Saas-Fee Advanced Course 28
Lecture Notes 1998

Springer
Berlin
Heidelberg
New York
Barcelona
Hong Kong
London
Milan
Paris
Singapore
Tokyo

Physics and Astronomy ONLINE LIBRARY

http://www.springer.de/phys/

B.W. Carney W.E. Harris

Star Clusters

Saas-Fee Advanced Course 28
Lecture Notes 1998
Swiss Society for Astrophysics and Astronomy

Edited by L. Labhardt and B. Binggeli

With 204 Figures and 25 Tables

Springer

Professor Bruce W. Carney
Department of Physics and Astronomy
University of North Carolina
Chapel Hill, NC 27599-3255, USA

Professor William E. Harris
Department of Physics and Astronomy
McMaster University
Hamilton ON L8S 4M1, Canada

Volume Editors:

Dr. Lukas Labhardt
Dr. Bruno Binggeli
Astronomisches Institut
Universität Basel
Venusstrasse 7
4102 Binningen, Switzerland

This series is edited on behalf of the Swiss Society for Astrophysics and Astronomy:
Société Suisse d'Astrophysique et d'Astronomie
Observatoire de Genève, ch. des Maillettes 51, CH-1290 Sauverny, Switzerland

---

Cover picture: (Left) Palomar 2, a globular cluster about 25 kiloparsec from the Sun in the outer halo of the Milky Way. *(Right)* A globular cluster in the halo of the giant elliptical galaxy NGC 5128, 3900 kiloparsec from the Milky Way (see Harris, p. 225)

ISBN 3-540-67646-5 Springer-Verlag Berlin Heidelberg New York

Library of Congress Cataloging-in-Publication Data applied for.

Die Deutsche Bibliothek – CIP-Einheitsaufnahme
Star clusters : with 25 tables / Saas Fee Advanced Course 28.
B. W. Carney; W. E. Harris. Swiss Society for Astrophysics and Astronomy. Ed. by L. Labhardt and B. Binggeli. – Berlin; Heidelberg; New York; Barcelona; Hong Kong; London; Milan; Paris; Singapore; Tokyo: Springer, 2001
(Saas Fee Advanced Courses ; 28)
(Physics and astronomy online library)
ISBN 3-540-67646-5

Springer-Verlag Berlin Heidelberg New York
a member of BertelsmannSpringer Science+Business Media GmbH

Printed in Germany

Typesetting: Camera-ready copy from authors/editors
Cover design: *design & production* GmbH, Heidelberg
Printed on acid-free paper SPIN 10769931 55/3141/xo - 5 4 3 2 1 0

Preface

Each year, the Swiss Society of Astrophysics and Astronomy (SSAA) organizes an Advanced Course on an astronomical subject of general interest. These annual courses have become known as the "Saas-Fee" Courses of the SSAA. They are held in a Swiss alpine resort in early spring, and last for one week. The subject chosen gets covered in complementary lectures by three invited scientists, all renowned experts in their respective fields. In accordance with the traditional format of the course every lecturer gives nine to ten lectures of 50 minutes.

The 28th Advanced Course of the SSAA took place in Les Diablerets from 30 March to 4 April 1998. It was devoted to observational and theoretical aspects of galactic and extragalactic star clusters. Up to now this very active field of modern astronomy has not been the subject of any of the previous Advanced Courses. We were very happy to attract three outstanding scientists who delivered brilliant, masterly interwoven lectures.

Star clusters within the Milky Way have played and continue to play a foremost role as laboratories for the study of stellar evolution and stellar dynamics. Bruce Carney lectured on "Stellar Evolution in Star Clusters" which included a detailed discussion of the chemical properties of the different stellar components and a critical presentation of the results obtained from various age estimators. Carlton Pryor lectured on "Structure and Dynamics of Globular Clusters" starting from quasi-static equilibrium models for clusters, stepping across the progressing stages of dynamical evolution, and ending up with describing the tidal destruction of globular clusters. Putting special emphasis on the meaning of radial velocity measurements of cluster stars, Pryor combined in an exemplary way the results of theoretical studies with those of observational work. William Harris gave a thorough review of the observed characteristics of the globular cluster systems of the Milky Way and other galaxies, concluding with a discussion of cluster formation scenarios. His first lecture was devoted to the basic principles of photometric methods and the photometry of non-stellar objects.

Almost 50 participants from 10 countries attended the course with great interest and enthusiasm. With them we were looking forward very much to the written versions of the lectures delivered. However, in spring 1999 it turned out to our great dismay that Tad Pryor's writing process was severely hampered by a wearisome health problem. Together with our authors we later had to decide to publish only two of the three expected contributions. We very much regret the omission of a significant part of the course and the subsequent delay in the preparation of the lecture notes. Bruce Carney and William Harris wrote two very comprehensive contributions to which they added the latest references in spring 2000. These lecture notes will certainly prove extremely valuable, both to students and researchers in the field.

It is a pleasure to thank our three lecturers for their great commitment and lively presentations which triggered many discussions during and after the lectures. The Saas-Fee Courses are financed in large part by the Swiss Academy of Sciences, to which the SSAA is attached. The kind hospitality provided by the hotel Eurotel-Victoria was also much appreciated.

Basel
August 2000

Lukas Labhardt
Bruno Binggeli

Table of Contents

List of Previous Saas-Fee Advanced Courses

!! 1997 Computational Methods for Astrophysical Fluid Flow
R.J. LeVeque, D. Mihalas, E.A. Dorfi, E. Müller

!! 1996 Galaxies Interactions and Induced Star Formation
R.C. Kennicutt, F. Schweizer, J.E. Barnes

!! 1995 Stellar Remnants
S.D. Kawaler, I. Novikov, G. Srinivasan

* 1994 Plasma Astrophysics
J.G. Kirk, D.B. Melrose, E.R. Priest

* 1993 The Deep Universe
A.R. Sandage, R.G. Kron, M.S. Longair

* 1992 Interacting Binaries
S.N. Shore, M. Livio, E.J.P. van den Heuvel

* 1991 The Galactic Interstellar Medium
W.B. Burton, B.G. Elmegreen, R. Genzel

* 1990 Active Galactic Nuclei
R. Blandford, H. Netzer, L. Woltjer

! 1989 The Milky Way as a Galaxy
G. Gilmore, I. King, P. van der Kruit

! 1988 Radiation in Moving Gaseous Media
H. Frisch, R.P. Kudritzki, H.W. Yorke

! 1987 Large Scale Structures in the Universe
A.C. Fabian, M. Geller, A. Szalay

! 1986 Nucleosynthesis and Chemical Evolution
J. Audouze, C. Chiosi, S.E. Woosley

! 1985 High Resolution in Astronomy
R.S. Booth, J.W. Brault, A. Labeyrie

! 1984 Planets, Their Origin, Interior and Atmosphere
D. Gautier, W.B. Hubbard, H. Reeves

! 1983 Astrophysical Processes in Upper Main Sequence Stars
A.N. Cox, S. Vauclair, J.P. Zahn

* 1982 Morphology and Dynamics of Galaxies
J. Binney, J. Kormendy, S.D.M. White

! 1981 Activity and Outer Atmospheres of the Sun and Stars
F. Praderie, D.S. Spicer, G.L. Withbroe

* 1980 Star Formation
J. Appenzeller, J. Lequeux, J. Silk

* 1979 Extragalactic High Energy Physics
F. Pacini, C. Ryter, P.A. Strittmatter

* 1978 Observational Cosmology
J.E. Gunn, M.S. Longair, M.J. Rees

* 1977 Advanced Stages in Stellar Evolution
I. Iben Jr., A. Renzini, D.N. Schramm

* 1976 Galaxies
K. Freeman, R.C. Larson, B. Tinsley

* 1975 Atomic and Molecular Processes in Astrophysics
A. Dalgarno, F. Masnou-Seeuws, R.V.P. McWhirter

* 1974 Magnetohydrodynamics
L. Mestel, N.O. Weiss

* 1973 Dynamical Structure and Evolution of Stellar Systems
G. Contopoulos, M. Hénon, D. Lynden-Bell

* 1972 Interstellar Matter
N.C. Wickramasinghe, F.D. Kahn, P.G. Metzger

* 1971 Theory of the Stellar Atmospheres
D. Mihalas, B. Pagel, P. Souffrin

---

* Out of print
! May be ordered from Geneva Observatory
Saas-Fee Courses
Geneva Observatory
CH-1290 Sauverny
Switzerland
!! May be ordered from Springer Verlag

Stellar Evolution in Globular Clusters

Bruce W. Carney

Department of Physics & Astronomy, University of North Carolina
Chapel Hill, NC 27599-3255, USA, bruce@physics.unc.edu

Abstract. 1. Summaries are given of the major stages of evolution for stars in globular clusters, including the nucleosynthesis networks, and their sensitivities to chemical abundances. Difficulties in comparing models with observations are described.

2. Unusual types of stars, including binaries, blue stragglers, and the "second parameter problem" are described. The blue stragglers appear to be the result of mass transfer, and either age or stellar interactions enhanced by high stellar densities are viable explanations for the second parameter, although age appears to be favored on Galactic scales.

3. RR Lyrae variables are discussed, including the relations between M_V and [Fe/H] and M_K and log P, and how the Oosterhoff classes are defined and differ. Some new results suggest that the two classes might differ in their Galactic histories and not only in metallicity.

4. Stellar populations, including the thick disk and the halo are defined. Possible subcomponents to the halo population are identified, as well as their origins. Part of the halo, that nearer the disk's plane, may share a common history with the disk, while the halo farther from the plane may be dominated by independent origin(s). The thick disk appears also to have arisen from a merger, based on an overlap in metallicity with the disk but has very different kinematics.

5. A simple model for the metallicity distribution function is described, as well as how metallicities and metallicity indicators are derived and calibrated. Results from recent analyses of high-resolution stellar spectroscopy agree well with each other, but disagree at intermediate metallicities with earlier results. New relations between [Fe/H] and the metallicity indicators ΔS and W' are derived.

6. Abundances of some special elements are described, including the "α" elements (O, Mg, Si, Ca, Ti) and the s–process and r–process elements, all of which are useful in the study of the Galaxy's nucleosynthesis history. Halo stars with unusual [α/Fe] ratios are identified and discussed. Unusual ratios for thick disk stars also suggest an origin separate from that of the disk. Lithium abundances in halo stars are also discussed briefly in terms of whether or not halo stars may be used to infer Big Bang nucleosynthesis abundances. Some new evidence for lithium production in low mass stars is presented.

7. Evidence for deep mixing in red giants is presented, including from the CN, ON, NaNe, and MgAl cycles associated with shell hydrogen burning. New evidence is also presented that indicates that helium may also be mixed into red giant photospheres. The possible role of rotation on mixing, mass loss, and horizontal branch morphology is summarized. Evidence for primordial variations in some elements within globular clusters is described, including CN variations in stars on cluster main sequences, and variable iron and r–process abundances in some cluster red giants.

8. The two methods for estimation of relative ages of globular clusters are reviewed: the turn-off luminosity derived from an assumed relation between horizontal branch and metallicity; and the color difference between the main sequence turn-off and the red giant branch at nearly equal metallicities. The most metal-poor globular clusters and field stars seem to show no discernible age differences, but an age spread begins to appear at intermediate metallicities, including in the outer halo. [O/Fe] and [α/Fe] ratios are compared with the relative age scales, suggesting that either some sub-populations (the disk clusters, for example) did not share a common chemical history with the oldest clusters, or that the SNe Ia timescale may be longer than 10^{10} years. (Or, as has been suggested recently, that the initial appearance of SNe Ia has more to do with metallicity than with age.)

9. The absolute ages of globular clusters are discussed in terms of the remaining uncertainties. Primary among them is the globular cluster distance scale and the apparent dichotomy from *Hipparcos* (and other) results for the luminosities of RR Lyrae variables. Differences between field and cluster RR Lyraes may be part of the cause, but the explanation fails at least two tests. Nonetheless, the ages of the globular clusters agree, at least within the large error bars, with the ages of field and cluster white dwarfs derived from deep luminosity functions and cooling theory, and from radioactive dating techniques.

1 Fundamentals of Globular Cluster Stellar Evolution

1.1 The Appeal

Globular clusters can entrance those fortunate observers with good optics, dark skies, and clear air. I first gazed at them with a 0.9–meter telescope on an island in the Chesapeake Bay during my military service, I have never tired of looking at them since that time over twenty-five years ago, and wondering about them. Indeed, those first glimpses into their depths afforded by both eyepiece and the literature led me to study globular clusters for my subsequent doctoral dissertation, and for much of my work since. And I am not alone: many fields of astrophysics involve globular clusters in one way or another.

Globular clusters, with thousands to millions of stars, were formed in our Galaxy's youth, and have not been created, at least here in the Milky Way, since then. As such, globular clusters and their field star counterparts are the stellar relics of a time long past, and are one of our primary means of learning the history of our Galaxy, and by extension, of disk galaxies in general. Dating of the oldest clusters provides cosmologists with a well-understood lower limit to the age of the Universe, while students of stellar evolution and stellar dynamics find the clusters to be unsurpassed, albeit apparently static, laboratories. My lectures will not address the problems of stellar dynamics, nor the wealth of globular cluster systems beyond our own Milky Way. One of my primary goals is to present our current knowledge of the relative and absolute ages of these systems and probes of the Galaxy's chronology. Another is to explore the chemical history revealed by spectroscopic studies, and what these studies also reveal about the complexities of stellar evolution. The history of the Galaxy is also one of dynamics, and we will explore those of the globular cluster system briefly, but more through the more numerous and better studied field stars. But first we need to understand some of the basic physics.

1.2 Basic Physics

Our understanding of stellar evolution comes from a comparison of "snapshots" in the form of cluster color–magnitude diagrams (or, occasionally, transformed into temperature–luminosity diagrams) with a series of calculations in which one solves, ironically, a series of static stellar models. The evolutionary series results from the static models by the small changes in chemical composition due to the energy generation by nucleosynthesis or in internal structure due to the transformation of gravitational potential energy into heat and radiation.

The static models are created by solving a set of four differential equations. Some are simple; some are not. We cast them here in their Eulerian mode (with respect to radius r). In the equations below, ϱ is the mass density, and we seek to solve the run of mass M_r, pressure P_r, luminosity L_r, and temperature T_r with radius. The simplest is the equation of the continuity of mass:

$$\frac{\mathrm{d}M_r}{\mathrm{d}r} = 4\pi\, r^2\, \varrho\,. \tag{1}$$

Almost as simple is that for the pressure gradient:

$$\frac{\mathrm{d}P_r}{\mathrm{d}r} = -\varrho_r \frac{G\, M_r}{r^2} - \varrho_r \frac{\mathrm{d}^2 r}{\mathrm{d}t^2}\,. \tag{2}$$

This is obviously not a static equation, and the last term is negligible during most phases of a star's life, and is therefore neglected. The result is the equation of hydrostatic equilibrium, which says nothing more than the pressure gradient must balance gravity. The third equation is that of the energy generation and the differential change in luminosity with respect to radius:

$$\frac{\mathrm{d}L_r}{\mathrm{d}r} = 4\pi\, \varrho_r\, r^2 \left(\varepsilon_r - T_r \frac{\mathrm{d}S}{\mathrm{d}T}\right)\,, \tag{3}$$

where ε_r is the energy generation rate per unit mass at radius r, and S is the entropy. A lot of physics is hidden within ε_r, as we shall see. The final equation is for the temperature gradient, which governs the transport of energy (and, of course, also the pressure gradient and ε_r):

$$\frac{\mathrm{d}T_r}{\mathrm{d}r} = -\frac{3\kappa\varrho}{ac} \frac{1}{T_r^3} \frac{L_r}{4\pi r^2}\,, \tag{4}$$

when the star manages to transfer all of its energy by radiative diffusion down the temperature gradient, or

$$\frac{\mathrm{d}T_r}{\mathrm{d}r} = \frac{\Gamma_2 - 1}{\Gamma_2} \frac{T_r}{P_r} \frac{\mathrm{d}P_r}{\mathrm{d}r} , \tag{5}$$

when the material is carrying the energy by convection. Again, there is a lot of physics subsumed within κ, the mass absorption coefficient (units are $\mathrm{cm}^2/\mathrm{gram}$), and the second adiabatic exponent, Γ_2. The latter, as might be expected from its use, relates changes in pressure and temperature:

$$\frac{dP}{P} + \frac{\Gamma_2}{1-\Gamma_2} \frac{dT}{T} = 0 . \tag{6}$$

When is the material convective? Generally, the condition for convective stability is given as:

$$\frac{\mathrm{d}T}{\mathrm{d}r} > \left(1 - \frac{1}{\Gamma_2}\right) \frac{T}{P} \left(\frac{\mathrm{d}P}{\mathrm{d}r}\right) . \tag{7}$$

Another useful way to look at this stability criterion, assuming a perfect gas law applies, is

$$\frac{\mathrm{d}\log T}{\mathrm{d}\log P} < 1 - \frac{1}{\gamma} + \frac{\mathrm{d}\log \mu}{\mathrm{d}\log P} . \tag{8}$$

This formulation makes it easy to see the importance of the mean molecular weight (i.e., ionization or dissociation) and the importance of γ,

$$\gamma = \frac{c_p}{c_v} = 1 + \frac{2}{f} , \tag{9}$$

where f is the number of degrees of freedom. The opacity, energy generation rate, entropy, and second adiabatic exponent are all complex functions of the mass density, temperature and chemical composition, usually divided coarsely into X, Y, and Z, the hydrogen, helium, and heavy element mass fractions, respectively. The pressure is also a function of these variables, but can be written simply in most cases applicable to globular cluster stars as the combination of gas pressure and radiation pressure

$$P = \frac{\varrho k T}{\mu m_\mathrm{H}} + \frac{1}{3} a T^4 , \tag{10}$$

where m_H is the mass of a hydrogen atom, and μ is the mean molecular weight. The last term in (10) is usually negligible for stars in globular clusters, but note the importance of μ. It is here (and in X, Y, and Z, to which μ is obviously related) that the static models predict a rate of change in a fundamental quantity. Using the incremental derivatives of these quantities and small time steps to alter them for inclusion into a new static model calculation is what provides the evolution of the model sequence. Further, "new physics" often seen in the literature generally is due to improvements in the calculation of P, κ, ε, S, and Γ_2.

The details of how static models are calculated is treated extremely well by Kippenhahn & Weigert (1990). This usually involves Lagrangian coordinates (mass, not radius, as the independent variable), and the solution of a fairly simple tri-diagonal matrix. What is important for students of globular clusters to recognize, however, is where the remaining problems lie, and which can, and no doubt do, introduce systematic uncertainties into our understanding. These fall into three broad categories.

Physical Problems. Equation (5) seems simple enough, but few models have used anything more sophisticated than the physically plausible and computationally straightforward "mixing length theory" developed by Böhm–Vitense (1958). The theory is extremely useful but is certainly not perfect. It also includes a free parameter, the ratio of the mixing length to the pressure scale height, as usually denoted as

$$\alpha_p = \ell_{\rm conv} \,/\, H_p \,. \tag{11}$$

The free parameter is a blessing and a curse. It is useful because it can be adjusted so that stellar evolution models can reasonably reproduce the current surface temperature, radius, and hence luminosity, of our Sun. One has some confidence thereby that α_p has been chosen well. (Note that some older stellar evolution models were not even subjected to such a calibration.) A value of α_p near 1.5 appears to work well for the Sun, and is therefore exported into the calculation of stellar evolution sequences for stars in globular clusters. The drawback to the use of α_p is, of course, that we often forget about it, and its effects on our interpretations of observational data. All normal main sequence stars and all red giants in globular clusters have convective envelopes, so convection is very important. Thus uncertainties in how to calculate the effects of convection introduce uncertainties into the outer layers of our model stars and evolutionary sequences.

Convection can also transport material into different radial regimes and thereby alter abundance ratios through nucleosynthesis. Convection is therefore intimately related to mixing, which as we will see remains a very important and as yet unsolved problem in stellar evolution. Finally, as stellar radii grow and luminosities increase, mass loss increases, and to degrees that may or may not have important consequences for subsequent evolutionary stages. Equations (1) – (10) are not the whole physical picture.

Mathematical Problems. The solution of the matrices is fairly straightforward, but four differential equations require four boundary conditions. Two are simple enough: $L_r = M_r = 0$ at $r = 0$. The outer two are less obvious. One might be tempted to employ $P_r = T_r = 0$ at $r = R$, where R is the stellar radius. Of course, a star with $T = 0$ at its surface does not radiate.

A common solution is to employ the "gray atmosphere" (i.e., no frequency dependence) solution, in which case one has

$$T = T_{\rm eff} \left(\tau + \frac{2}{3} \right) , \tag{12}$$

where τ is the optical depth, $\tau = \int \kappa \, \sigma \, dx$, and, of course,

$$L = 4\pi \, R^2 \, \sigma \, T_{\rm eff}^4 \, . \tag{13}$$

An even more correct approach is to fit a model atmosphere to the stellar interiors solution, as described by Kippenhahn & Weigert (1990). The most important point of all this is that the uncertainties introduced primarily affect the outer layers of the model star.

Observational Problems. The model calculations are done with an adopted chemical composition, denoted by X, Y, and Z. Since $X + Y + Z = 1$ by definition, we need to derive only two of these. Nature has not made our job easy here, however. The helium mass fraction is especially important in the estimation of μ, for example, but it is not directly measurable in globular cluster stars in general, and in those cases where the stars are hot enough for helium lines to be seen, the effects of gravitational settling or diffusion appear to be present (see Glaspey et al. 1989 and references therein). We must resort to indirect methods, including the values from metal-poor extragalactic H II regions (see Pagel et al. 1992; Pagel & Kazlauskas 1992), standard Big Bang nucleosynthesis calculations (see Wheeler et al. 1989), and the so-called "R method" described later. In all cases, agreement is quite good, with $Y \approx 0.23$. As for Z, over half the heavy element atoms in stars are oxygen, and oxygen also proves difficult to measure, compared to the less important iron peak elements. We discuss this in a later section.

A static model or an evolutionary sequence provides us with luminosity L and effective temperature $T_{\rm eff}$. What we observe are apparent magnitudes and colors, which are related to L and $T_{\rm eff}$, of course, but not in a manner easy to define. Apparent magnitudes must be corrected for the effects of interstellar absorption, distance, and the conversion from absolute magnitude in some bandpass into apparent bolometric magnitude via the "bolometric correction". As we will see, distance remains one of the key problems. Apparent colors must be corrected for interstellar extinction, and then converted into $T_{\rm eff}$ using additional corrections for metallicity and even gravity effects since line blanketing and continuum opacity can be affected by both. Consider Figs. 1 and 2, which show the filter transmission functions for the Johnson $UBVJK$ and the Cousins RI systems matched to the solar flux spectrum. Line blanketing plays a large role at shorter wavelengths and less so at longer wavelengths. Cooler stars are better observed at longer wavelengths, near their flux peaks, but hotter and fainter stars are more easily observed in the optical regime.

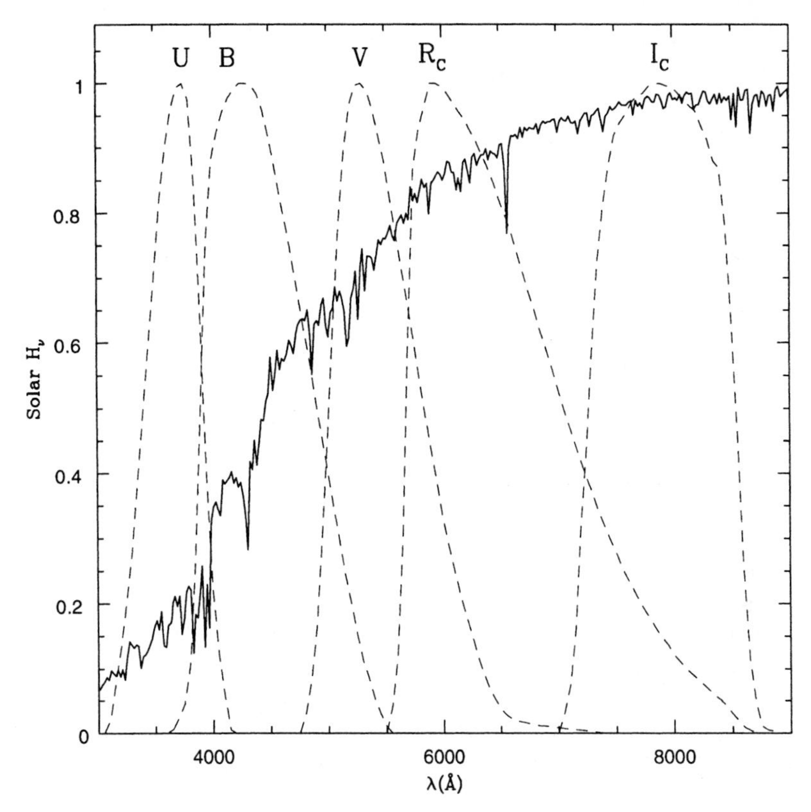

Fig. 1. The solar spectrum and $UBVR_CI_C$ bandpasses

The color–temperature scale is, in principle, calibrated most directly by measuring the angular diameters of stars and their fluxes over all wavelengths. One may see directly from (13) that dividing by the distance squared, there is a relation between the apparent bolometric luminosity measured outside the Earth's atmosphere, the angular diameter, and the effective temperature:

$$T_{\text{eff}} \propto \theta^{\frac{1}{2}}\, \ell^{\frac{1}{4}} \ . \tag{14}$$

Unfortunately, no stars in the temperature and metallicity ranges of globular cluster stars have had angular diameters determined with sufficient precision, so color–temperature calibrations have been done using three different approaches. All are secondary, and all depend on model stellar atmospheres.

Synthetic Spectra. In principle, the surface flux distributions derived from model stellar atmospheres may be passed through filter transmission functions (and aluminum mirror reflectivity functions and detector sensitivity functions) to determine synthetic colors and bolometric corrections. The zero point shifts

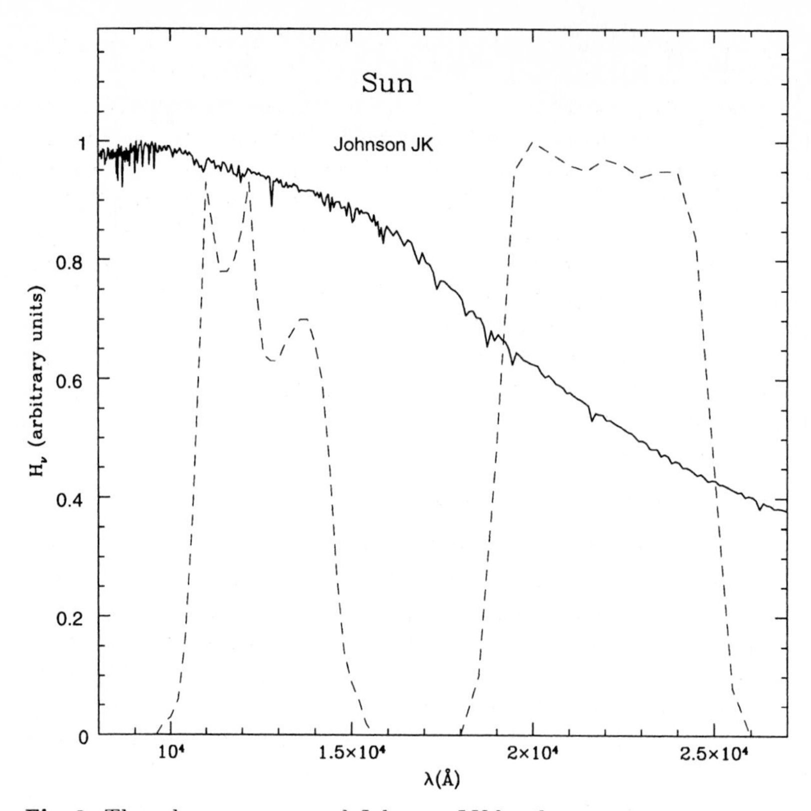

Fig. 2. The solar spectrum and Johnson JK bandpasses

to bring these on to the empirical scale may be accomplished using carefully selected standard stars, such as Vega. A good example of this method for the Strömgren photometric system applied to main sequence stars is given by Edvardsson et al. (1993). A similar example for metal-poor red giants is given by Cohen et al. (1978). Buser & Kurucz (1978, 1992) give synthetic colors for a wide variety of color indices, gravities, and chemical compositions.

Surface Flux Distributions. Rather than passing the model surface flux distribution through filter transmissions functions, it is sometimes more direct to compare the computed surface flux distributions with spectrophotometry, the measurement of stellar flux in narrow, generally weak-lined regions of stellar spectra. This is the approach taken by Peterson & Carney (1979) and Carney (1983). Its limitation is the relatively small number (under 100) of metal-poor stars with spectrophotometric data of high quality.

Infrared Flux Method. The ratio of the stellar flux at some point out on the Rayleigh–Jeans tail of the flux distribution to the integrated stellar stellar flux is a good temperature indicator, and relatively insensitive to chemical composition. This can be seen relatively easily from the blackbody equation, where the total stellar flux rises as $T_{\rm eff}^4$, while the flux per unit wavelength on the Rayleigh–Jeans tail rises only as $T_{\rm eff}^{1.6}$ at near-IR wavelengths. Broadband filters are often employed, and the major problem is the conversion of such magnitudes into $\rm erg\,cm^{-2}\,s^{-1}\,Å^{-1}$. Saxner & Hammarbäck (1985), Bell & Gustafsson (1989), and Blackwell et al. (1990) have applied the method to metal-rich stars, while Alonso et al. (1996; 1999a,b) have extended the work to metal-poor dwarfs and giants, and from such temperature estimates have produced color–temperature relations for a variety of color indices.

Fortunately, the results from the surface flux distributions and the Infrared Flux Method (IRFM) agree well (Alonso et al. 1996), as do the angular diameters and the IRFM results (Alonso et al. 1999b).

1.3 The Major Stages of Stellar Evolution

Figure 3 identifies the major stages and timescales for the stars in globular clusters. The main sequence and red giant models are based on new models from Yale (see Green et al. 1987), and supplied kindly by Dr. Sukyoung Yi, while the horizontal branch and asymptotic giant branch stages are taken from Yi et al. (1997). These replace the older "Yale Isochrones" (Ciardullo & Demarque 1977.) In all cases, the adopted chemical compositions are very similar. In the case of the main sequence and red giant stages, Fig. 3 indicates some of the timescales, from which it is clear that, as expected, the timescales drop quickly as the luminosity rises. Indeed, the timescale from the main sequence turn-off point at core hydrogen exhaustion to the tip of the red giant branch, where helium burning is ignited in degenerate conditions, is so brief that the range in masses is quite small at a constant age. For a cluster with an age of 10 Gyrs, the turn-off mass for $Z = 0.001$ is 0.794 $\rm M_\odot$, while the RGB tip mass is 0.826 $\rm M_\odot$. For an age of 15 Gyrs, the values are 0.883 $\rm M_\odot$ and 0.924 $\rm M_\odot$, respectively. The dots for the horizontal branch (HB) and asymptotic giant branch (AGB) stages indicate spacings of only 10^7 years, so it is easy to see the luminosities at which such stars spend most of their lives.

Two other items are shown in Fig. 3. The horizontal line at the top represents the approximate track of post-AGB evolution, based on the luminosities of a few such stars, as discussed later in this section. The dashed diagonal lines are approximations to the hot and cool limits to the instability strip. It has been assumed that the width of the strip matches that of the RR Lyrae instability strip, and the slope of the cool edge was set by the approximate range in $(B-V)_0$ of SX Phe variables in globular clusters (Nemec et al. 1994). The colors have been converted to approximate temperatures using the synthetic colors computed by Buser & Kurucz (1978, 1992). The hot edge of the lower instability strip for metal-poor stars is not easy to estimate in any case since if mass transfer is the cause of SX Phe variables (discussed in Sect. 2), a natural hot limit is set by stars with twice the mass of the main sequence turn-off. All

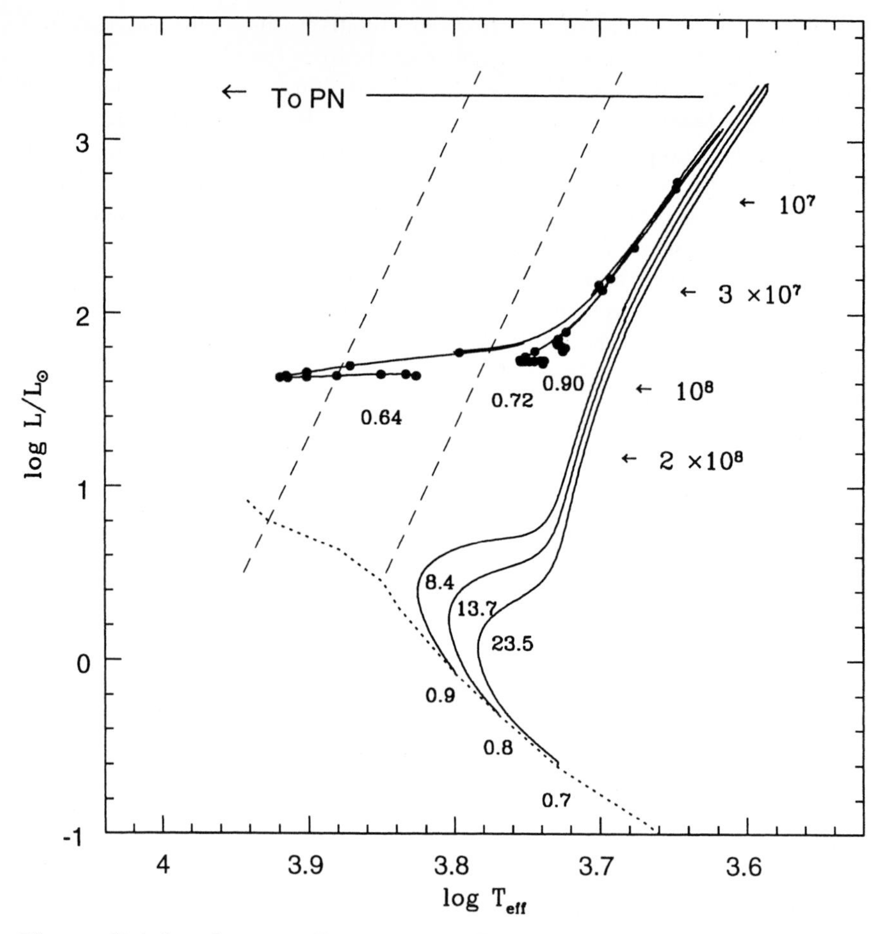

Fig. 3. Globular cluster stellar evolution. The *dotted line* is the zero-age main sequence. Turn-off ages in Gyrs are given for three main sequence masses (of 0.7, 0.8, and 0.9 $M_\odot$). Times remaining to the He shell flash are given along the RGB. Zero-age HB masses are given, and each *dot* signifies 10^7 years of evolution. The *dashed lines* denote the approximate edges of the instability strip

globular cluster SX Phe variables, as well as high-luminosity ($P > 20$ days) Cepheids in clusters lie within the strip as shown in Fig. 3.

Nucleosynthesis. Throughout the stages depicted in Fig. 3, masses are small enough that the nucleosynthesis occurs within the context of hydrogen and helium burning only. On the main sequence, globular cluster stars are primarily converting hydrogen into helium via the proton–proton chain. The details of this nucleosynthesis are given in many textbooks, but it is worth

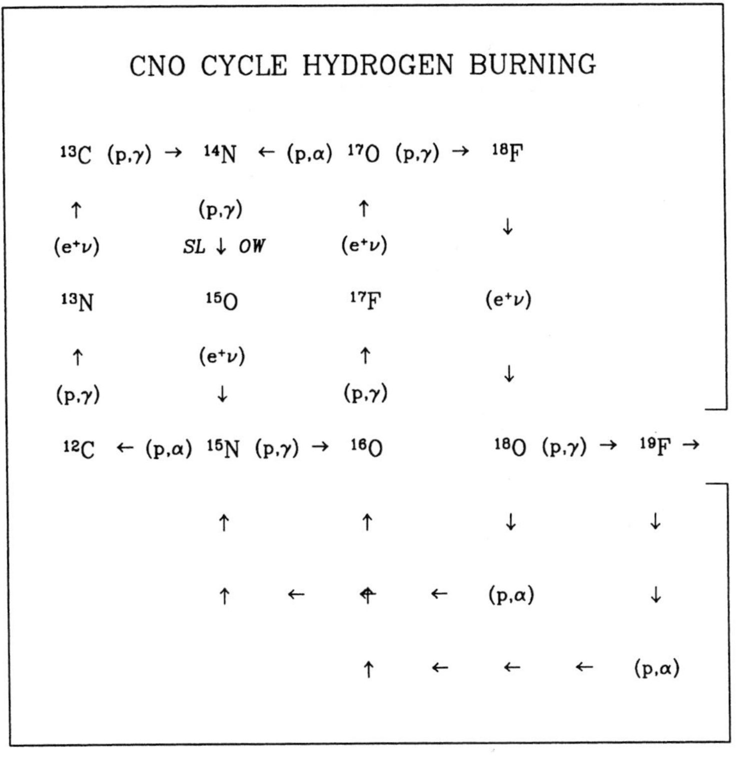

Fig. 4. The major chains of the CNO cycle

concentrating on at least the "pp I" chain:

$$^1\mathrm{H} + {}^1\mathrm{H} \rightarrow {}^2\mathrm{H} + \mathrm{e}^+ + \nu \,, \tag{15}$$

$$^2\mathrm{H} + {}^1\mathrm{H} \rightarrow {}^3\mathrm{He} + \gamma \,, \tag{16}$$

$$^3\mathrm{He} + {}^3\mathrm{He} \rightarrow {}^4\mathrm{He} + 2\,{}^1\mathrm{H} \,. \tag{17}$$

The first reaction is the most important since it is the slowest reaction, involving as it does the weak nuclear force, and therefore sets the speed of the entire process and its temperature sensitivity. Since the reactions are all two-body captures, the reaction rate depends on the square of the density, and the temperature sensitivity is modest:

$$\varepsilon(\mathrm{pp\ chain}) \propto \varrho^2\, T^4 \,. \tag{18}$$

However, as the core hydrogen nears exhaustion, the temperatures rise sufficiently that the CNO cycle can begin to operate. This is a much more complicated cycle. While it mostly uses the CNO nuclei as catalysts to convert hydrogen into helium, there is leakage out of these cycles, as shown in Figs. 4 and 5. One can see immediately that much larger Coulomb barriers are involved than in the proton–proton chain, and generally the nucleosynthesis

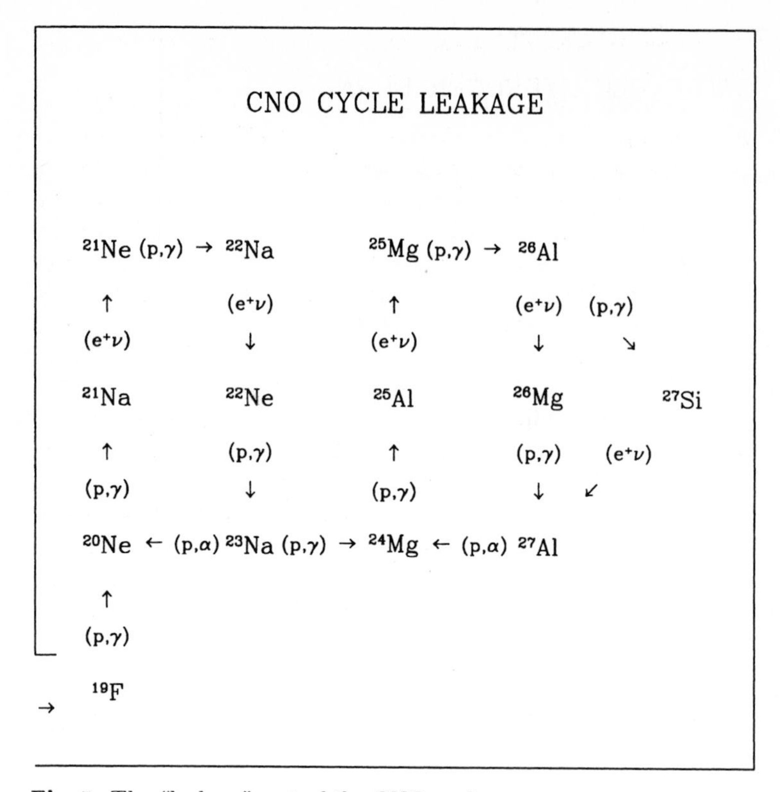

Fig. 5. The "leakage" out of the CNO cycle

rate is set by the slowest reaction: the capture of a proton into a ^{14}N nucleus. At temperatures near 25×10^6 K, $\varepsilon(\mathrm{CNO}) \propto \varrho^2 T^{16.7}$. Two points in Fig. 4 deserve special note. First, the $^{14}\mathrm{N(p,}\gamma)^{15}\mathrm{O}$ reaction is very slow, causing a "bottleneck" in the first cycle (called the CN cycle). Second, less than one ^{15}N nucleus in a thousand undergoes the (p,γ) reaction, so there is not a great deal of "leakage" into the second cycle (called the ON cycle). Hence its rate is also set by the speed of the $^{14}\mathrm{N(p,}\gamma)^{15}\mathrm{O}$ reaction.

When hydrogen has been exhausted, temperatures and densities may reach sufficiently high values in globular cluster stars for helium burning to occur. This is basically a three-body process,

$$3\,^4\mathrm{He} \rightarrow \,^{12}\mathrm{C}\,, \tag{19}$$

with the possibilities of many attendant reactions which may contribute little to the energy budget but produce new heavy elements through the effect of slow neutron capture on iron peak nuclei. All of this is well described by Pagel (1997).

Main Sequence Evolution. The main sequence phase is the lowest luminosity, and hence the longest, stage in a star's life. The low luminosity is the result of the relatively low Coulomb barriers and the relatively high efficiency of the hydrogen fusion process, which liberates almost 85% of the total energy per nucleon in the possible conversion of hydrogen all the way to iron, which is the nucleus with the highest binding energy per nucleon and therefore the end point of energy-generating nucleosynthesis. The low luminosities of main sequence stars makes their study in globular clusters difficult, yet their importance is fundamental. Main sequence fitting is one of the key methods by which cluster distances may be estimated, and the luminosity of the turn-off is the primary (and best understood) means of estimating absolute and relative cluster ages. It is therefore vital that we understand what affects the location of the main sequence in the luminosity–temperature and color–magnitude diagrams and therefore how well we can trust our conclusions regarding distances and ages.

An excellent beginning is Eddington's (1926) classic work on the internal workings of stars, along with that of Chandrasekhar (1957) and Clayton (1968). Kippenhahn & Weigert (1990) is an excellent modern work including the details of the numerical solution to the equations of stellar structure. For a model star in which the opacity, κ, is related to density, ϱ, and temperature, T, via the "Kramer's opacity",

$$\kappa = \kappa_0 \, \varrho T^{-3.5} \, , \tag{20}$$

one may derive (see Clayton equation 6-60)

$$L_{\rm MS} \propto \mu^{7.5} \, M^{5.5} / \kappa_0 \, , \tag{21}$$

where M is the mass, κ is the opacity, and μ is the mean molecular weight:

$$\mu = \frac{4}{3 + 5X - Z} \approx \frac{4}{8 - 5Y} \, . \tag{22}$$

The Kramer's opacity is only an approximation, and applicable to lower mass stars. Nonetheless, for low-mass stars, which are all that is left within globular clusters, (21) is a very useful tool. For example, as had been known for a long time, the luminosity is a high power of the stellar mass, and hence the luminosity of the turn-off is an age indicator because it is so sensitive to the stellar mass. However, both μ and κ play key roles. As one can see from (22), it is the helium mass fraction that primarily determines μ, and hence the precision of our estimates of the absolute and relative ages of star clusters is limited by our knowledge of Y. Figure 6, taken from the models of VandenBerg & Bell (1985), shows this graphically for a low metallicity composition and an old age. Note that the loci diverge at higher masses, which is due to the higher luminosity resulting in faster evolution, which in turns leads to higher luminosity.

Of course, the measurable plane is not, regrettably, luminosity *vs.* mass, but luminosity *vs.* temperature. Higher luminosity requires a higher surface

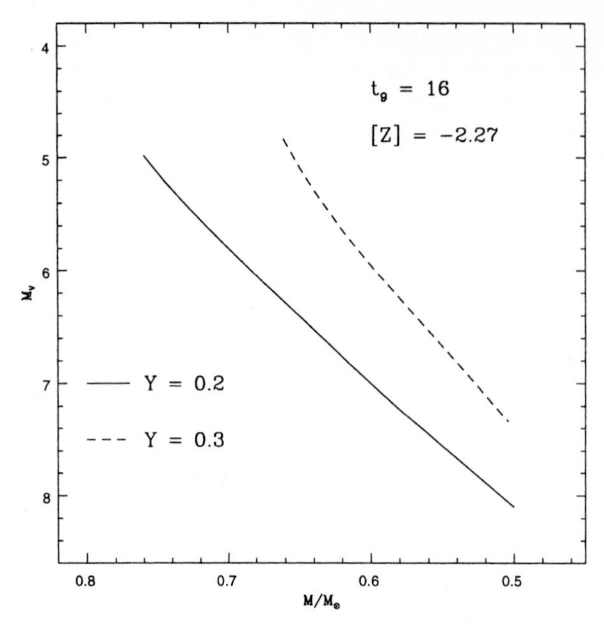

Fig. 6. The effects of helium abundance on absolute visual magnitude

temperature (and larger radius). The consequences, shown in Fig. 7, appear to be a reversal of the trend seen in Fig. 6. In the observational plane, the stars with higher helium abundances have *lower* luminosities at equal color indices. The higher luminosity at fixed mass has resulted in a shift to higher temperature (and a bluer color index) and in the observational plane, and this temperature shift overcomes the luminosity shift.

Note also in Fig. 7 the strong effect of the heavy element mass fraction, Z, despite the low metallicities. Such small changes in Z do not affect μ, but they still affect κ, and this plays two important roles. On the less evolved lower main sequence, (21) predicts that a lower Z, which results in a lower κ, will result in a higher luminosity. As we have seen, however, this also requires a higher surface temperature and the net result is that at a fixed temperature/color index, the lower metallicity isochrone is fainter. The second effect is an augmentation of the first. At a fixed luminosity (which recall is set by the conditions in the stellar interior), a lower opacity means a smaller radius, which requires a higher temperature to emit the fixed luminosity. One must always remember that what we observe is dictated by both the stellar luminosity, which is created within the stellar interior, and by the stellar photosphere, which, if it is in pressure equilibrium, adjusts its temperature and radius to emit that luminosity.

Why is the lower metallicity main sequence turn-off both hotter and brighter? We answer this by first considering the Y effects. The higher helium abundance stars are more luminous, and thus deplete their energy reserves

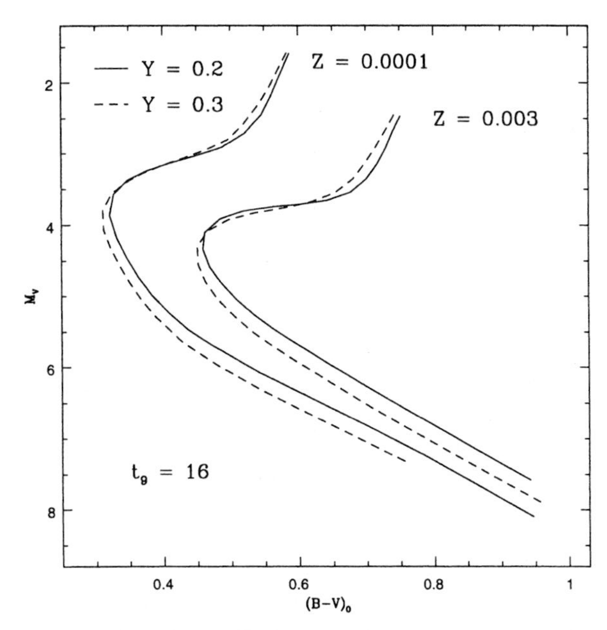

Fig. 7. The effects of helium and heavy element abundances in the color–magnitude diagram

sooner. Thus the masses of the stars at the turn-off points are lower for higher Y, and so are the luminosities. According to VandenBerg & Bell (1985; whose isochrones were used to produce Figs. 6 and 7) for $t_9 = 16$ and $Z = 0.0001$, the turn-off masses at $Y = 0.2$ and 0.3 are 0.810 and 0.685 $M_\odot$, respectively, while at $Z = 0.003$, they are 0.854 and 0.720 $M_\odot$, respectively. But the higher luminosity at the higher helium abundance and the resultant higher surface temperature almost compensate for the lower masses, so the $Y = 0.2$ and 0.3 turn-offs and subgiant branches almost overlap. The effects of Z may now also be understood: increasing Z increases the opacity, which lowers the stellar luminosity, which means that stars of a given mass live longer. Thus the turn-off mass is higher for a higher metallicity, as seen above. Despite the high power of the mass sensitivity in (21), the small mass difference means that the mass sensitivity is overcome by the larger opacity effect (which scales roughly with Z). Therefore the higher metallicity isochrones have lower luminosities and, hence, lower effective temperatures at the main sequence turn-off.

Finally, while we have looked separately at the effects of the helium and heavy element mass fractions, Y and Z, one must keep in mind two more points. First, the abundances of helium and the heavy elements in stars depend on the Big Bang nucleosynthesis plus the enrichment of the interstellar medium from preceding generations of stars. Since the manufacture of helium precedes the production of heavy elements in any given star, we must expect that on average, increasing heavy element abundances are accompanied by increases in the helium abundance,

$$\Delta Y = f(\Delta Z)\ . \tag{23}$$

Second, not all heavy elements produce the same effects on the opacities, especially in stellar photospheres, and, as we will see, metal-poor stars do not show the same elemental abundance ratios as does our Sun.

Red Giant Branch Evolution. When the core hydrogen is exhausted, the core contracts and heats up. The temperature at its outer edge eventually becomes high enough that hydrogen fusion (CNO cycle) begins in the lower density regions just beyond it, in a relatively thin shell. The increasing temperature leads to higher luminosity, but at first much of this extra energy is used against the stellar potential energy and the star expands in size, cooling off its surface layers as it does so. This is the subgiant branch, where the bolometric luminosity remains roughly constant while the star grows and reddens. Once the star has cooled to the point where most of its outer layers are convective, the star cannot cool very much further and remain in pressure equilibrium (the "Hayashi track"). The location of the Hayashi track in temperature has some weak sensitivity to mass and greater sensitivity to opacity, hence metallicity. Assuming a Kramer's opacity, one may derive the following relation between temperature, mass, luminosity, and the zero point of the Kramer's opacity, κ_0, which depends on the overall metallicity [see (20)]

$$T_{\rm eff} \propto \frac{M^{0.2}\,L^{0.05}}{\kappa_0^{0.1}}\ . \tag{24}$$

The increasing luminosity of a red giant requires a larger radius and the star grows to become a "mature" red giant. Of course, as the luminosity rises, the speed of the star's evolution increases, and Fig. 3 shows that the time remaining until the ignition of helium in the inert core (at the tip of the red giant branch) decreases quite rapidly. What happens inside the star has been depicted nicely by Kippenhahn & Weigert (1990), based on the work of Thomas (1967), and their Fig. 32.3 is reproduced here as Fig. 8. While the model star is more massive than typical globular cluster stars, the basic processes are identical. The most intense region of hydrogen burning is confined to the inner 10%, increasing to 20%, of the stellar mass, while some degree of burning extends out to almost 50% of the stellar mass. Once the core burning phase ends and the shell burning phase begins, the star expands, cools, and the convection zone deepens. It reaches well into the regions where hydrogen burning has been active, and therefore at this point we expect to begin to see the first signs of the products of nucleosynthesis showing up in the red giant photosphere.

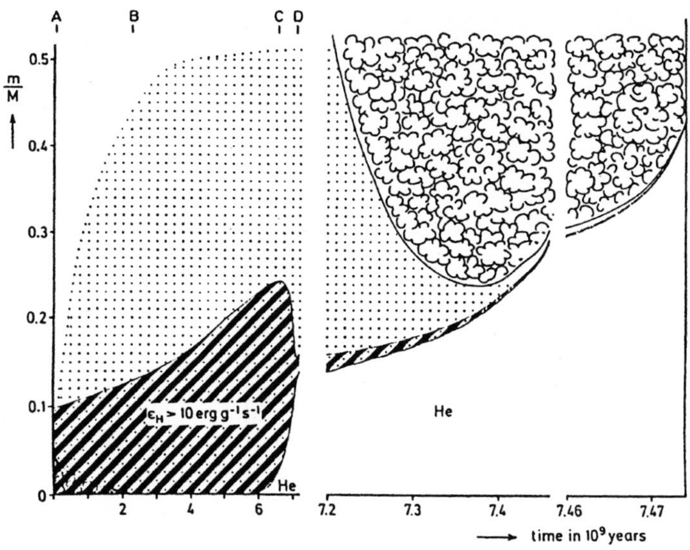

Fig. 8. The evolution of the hydrogen burning and convective envelope in a star of $1.3 M_\odot$, from Thomas (1967). The *diagonal* stripes indicate regions of strong hydrogen burning, while the *dotted* regions indicate weaker burning. The *cloudy* region represents convective regions

An interesting consequence of the interplay between the convection zone and the hydrogen burning shell can be inferred from the useful homology relation for red giant luminosities given by Kippenhahn & Weigert (1990):

$$L_{\mathrm{RGB}} \propto \mu^{\frac{22}{3}} M_c^7 R_c^{-\frac{16}{3}} . \tag{25}$$

The mean molecular weight does not change significantly since the core is mostly helium, but its increasing mass and radius largely determine the stellar energy output. However, the mean molecular weight does play an interesting role at one point. The convective zone plays the role of a giant mixer. Hydrogen burning leads to an increase in μ, but the deepening convection zone mixes that material with unprocessed material with a lower μ value. Thus when the hydrogen burning shell reaches the point where the convection zone had obtained its deepest penetration (and thereby lowering μ), the stellar luminosity drops. Careful inspection of stellar evolution tracks and isochrones appropriate to globular cluster stars reveals such a drop in luminosity at M_V values comparable to that of the horizontal branch. Lower luminosities imply slower evolution, and hence an increase in the number of stars seen per unit luminosity. The effect may be seen in observational data in terms of luminosity functions. Fusi Pecci et al. (1990) and Sarajedini & Forrester (1995) have shown a nice set of red giant branch luminosity functions, and in a number of cases an excess of stars due to the slower evolution is indeed seen at the

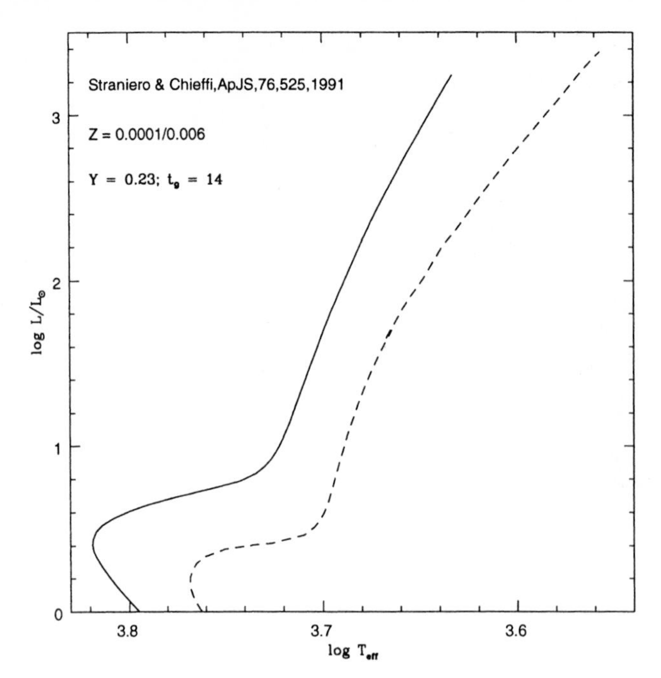

Fig. 9. The helium flash luminosity in the theoretical plane

expected luminosity level. It is a nice testimony to the general validity of the stellar evolution models, including the treatment of convection.

Red giant branch evolution and shell hydrogen burning terminate abruptly when the degenerate helium core becomes hot enough for the "triple–α" process to commence. Because ignition occurs under degenerate conditions, it is extremely difficult to model the event. Nonetheless, it is somewhat easier to estimate *when* it occurs. Buzzoni et al. (1983) have provided a useful approximation:

$$\log L_{\rm flash} \approx (0.75 - Y)\, M_{\rm RGB} + 0.09 \log Z - 1.12\, Y + 3.93\,. \tag{26}$$

As might be guessed due to its effects on the mean molecular weight, the helium mass fraction is fairly important while that of the heavy element mass fraction is minor. Nonetheless, we do not expect significant variations in Y among the most metal-poor stars, so it is worth concentrating on the effects of Z. Figure 9 compares the 14 Gyr isochrones for $Z = 0.0001$ and 0.006 at $Y = 0.23$ (from Straniero & Chieffi 1991). As we saw in Fig. 7, the more metal-rich track is, generally speaking, shifted to cooler temperatures, mostly due to the opacity effects. The more metal-rich track also achieves a higher luminosity at helium ignition, consistent with (26). But comparisons of theory with observations must be made carefully, as shown in Fig. 10. Here the same models have been transformed into the observational plane M_V *vs.* $B-V$, and now the more metal-rich sequence appears to have a *lower* luminosity. This

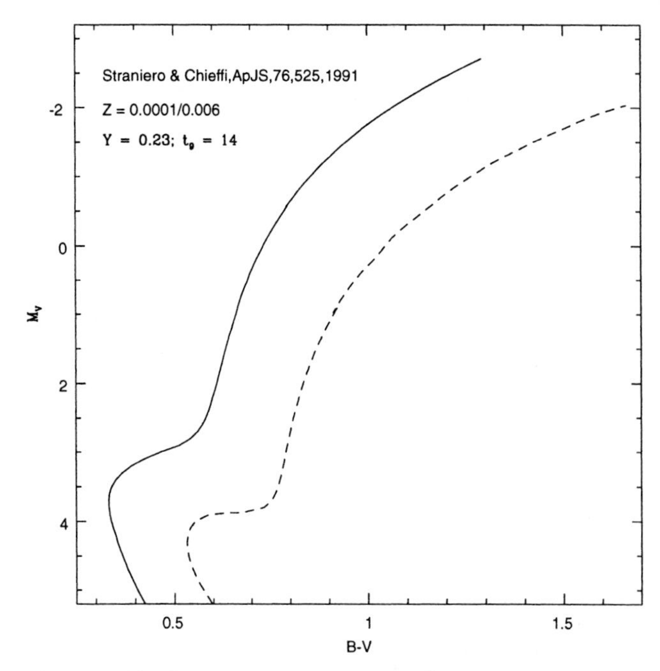

Fig. 10. The helium flash luminosity (represented by absolute visual magnitude) in the observational plane

is due primarily to the fact that the cooler temperatures seen in Fig. 9 for the higher metallicity stars result in larger bolometric corrections. The concomitant increased line blanketing also leads to larger bolometric corrections. In other words, M_V becomes a less reliable indicator of bolometric luminosity at the red giant branch tip as the metallicity increases. Fortunately, there are bandpasses where these effects almost nullify one another and the absolute magnitude of the red giant branch tip becomes almost independent of metallicity. Da Costa & Armandroff (1990), Freedman et al. (1991), Lee et al. (1993), and Madore et al. (1997) show in particular that the Cousins I bandpass satisfies this criterion, so that M_I of the red giant branch tip may be a good "standard candle", relatively insensitive to metallicity.

The alert reader may have noticed that Figs. 3 and 7 suggest that the color difference between the main sequence turn-off and the base of the red giant branch is fairly sensitive to age and not as sensitive to Y. We will see later that this may be used effectively as a means to compare relative ages of comparable metallicity clusters, but the detailed comparisons between observational data and theoretical models is difficult. As noted already, the conversion of color indices into temperatures and bolometric corrections is a problem. But there is more.

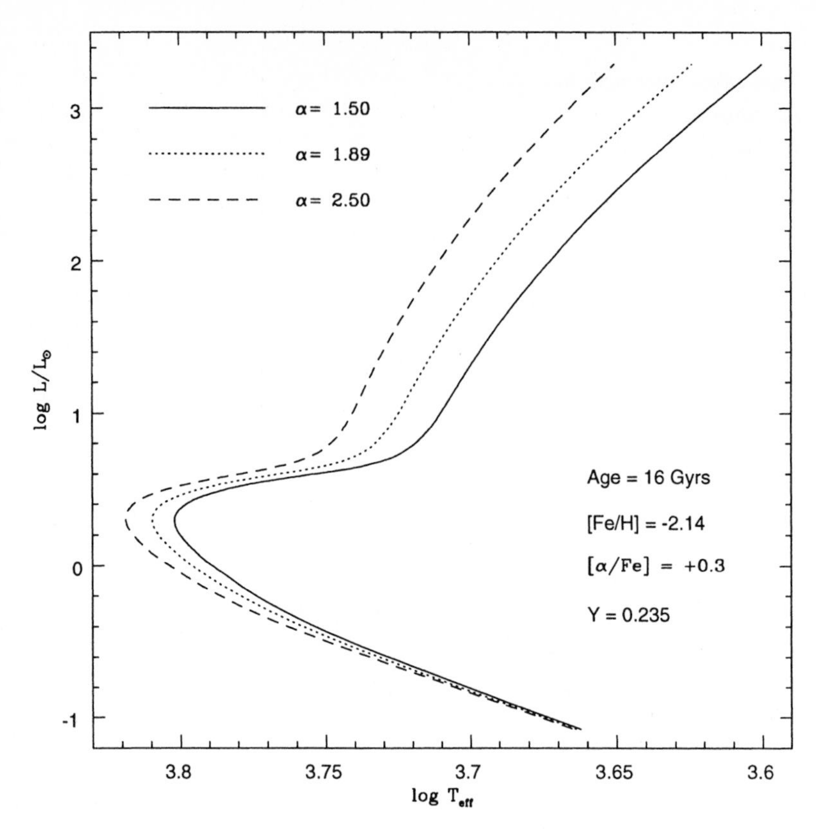

Fig. 11. The effects of the mixing length parameter α, using isochrones computed by D. VandenBerg (private communication)

First, as we have noted, convection is handled by mixing length theory, and α_p, the ratio of the mixing length to the pressure scale height, is more or less a free parameter. It can be fixed by comparing models with the Sun, but the validity of its extension to other types of stars is less obvious. Further, the more convective a star is, the more important that α_p value becomes. This is best illustrated in Fig. 11, following VandenBerg (1983). As expected, the effects are largely a function of the effective temperature. The luminosities of the non-convective cores of low-mass stars are not sensitive to the value of α_p, but the stellar radius is quite sensitive since the outer envelopes are convective. Increasing α_p decreases the stellar radius because it forces a steeper temperature gradient to carry the flux. To emit the same total luminosity, the surface temperature must rise to compensate. And the effects are greater for the cooler, more convective stars. Thus the predicted temperature or color difference between the main sequence turn-off and the base of the red giant branch depends on how convection is handled. The *observed* differences do not depend, of course, on the theoretical methodology, assuming that the physics is the same everywhere and that secondary effects are negligible.

One secondary effect may be important, however. We have seen that the heavy element abundances are important due to their opacity effects, and have also noted that metal-poor stars in globular clusters and the field do not show solar elemental abundance ratios. R. Rood noted the importance of non-solar abundance ratios long ago, and the effects have been shown fairly recently in a nice fashion by Salaris et al. (1993). Figure 12 is related to their work, and is supplied by Don VandenBerg. It shows the relative effects of "high-temperature" and "low-temperature" opacities on the main sequence and red giant branches. Oxygen is a "high-temperature" opacity source, whereas magnesium, calcium, and iron are "low-temperature" opacity sources. Note that increasing oxygen alone shifts the main sequence and turn-off due to its important role in the interior opacity. However, its has little or no effect on the position of the red giant branch. The important continuum opacity source in cool stellar atmospheres is H^-, and at cool temperatures the electrons do not come from hydrogen but from the metals, even though they may be depleted by one or two orders of magnitude compared to the Sun. The ionization potential of oxygen is high, very similar to hydrogen in fact, so increasing the oxygen abundance does not lead to an increase in the electron number density and, hence, to an increase in the continuum opacity that would lead to a larger radius and hence lower temperature at fixed interior luminosity. Increasing *all* the "α" elements, which are the primary electron donors, does lead to such an increase in the H^- opacity.

In conclusion, the comparisons of observed red giant branch sequences with theory are encouraging given the evidence for mixing (to be discussed later) and the "bump" in the luminosity functions, but care is required in understanding the limitations of such comparisons. Convection and chemical composition play important roles, and, as we will see, mixing is apparently much more extensive than current models predict.

Horizontal Branch Evolution. Following the ignition and stabilization of core helium burning and some shell hydrogen burning, low mass stars as are found in globular clusters settle into what is traditionally called the "horizontal branch" (HB) stage of evolution. The name arises, as may be seen approximately in Fig. 3, because the stars have comparable luminosities. Observationally, the wide range of temperatures seen along the HB leads to a very large range in bolometric correction, and, hence, very large differences in the absolute and apparent V magnitudes. Figure 13 shows a cluster whose HB extends to such high temperatures that the hot side "droops" to very faint magnitudes due to this effect.

The horizontal branch is often subdivided into three sections: based on the cool and hot edges of the instability strip. Red HB (RHB) stars extend almost to the red giant branch, while blue HB (BHB) stars extend to very

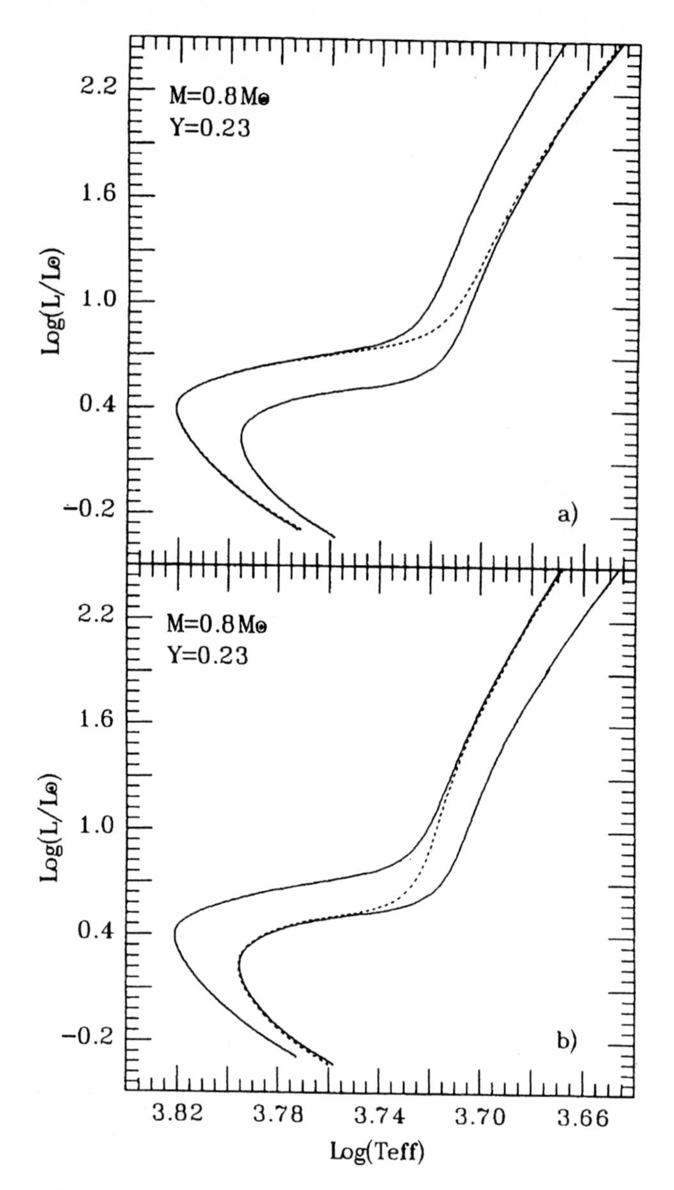

Fig. 12. The effects of interchanging opacity tables for $Z = 10^{-3}$ and 10^{-4} with each other below temperatures of 12,000 K. This approximately represents the effects of the α elements on the photospheres. In the upper panel one sees increasing the low-temperature opacity of a metal-poor model does not affect the main sequence much, but does move the red giant branch track to one of similarly high metallicity. In the lower panel, the low-temperature opacities are reduced, and the red giant branch track shifts to follow the low-metallicity track. The figure was supplied by Dr. D. VandenBerg

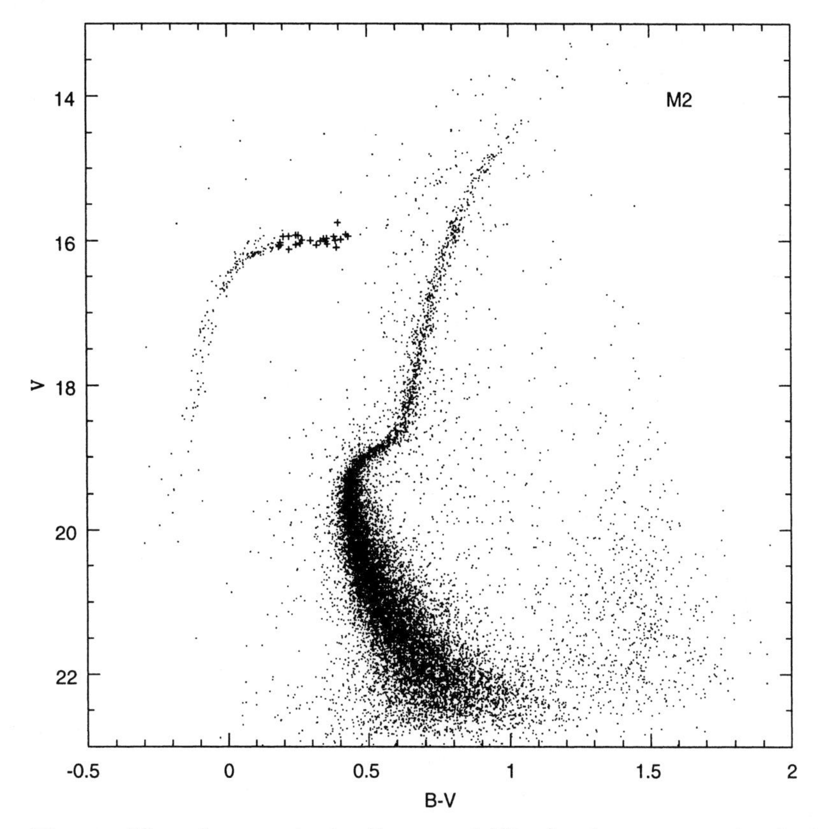

Fig. 13. The color–magnitude diagram of M2, showing an extensive horizontal branch. The RR Lyrae variables are denoted (*plus signs*)

high temperatures. Stars within the strip are RR Lyrae variables. At low enough temperatures, the pulsation mechanism is damped by convection. As one crosses the strip from lower to higher temperatures, one first encounters long-period, low-amplitude variables, then shorter period and larger amplitude stars. The reason for the change is simple enough: with the luminosities roughly constant, a higher temperature means a smaller radius and, hence, a higher stellar density. Since

$$P \propto \varrho^{-\frac{1}{2}} , \tag{27}$$

the periods become shorter as the surface temperature rises. The larger amplitudes result from the decreasing depths of the hydrogen and helium ionization zones. There is a transition period and temperature at which the pulsation changes from the fundamental mode to the first overtone mode, and these stars are readily discerned from their positions in period–temperature planes as well as from their light curves. The fundamental mode pulsators, called RR_{ab} variables, show asymmetric light curves, while the first overtone

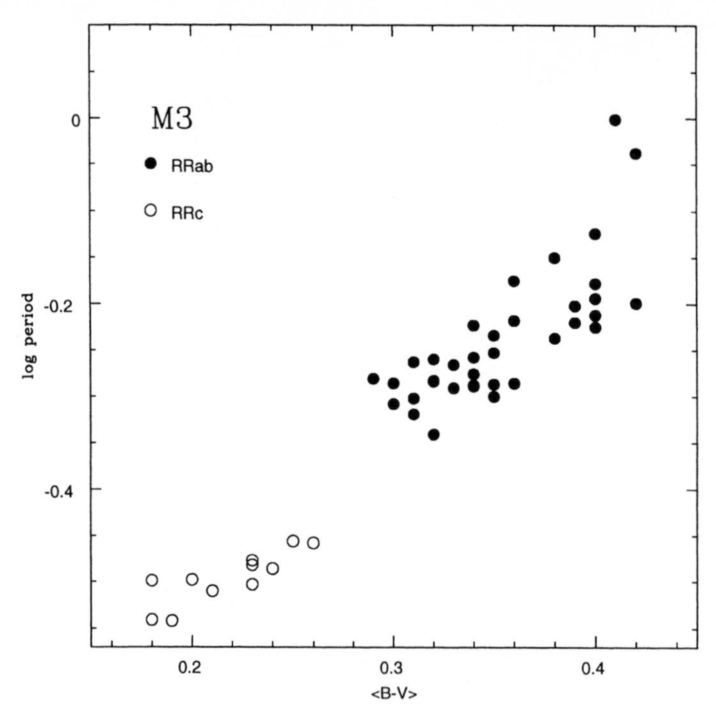

Fig. 14. M3 RR Lyrae periods *vs.* magnitude-averaged $B-V$. Note the clear separation of fundamental (RR_{ab}) and hotter, denser, first overtone pulsators (RR_c)

pulsators, called RR_c variables, show sinusoidal light curves. There are a few variables pulsating in both modes, known as RR_d variables (see Clement et al. 1993 for a recent summary). Figure 14 shows the magnitude-averaged $B-V$ color index plotted against the logarithm of the pulsation period for a well-studied subsample of RR Lyraes in the globular cluster M3 (NGC 5272), with data taken from Sandage (1990a). Figure 15 shows the variation of amplitude in the B bandpass with the logarithm of the period. For more details, Smith's (1995) excellent book should be consulted. One interesting oddity is that the mean periods of the RR_{ab} variables in the Galaxy's globular clusters and in its metal-poor field stars seem to cluster around two values: 0.55 and 0.65 day, called Oosterhoff I and Oosterhoff II clusters, after the astronomer who first drew attention to this phenomenon (Oosterhoff 1939). The dichotomy may be more than a simple oddity, as we discuss later.

The luminosity of these stars is determined primarily by the mass of the helium core, and secondarily by the initial helium abundance. Renzini (1977) gives a relation derived from the models of Sweigart & Gross (1976), determined at the point where $\log T_{\rm eff} = 3.85$:

$$\log L(\mathrm{HB}) \propto 3.2 M_{\rm core} + 2.0 Y - 0.04 \log Z \; . \qquad (28)$$

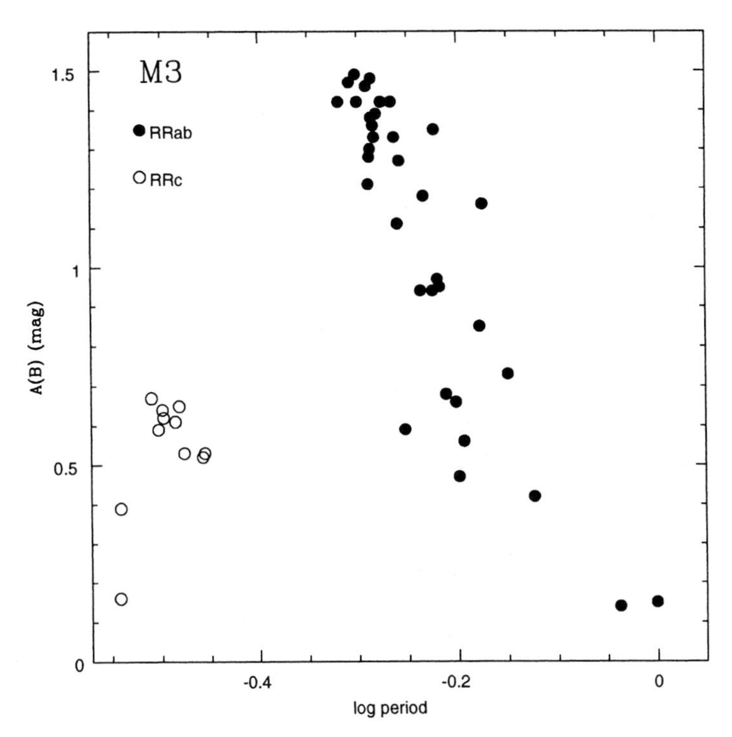

Fig. 15. M3 RR Lyrae periods *vs.* B pulsational amplitudes. Note the rise from low-amplitude pulsation at the cool/red/long period edge of the distribution and the rise to large amplitudes at higher temperatures and densities

Note that increasing the helium abundance increases the luminosity of the HB, while it diminishes the luminosity of the main sequence turn-off (see Fig. 7). This gap between the HB and turn-off luminosities has the advantage of being measured straightforwardly, as well as being insensitive to reddening. In his excellent review of single and binary star evolution, Iben (1991) gives an approximate relation:

$$\log t_9 \approx 1.146 + 1.12(\delta - 1.4) - 1.98(Y - 0.23) - 0.084(\log Z + 3) , \quad (29)$$

where δ is the V magnitude difference in the HB and turn-off luminosities. (See Iben 1971 for several other instructive relationships.) As expected, the gap between the horizontal branch and turn-off luminosities is quite sensitive to Y, but is also sensitive to age. Accurate relative ages of globular clusters can thus be determined using this method, although one must, as always, take care to make certain the clusters' chemical compositions are well understood. Both helium and heavy element abundances must be studied. The helium abundance dominates the heavy element abundance in determining the luminosity of the horizontal branch, while the heavy element abundance is most important in the luminosity of the main sequence turn-off.

Buzzoni et al. (1983) have derived an approximate relation for the lifetime of HB stars:

$$\log t_{\mathrm{HB}}[\mathrm{years}] = 9.51 - 0.22 M_{\mathrm{HB}} - 2.58 M_{\mathrm{core}} - 0.21 Y + 0.01 \log Z \; . \quad (30)$$

As expected from the lower efficiency of the "3α" process, and the higher stellar luminosities, HB stars have much shorter lives than main sequence stars, and hence are far fewer in number. However, red giant stars are also short-lived and thus are few in number, and their numbers relative to horizontal branch and main sequence stars may be measured with reasonable precision in clusters. As Iben (1968) first noted, the relative lifetimes of HB and RGB stars are sensitive to the helium abundance, Y, and so counts of their relative numbers can in principle be used to estimate Y, or at least relative Y values, within clusters. Buzzoni et al. (1983) give a simple relation:

$$Y = 0.380 \log R + 0.176 \; , \quad (31)$$

where R is the number of horizontal branch stars divided by the number of RGB stars more luminous than the HB. (Note that exclusion of post-HB, or asymptotic giant branch stars, must be done, or included in an alternative ratio denoted as R'.) The metal-poor clusters appear to have very similar R values, and hence very similar helium abundances (Buzzoni et al. 1983; Caputo et al. 1987).

While the core mass determines the luminosity of the horizontal branch, it is the total mass, or, alternatively, the envelope mass, that determines the location of core helium–burning stars along the HB, as indicated in Fig. 3. We have seen that for fixed Y, the mass of the main sequence turn-off stars is a function of metallicity, with more metal-rich stars having higher masses at a fixed age. One therefore expects that the more metal-rich clusters would be populated by more massive, hence redder HB stars. Metallicity is thus expected to be the "first parameter" that affects the distribution of stars along the HB. This may be seen observationally in Fig. 16. Cluster metallicities, taken from Zinn (1985), are plotted against the quantity $(B-R)/(B+V+R)$, where B, V, and R refer to the numbers of stars on the BHB in the instability strip, or on the RHB, respectively. (See also the discussion by Fusi Pecci et al. 1993 for alternative means of quantifying the HB color distribution.) Notice that we have restricted the range of clusters' Galactocentric distances, for reasons which will be made clear later. Naively, one might expect that since the red giant stage is so much shorter than the main sequence stage, the total range of stellar masses populating the HB would be small, and hence the spread in color *within* any one cluster should also be small. This is often not the case, as Fig. 13 shows. It appears that there is a stochastic process that varies the amount of mass lost by each star during its evolution to the tip of the red giant branch. Higher mass loss leads to BHB stars; lower mass loss produces RHB stars. How much mass is actually lost from the main sequence turn-off to the horizontal branch stage is very hard to predict.

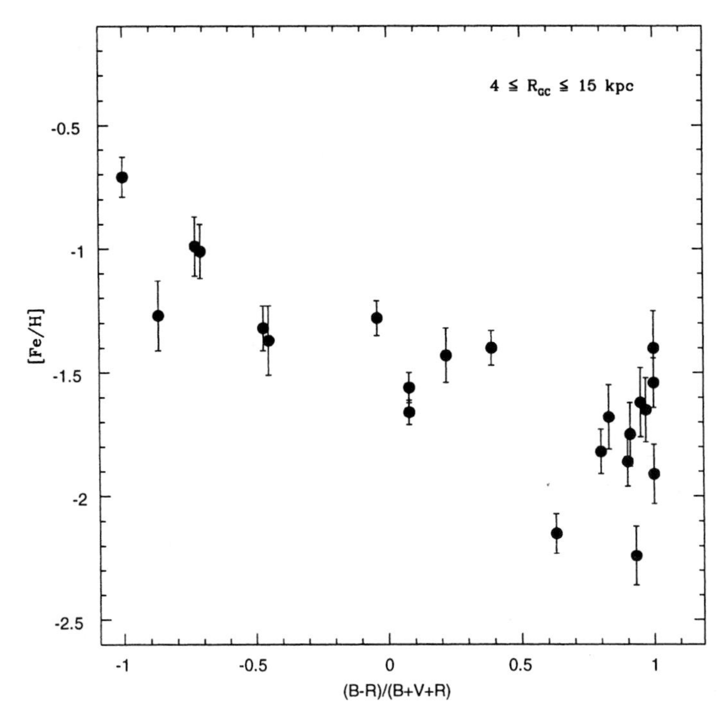

Fig. 16. The color classification of globular cluster horizontal branches *vs.* metallicity. Note that we have restricted the sample to include only clusters lying between 4 and 15 kpc of the Galactic center

There is one model-dependent means available, however, to estimate the masses of some HB stars, specifically the RR_d variables. The "Petersen diagram" plots the ratio of the first overtone to fundamental periods, P_1/P_0 (or Π_1/Π_0 when theoretical values are used) vs. P_0 (or Π_0). Cox (1991) and Clement et al. (1993) summarize some recent results, from which it appears that cluster RR Lyrae variables have $M \approx 0.7 M_\odot$ in both the metal-rich and the metal-poor globular clusters. Taken at face value, the RR Lyrae stars appear to have lost about $0.1 M_\odot$ since they left the main sequence.

AGB & Post-AGB Evolution. The stages of stellar evolution following core helium exhaustion are called the asymptotic giant branch (AGB) and the post-AGB. Shell burning is all that remains prior to the (possible) expulsion of the outer envelope and the formation of a planetary nebula, followed by the emergence of a white dwarf stellar remnant. One can see the AGB stages depicted in Fig. 3, and also that these stages are very brief.

Brevity does not mean, however, that these stages are uninteresting, as Renzini's (1977) review shows. The stars are luminous and cool, and therefore as in their earlier incarnations as red giants, the questions of mixing and mass loss recur. Since the stellar luminosities are again governed by the core mass and AGB stars have higher core masses than RGB stars, we expect the AGB to extend to higher luminosities. But since no stars more luminous than the RGB tip are seen in clusters, mass loss must play a significant role. The nucleosynthesis occurring within the shells is not even steady. The helium shell actually undergoes a series of flashes near the end of the AGB stage, with inter–flash periods measured in only thousands of years. The shell flashes may also promote even more vigorous mixing than had been the case during the RGB phase, and it is possible that the convective shell that develops around the helium-burning shell may reach the hydrogen-burning shell. The importance of this is that the temperatures are certainly high enough to promote vigorous hydrogen burning as well if a supply of protons can be found. Particularly important in the helium-burning stage is the $^{13}C(\alpha,n)^{16}O$ reaction. The neutron flux may become high enough to produce significant enhancements of neutron capture, particularly slow neutron capture, elements, and the vigorous mixing may bring these products to the surface layers. We will comment on the search for such elements later.

Following the final shell flash, a star will begin to shrink in size while it burns the last of its available hydrogen fuel, and will cross the luminosity–temperature plane, as shown schematically by the line in Fig. 3. Howard Bond has drawn attention to these stars over the years, both because of their interesting chemistry, discussed later, and their high and fairly well-defined luminosities. He has in particular pointed out (Bond 1997) that the very large bolometric corrections at the highest temperatures means that such stars may be found at visual magnitudes if they are as late as A to F in spectral type. His studies of field examples and his search for such stars in globular clusters indicate fairly constant luminosities, $M_V \approx -3.4$. These are comparable to Population I Cepheids and so could provide us with a new "standard candle" to measure extragalactic distances, especially in elliptical galaxies where the normal Cepheids are absent.

Planetary nebulae have been found in globular clusters, including one in M15 found by Pease (1928), another in M22 found by Gillett et al. (1988), and two more found by Jacoby & Fullton (2000) in their thorough survey of all globular clusters (one each in Palomar 6 and NGC 6441). As Jacoby et al. (1997) discuss, the frequency of planetary nebulae in globular clusters is lower than expected. Since they argue that a central star mass of at least $0.55M_\odot$ is required to produce a detectable planetary nebula, it may be that most globular cluster stars are not massive enough at the end of their AGB evolutionary phase to produce a planetary nebula. In fact, Richer et al. (1997) have argued that the white dwarf sequence they detected in the nearby globular cluster M4 is consistent with a mean mass of only $0.51M_\odot$. The question then becomes inverted, and one must ask why there are any planetary nebulae in globular clusters, and why the post-AGB stars found by Bond appear to have masses large enough to lead to planetary nebulae. Jacoby et al. (1997) invoke mass transfer within binary systems, which illustrates the point made many times that it is unwise to assume that all stars in clusters evolve as single objects. Note as well that if this explanation for the paucity of planetary nebulae in globular clusters is correct, then the mass lost during the AGB and post-AGB stages may reach $0.2M_{\odot}$.

2 Unusual Features in the H-R Diagram

The ideal case of single stars, uncomplicated by rotation, magnetic fields, mixing, and mass loss, has provided us with an excellent perspective with which to understand what is revealed in color–magnitude diagrams and in the spectroscopy of globular cluster stars. Stellar evolution models explain what we see, and, as will be discussed later, may be put to use to estimate the relative and absolute ages of clusters, and thereby give us the framework of the Galaxy's early history. But improving our understanding of reality proceeds in part by scrutinizing the differences between our models and what we see. Indeed, pathology is often the primary means by which we understand how things work. Here we take up three topics of pathology in the color–magnitude diagram, and in a later section we take up unusual abundance patterns derived via stellar spectroscopy to explore the internal workings of globular clusters stars.

2.1 Binary Sequences

Cecilia Payne–Gaposchkin used to remark that "three out of every two stars is a binary", and if one undertakes a census of all the "stars" to which we have assigned names, the number of individual stars will significantly exceed the number of names. In globular clusters the situation becomes even more interesting since binary systems can be created, altered, or destroyed by interactions with their nearby neighbors. Indeed, globular clusters have proven outstanding laboratories to test models of the dynamical evolution of stellar systems. There is a very thorough review by Hut et al. (1992), in which the whole range of binary systems in globular clusters is reviewed. We will concentrate here on two questions. How do binary systems affect the color–magnitude diagram and what might we infer from it? What are the probable primordial binary frequency and the distribution of orbital parameters?

Binaries and the Color–Magnitude Diagram. As we have seen, the evolutionary stages following the main sequence are rapid, due to the high luminosities and the reduced efficiency per mass of the helium-burning stages. Therefore, for a relatively flat distribution of the probabilities of secondary masses, the secondary is most likely to be a main sequence star. A smaller possibility is that the companion is a stellar remnant such as a white dwarf or a neutron star. An even smaller possibility, given the small range in progenitor stellar masses among post-main sequence stars, is that the companion to a post-main sequence star is another post-main sequence star. The addition of the light from a secondary star is thus not likely to affect significantly the luminosity or color of stars on the upper red giant branch, the horizontal branch, or the asymptotic giant branch. It is the main sequence where the effects of a secondary companion are most pronounced, and, arguably, most important. The reason for this is that for a pair of main sequence stars, the secondary is always no brighter and no bluer than the primary. The addition of the secondary's light thus causes a combined light system that is always brighter and either the same color or redder than the primary. Two stars with the same V magnitude and $B - V$ color index will appear as one star with the same $B - V$ value, but a V magnitude enhanced by 0.75 mag. Fainter and cooler secondaries create combined light systems that lie less than 0.75 mag above the single-star main sequence. This has the effect of appearing to broaden the main sequence in $B-V$ at equal M_V or in M_V at equal $B-V$. If one tries to determine the distance to a cluster whose main sequence has been so broadened to a cluster whose main sequence which has been "cleaned" of binaries by either careful photometry, careful spectroscopy, or by dynamical processes within the cluster itself, a systematic error may result if one tries to fit the mid-points of the color–magnitude of the clusters' main sequences rather than the faint/blue edges. The mid-points will be affected by binaries, after all, whereas the faint/blue side should reflect the single-star distribution.

Are such effects seen in globular clusters? They certainly are, and a particularly nice example is the discussion by Bolte (1992) of the main sequence of NGC 288. His Fig. 12 shows how secondaries will affect the distribution of stars in the color–magnitude diagram, and his data seem to show a broadening of the main sequence that reaches a point 0.75 above its faint/blue side. Bolte also shows a nice method to try to demonstrate the presence of binaries quantitatively. At equal V magnitudes, he shows the distribution of stars' colors relative to a model of Gaussian distributions in color. Subtraction of those Gaussians clearly reveals the excess on the red side and the absence of such an excess on the blue side, consistent with the combined light binary hypothesis, as seen in Fig. 17. So NGC 288, and many other globular clusters that have been studied carefully, show evidence for binary systems.

Primordial Binary Frequencies & Orbital Parameters. One must remember that binaries come in two broad categories: optical and physical. The former are chance superpositions, which are perhaps common in the dense globular clusters. The physical binaries are true binaries that orbit about one another, but the skeptic in us will not permit us to call any one system an optical binary and another a physical binary unless there are orbital data available. At the magnitudes of globular cluster main sequences, radial velocity data are difficult to obtain. And as mentioned already, dynamical processes may alter the frequency of binaries and their orbital elements. While

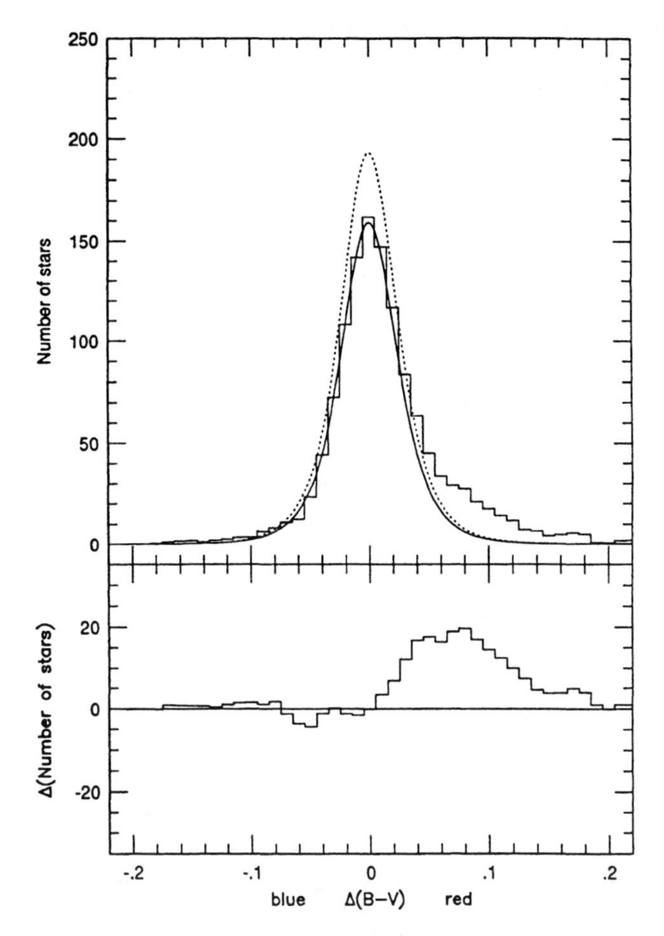

Fig. 17. The color distribution observed on the main sequence of NGC 288, as observed, and then with the subtraction of a Gaussian representing observational errors. The residual, redder than the main sequence locus, represents the contributions of binary systems. Taken from Bolte (1992)

one might attempt to circumvent this second problem by studying low density and dynamically unevolved clusters, one cannot escape the first problem: the stars to be studied are faint.

A possible solution, then, is to study the lowest density environment: the field stars. Since the mass of the halo field population exceeds that of the clusters by a factor of about one hundred, field stars are found much closer to the Sun, and are much more easily studied. Since roughly 1980, Dave Latham of the CfA, Tsevi Mazeh of the University of Tel Aviv, and I, plus several other colleagues, have been studying two samples of stars that bear on the question of binary frequency and orbital parameters for the metal-poor population in our Galaxy. The first and by far the largest program is a photometric and spectroscopic study of 1450 stars selected from the Lowell Proper Motion Catalog. Radial velocities have been obtained using the echelle–Reticon systems of the CfA attached to the 1.5–meter Wyeth reflector at Oak Ridge Station in Harvard, Massachusetts, the 1.5–meter Tillinghast reflector atop Mt. Hopkins, Arizona, and the MMT, also atop Mt. Hopkins. As of July 1998, we have over 28,000 radial velocities for these stars, or about 20 velocities per star. This statistic is slightly misleading, however, because stars with variable radial velocities are observed much more often in an attempt to derive their orbital parameters. But our goal is to have at least ten or so velocities for each star covering a period of at least 3000 days. Some stars have been observed for over twice that long. The radial velocity precision is about 1 km/sec, although it varies slightly with metallicity. To date, 160 single-lined spectroscopic binary orbits have been solved (Latham et al. 1999), as have 34 double-lined spectroscopic binary orbits (Goldberg et al. 1999). (Eighty of these orbits have been published previously: see Latham et al. 1988, 1992.)

The second program is a continuation of the study reported on by Carney & Latham (1986). They observed a number of metal-poor field red giants, all of which had been selected without kinematics bias. (This is therefore unlike the obvious kinematic bias in the above sample.) The radial velocity monitoring has continued. The 90 red giants are all metal-poor, with [Fe/H] ≤ -1.2, as best as we can tell from photometric indicators. Again, we have roughly 20 velocities per star, and, again, there is a bias for more velocities for the velocity variable stars. All stars have at least 8 velocities with coverage spanning about 15 years. To date, we have single-lined orbital solutions for 13 systems. (As expected from the remarks in the prior section, comparable luminosity/mass double-lined spectroscopic binary systems are expected to be very rare among evolved stars, and we have not found any.)

The binary frequency for systems with periods less than 3000 days is $14.5 \pm 4\%$ for the metal-poor red giants. For the 318 stars in the proper motion sample with [Fe/H] ≤ -1.5, there are 44 binaries, for a frequency of $14 \pm 2\%$, whereas for the 484 stars with [Fe/H] ≤ -1.0, there are 65 binary systems, for a frequency of $13.5 \pm 1.5\%$. The binary frequencies are thus all very similar for the metal-poor dwarfs and giants, although it should be recalled that this covers only these relatively short period systems. We can say little at this point about wider pairs with longer periods.

The orbital parameters of the systems studied may be used to address a number of issues, including, for example, the distribution of secondary masses using the mass ratios from the double-lined systems and the mass function and some statistical treatment of the distribution of orbital inclinations for the single-lined systems. One of the more relevant results (for these lectures) is the distribution of orbital eccentricities. As Fig. 18 shows, most of the binaries in our sample, and in the subsamples dealing with only the most metal-poor stars, have fairly eccentric orbits. The mean is around 0.4. The few circular

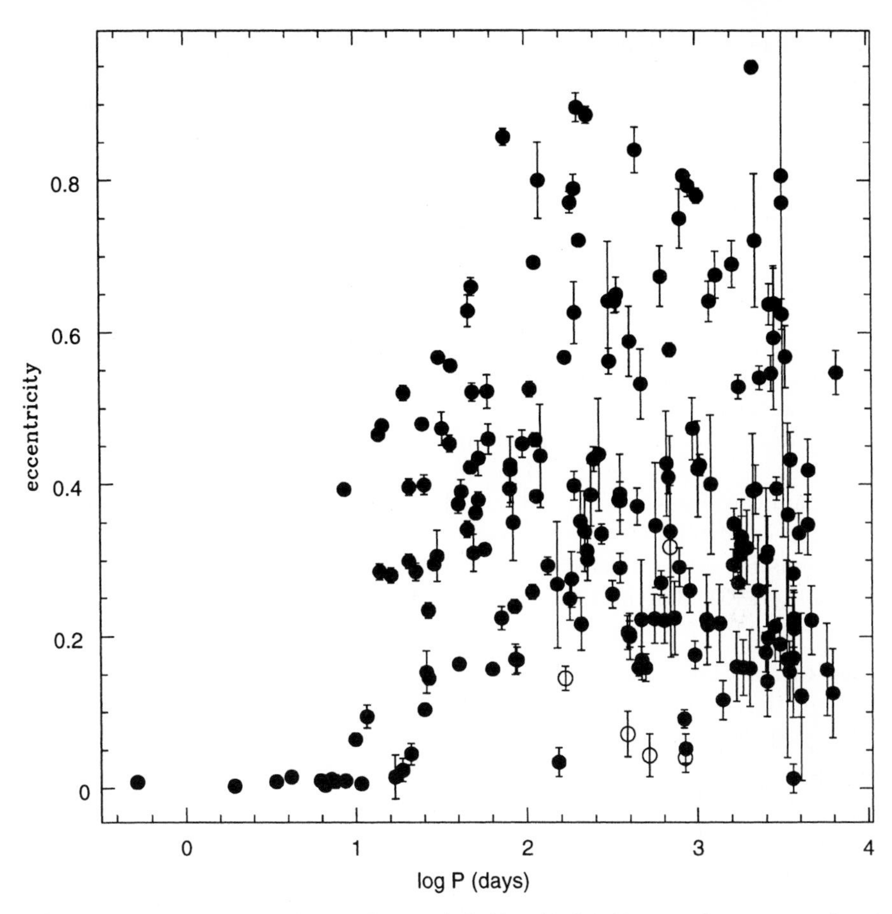

Fig. 18. Orbital eccentricity *vs.* log period (days) of field stars from a sample of proper motion stars. *Open circles* are blue stragglers

orbits tend to be found among the shortest period systems, which can be understood readily as the effect of tidal circularization. Close binaries exert tidal stresses on each other, causing them to first become co-rotating, tidally locked on one another. The orbits are also affected, slowly becoming circular. Since the effect depends on a high power of the ratio of the separation to stellar radii, the close systems circularize first. As time passes, increasingly wide systems become circularized, so that among the metal-poor field dwarfs, essentially all binary systems with periods of 20 days or less are circular (see Latham et al. 1992). The small number of longer period systems with near-circular orbits seen in Fig. 18 tend to be unusual in other regards, as we discuss below.

2.2 Blue Stragglers

"Blue stragglers" were identified by Sandage (1953) when he obtained photometry reaching well below the main sequence turn-off for the globular cluster M3. A small number of stars appeared to define an extension of the main sequence past the turn-off to brighter magnitudes and bluer color indices. Since these stars seemed to have not followed the expected evolutionary timescale, but rather "straggled" behind, the term is both felicitous and descriptive.

Blue stragglers are of particular interest for two primary reasons. First, and most obviously, why do they exist? Is there something wrong with our understanding of stellar evolution and if so, what? Blue stragglers offer us an exercise in stellar pathology. We concentrate on this issue here, arguing for the importance of mass transfer in binary systems. This relates to the second great interest in blue stragglers. As binary systems, which may be hardened, softened, created, and destroyed in dense stellar environments, they are probes of dynamical processes. This section of the lectures is not intended to discuss this aspect of blue stragglers, fascinating as it is. One should consult the numerous reviews of blue stragglers, including Stryker (1993), Livio (1993), Trimble (1993), Bailyn (1995), and Leonard (1996).

Review of Basic Properties. Bailyn (1995) has provided a particularly illuminating Hertzsprung–Russell (H–R) diagram (his Fig. 1, supplied by Dr. Ata Sarajedini) wherein the known globular cluster blue stragglers are plotted in M_V *vs.* $(B-V)_0$ as well as illustrative metal-poor and metal-rich isochrones. We reproduce the diagram here as Fig. 19. Blue stragglers extend from the main sequence turn-off to magnitude levels consistent with about twice that mass (consistent with the limit of mass transfer, although a few stragglers might exceed this limit). Further, while many blue stragglers are on or near the zero-age main sequence, many more appear to lie about a magnitude above it. Some evolution of the primary seems to have occured, therefore, in most cases, either before the putative mass transfer occured, or because of it. While recognized relatively easily using traditional UBV photometry, Ferraro et al. (1997) have shown the much greater power of ultraviolet photometry in identifying such stars.

The blue straggler domain extends into the instability strip, and in consequence a number of them pulsate. Because they are roughly main sequence stars, their densities are high and their pulsation periods are short ($P \propto \varrho^{-\frac{1}{2}}$). Typical values are an hour or less. These are known as SX Phe variables, after their field star prototype. A particularly good discussion of such variables is provided by Nemec et al. (1995) in their study of such stars in NGC 5053. This cluster is low density, which has two advantages. First, crowding is not a major problem. Even without the *Hubble Space Telescope (HST)*, stars may be studied right into the cluster core. Second, encounters with other stars or binaries have been less frequent than in other clusters, so the blue straggler phenomenon itself may be studied more readily. The variability is more than

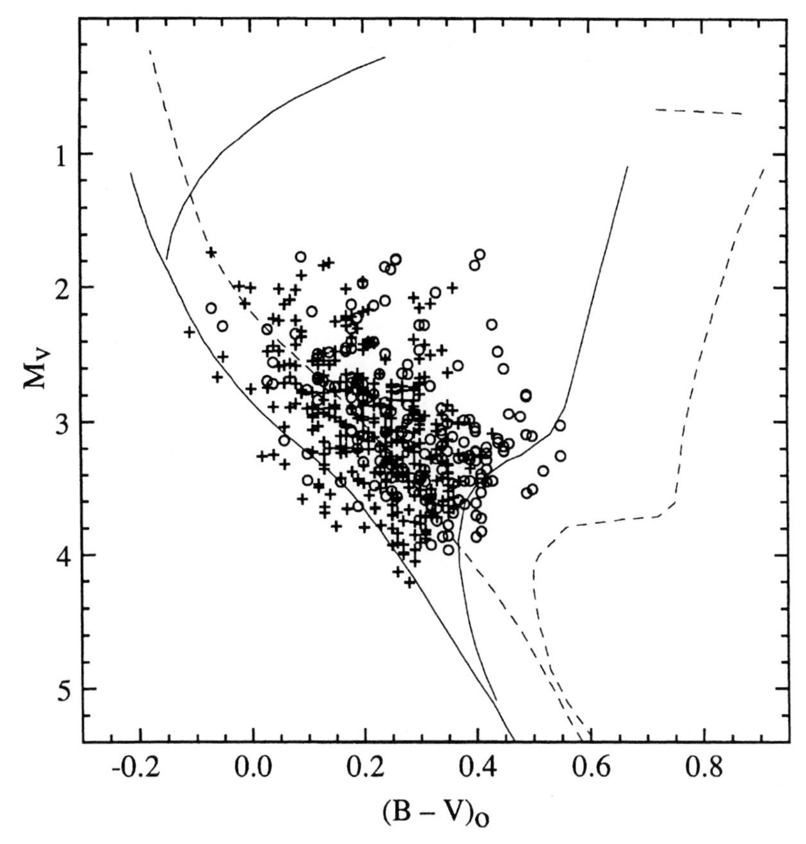

Fig. 19. H–R diagram of 512 globular cluster blue straggler stars. *Open circles* represent clusters with [Fe/H] > −1.6 while *plus signs* represent those with [Fe/H] ≤ −1.6. The blue stragglers in metal-poor clusters appear to be bluer on-average than those in metal-rich clusters. The *solid* and *dashed lines* indicate zero-age and current age tracks for clusters like M92 and 47 Tuc, respectively. Diagram supplied by Dr. A. Sarajedini

a curiosity: it can be a tool. Since the period is related to the density, hence the mass and radius, hence mass, luminosity, temperature, and chemical composition, we may use measured parameters to infer otherwise unobservable ones, such as mass. Nemec et al. (1995) performed such an analysis relative to the cluster's RR Lyrae variables, and assuming they have masses of 0.75 $M_\odot$, the blue straggler masses are 1.2 to 1.8 $M_\odot$, assuming they are pulsating in the fundamental mode. The blue stragglers are indeed more massive than the main sequence stars. The high masses are also consistent with two other forms of mass estimation. First, stars that are heavier than normal stars, be they blue stragglers or binaries or both, will sink to the cluster's central regions over time. Dynamical models of such processes, along with measurements of the degree of central concentration, have provided typical blue straggler mass estimates of 1.2 $M_\odot$ in NGC 5053 (Nemec & Cohen 1989). This does not estimate masses for exactly the same stars as the pulsational masses since only the hotter and presumably more massive blue stragglers occupy the instability strip. Second, Shara et al. (1997) have obtained a spectroscopic estimate of the gravity of a blue straggler in 47 Tuc, along with the luminosity and temperature. From these the star's mass was estimated to be 1.7 ± 0.4 $M_\odot$.

In general, blue stragglers have been found to be concentrated to the central regions of clusters, consistent with dynamical evolution and the greater chances for encounters that would lead to mass transfer binaries. M3 is an interesting exception in that Ferraro et al. (1993) found that the density of blue stragglers relative to other stars is indeed high in the central regions, drops to low levels as one moves toward the outer regions, then rises again about 5′ from the core. One explanation for this dichotomy is that we are seeing the enhanced production of blue stragglers in the cluster core, either during or subsequent to the cluster's formation, coupled with dynamical evolution slightly outside the core. The less dynamically evolved outer parts of the cluster would then include the "primordial" blue stragglers, which may be the easiest to study to determine their cause. But the relatively faint magnitudes make these studies difficult and so once again, we turn to field stars.

Mass Transfer. A variety of explanations for blue stragglers have been advanced over the years. Like most mysteries, much of the evidence is circumstantial and one obtains a solution in the Holmesian fashion of eliminating the impossible, leaving the truth, however improbable, behind. We hope. Delayed star formation in globular clusters can be eliminated since their high velocities and passage through the Galactic plane every 10^8 or so years would sweep them clean of gas. Pulsationally-driven mass loss (Willson et al. 1987) reverses the mystery, suggesting that blue stragglers are normal and that what appear to be normal main sequence stars are merely the remnants of once more massive stars. This hypothesis cannot explain the presence of the fragile element lithium in the normal main sequence stars if their surfaces were once buried deep inside the more massive original star. (Lithium is destroyed easily by proton capture at temperatures near two million degrees.) Besides, the SX Phe instability strip is far from the main sequence turn-offs now observed. Therefore we assume that the blue stragglers are abnormal and inquire why. Internal mixing within stars could provide the central regions with additional hydrogen fuel and prolong the core hydrogen-burning stage, as pointed out by Wheeler (1979a,b) and Saio & Wheeler (1980). Blue stragglers are observed to be deficient in lithium (Hobbs & Mathieu 1991; Pritchet & Glaspey 1991; Glaspey et al. 1994), and mixing is consistent with these observations. The only argument against mixing is the observation by Mathys (1991) that two blue stragglers studied in the old open cluster M67 are deficient in lithium but enhanced in s–process elements. Core hydrogen burning cannot readily

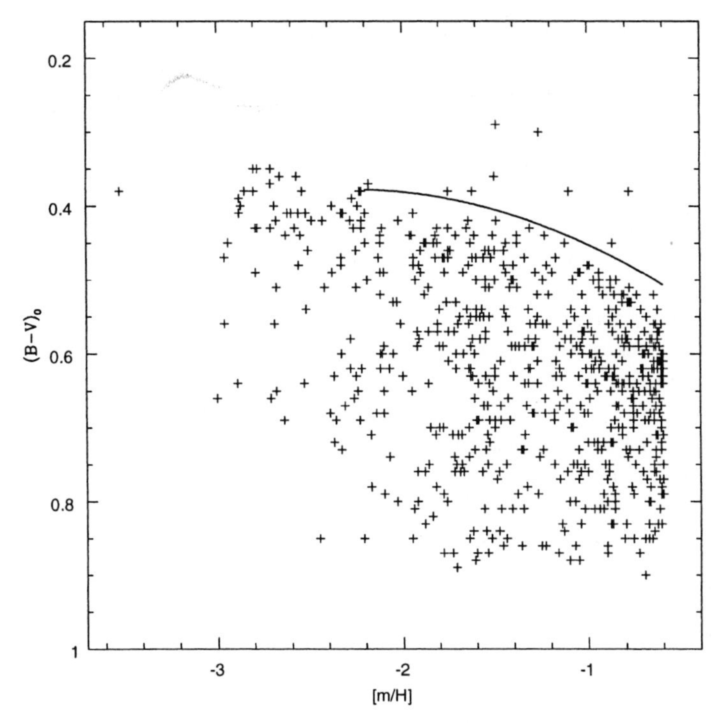

Fig. 20. De-reddened $B-V$ values *vs.* metallicity for stars from the proper motion survey of Carney et al. (1994)

explain this last observation since it does not provide a neutron source. The default and generally preferred model for blue stragglers, or at least the one that may apply to most of them, is mass transfer within a binary system. This can explain the large masses, the lithium deficiencies, and, if the donor was sufficiently evolved, the s–process enhancements. But further tests are needed, including studies of binary frequencies and orbital parameters, and, again, the field stars are most easily studied.

We (Carney et al. 1999a) have approached this by first trying to learn how to identify blue stragglers in the field population. Figure 20 shows the derreddened $B-V$ color index for all stars in our survey of proper motion stars with metallicity $[m/H] \leq -0.6$. The line is defined by the same color index of globular clusters, and we identify all nine stars above this line as blue straggler candidates. The first question we ask is if the binary frequency is different for these stars than those below the line. Of the 711 stars in the survey with such low metallicities, there are 114 single-lined, double-lined, or multiple systems, for a binary frequency of 15.5±1.5%. Of the nine candidate blue stragglers, five are binaries with orbital solutions, for a binary frequency of 56±25%. This is a lower limit to the true blue straggler binary frequency, however. The division between the blue straggler candidates and the normal stars is a bit uncertain.

Two of the stars are extremely close to the line. Small errors in photometry might have moved them into this domain. Second, the Galaxy appears to have accreted a small galaxy which included a fairly young but very metal-poor population (Preston et al. 1994). We (Carney et al. 1994) estimated that one or two of our blue straggler candidates may belong to this population, and thus would not be *bona fide* blue stragglers. (Although they would be explained, in a sense, by delayed star formation, except that it happened outside the Milky Way.) Finally, two of the stars actually show long-term trends in their radial velocities. If they are binaries, we have not observed them long enough yet to compute their orbits. And it is the orbits themselves that speak about the mass transfer mechanism. Figure 18 shows the five blue stragglers with binary orbital solutions occupy an unusual location in the eccentricity *vs.* $\log P$ plane. Four of the five have quite low eccentricities, with $e < 0.15$. This can be understood within the mass transfer model. The initially more massive star begins to transfer mass, and as the mass ratio decreases and approaches unity, the separation decreases, increasing the power of tidal interactions. If the mass transfer continues and the original secondary becomes the more massive star in the system, the separation increases, leading to longer periods, but with a nearly circular orbit. Perhaps the best evidence in support of this model is that several classes of unusual stars, enriched in carbon which was manufactured during helium-burning stages of evolution (i.e., AGB stars), also show high binary frequencies and long-period orbits with small orbital eccentricities. McClure & Woodsworth (1990) found essentially all of the Ba II and CH stars to be binaries. The eight CH stars with orbital solutions show periods of 328 to 2954 days, $e < 0.18$, and $\langle e \rangle = 0.04$. Half of the 16 Ba II binary stars' orbits have $e < 0.1$, and all have low eccentricities compared to normal G and K giants, and their periods range from 70 to 4400 days. Udry et al. (1998a,b) have obtained orbital solutions for 26 strong barium stars. Because there may be more than one mode of mass transfer (Han et al. 1995), we restrict our analysis to stars with orbital periods of less than 2000 days. Including HD 121447 (Jorissen et al. 1995), all 17 stars have orbital eccentricities smaller than the 0.23 value that is typical of normal giant stars in binary systems (Boffin et al. 1992). The mean eccentricity for the strong barium star binaries is $\langle e \rangle = 0.074 \pm 0.018$ ($\sigma = 0.072$). As for subgiant CH stars, McClure (1997) has found that nine of the ten he has been studying are binaries, and six have orbital solutions. The periods are long, 878 to 4140 days, and the eccentricities are small, ranging from 0.09 ± 0.08 to 0.16 ± 0.08. Finally, the first dwarf carbon star to be studied, G77-61, has proven to be a binary with a near-circular orbit and a period of of 245 days (Dearborn et al. 1986). Clearly such long periods and relatively low eccentricities are related to mass transfer, and therefore similar orbital properties for the metal-poor field blue stragglers also suggest mass transfer as the cause of the phenomenon. Indeed, there is one more blue straggler binary worth mentioning: CS 22966-043, an SX Phe variable which Preston & Landolt (1998) have found to be a binary with a period of 431 days and an orbital eccentricity of 0.10. One final test is the secondary mass. If the mass transfer was slow enough to permit the donor star's core to grow more or less normally, the remnant white dwarf should have a nearly normal mass. Using the mass function and the assumption of randomly-oriented orbits, McClure & Woodsworth (1990) found the Ba II and CH stars have secondaries with $M_{\rm sec} \approx 0.6$ $M_{\odot}$, as did McClure (1997) in his study of subgiant CH stars. The five metal-poor field blue stragglers also show $M_{\rm sec} \approx 0.6$ $M_{\odot}$ (Carney et al. 1999).

2.3 The Second Parameter

We have seen that where stars appear on the horizontal branch depends primarily on their mass, or, since their core masses are quite similar, on their envelope mass. Further, since the total lifetimes are relatively short compared to main sequence lifetimes, the total range in masses of stars populating the horizontal branch should be relatively small. Since the metallicity is a primary factor in the masses of stars at the main sequence turn-off, hence at the red giant branch tip, and hence on the horizontal branch, it is the "first parameter". If all stars in a cluster evolved exactly in the same fashion, the total range in the color distribution along a horizontal branch should be small, and that color should correlate with metallicity. But it is not that simple.

The color distribution of stars in horizontal branches is small in some clusters, as expected, such as the metal-rich 47 Tuc, with a predominantly red horizontal branch, or in metal-poor NGC 288, with a predominantly blue horizontal branch. But many clusters show a spread of stars all across the horizontal branch, such as M3 or M5. And some are truly unusual: NGC 2808 has a strong concentration of stars on both the red and blue sides of the instability strip — its color distribution is bimodal. Apparently many stars in these clusters begin their horizontal branch stage with a wider range of masses than expected. Some sort of stochastic process appears to be operating that alters the amount of envelope mass lost during the red giant branch stage. Binaries could be partly to blame. Rotation may also be a cause since if it is strong enough to prolong the red giant evolution, the high luminosities and low gravities would lead to increased mass loss. This problem is not yet solved, despite a large number of earnest efforts. However, this is not the usual definition of the "second parameter".

The "second parameter" (and I will drop the quotation marks henceforth) was first noted by Sandage & Wildey (1967) in their study of the distant and intermediate metallicity globular cluster NGC 7006. It has a metallicity very similar to that of M3 and M13, but the three clusters have very different horizontal branch morphologies. Using metallicities from Zinn (1985) and the color distribution indicator $(B-R)/(B+V+R)$ (Lee et al. 1994) shows the following values: NGC 7006 (−1.59; −0.11); M3 (−1.66; +0.08); M13 (−1.65; +0.97). This is the hotly debated second parameter: it is not the spread within any given cluster, but the variation from one cluster to another, even though

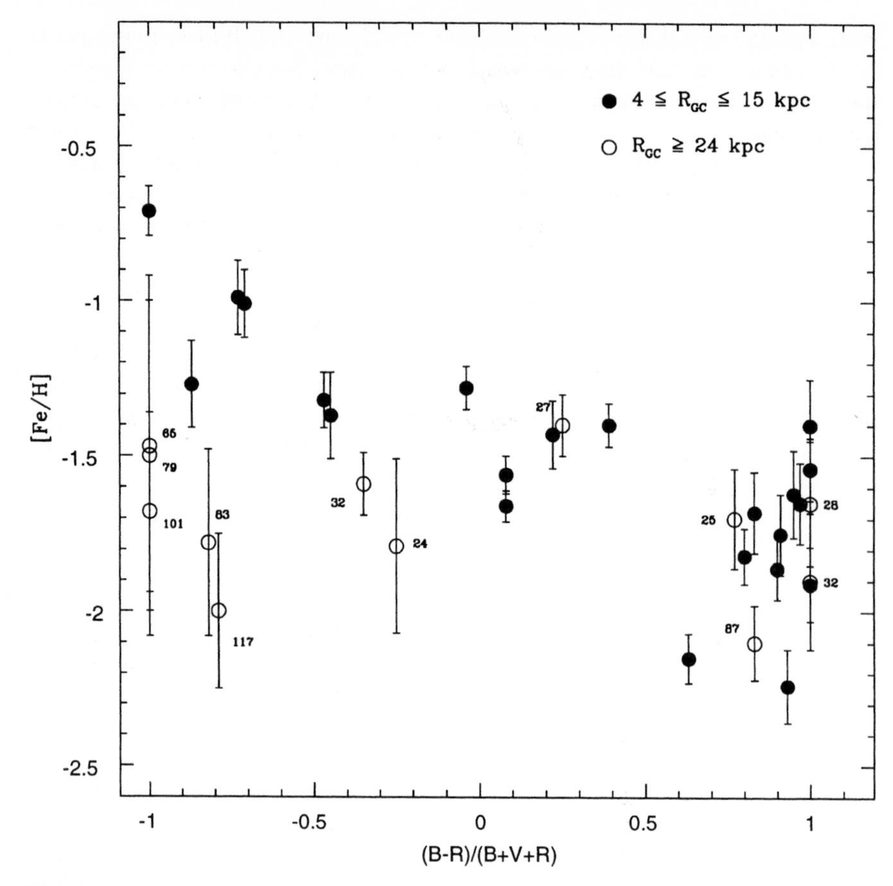

Fig. 21. The colors of local and distant globular clusters *vs.* metallicity. The numbers next to the *open circles* are the Galactocentric distances in kpc

the two phenomena may share a common cause. The second parameter is of obvious interest for students of stellar evolution, as discussed by, among others, Renzini & Fusi Pecci (1988) and D'Cruz et al. (1996), but here we will concentrate on what it tells us about globular clusters and what it might reveal about the history of the Galaxy rather than of individual stars. Its importance was stressed most eloquently in the discussion by Searle & Zinn (1978). Their analysis of the metallicities of globular clusters failed to show any trend with Galactocentric distance. However, many/most of the most distant globular clusters have horizontal branches that are much redder than those of local but similar metallicity clusters. The second parameter is a global phenomenon. Figure 21 shows this graphically. So the second parameter has significant Galactic aspects, and may thus be a vital clue in our study of the formation of the globular cluster system. The extensive debate may be summarized as one of the relative importance of hereditary influences such as chemical composition, age, and rotation *vs.* environmental influences. Searle & Zinn (1978) argued that a plausible explanation for the second parameter phenomenon was age. Younger clusters will have higher mass stars populating the horizontal branch, and hence will have redder horizontal branches. Thus the global behavior of the second parameter suggests that the outermost globular clusters would be younger than the inner clusters.

Lee et al. (1994) have provided an excellent summary and evaluation of the possible hereditary causes of the second parameter, including differences in age, abundances of helium and the CNO elements, and core rotation. Their work is based on model horizontal branches computed using a Gaussian formulation of variable mass loss to model the observed dispersion along the horizontal branch seen *within* clusters. But the *mean colors* of the horizontal branches do change as the above parameters are varied. Differences in helium abundance at fixed metallicity are unexpected since nucleosynthesis must manufacture helium prior to manufacturing the heavy elements. And Lee et al. (1994) are able to rule it out as the cause of the second parameter effect. Recall that increasing the helium abundance increases the mean molecular weight, so a star's luminosity increases, and hence at the current age of the Galaxy, a cluster with a higher helium abundance, other things being equal, will have lower mass stars populating the horizontal branch, so it will be bluer. Another observational consequence is that the fundamental pulsational periods of the RR Lyrae variables will increase at fixed temperature due to higher luminosities (recall the effects of helium abundance on the core mass, which sets the horizontal branch luminosity). This may be tested using period shift analysis, in which one compares stars in different clusters at equal temperature. The expression most often used in these comparisons is that of van Albada & Baker (1971):

$$\log P_0 = 0.84 \log L - 0.68 \log M - 3.48 \log T + 11.497 . \tag{32}$$

Assuming similar masses, this simply restates the idea that at equal temperatures a star with a larger radius, hence larger luminosity, hence lower density, will have a longer pulsational period. The changes wrought by increasing the helium abundance are not seen in period shifts between the different RR Lyrae variables. Therefore differences in helium abundances appear to be ruled out as the cause of the second parameter.

Core rotation also appears to be ruled out by this sort of analysis since core rotation is thought to prolong the red giant branch stage, which extends the duration of mass loss (making the horizontal branch bluer), and also increasing the core mass. Once again the larger core mass and higher luminosity are not consistent with the modest period shifts seen in the clusters' RR Lyraes.

The CNO abundances are probably ruled out on the basis of the comparison between NGC 288 and NGC 362. The horizontal branch morphologies are very different, with that of NGC 288 being very blue [$(B-R)/(B+V+R) = +0.95$], while NGC 362's is very red (-0.87). According to Gratton (1987b) and Carretta & Gratton (1997), their metallicities are very similar, [Fe/H] =

-1.07 and -1.15, respectively, while Gratton (1987a) quotes [O/Fe] = +0.6 and +0.2, respectively. Using a larger sample, Dickens et al. (1991) confirmed the equivalence of [Fe/H] and [CNO/Fe] in the two clusters. Enhanced CNO abundances redden the horizontal branch. Thus taking the [Fe/H] and [O/Fe] results at face value, NGC 288 should have the redder horizontal branch, contrary to what is observed. Before accepting the conclusion that CNO abundances (or other elemental abundances) are not the cause of the second parameter, one should note that this applies to only these two clusters, and may not be representative of whole globular cluster system. As we discuss in a later section, many clusters' evolved stars show signs of deep mixing, including depletion of oxygen as it is transformed into nitrogen (see Fig. 4). A more robust indicator of the initial oxygen enhancements in these two clusters is [Si/Fe] since silicon is not so easily destroyed and in halo dwarfs appears to be very well correlated with oxygen abundances when mixing has not occurred. According to Gratton (1987b), [Si/Fe] = $+0.30 \pm 0.03$ in NGC 288 and $+0.09 \pm 0.08$ in NGC 362, again pointing out that abundances do not seem to explain the large difference in horizontal branch morphology. The issue of chemistry is, however, particularly valid for another second parameter pair discussed by Lee et al. (1994): Ruprecht 106 and NGC 6752. The latter appears to have normal enhancements of the "α" elements, which includes oxygen (see Carney 1996 for a review of cluster abundances), while Ruprecht 106 has, in fact, roughly solar values of [α/Fe], meaning it is depleted relative to normal halo values.

It should not be surprising that arguments have been advanced in favor of environment playing a potentially key role in the second parameter, and in particular through stellar encounters as inferred from the central density ϱ_0. The basic idea is that stellar encounters may foster envelope loss in distended red giants, leading to bluer horizontal branch stars. Perhaps the clearest illustration of the possible importance of central density was provided by Buonanno et al. (1997), in particular their Fig. 1. They showed the cumulative distribution of clusters, which is simply the running sum of the total fraction of clusters studied (beginning at zero and rising to unity when the entire sample has been included), as a function of both the metallicity and the horizontal branch color, defined anew as $B2/(B+V+R)$, where $B2$ is the number of horizontal branch stars with $(B-V)_0 \leq -0.02$. The clusters were divided into two samples according to whether $\log \varrho_0$ [$M_\odot$/ pc^3] is greater or less than 3.0. In the case of [Fe/H], the cumulative distributions of the high- and low-density clusters were a good match. But in the case of $B2/(B+V+R)$, the higher-density clusters showed a strong preference for blue horizontal branches. Buonanno et al. (1997) also attempted to refine the analysis beyond a crude division into two density regimes. They compared $B2/(B+V+R)$ with another horizontal branch indicator, $(B-V)_{\rm peak}$, as defined by Fusi Pecci et al. (1993). This essentially measures the peak of the distribution in color along the horizontal branch. As expected, for clusters with blue or bluish ho-

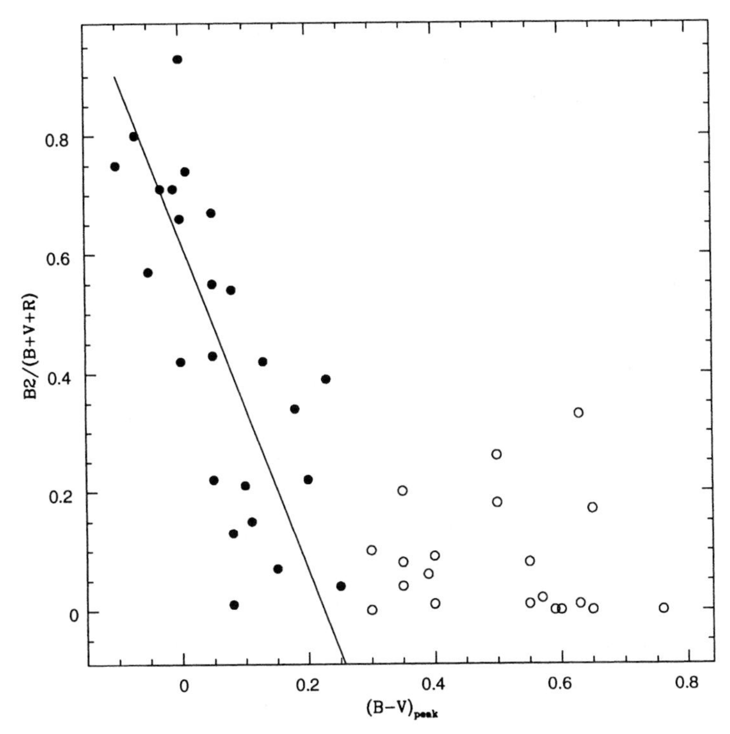

Fig. 22. A plot of $B2/(B+V+R)$ *vs.* $(B-V)_{\rm peak}$, following Buonanno et al. (1997). *Dots* and *circles* represent clusters employed and not employed in determining a linear relation between the two quantities. The *line* gives the linear least squares bisector fit

rizontal branches, the two parameters are related. As $(B-V)_{\rm peak}$ approaches -0.02, $B2/(B+V+R)$ changes dramatically. Buonanno et al. (1997) did a linear regression between the two parameters, and then compared the difference between a cluster's *observed* value of $B2/(B+V+R)$ with the value *predicted* by the regression and then plotted that residual difference *vs.* the cluster's central density. A clear trend was seen: bluer horizontal branches (larger value of $B2/(B+V+R)$ than predicted) correlated with $\log \varrho_0$. All is not so simple, however, and this exercise illustrates the importance of proper care with linear regressions. A glance at Fig. 2 of Buonanno et al. (1997) shows that the linear fit is not ideal: there are more points above the linear fit at one end, and more below it at the other. This is a classic problem resulting from a linear regression of only one variable on the other when both have significant uncertainties. Sarajedini et al. (1997) noted this, and did the inverse fit: they used $B2/(B+V+R)$ as the independent variable rather than $(B-V)_{\rm peak}$. A plot of the residual *vs.* $\log \varrho_0$ showed no significant trends. But this approach also neglects the errors in both variables. In Fig. 22 we

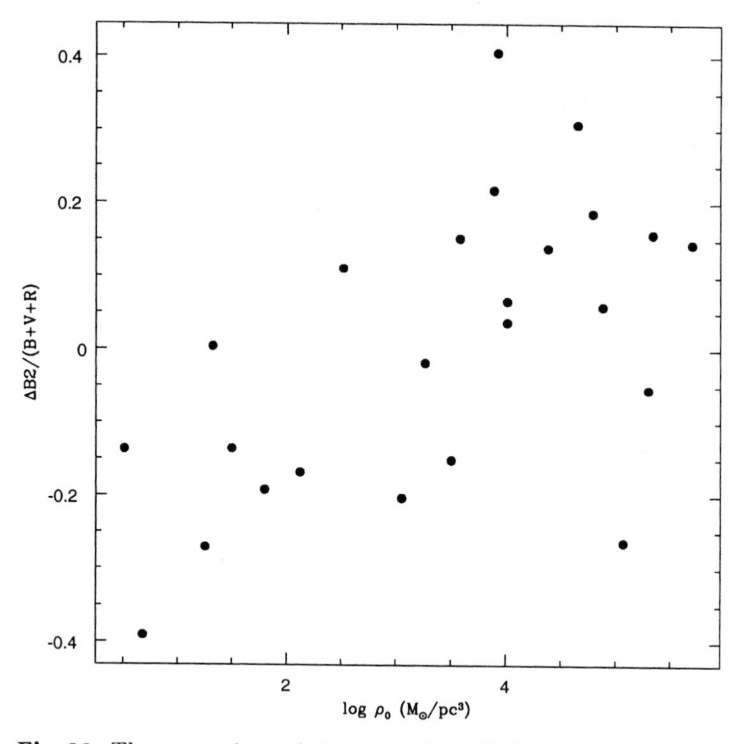

Fig. 23. The comparison of the residuals in $B2/(B+V+R)$ *vs.* $(B-V)_{\rm peak}$ as a function of cluster density, following Buonanno et al. (1997). The difference here is that we have used a linear least squares bisector to determine the slope of the relation for $(B-V)_{\rm peak} < 0.30$

show the data from Buonanno et al. (1997) but with a fit that takes both sets of errors into account: we have done a linear least squares bisector approach (see Isobe et al. 1990) for the clusters with $(B-V)_{\rm peak} < 0.30$, and we show the residuals *vs.* $\log \varrho_0$ in Fig. 23. As found by Buonanno et al. (1997), there is an apparent trend between the "blueness" of the horizontal branch colors and the clusters' central densities. However, it is hard to claim that higher densities lead to dramatically bluer horizontal branch colors and that cluster density and the implied dynamical effects are the major causes of the second parameter effect. There is no obvious trend for $\log \varrho_0 > 2.5$, and almost the entire range of variations seen in $\Delta[B2/(B+V+R)]$ is apparent in the highest-density clusters. The apparent trend in Fig. 23 is caused by the lowest-density clusters, *not* the highest. And here one must be very careful because the clusters we see are survivors of tidal destruction mechanisms in the Galaxy. The lower density clusters can survive only if they remain relatively far from the Galaxy's dense central regions. Thus we must ask if the apparent trend in Fig. 23 might be caused by a *global* trend (low-density clusters have larger Galactocentric distances) rather than *internal* dynamical effects. Consider the five clusters with $\log \varrho_0 < 2.0$ and $\Delta[B2/(B+V+R)] < 0$. Their Galactocentric distances are 11, 93, 16, 16, and 23 kpc, for a mean of 32 kpc and a median of 16 kpc. The other 19 clusters have a mean Galactocentric distance of 8.6 kpc and a median of 7 kpc. In other words, Fig. 23 is consistent with the idea that there is no relation between horizontal branch and color and cluster central density, but that there is a a *global* second parameter effect. The trend in Fig. 23 is due to primarily the inclusion of distant clusters which preferentially manifest the second parameter effect and whose densities are lower than average. Whether younger ages are the cause of the unusually red horizontal branches for these clusters (NGC 288, NGC 2419, NGC 5053, NGC 5466, and NGC 7492) requires careful analyses of their ages relative to other clusters with similar metallicities. We discuss some of these clusters in Sect. 8.

Thus age remains a plausible cause for the second parameter. Perhaps the most suggestive piece of evidence that age is the dominant cause of the second parameter is Fig. 8 of Lee et al. (1994), reproduced here as Fig. 24. Cluster density, indicated by the central density ϱ_0 or the concentration class parameter, c, shows no variation with the horizontal branch type as a function of Galactocentric distance. But as Searle & Zinn (1978) noted, the horizontal branch type is observed to depend on metallicity and on Galactocentric distance. One might expect that age would vary with distance in that the overall gas density was probably higher in the central regions of the young Galaxy, and that this would expedite star formation. We return to this idea in Sect. 8.

3 RR Lyrae Variable Stars

There are many types of variable stars in globular clusters, from the blue straggler SX Phe class to Miras (in metal-rich clusters) and RV Tau variables, plus RR Lyr variables on the horizontal branch and Type II Cepheid variables above it. For a good summary of the classes and lists of such stars within globular clusters, one should consult the 4th edition of the "Catalog of Variable Stars in Globular Clusters", available from the University of Toronto. Nemec et al. (1994) also give an excellent summary. For a very thorough discussion of RR Lyraes, I recommend the book by Horace Smith (1995).

We concentrate here on the RR Lyrae variables because of their special importance in estimating distances to clusters, to the Galactic Center, and to other galaxies in the Local Group. The absolute magnitudes of the RR Lyraes are also of fundamental importance in the estimation of relative and absolute ages of globular clusters. Figure 3 shows why. The age of a cluster is determined from the luminosity of the main sequence turn-off. If we know the luminosity of the RR Lyraes in a cluster, and if we then measure the luminosity, or (in observational terms) the magnitude difference between the RR Lyrae and the turn-off, we have thereby measured the main sequence

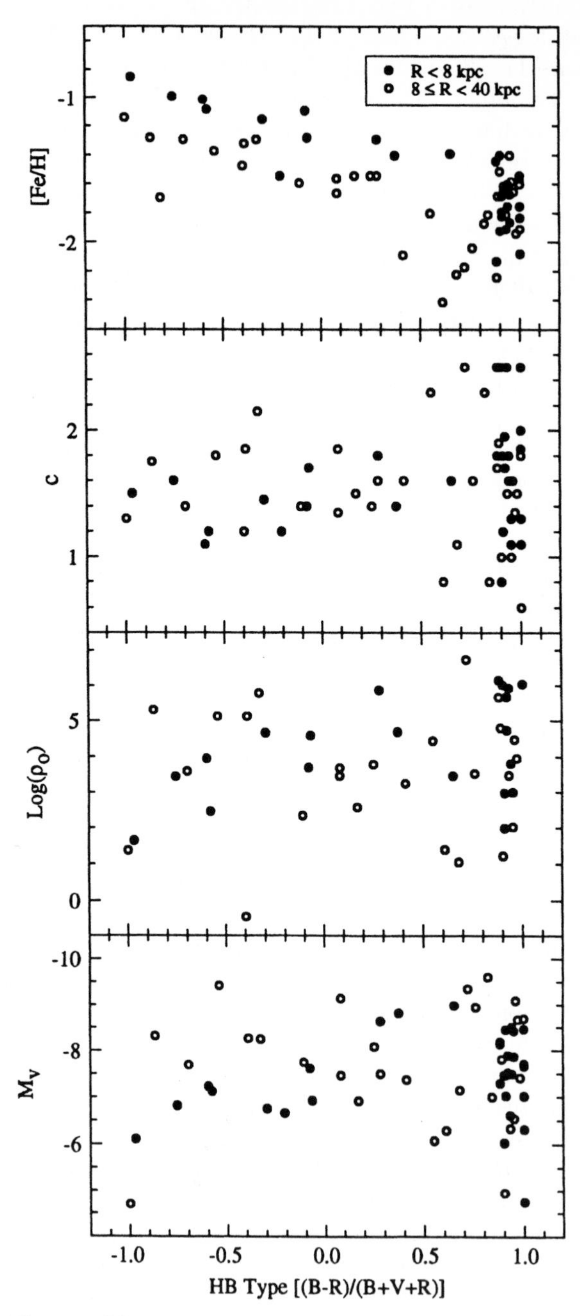

Fig. 24. The variations of the horizontal branch type with various parameters, divided into two distance regimes, taken from Lee et al. (1994)

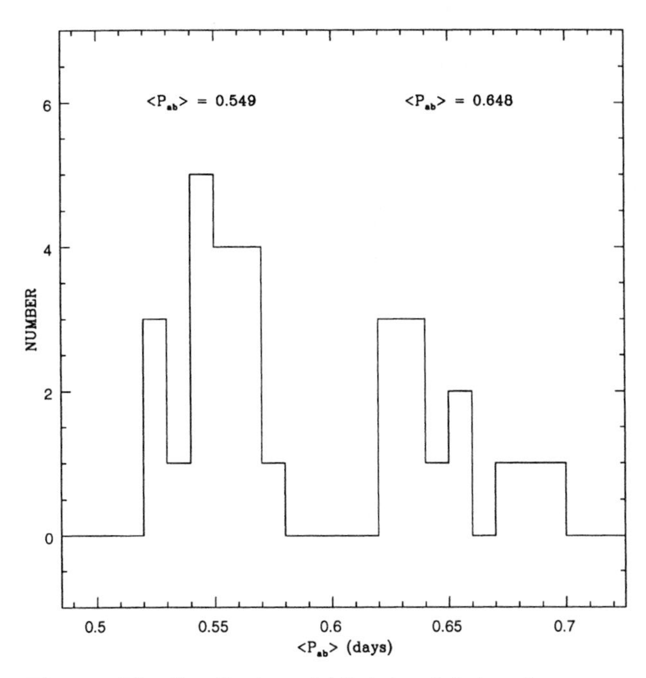

Fig. 25. The distribution of $\langle P_{\rm ab} \rangle$ for globular clusters

turn-off luminosity. This $\Delta V_{\rm TO}^{\rm HB}$ parameter has another very great advantage in that it is independent of reddening. This is the essence of (29).

3.1 Oosterhoff Classes

Although I have referred to the division of the Galaxy's globular cluster RR Lyrae populations into the Oosterhoff I and II classes as an oddity, it may actually prove to be a profound difference. Sandage (1982) has stressed this in particular in analyses of the relative luminosities of the two classes of RR Lyraes. *Why* are there two classes? How and why do their luminosities differ? Figures 25 and 26 show the distributions of the mean $\rm RR_{ab}$ periods $\langle P_{\rm ab} \rangle$ of the cluster and field star samples, showing both are bimodal. Figure 27 shows a richness estimate *vs.* cluster metallicity from Carney et al. (1992), based on the $N_{\rm RR}$ estimate of Suntzeff et al. (1991) (the number per unit cluster luminosity) and $V/(B+V+R)$, which is the fraction of horizontal branch stars which are RR Lyraes. This again emphasizes the Oosterhoff dichotomy, and in particular the apparent discontinuity in the frequency of RR Lyraes around $[\rm Fe/H] \approx -1.6$ that also appears to separate the Oosterhoff I and II classes.

Another reason why this classification and dichotomy may be important is that it appears to be common only to the Milky Way. The RR Lyraes in neighboring dwarf galaxies have different values of $\langle P_{\rm ab} \rangle$ than *either* of

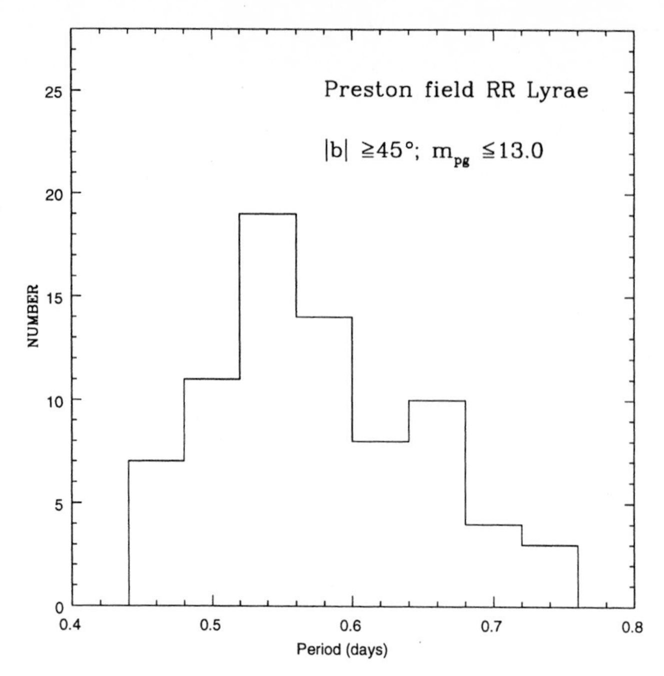

Fig. 26. The distribution of $\langle P_{\mathrm{ab}} \rangle$ for $\mathrm{RR_a}$ Lyraes taken from Preston (1959)

the Oosterhoff I and II classes. The mean periods of the neighboring dwarfs studied to date are: Carina (0.62 day; Saha et al. 1986); Draco (0.61 day; Nemec 1985); LMC (0.58 day; Hazen & Nemec 1992; Alcock et al. 1996); Leo II (0.59 day; van Agt 1973); Sculptor (0.60 day; Goldsmith 1993); and Ursa Minor (0.64 day; Nemec et al. 1988). So why is the Milky Way so special?

3.2 RR Lyrae Luminosities

If the luminosity of the horizontal branch is a function of the helium and heavy element mass fractions, Y and Z, and if changes in Z are linearly related to changes in Y, we would expect that

$$M_V(\mathrm{RR}) = a\,[\mathrm{Fe/H}] + b\,. \tag{33}$$

If we have a collection of $\Delta V_{\mathrm{TO}}^{\mathrm{HB}}$ data, then the slope, a, determines the relative cluster ages because it determines the relative turn-off luminosities. The zero point, b, determines the absolute ages. Relative and absolute ages are discussed later, as is the zero point. Here we focus on the four methods used to estimate the slope of the relation.

Theory. The modelling of horizontal branches is a serious challenge, involving not only introducing sufficient variation in mass loss to populate the

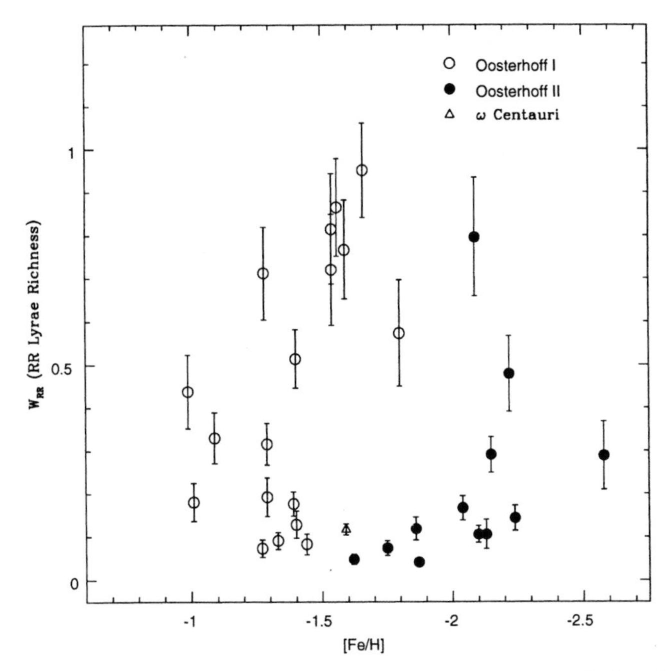

Fig. 27. A measure of the richness of RR Lyraes in globular clusters *vs.* metallicity

horizontal branch for any cluster in a manner consistent with observations, but also to allow for evolutionary effects. Theoretical identification of the zero-age horizontal branch is generally not sufficient because it is not what is measured. Mean magnitudes of the variables are the most reliable observable quantity. The models of Lee et al. (1990) satisfy these demands, and they have obtained the following results.

$$M_{\mathrm{bol}}(\mathrm{theory}) \propto 0.20\,[\mathrm{Fe/H}]\,, \tag{34}$$

and, once the bolometric corrections have been included,

$$M_V(\mathrm{theory}) \propto 0.17\,[\mathrm{Fe/H}]\,. \tag{35}$$

Baade–Wesselink Method. The Baade–Wesselink method is simple in theory and difficult in practice, like so many other things in life. Its basis is simply the definition of effective temperature, (13). As noted already, dividing both sides of the equation by d^2 leads to (14), where θ is the angular diameter. We can measure the changes in the apparent luminosity, $\Delta\ell$, from the changes in brightness of the variable star, and likewise we can, in principle, measure the changes in T_{eff} from changes in color. Thus we can estimate the changes in the angular diameter. From spectroscopy, we may estimate the changes in linear diameter or linear radius, and by comparing that to the changes in angular diameter, we determine the distance. The change in radius, ΔR, is found from

$$\Delta R = \int p\,(v_{\rm rad} - \gamma)\,dt\,, \tag{36}$$

where γ is the star's systemic velocity, obtained by integrating the radial velocity over the pulsational cycle. The factor p is the projection factor, which converts *radial* velocity into *pulsational* velocity. The two are not the same, of course, because the whole star pulsates, but our unresolved observations of the stellar disk includes the disk center, where $v_{\rm rad} = v_{\rm pul}$ as well as the stellar limb, where no radial velocity variations are detected. The projection factor thus depends on geometry, and a good knowledge of limb darkening.

To make the Baade–Wesselink analysis effective, three conditions must be satisfied. First, the color index used to measure the temperature should be as sensitive as possible, which means the longest wavelength baseline possible. Further, because one wants to compare derived luminosities for stars with very different metallicities, the color index employed should have minimal sensitivity to metallicity. And the flux that comes through the two bandpasses that define the color index should form at similar layers within the star throughout its pulsational cycle. Otherwise the color–temperature relation winds up being sensitive to the $T - \tau$ relation rather than only the effective temperature (τ is the optical depth, defined in Sect. 1). For RR Lyraes, the optimal color index has been found to be $V - K$, as proposed initially by Longmore et al. (1990). Second, the changes in the apparent luminosity should be measured at a wavelength that is less sensitive to changes in temperature and more sensitive to changes in radius. Bandpasses near the flux peak are not good in this regard because a small change in temperature can produce large changes in the flux. One wants to separate $\Delta \ell$ and $\Delta T_{\rm eff}$ as cleanly as possible. The K bandpass is nearly ideal, lying on the Rayleigh–Jeans tail of the flux distribution. Finally, one also wants the spectral lines studied to derive the changes in linear radius to form at a constant optical depth, or at least separated from the continuum-forming regions by a constant amount through the pulsation cycle. Otherwise systematic errors will creep in as different layers, moving at differing velocities, will determine $v_{\rm rad}$ and, hence $v_{\rm pul}$. Jones et al. (1987) discussed this last point at length.

Even with great care, the results from Baade–Wesselink analyses are subject to three major cautions. First, the projection factor p cannot be determined empirically. While it is not likely to vary from star to star much, the value selected will have a direct effect on the derived distances and luminosities. This is a primary uncertainty in the absolute luminosities derived from Baade–Wesselink analyses. However, it does not affect the *relative* distances and luminosities, so the Baade–Wesselink method is still a powerful means of measure the slope, a. Second, the slope of the $V - K$ *vs.* $T_{\rm eff}$ relation appears well-determined from synthetic colors, derived from model atmospheres (see the discussion by Jones et al. 1992), but the zero point is still somewhat

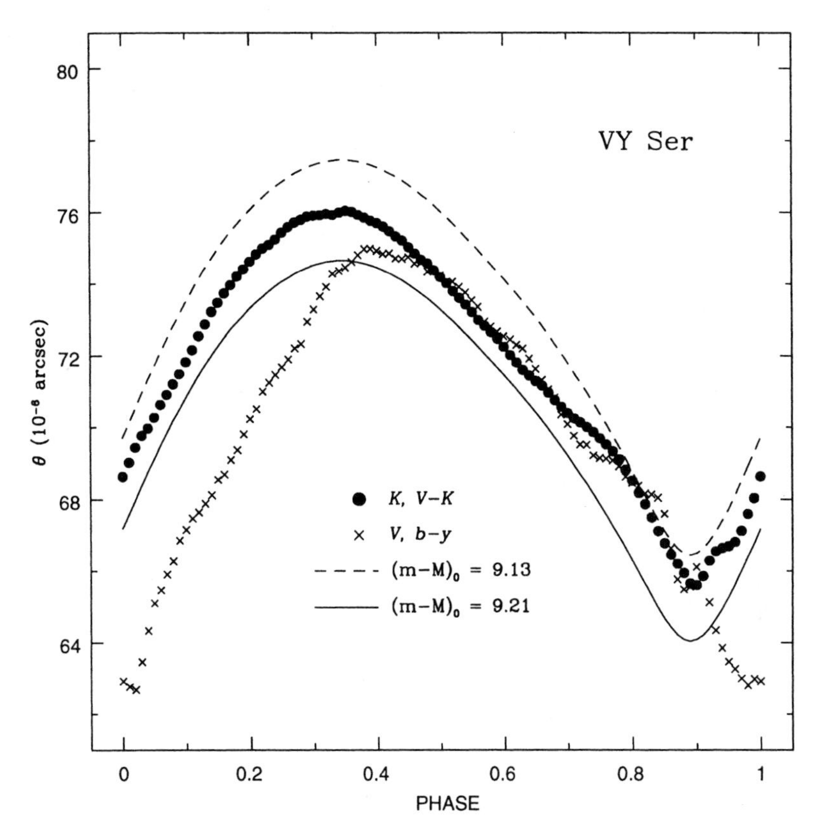

Fig. 28. The angular diameter over the pulsation cycle derived from radial velocities (*lines*) *vs.* derived from optical and infrared photometry

uncertain. It is usually set by comparing the theoretical values with a star whose effective temperature has been well determined (by measuring θ and the apparent bolometric luminosity). The uncertainty in the zero point of the color–temperature relation affects the absolute distances and luminosities, but its well-determined slope leads to reliable values for a. Finally, and this bears on the discussion in the following section, blue colors appear to be unreliable. Part of this is no doubt to the requirement that the flux peak be avoided. But part of it may be due to low-level emission in the expanding photospheres of RR Lyraes. The concern about blue magnitudes shows up most clearly in Fig. 28. Jones et al. (1988) analyzed the photometry and velocities for the metal-poor field RR Lyrae VY Ser, and Fig. 28 shows the angular diameters derived from the radial velocities adopting two different distances compared with those derived from the optical and infrared photometry, respectively. The optical photometry leads to serious mismatches between the varying angular diameters, while the infrared photometry provides a good match over

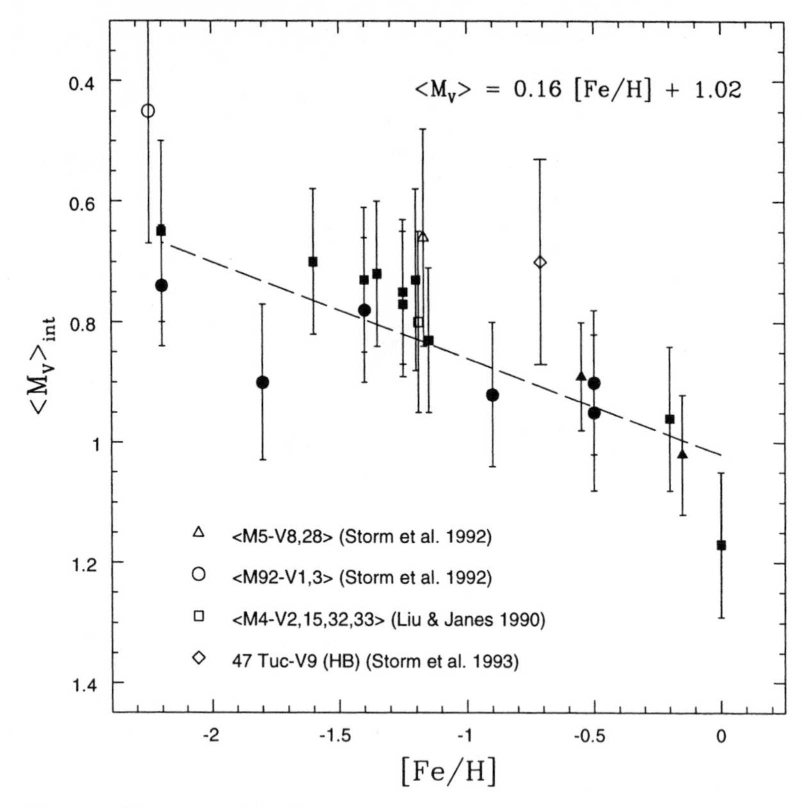

Fig. 29. The results from Baade–Wesselink analyses of field and cluster RR Lyraes obtained using infrared photometry

the full pulsational cycle. What this means, in turn, is that the $B-V$ and $b-y$ color indices are probably good indicators of $T_{\rm eff}$ only during parts of the pulsational cycle, especially those following maximum radius. This means that such color indices are not good during the rise to maximum radius, nor when they are used by averaging over the pulsational cycle.

Figure 29 summarizes the results from analyses of both field and cluster RR Lyraes (the former from Carney et al. 1992; the latter from Liu & Janes 1990a,b and Storm et al. 1994a,b). Note that the value for 47 Tuc is not for the highly evolved variable V9, but rather the implied luminosity of the cluster's horizontal branch. Within the limits of the scatter, there appears to be a linear relation between M_V(RR) and [Fe/H], with a slope of 0.16 mag/dex. A similar figure but using $M_{\rm bol}$(RR) leads to a slope of 0.21 mag/dex. How does one test if this is plausible, aside from comparing the results with those obtained by other methods? It turns out that there is a relatively well-defined relation for RR Lyraes between M_K and $\log P$. One expects such a relation for the following reason. The RR Lyraes have roughly constant luminosities within

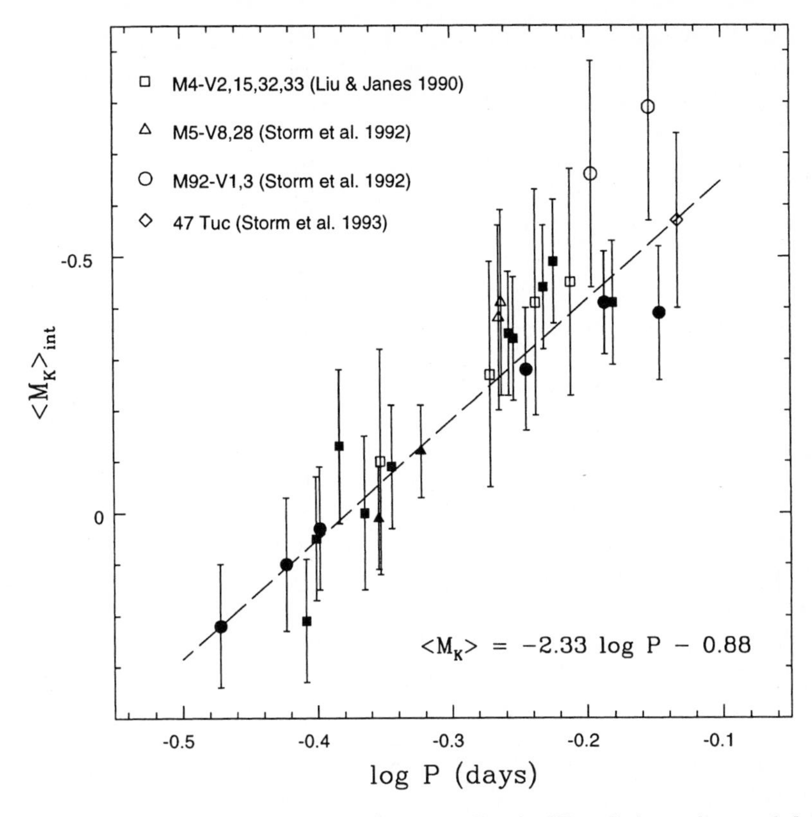

Fig. 30. The M_K *vs.* $\log P$ results from Baade–Wesselink analyses of field and cluster RR Lyraes obtained using infrared photometry

a cluster, which is of course what makes them so interesting. Since $V-K$ is an excellent temperature indicator, so is $M_V - M_K$. Since M_V is roughly constant, M_K is proportional to $T_{\rm eff}$. Since the luminosity is constant, the temperature and radius are highly correlated, so M_K is also proportional to the stellar radius. Since the masses are very similar for all RR Lyraes within a cluster (and from cluster to cluster, as we have seen), M_K is related to the density and hence to the pulsation period. Theory predicts a slope of -2.22 in the M_K *vs.* $\log P$ relation (Longmore et al. 1990), and this is supported by observations in globular clusters. In the cluster ω Centauri, where there is a wide range of [Fe/H], the slope is observed to be -2.28 ± 0.07 (Carney et al. 1992). In the six globular clusters in which more than 20 RR Lyraes have been studied, the slope is -2.31 ± 0.06, using data from Longmore et al. (1990). Thus if the Baade–Wesselink method has produced correct relative distances of the field and cluster RR Lyraes, a similar slope should be found. (Note t[illegible] here the variable V9 in 47 Tuc is plotted by itself: its agreement sug[illegible] lower-than-average mass, according to Storm et al. 1994b.) Figure [illegible] that the derived slope is −2.33, in excellent agreement. This indicates that the Baade–Wesselink method yields reliable *relative* distances and hence a reliable value for the slope of the M_V *vs.* [Fe/H] relation.

Period Shift Analyses. Equations (27) and (32) provide the basis for estimating relative luminosities of RR_{ab} stars. We may include the first overtone RR_c variables if we "fundamentalize" their periods, following van Albada & Baker (1971),

$$\log \frac{P_0}{P_1} = 0.095 - 0.032 \log \frac{M}{M_\odot} + 0.014 \log \frac{L}{L_\odot} + 0.09 \log \frac{6500}{T_{\rm eff}} . \quad (37)$$

For typical RR Lyrae masses of 0.7 $M_\odot$, $L \approx 45\ L_\odot$, and $T_{\rm eff} \approx 7200$ K for RR_c variables (see Carney et al. 1992), one may change $\log P_1$ into $\log P_0$ by adding about 0.12.

As described by Sandage (1990b), the essence of the period shift analysis is to compare variables at equal temperatures. Then differences in $\log P_0$ are directly related to differences in $\log L/M^{0.81}$. Since we have seen that the mass differences are quite modest, relative luminosities may be derived readily within a globular cluster, which may be compared to what is actually observed, as discussed later in the case of M2. More important, comparisons may be made between clusters or between field stars. Sandage (1990b) undertook the first extensive comparison, using pulsation periods, light curves, and photometry for variables in ten globular clusters, as well as a large sample of field stars taken from Lub (1977). One of the key problems in this sort of analysis is how to define the variables' mean temperatures, or more exactly, the equilibrium temperatures (the temperature the star would have were it not pulsating), of the variables. Originally, Sandage (1982) employed the blue pulsational amplitude, which, as Fig. 15 shows, is probably a good indicator of relative temperatures within a globular cluster, with the added benefit of being a measurable quantity that is independent of reddening. The period shift is then apparent in Fig. 31, where we compare the data for M3 variables from Carretta et al. (1998) with those in M2 from Lee & Carney (1999a). The two clusters have very similar metallicities. At equal A_B values, the M2 variables have longer periods, and, by inference, lower densities, larger radii, and greater luminosities. Because the relation between A_B and $T_{\rm eff}$ may have some metallicity sensitivity, Sandage (1990b) utilized the magnitude-weighted mean $\langle B - V \rangle_{\rm mag}$ color index and a model atmosphere-based calibration of $B - V$ *vs.* $T_{\rm eff}$. Neglecting the effect of differing masses, Sandage (1990b) found the $M_{\rm bol} \propto 0.35$[Fe/H] for the clusters. Similarly, the field stars led to $M_{\rm bol} \propto 0.25$[Fe/H]. The cluster result certainly does not agree well with the Baade–Wesselink analyses.

There are two potentially significant problems with the period shift analyses, illustrating again the subtleties that sometimes emerge in astronomy.

First, we have noted that the Baade–Wesselink analyses suggest that blue bandpasses are not a reliable measure of the temperature during at least

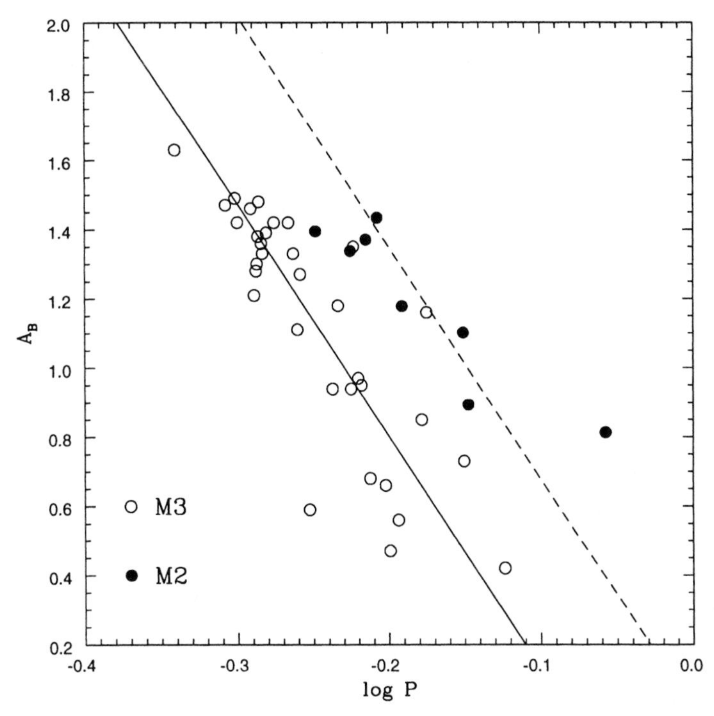

Fig. 31. The period shift of M2 *vs.* M3. Note that at equal amplitude and (probably) temperature, the M2 variables have longer periods, hence lower densities, larger radii, and higher luminosities

part of the pulsation cycle of RR Lyraes, and hence that $\langle B-V\rangle_{\rm mag}$ may not accurately reflect $\langle T_{\rm eff}\rangle$, even if the $B-V$ *vs.* $T_{\rm eff}$ relation is accurate. A new temperature indicator is required, and if possible, one that is independent of reddening. Carney et al. (1992) used the results from the Baade–Wesselink method and their infrared photometry to derive $\langle T_{\rm eff}\rangle$ for a variety of field stars. They then derived a relation between Θ ($= 5040/T_{\rm eff}$) and three reddening-free parameters:

$$\Theta = 0.261 \log P_0 - 0.028 A_B + 0.013\,[{\rm Fe/H}] + 0.8910\,. \qquad (38)$$

The utility of this relation may be tested by undertaking a period shift analysis *within a single cluster* and comparing what the period shift predicts in terms of $\Delta M_{\rm bol}$ and comparing that to the *observed* differences in V. Equation 38 produces an excellent correlation between the predicted values of $\Delta M_{\rm bol}$ and the observed values of $\Delta V = \Delta M_V$ in M3 (Carney et al. 1992). Use of this revised temperature calibration applied to nine globular clusters results in $\Delta M_{\rm bol} \propto 0.18[{\rm Fe/H}]$, half the slope found by Sandage (1990b) and in agreement with the Baade–Wesselink results. Note that the variable V10 in M2 has an unusually long period compared to other variables in the cluster.

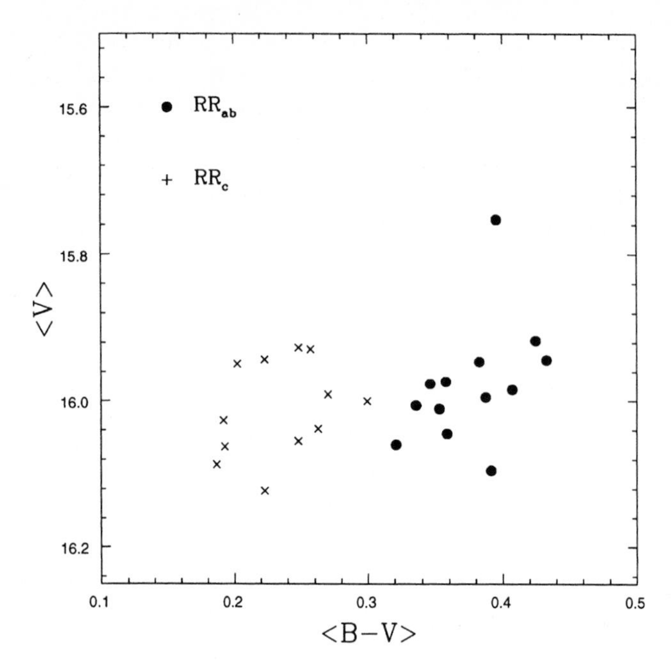

Fig. 32. The mean V magnitudes of M2 RR Lyraes. *Crosses* are RR_c variables, while *dots* are RR_{ab} variables. V10 is the brightest RR_{ab} variable

Figure 32 shows that the mean V magnitude of V10 is indeed brighter than the rest of the variables in the cluster.

The second subtle problem has to do with the field RR Lyraes. What we are trying to establish is the relation between M_V and [Fe/H] for the RR Lyraes as a population. Of course, any sample selected for study runs the risk of bias, favoring stars that may be more highly evolved and hence brighter. SS Leo, for example, was noted by Carney et al. (1992) to probably be such a star, based on its unusually long period for either its blue amplitude or its derived mean temperature. The problem with such stars is that they are (a) short-lived, hence (b) uncommon, hence (c) not representative of the bulk of the field RR Lyrae population. They are also brighter than average, and may therefore distort the results. While a randomly selected sample of field stars is probably representative, the Lub (1977) sample is neither random nor representative of the field population. It was selected to fully sample stars in the period–metallicity plane, and hence includes longer period stars than would be obtained from a random sample. Carney et al. (1992) therefore chose to work with a more randomly selected sample of field stars, that of Suntzeff et al. (1991). The period shift analysis of these stars led to $\Delta M_{\rm bol} \propto 0.20\,[\rm Fe/H]$.

Subtle Biases? Feast (1997) argued that various biases might affect the selection of the field RR Lyraes selected for study, and that therefore the analysis should be done by fitting [Fe/H] to M_V(RR) rather than the other way around, as is usually the case. Upon taking the metallicity as the dependent variable and inverting, Feast found a very steep M_V(RR) *vs.* [Fe/H] relation, 0.33 mag/dex. Fernley et al. (1998b) re-evaluated Feast's claim, noting that the suspected biases were not present in the selection of the stars for study, and that a maximum likelihood regression, allowing for the errors in *both* [Fe/H] and M_V(RR) results in:

$$M_V(\mathrm{RR}) = (0.20 \pm 0.04)\,[\mathrm{Fe/H}] + (0.98 \pm 0.05)\,. \qquad (39)$$

This is not too dissimilar from the detailed review of various methods by Carney et al. (1992), who found

$$M_V(\mathrm{RR}) = (0.15 \pm 0.01)\,[\mathrm{Fe/H}] + (1.01 \pm 0.08)\,. \qquad (40)$$

At [Fe/H] = −2.0 and −1.0, these two relations differ in the sense of (39) minus (40) by +0.13 mag and +0.08 mag, respectively.

M31 Globulars. Perhaps the most reliable approach to solving M_V(RR) *vs.* [Fe/H] is to study the globular clusters in M31. The horizontal branch is very faint, $V \approx 25$, and crowding is extreme, but *HST* has been successful in providing the necessary photometry. Cluster metallicities were estimated using the colors of the red giant branches. Relative V_0 magnitudes of the clusters' horizontal branches then yield relative M_V values. Fusi Pecci et al. (1996) have summarized the available data, and concluded that the slope of (33) is $a = 0.13 \pm 0.07$ mag/dex.

Summary. Table 1 summarizes the various results for the slope a. With the revised results from the period shift analysis, it appears that the slope is well determined and that there may be a simple linear relation between M_V(RR) and [Fe/H].

Table 1. Summary of results for slope of absolute magnitudes *vs.* [Fe/H]

Technique	$\frac{\Delta M_{\mathrm{bol}}}{\Delta[\mathrm{Fe/H}]}$	$\frac{\Delta M_V}{\Delta[\mathrm{Fe/H}]}$
Theory	0.20	0.17
Baade–Wesselink	0.21	0.16
Period shift (clusters)	0.18	....
Period shift (field)	0.20	....
M31 clusters	....	0.13

3.3 Oosterhoff Classes *Redux*

The above analyses have been done in rather traditional astronomical practice: we define two variables and seek a linear relation. But is it this simple? Have we conveniently ignored the evidence cited already that the Oosterhoff dichotomty may be discontinuous rather than continuous? Jaewoo Lee and I have worried about this, and have completed a study of the globular cluster M2, which belongs to the Oosterhoff II class despite having a metallicity (−1.62) very similar to several Oosterhoff I clusters. We (Lee & Carney 1999a) have analyzed the light curves of 30 RR Lyraes in M2 (13 of which are newly discovered) and compared them *via* a period shift analysis to the variables in M3, whose metallicity, −1.66, is essentially identical to that of M2, according to Zinn (1985). Figure 31 shows the results. The M2 variables have longer periods than do those of M3 at similar temperatures. The period shift analysis leads to a difference of 0.23 mag in $M_{\rm bol}$. Figure 13 shows the color–magnitude diagram of M2 obtained from the 11 best-seeing BV pairs with long exposure times and the 7 pairs with short exposure times and analyzed all of them simulataneously using Peter Stetson's powerful program ALLFRAME (Lee & Carney 1999b). We have done a main sequence fit to that of M3, using data for its main sequence from Ferraro et al. (1997) and for its RR Lyrae variables from Carretta et al. (1998). The result is that the M2 variables are brighter than those in M3 by 0.17 mag in M_V, consistent with the period shift analysis, which appears again to be reliable. M2 and M3 form yet another of the many "second parameter" pairs, with M2 having a much bluer horizontal branch $[(B-R)/(B+V+R) = 0.96$ *vs.* $0.08]$, and, again, we wonder why. The difference in the luminosities of the RR Lyraes, and the difference in the richness values (M2: 0.05; M3: 0.95) can be understood if the RR Lyraes in M3 are mostly near their zero-age horizontal branch position, while those in M2 are largely evolved from the blue horizontal branch. Further, it appears that the differences are caused by differences in age. Figure 33 shows the main sequence and lower giant branch domains after matching the clusters 0.05 mag redward of the turn-off, following the methodology developed by VandenBerg et al. (1990). We have also superposed isochrones. It appears that M2 is roughly 2 Gyrs older than M3, which would explain its bluer horizontal branch, and hence its brighter RR Lyraes. The observed period change rates observed for the RR Lyraes in the two clusters support this idea (Lee & Carney 1999b). Those in M3 do not have detectable period change rates, on average, consistent with the slow evolution near the zero age horizontal branch, while those in M2 are lengthening, consistent with evolution toward lower densities, hence larger radii, hence cooler temperatures.

The conclusion that the RR Lyraes in Oosterhoff II clusters have longer periods than those in Oosterhoff I clusters because the former clusters are older than the latter is not secure, simply because we have relied upon only one cluster pair. However, it suggests that some care may be needed in the

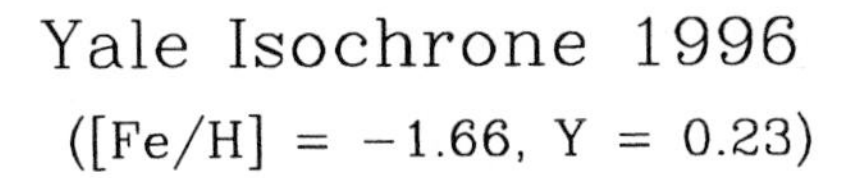

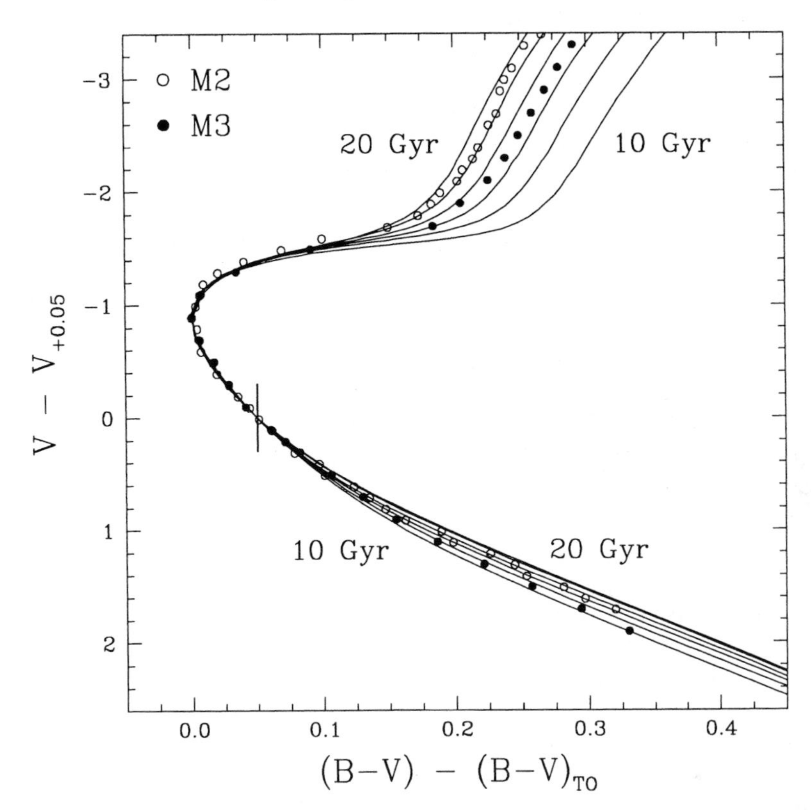

Fig. 33. A comparison of the main sequence turn-off regions of M2 and M3

application of (39) and (40) to *all* clusters without being aware of the evolutionary status of the clusters' RR Lyrae variables.

4 Stellar Populations

While our topic is globular clusters, it is prudent to extend the discussions to their field star counterparts, as we in fact have done already. Globular clusters have the great advantage of being laboratories where the relative luminosities, temperatures, and gravities of stars are easily determined. Clusters are therefore primary means for studying the effects of stellar evolution and estimating ages to a reasonable level of precision. One drawback to clusters, however, is that they are rather distant. Generally only the brightest stars may be studied spectroscopically, and proper motions, and hence three-dimensional space velocities are available for a small number of clusters. (Good summaries of the clusters' proper motions are given by Cudworth & Hanson 1993; Dinescu et al. 1997,1999; and Odenkirchen et al. 1997.) Field stars are much more numerous than cluster stars. In fact, the total mass in metal-poor field stars exceeds that in clusters by a factor of at least 50 (Carney et al. 1990a; Suntzeff et al. 1991), and since clusters each have rather a large number of stars, the total number of individual field stars exceeds the total number of clusters by a factor of about 10^7. While they probably have not sampled a larger number of histories (i.e., formed within the Milky Way or been accreted into it), the nearer field stars are closer to us than the nearest clusters. The field stars therefore provide us with a more convenient pool of targets to study the relationship(s) between chemistry and kinematics, on which we focus in this section.

Before proceeding, some definitions are required. The radial and tangential velocities may be combined to yield a space velocity, but it is more convenient to define a new coordinate system, one related to the Galactic plane and center. The U, V, and W velocities are orthogonally directed toward the anticenter ($\ell = 180°$, b = 0°), the direction of the Local Standard of Rest ($\ell = 90°$, b = 0°), and the North Galactic Pole (b = 90°). The V velocity is clearly related to orbital angular momentum. Since the circular orbital velocity at the Sun's distance from the Galactic center is thought to be close to 220 km s^{-1}, stars with V ≤ -220 km $s^{-1}$are on retrograde orbits.

From the U, V, and W velocities, and an adopted mass model for the Galaxy, one may compute Galactic orbits (see Carney et al. 1990b, 1994; Allen & Santillan 1991; Schuster & Allen 1997). The key resulting quantities here are the mean apogalacticon distance, $R_{\rm apo}$, the mean perigalacticon distance, $R_{\rm peri}$, and the mean maximum distance from the Galactic plane, $|Z_{\rm max}|$. From these one may also compute an orbital eccentricity:

$$e = \frac{R_{\rm apo} - R_{\rm peri}}{R_{\rm apo} + R_{\rm peri}} . \tag{41}$$

4.1 How to Identify a Stellar Population

The term "stellar population" is used frequently in the literature, as are specific examples, including the "disk", "thick disk", "bulge", and "halo". These may be related to one another in that they share a continuous history, in which case we are in fact referring merely to *stages* in a process, or they may in fact refer to *separate*, perhaps even unique, histories. This is one of the big questions for studies of stellar populations. Therefore, we need to wrestle first with a proper definition. I take the term "population" to refer to *a mixture of stars and gas that share a common chemical and dynamical history.*

Perhaps the most common means of labelling a stellar population is by its metallicity. There is a tendency to associate a star with [Fe/H] = −1.5 with the Galactic halo. But what if the Galactic halo was produced by a series of

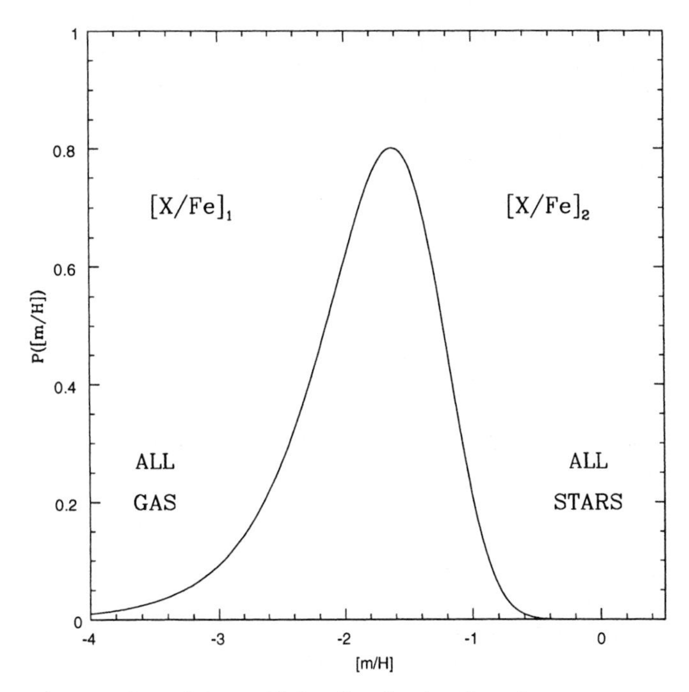

Fig. 34. A model metallicity distribution function

mergers of dwarf galaxies with the Milky Way, while the stars ancestral to the Galactic disk also went through a metal-deficient stage? In that case we would invoke kinematics in deciding whether the star was most likely associated with the dynamically hot halo or the dynamically cool disk. Even this is difficult because the velocity dispersions are large, in general, and at low metallicities, it becomes very difficult to assign any one star or cluster to a "proto–disk" stage of Galactic evolution or to a merger remnant. One must also recall that the dispersion in [Fe/H] may be small within a given globular cluster (ω Cen being an exception), but it is large among the field stars. Nonetheless, this dispersion may be used to advantage. In Fig. 34 we show a very simple model of chemical evolution. It assumes a "closed box", so that there is no inflow and outflow, and that the heavy elements synthesized in supernovae are returned to new generations of stars promptly. (See the next section for some details.) What one sees in general is that when most of the mass is still in gas, before many supernovae have exploded, the probability of finding a star with a low metallicity is very small. As more of the gas is converted into stars and enriched in heavy elements, the probability of finding a star in the ensemble with a higher metallicity is increased, until more than half the gas has been converted into stars. At that point the probabilities of finding stars with higher metallicities decline. The result of this simple chemical evolution model is the curve shown: sharply peaked with long tails at higher and lower metallicities. This metallicity distribution function (MDF) is a hallmark of a stellar population. The mean metallicity in this simple model is set by the "yield", the ratio of heavy elements produced by supernovae to that locked away in stellar remnants and long-lived low mass stars that we now may study. As the chemical enrichment proceeds, the elemental abundance ratios, [X/Fe], may change. Type II supernovae, created during the deaths of massive stars, dominate the early stages, but if the chemical evolution timescale is long enough, other types of supernovae, such as Type Ia, may begin to contribute to the nucleosynthesis, and they may produce a distinctly different heavy element abundance profile. We return to this in a later section.

As noted already, one of the characteristics of the halo population is that it is metal-poor: the peak in the [Fe/H] distribution is near -1.6 (Laird et al. 1988). The halo is also old, and because we believe that the overall heavy element abundances have increased with time, so we assume that the halo is metal-poor because it is old. But this is not wholly the truth. This may explain why the most metal-poor stars in the halo have lower [Fe/H] values than the most metal-poor stars in the disk, assuming the disk's oldest stars are younger than those in the halo, but it does not easily explain why the mean metallicities are different. The mean is set, at least in this simple model, by the yield, which depends primarily on the stellar initial mass function. There is no obvious reason why the halo should produce fewer supernovae per unit mass than stars forming in the disk, and hence no reason to expect, in a closed system, that the halo would end up more metal-deficient than the disk. The answer to this conundrum lies in our assumption of a closed system. The halo is not and probably never has been a closed system, so gas is lost from it rather than being retained to form new stars. This mass loss reduces the yield, and hence the mean metallicity. Thus the halo is metal-poor not because it is old, but because it could not retain its gas. Of course, if the halo is the ancestor of the disk, it does make some sense to associate low metallicities with great ages: it is the entire Galaxy itself that is then the "closed system".

Where did the halo gas go? There are two obvious repositories: the bulge and the disk. As we will see, the halo has very little net angular momentum, so its gas is more likely to find its way to the Galaxy's central regions. Two pieces of evidence suggest that the bulge is, in fact, the repository for this gas, and that the halo and bulge populations share common histories, at least in part. The first argument is one of consistency. The estimated mass of the bulge, $\approx 2 \times 10^{10} M_\odot$ (Blanco & Terndrup 1989), is consistent with the estimated gas mass lost from the halo ($\approx 3 \times 10^{10} M_\odot$; Carney et al. 1990b). The second argument is more direct: the angular momentum distribution of the halo and bulge populations are similar and unlike that of the disk (Wyse & Gilmore 1992), as shown in Fig. 35. The bulge is more metal-rich than the halo because it is deep in the Galaxy's gravitational potential, so that gas could not easily escape, and the transformation of gas into stars went essentially to completion, and with a mean metallicity more comparable to

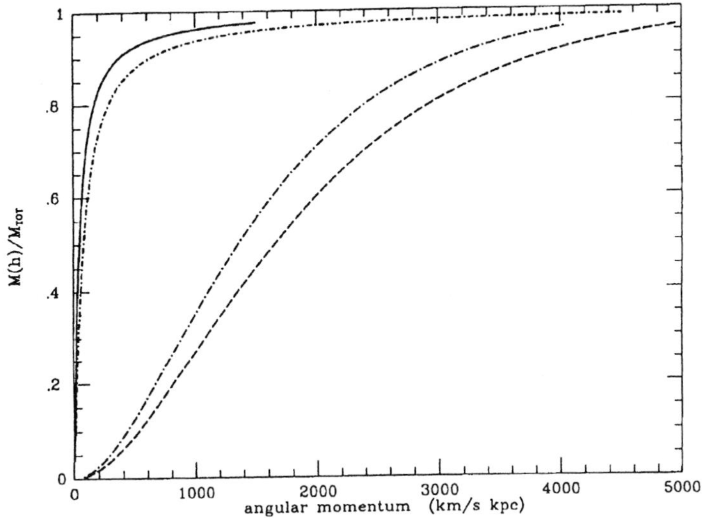

Fig. 35. The angular momentum distributions of the halo, bulge, and disk populations, according to Wyse & Gilmore (1992). The *solid line* is the normalized angular momentum distribution for the bulge, the *short dash-dot curve* is for the spheroid/halo, the *long dash-dot curve* is for the thick disk, and the *dashed line* is for the thin disk

that of the Galactic disk. The mean metallicity of the bulge appears to be slightly less than solar, $\langle[\mathrm{Fe/H}]\rangle \approx -0.25$ (McWilliam & Rich 1994).

The power of the MDF to identify stellar populations is particularly apparent in Fig. 36, taken from Zinn (1985). Two distinct MDFs are discernible, one with a peak [Fe/H] of ≈ -1.5, and the other with ≈ -0.5. The clusters are distributed very differently in the Galaxy as well. The metal-rich population, often associated with the "thick disk", noted in star count analyses by Yoshii (1982) and Gilmore & Reid (1983), is, as its name implies, confined to the disk. The metal-poor population extends to distances far from the plane of the disk. Further, some of the metal-poor clusters appear to be concentrated in the plane, which suggests that either the disk's gravitational potential has a strong influence on the clusters' distribution, or that some of the metal-poor clusters are the low-metallicity tail of the disk population of clusters. There appear to be two histories, probably separate, among the Galaxy's globular clusters, but the questions we ask are similar in both cases. Are either the halo or the thick disk ancestral to the Galaxy's disk population? Or are either remnants of a merger event?

4.2 The Halo Population(s)

Two Independent Histories? Is the halo population monolithic, with a single history that can be seen in a continuous relation between chemistry and

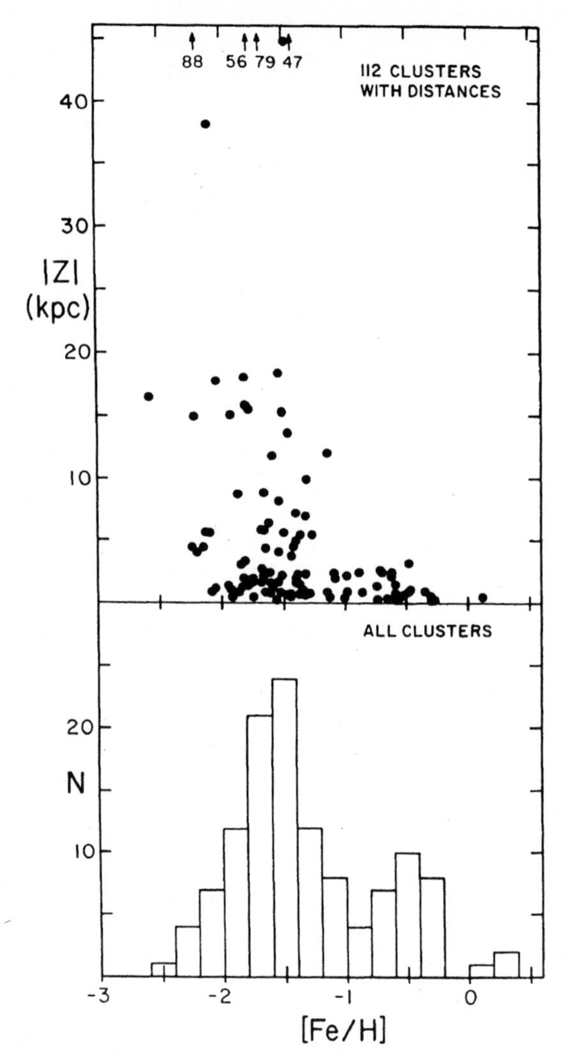

Fig. 36. The distributions of [Fe/H] and distance from the plane of the Galaxy's globular clusters, according to Zinn (1985)

kinematics? Or does it represent the merger of a variety of independent "proto-Galactic fragments" with independent chemical and dynamical histories? In essence, we ask of the halo what is asked of the stars in the globular cluster ω Cen: is there one history producing the observed range in metallicities, or is there more than one? (Norris et al. 1997 have argued the latter for the cluster.)

The "monolithic" model was the conclusion of Eggen et al. (1962; hereafter ELS), who began the modern era of the study of the Galaxy's formation and evolution. They made four major points. **1.** There are few, if any, stars with both low velocities and low metallicities. **2.** There are few, if any, stars with high velocities and high metallicities. **3.** Related to these two points, there is a monotonic relation between metallicity and kinematics, implying a single history. **4.** The high Galactic orbital eccentricities of the most metal-poor, presumably oldest, stars indicate that the Galaxy began in a state of "collapse", implying a formation timescale of $\approx 2\times10^8$ years, the approximate free-fall timescale. The beauty of the conclusion is not only that there are supporting observations, but that it makes intuitive sense, and it finds a close analogy in the formation of our Solar System. But has the Galaxy's evolution really been monolithic, or could accretion or at least less incoherent evolution have played a role? The ELS study involved stars selected from two catalogs, one of relatively nearby stars, hence without much bias, and the other from a catalog of high-velocity stars, and hence considerable kinematical bias. With two samples, one must ask how an MDF might filter the observations and affect what we see. If the local sample, with predominantly low velocities, is not a large enough sample, one will not easily find the metal-poor tail in the distribution. This might explain point 1. Similarly, a high-velocity sample, even if studied without a bias in metallicity, may not readily reveal the metal-rich tail (point 2). Thus a merger of two samples could lead to point 3 by means of subtle MDF-induced biases.

The final point, predicting a small and essentially non-measurable age spread, is one of the key points in the alternative view presented by Searle & Zinn (1978; hereafter SZ). In a coherent or monolithic evolutionary process, a metallicity gradient should develop if the Galaxy contracted on a timescale long enough for supernovae ejecta to be incorporated into new generations of stars. The ELS model envisions a rapid process; too rapid, perhaps for a detectable metallicity gradient to be created. SZ made a careful study of globular cluster metallicities as a function of Galactocentric distance, R_{GC}, and found no evidence for a gradient. This may be seen in Fig. 37, which includes a recent updated list of globular cluster metallicities, restricted to those with $[Fe/H] \leq -0.8$ to eliminate the thick disk population. Also included are studies of field RR Lyraes from Suntzeff et al. (1991), which yield a radial metallicity gradient identical to that found by SZ: -0.004 dex/kpc. Finally, Carney et al. (1990a) studied local field stars with high proper motions. Restricting the sample to stars with $V \leq -150$ km s^{-1}, they found -0.008 dex/kpc. Figure 37 has been updated in that it uses the larger database from Carney et al. (1994), and the gradient is nominally -0.019 ± 0.006 dex/kpc, although it is clear from the figure that the gradient is very weak beyond apogalacticon distances of 10 kpc. A strong metallicity gradient does not appear to exist in the outer halo. However, as we have discussed, SZ found that the horizontal branch morphology depends on R_{GC}, as Fig. 21 shows. If the second para-

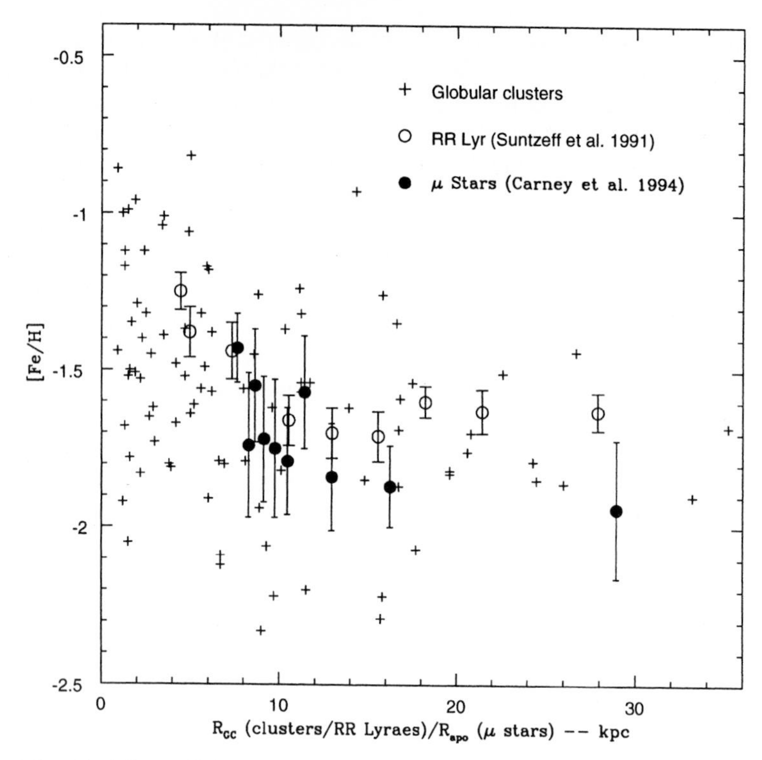

Fig. 37. The radial metallicity gradient of the halo population from three different studies

meter is primarily age, then there is apparently a large enough age spread to have permitted the formation of a metallicity gradient. Its absence suggests that the evolution of the halo, and the outer halo in particular, has been more incoherent, even chaotic, rather than coherent. SZ argued that the halo was assembled from "proto-Galactic fragments". These formed into stars and clusters more slowly in the outer halo; hence their younger ages and redder horizontal branch colors. The lack of a metallicity gradient indicates they evolved more or less independently of each other.

Thus one of the key questions about the formation and evolution of the Galaxy's halo is which model more closely applies. Aside from metallicity gradients, one may return to the initial ideas of ELS and re-explore the relationship between chemistry and kinematics. This was done by Sandage & Fouts (1987), who found a steady trend in V velocity (i.e., angular momentum) with metallicity, as inferred from the photometric metallicity indicator $\delta(U-B)_{0.6}$ (see Sandage 1969a,b for the definition and Carney 1979 for the calibration). The steady trend was criticized by Norris & Ryan (1989), who noted that the apparent trend could be created when none exists due to the decreasing precision of the ultraviolet excess as a metallicity indicator at

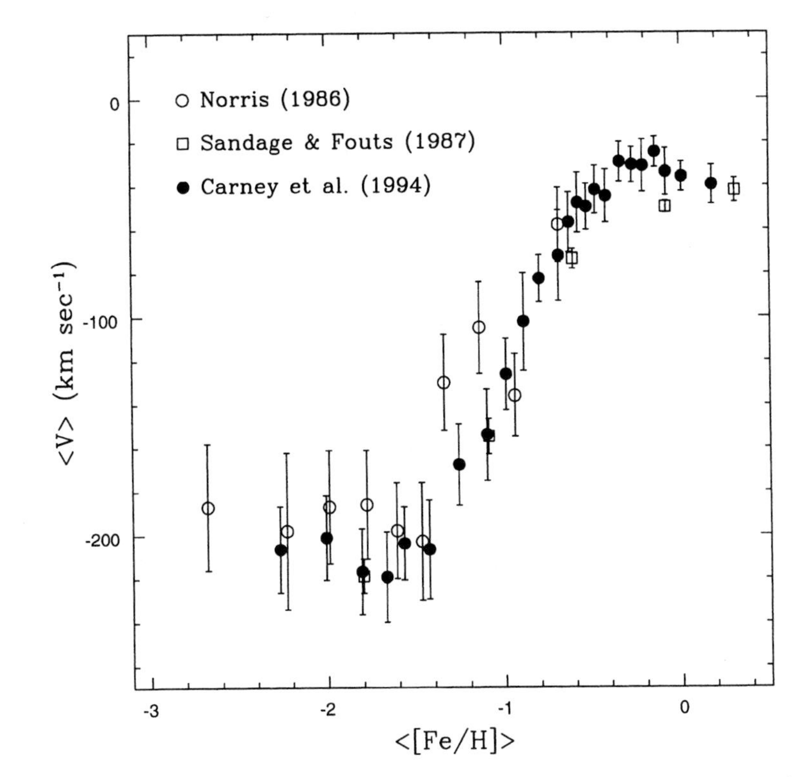

Fig. 38. The mean V velocity values *vs.* metallicity from three independent studies

the lower metallicity levels and the use of two separate catalogs and hence two separate sets of selection biases. Further, Norris (1986) employed a kinematically unbiased sample and found that the mean V velocity is constant for $[Fe/H] \leq -1.4$. Carney et al. (1990b) found a similar result from their study of field stars selected from proper motion catalogs. In Fig. 38 we show the results from these studies, except that of Carney et al. (1990b) has been updated using the data from Carney et al. (1994). The lack of a relation between chemistry and kinematics for stars with $[Fe/H] < -1.4$ is at odds with the ELS data and conclusions, but only barely. If the time required for the mean metallicity to rise to $[Fe/H] = -1.4$ in a coherently-evolving galaxy is short enough, we would not expect such a correlation. As discussed in Sect. 8, there is, in fact, no evidence that the most metal-poor clusters have detectable age differences, while the evidence in the intermediate metallicity domain, around $[Fe/H] = -1.4$, does suggest some age spread. Of somewhat greater concern should be that the apparent change in the kinematics appears quite near the peak in the halo MDF. That is, the relative number of halo stars and clusters declines relatively rapidly at higher metallicities, and "contamination" of a mean velocity by the metal-poor tails of the much more numerous

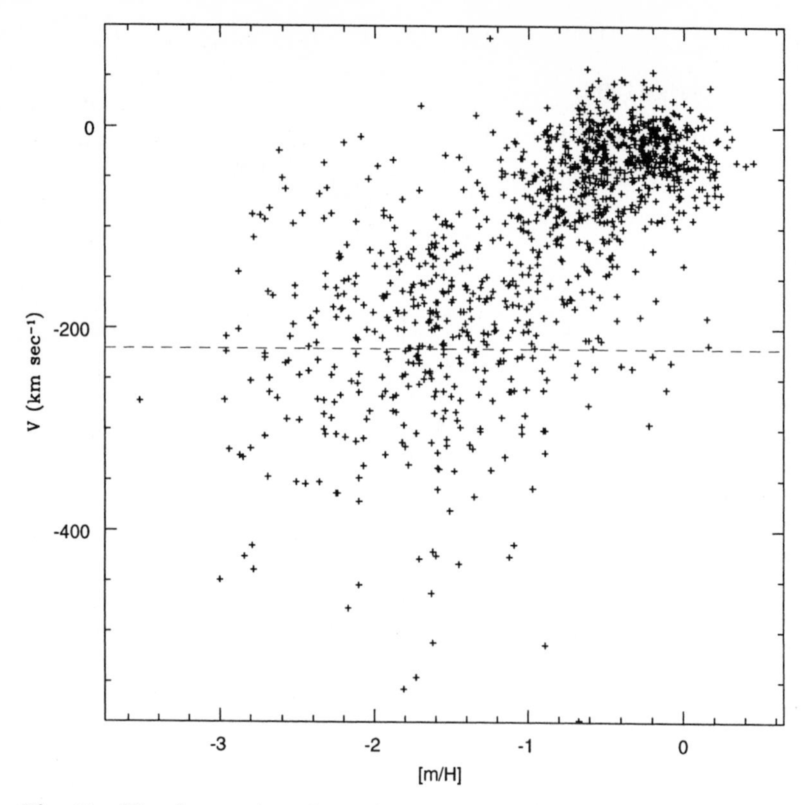

Fig. 39. The data points from Carney et al. (1994) used to produce part of Fig. 38. The *dashed line* is the approximate division between prograde and retrograde Galactic orbits

thick disk and disk stars may create an apparent trend where none exists. In these cases, *it is always better to look at the individual data themselves*, which we show in Fig. 39. The dynamically hot (i.e., high-dispersion) halo population fills much of the figure, with the disk populations confined to the low-velocity, metallicity-rich quadrant. There is no clear sign of a coherent or monolithic transformation from a high-velocity, metal-poor population into the disk. Similar results were found by Layden (1995) in his analysis of the motions of over 300 field RR Lyrae variables, and by Chen (1998) in his re-analysis of our proper motion sample. In fact, Chen (1998) suggested that large-scale velocity "substructure" is present in the Carney et al. (1994) data. The substructure represents apparent clusterings in velocity space, and was also found by Majewski et al. (1996) in their study of the Majewski (1992) data. The groupings appear to be real, but the velocity dispersions are very large, so these do not represent the "moving groups" suggested by Eggen (1977, 1978, 1987), Doinidas & Beers (1989), Arnold & Gilmore (1992), and

Kinman et al. (1996), which have very small velocity dispersions. The moving groups could be remnants of small "fragments" in the SZ model, but they could also represent debris from the tidal destruction of the Galaxy's own globular clusters, as suggested by Aguilar et al. (1988), Murali & Weinberg (1997), and Gnedin & Ostriker (1997). The higher velocity dispersion kinematic substructure, however, is very much harder to explain within a model wherein the halo and disk populations are different stages in a single evolutionary process.

The ELS model is not confirmed, but one must be very careful to not overinterpret these results, either. There remains room for both the ELS and SZ viewpoints to be accomodated. One must remember that the data employed in Fig. 39 are biased kinematically. Stars with lower velocities have lower probabilities of being included in the sample. In other words, there remains a bias against low-velocity, low-metallicity stars.

The idea of two different populations or separate histories for the metal-poor stars is not new. Hartwick (1987) argued that models of the dynamics of RR Lyrae variables with [Fe/H] < -1 required two components, one spherical and a more flattened ($c/a \approx 0.6$) inner component that is dominant at the solar Galactocentric distance. Sommer–Larsen & Zhen (1990) studied a kinematically unbiased sample of 118 stars that are even more metal-poor: [Fe/H] ≤ -1.5, and again found that to explain the kinematics required two components. Their fit indicated that most of the mass, about 60% in the solar neighborhood, included a spheroidal component ($c/a = 0.85 \pm 0.12$), while the rest was distributed closer to the disk, with $|Z| < 3$ kpc, and which was not flattened by rotation alone. The difference in the dominance of the spheroidal *vs.* the flattened component in these two studies may reflect the relative metallicities, with the spheroidal one being more metal-poor. A third sample of field stars, those identified as blue horizontal branch stars by Kinman et al. (1994) also appears to show spheroidal and flattened components. The metallicity dependence is, of course, somewhat harder to discern because metallicity is, after all, the first parameter.

Do the metal-poor globular clusters also divide into two components? Zinn (1993) and Da Costa & Armandroff (1995) have suggested that they do. They used the horizontal branch morphology (see Fig. 21) to identify "young halo" and "old halo" clusters, under the assumption that age is the dominant second parameter. This seemed to be confirmed by age differences derived by Chaboyer et al. (1996), who used $\Delta V_{\rm TO}^{\rm HB}$ values for a large number of clusters, and a horizontal branch luminosity-metallicity relation very similar to that derived from the Baade–Wesselink analyses, to find a significant mean age difference between these two subsamples of metal-poor clusters. There are three quantities that may be measured, to varying degrees of precision, from globular clusters that bear on the question at hand: did they form as part of a coherent collapse/contraction of the proto-Galaxy, or were they formed in "fragments" which evolved independently but eventually became part of the

Galaxy? **1.** Does either show a metallicity gradient, as would not be expected from the independent fragments model and which *might* be seen in the coherent evolution model, if sufficient time elapsed? Zinn (1993) noted that the "young halo" clusters showed no signs of a radial metallicity gradient, whereas the "old halo" clusters do. The mean metallicity of the "old halo" clusters declines from -1.44 ± 0.06 for $R_{GC} < 6$ kpc to -1.80 ± 0.07 for $6 \leq R_{GC} \leq 15$ kpc, to -1.93 ± 0.10 for $15 \leq R_{GC} \leq 40$ kpc. **2.** The angular momentum, signifying a relation to the disk populations, may be estimated roughly by using the clusters' radial velocities. Using only one component of motion (radial velocities) results in large uncertainties, but Zinn (1993) found that the 19 "young halo" clusters showed an apparent retrograde rotational velocity. Assuming that the local circular Galactic orbital velocity, Θ_0, is 220 km s^{-1}, he found $\langle v_{rot} \rangle = -64 \pm 74$ km s^{-1}. Da Costa & Armandroff's sample of 21 clusters yielded -46 ± 81 km s^{-1}. For the "old halo", the net rotation was prograde. Zinn (1993) found $\langle v_{rot} \rangle = +75 \pm 39$ km s^{-1} for 24 clusters with $6 \leq R_{GC} \leq 40$ kpc, while Da Costa & Armandroff (1995) found $+40 \pm 41$ km s^{-1} for 27 clusters in the same distance range. The error bars are, unfortunately, large, but the differences appear to be real. **3.** Using only a component of the radial velocities to estimate the V velocity or the net rotational velocity v_{rot} is not an optimum use of the data, so Zinn (1993) also computed the dispersion in the line-of-sight velocity, σ_{LOS}. This is a measure of how "hot" the kinematics are, with higher velocities being more typical of a spheroidal component, and lower velocities being more typical of a disk component. Zinn (1993) and Da Costa & Armandroff (1995) found σ_{LOS} values of 149 ± 24 and 163 ± 25 km s^{-1} for their 19 and 21 "young halo" clusters, respectively, but 99 ± 14 and 115 ± 16 km s^{-1} for the 24 and 27 "old halo" clusters, respectively. These differences are more significant and less sensitive to the positions of the clusters relative to the Sun (which determines how much of the radial velocity component contributes to the V or v_{rot} velocities). Rather than rely on "young halo" and "old halo", we may also consider the differences between the Oosterhoff I and Oosterhoff II classes. Carney (1999) and Lee & Carney (1999b) found $\langle v_{rot} \rangle = -76 \pm 53$ and $+42 \pm 58$ km s^{-1}, respectively, while the line of sight velocity dispersions are 130 ± 30 and 112 ± 33 km s^{-1}, respectively. Bill Harris has pointed out the large influence of NGC 3201 in these calculations. Removing it from the Oosterhoff I results leads to $\langle v_{rot} \rangle = -40 \pm 53$, which still differs significantly from the results for the Oosterhoff II clusters. (And removing a single cluster introduces a biased result in any case.)

As suggested in Fig. 34, a prolonged period for the transformation of gas into stars might lead to differing element-to-iron ratios. As we discuss in Sect. 6, the "α" elements, including oxygen, silicon, magnesium, calcium, and titanium, appear to be measures of the relative dominance of SNe II *vs.* SNe Ia. Wyse & Gilmore (1988, 1993) and Smecker–Hane & Wyse (1992) have argued further that the change from enhanced levels of [α/Fe] to near-solar values is a chronometer if the timescale required for the appearance of

SNe Ia is about 10^9 years. The idea here is that if one finds [α/Fe] values that are constant, then the timescale is less than 10^9 years. Carney (1996) reviewed the available [α/Fe] data for globular clusters, and found that the three disk clusters, the three "young halo", and the fourteen "old halo" clusters studied all have [α/Fe] $\approx +0.3$. Does this mean they all have the same age? Not exactly. The conclusion that they have similar ages would be based on two assumptions: (a) that the SNe Ia begin to appear after 10^9 years; and (b) that these clusters have shared a common history. But the second point is, of course, the question under discussion. In principle, all the change in the [α/Fe] tells us is the *duration* of the chemical enrichment process, as pointed out by Carney et al. (1990b), and discussed at greater length by Gilmore & Wyse (1998). If gas cloud A begins to form stars at time zero, and requires, say, 3 Gyrs to complete it, then its most metal-poor stars will show only SNe II abundance patterns, with [α/Fe] > 0, while its most metal-rich stars will have very substantial SNe Ia contributions, and [α/Fe] ≈ 0. If gas cloud B originates independently of A, but begins its star formation after that in A has ended, B will nonetheless have its metal-poor stars dominated by SNe II ejecta, and [α/Fe] > 0. The point is that the metal-poor stars produced by clouds A and B would both have enhanced [α/Fe] ratios, despite having very different ages. Thus the equivalence of [α/Fe] ratios in the "young halo", "old halo", and disk clusters can be explained by either (a) no age differences and common histories; (b) age differences but independent origins; or (c) incomplete understanding of the SNe II *vs.* SNe Ia timescales.

Before continuing on to a discussion of signs of discrete accretion events, it is worth returning to the field star data one more time to explore the possible dual histories of the metal-poor stars in the solar neighborhood. One particularly intriguing piece of evidence was found by Majewski (1992). His study of the proper motions and metallicities for a *complete* sample of stars in the direction of the North Galactic Pole indicated that the stars more than 5 kpc from the plane showed a net retrograde rotation: $\langle v_{\mathrm{rot}} \rangle = -55 \pm 16$ km s^{-1}. Chen (1997) re-analyzed Majewski's (1992) data and also found a retrograde signature, although not quite as extreme, $\langle v_{\mathrm{rot}} \rangle = -31 \pm 10$ km s^{-1}. Rodgers & Paltoglou (1984) had previously drawn attention to the apparent net retrograde rotation of globular clusters with [Fe/H] ≈ -1.5, and van den Bergh (1993) made essentially the same point, noting that the more metal-rich Oosterhoff I globular clusters show a greater tendency for retrograde motion. A net retrograde rotation is a pretty convincing argument for an origin largely independent of the Galactic disk, which has very positive net angular momentum, of course, or any other sample with a net prograde rotation, such as the "old halo" clusters discussed above. The "old halo" and "young halo" discussion then is transformed into one of "young halo" (possible retrograde rotation) and "old halo" (prograde rotation), respectively. Majewski's (1992) result recasts the question, at least for the field stars, into a "high halo" and a "low halo". The former may show retrograde rotation and may therefore

Table 2. Summary of net rotation velocities and metallicity gradients for "low halo" and "high halo" samples

Criterion	No.	$\langle v_{\rm rot} \rangle$	$\langle$ [m/H] $\rangle$	$\frac{\Delta[\rm m/H]}{\Delta R_{\rm apo}}$
		$\lvert Z_{\rm max} \rvert \geq 4$ kpc		
unweighted	45	-30 ± 16	-1.95	-0.005 ± 0.006
weighted	45	$+15 \pm 4$	-1.67	-0.019 ± 0.009
		$\lvert Z_{\rm max} \rvert \leq 2$ kpc		
		— *unweighted* —		
$[\rm m/H] \leq -1.5$	150	$+27 \pm 7$	....	....
$e \geq 0.85$	97	$+12 \pm 6$	-1.65	-0.030 ± 0.017
$(\rm U^2 + V^2)^{\frac{1}{2}} \geq 200$	51	$+44 \pm 13$	-1.62	-0.036 ± 0.018
		— *weighted* —		
$[\rm m/H] \leq -1.5$	148	$+104 \pm 6$	-2.00	....
$e \geq 0.85$	79	$+14 \pm 3$	-1.62	-0.023 ± 0.019
$(\rm U^2 + V^2)^{\frac{1}{2}} \geq 200$	52	$+74 \pm 9$	-1.60	-0.039 ± 0.022

be associated with the "young halo" while the "low halo", showing prograde rotation, would be another name for the "old halo". In studying field stars, the problem is fairly easy in terms of "high halo": one either finds stars far from the plane, as Majewski did, or finds stars closer to the Sun, but whose orbits carry them far from the plane, with data supplied, for example, by Carney et al. (1994). But how does one identify a metal-poor population closer to the plane, where the disk populations dominate? One may, of course, invoke a metallicity criterion, using stars no more metal-rich than the peak in the MDF of the halo population, say $[\rm m/H] \leq -1.5$. This is an effective criterion, but it hampers the estimation of a metallicity gradient for obvious reasons. Use of the V velocity as a criterion is also imprudent if one wishes to establish the prograde or retrograde rotation of the sample. Carney et al. (1996) therefore employed two other criteria: $(\rm U^2 + V^2)^{\frac{1}{2}} \geq 200$ km s^{-1}, and $e \geq 0.85$. The virtue of these two criteria is apparent when one considers the resultant MDFs: both show the same shape as Fig. 34, with a peak at $[\rm m/H] \approx -1.7$, and wings extending to either side, as may be seen in Figs. 40 and 41. With these criteria, we estimate the net rotational velocity of the "high halo" ("young halo") and "low halo" ("old halo") samples, as well as their radial metallicity gradients. The results are summarized in Table 2. There *do* appear to be differences between the two samples, and in the manner suggested by Zinn (1993) and Da Costa & Armandroff (1995). The "high-halo" stars are in retrograde rotation, and have no radial metallicity gradient, suggesting an

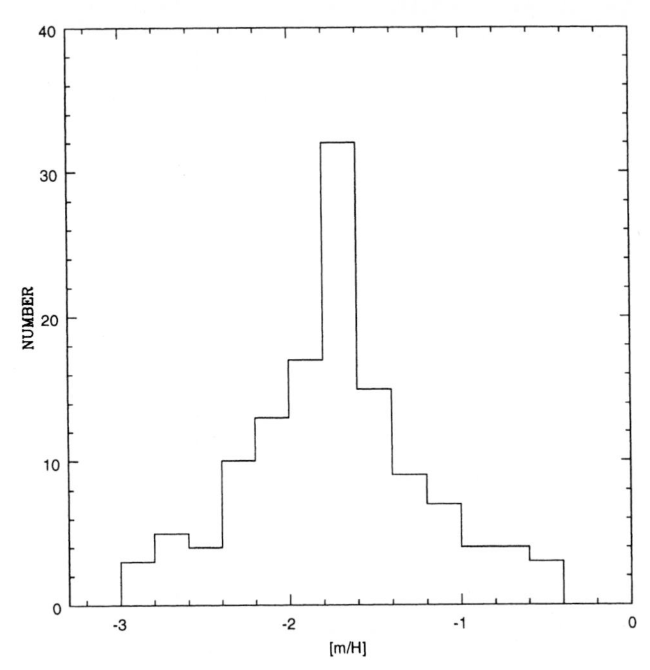

Fig. 40. The metallicity distribution function (MDF) for stars with $(U^2 + V^2)^{\frac{1}{2}} \geq$ 200 km s^{-1}

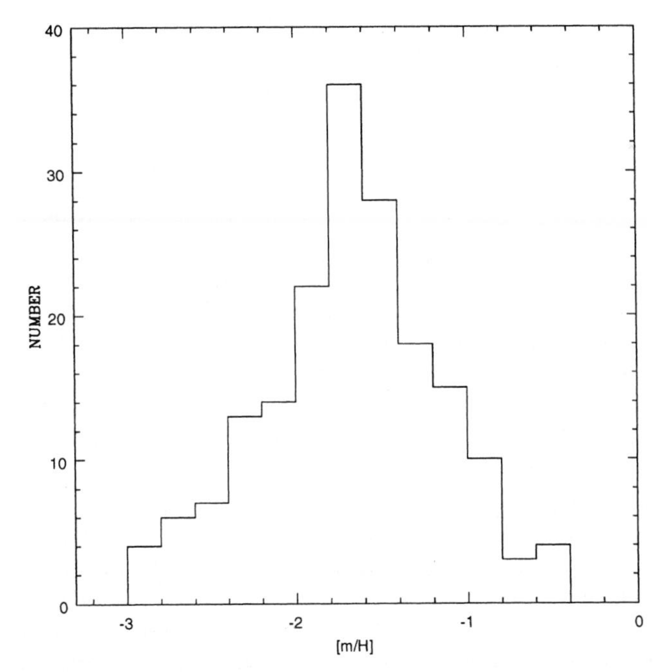

Fig. 41. The metallicity distribution function (MDF) for stars with $e \geq 0.85$

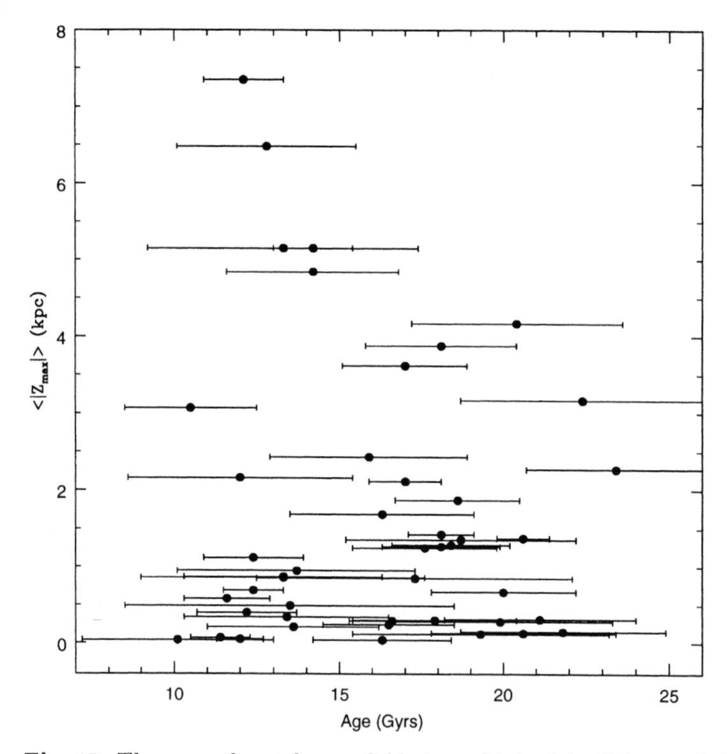

Fig. 42. The ages of metal-poor field stars obtained by Schuster & Nissen (1989b) plotted *vs.* mean maximum distance reached from the Galactic plane on their calculated orbits

origin that is related to the SZ model or to accretion. The "low halo" appears to be in prograde rotation and may have a radial metallicity gradient. Further, the "high halo" appears to have a different mean metallicity, itself suggestive of a separate history. If we employ a preliminary version of the weighting algorithm (discussed below), we see that the retrograde rotation for the "high halo" becomes closer to zero net rotation, consistent with a series of merger events, and that the differences between the "high halo" and "low halo", in terms of rotation, mean metallicity, and radial metallicity gradients, persist.

Can we say anything about ages? Figure 42 shows the stars within the study of Carney et al. (1996) whose ages were estimated using Strömgren photometry by Schuster & Nissen (1989b). Their age estimation technique works for field stars because an isochrone, normally thought of as a relation between temperature and luminosity, is also a relation between temperature and surface gravity, which is what the Strömgren system was designed to measure. While ages are not available for many stars, the figure suggests that the "high halo" is younger than the "low halo", as we had expected. Further, there is a hint that the age distribution of metal-poor field stars might even

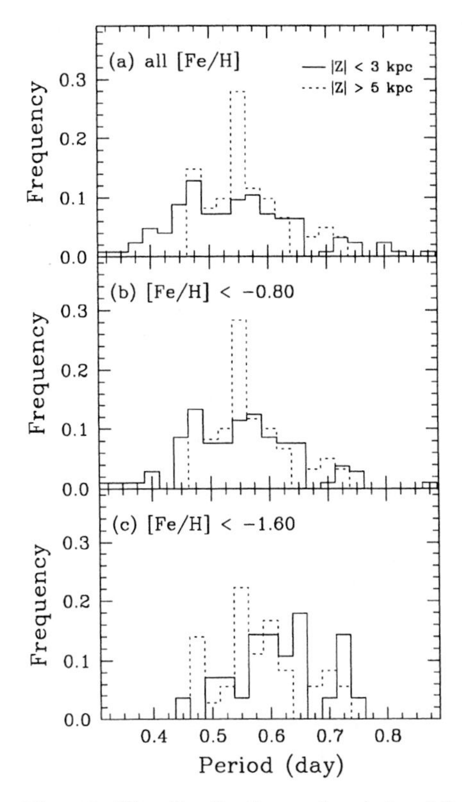

Fig. 43. The distributions of periods of field RR Lyraes, divided into three metallicity domains and for two ranges in distance from the plane

be bi-modal. The mean metallicities of the five stars which reach the largest distances from the Galactic plane is −1.91, comparable to the 34 other stars, −1.85, but the mean ages are quite different, 13.2 ± 0.8 Gyrs *vs.* 15.3 ± 0.3 Gyrs (the errors are those of the mean).

We have seen that the Oosterhoff I and II RR Lyrae variables may arise from populations of different ages and histories, with the Oosterhoff I ($\langle P_{\rm ab} \rangle =$ 0.55 day) younger than those in the Oosterhoff II class ($\langle P_{\rm ab} \rangle = 0.65$ day). If this age difference is true, and as Fig. 42 suggests, that the "high halo" is younger than the "low halo", then we should see a difference in the distributions of the field RR Lyrae variables' periods as we move to larger distances from the plane, with the Oosterhoff I variables present at all heights and the Oosterhoff II variables confined more closely to the plane. Lee & Carney (1999b) have used the unbiased studies of field stars from Preston (1959), Butler et al. (1982), Kinman *et al.* (1982, 1984, 1985), Suntzeff et al. (1991, 1994), Kinman (1998), Suntzeff (1998), Saha (1984), and Saha & Oke (1984). Figure 43 shows the distributions of field RR Lyrae variables' periods in three metallicity regimes for stars within 3 kpc of the plane and more than 5 kpc from the plane. The key point is that the Oosterhoff I peak is seen in all metallicity domains and both close to and far from the plane, while the Oosterhoff II variables dominate at low metallicities and only close to the plane.

Accretion. If there are really two independent histories for the Galaxy's metal-poor stars, one involving a coherent ELS-like collapse or contraction, and the other involving independent "fragments" as envisioned by SZ, then we are in essence now forced to argue that accretion has played a significant role in the Galaxy's early evolution. Is there more direct evidence for distinct accretion events?

The Sagittarius dwarf, discovered by Ibata et al. (1994), is clearly such an event in progress. And as Da Costa & Armandroff (1995) and others have noted, four of the Galaxy's globular clusters (M54, Terzan 7, Terzan 8, and Arp 2) actually belong to Sagittarius.

Another sign, albeit less obvious, for a discrete accretion event is found in the work of Preston et al. (1994). The enormous efforts by George Preston to identify weak-lined stars down to $B = 15$ mag using photographic plates, an interference filter, and an objective prism on modest aperture Schmidt telescopes yielded several very interesting subsamples. Figure 44 shows the color–color plane for some of their stars. The stars with $0.15 \leq (B-V)_0 \leq 0.40$ merit careful attention. Redward of that limit, and with $(U-B)_0 \leq -0.1$, are the metal-poor halo dwarfs, with colors redder than globular cluster turn-offs (see Fig. 20). Stars in this bluer color interval are then, like blue stragglers, bluer than the turn-offs. Further, lower gravities shift stars to redder $(U-B)_0$ colors, while lower metallicities shift them blueward. The stars plotted as filled squares are thus either blue stragglers or stars belonging to a young, metal-poor population. Preston et al. (1994) used blue straggler statistics from globular clusters to estimate that only about 10% of these stars are likely to be blue stragglers: the remainder are, as noted in Sect. 2, likely to be truly younger. Further, Preston et al. (1994) drew a comparison with the Carina dwarf galaxy, a large fraction of whose stars may be comparably young (only a few Gyrs: see Smecker–Hane et al. 1994). (It is worth noting that the Sagittarius dwarf also shows signs of episodic star formation: see Sarajedini & Layden 1995.) The kinematics of these stars supports a merger origin in that they do not agree with those of either the halo or the thick disk. The velocity dispersions are isotropic, about 90 km s^{-1} in all three components, and the mean V velocity is about -90 km s^{-1}.

Is much of the halo due to mergers of such large fragments as Sagittarius or Carina? This seems unlikely, as Unavane et al. (1996) have argued. If a significant fraction of such victim galaxies' stars are young, then we would expect to find a large number of blue straggler-like stars in studies of nearby metal-poor stars, quite contrary to what is seen in Fig. 20. On the other hand, this argument is weakened if the bulk of these mergers occurred early in the

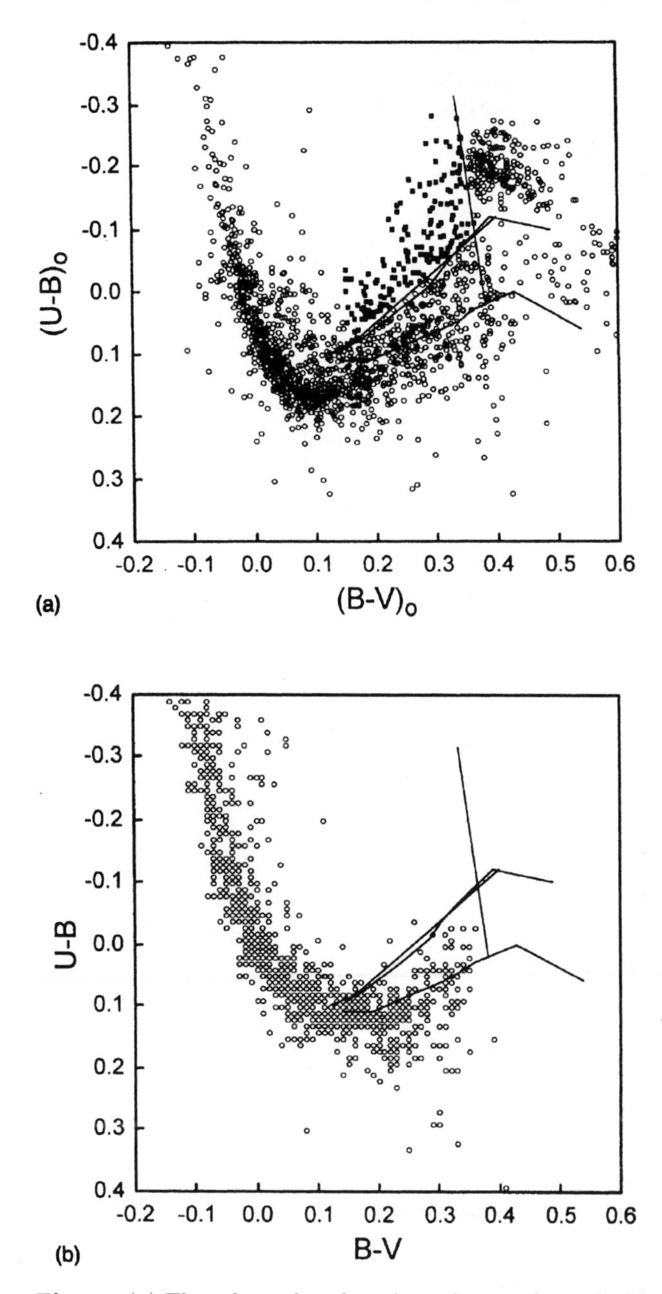

Fig. 44. (**a**) The color–color plane from the sample studied by Preston et al. (1994). The stars plotted as *filled squares* are bluer than the nominal halo population main sequence turn-off (represented by the almost *vertical line*), and with high gravities, judged by their position above the mean locus. (**b**) Bright Star Catalog data for B8 – F0 stars with $V \leq 6.0$ and luminosity classes III, IV, and V

history of our Galaxy (and those of the victims) since the stellar remnant we would now see would be only old and metal-poor. A thousand Draco or Ursa Minor dwarfs could provide the total halo stellar population. Such a large number of small galaxies merging early on, however, begins to sound more like the SZ "fragments" model rather than discrete merger events.

There is another method of identifying stars that probably originated outside the Galaxy's chemical "sphere of influence": study the [α/Fe] ratios and see if there are unusual stars or clusters. We address this in Sect. 6.

4.3 The Thick Disk Population

Majewski (1993) has written an excellent summary of the study of the thick disk population, which apparently includes a number of globular clusters, according to the MDF seen in Fig. 36. The radial velocity data of the metal-rich disk clusters (Da Costa & Armandroff 1995), and the space motions for a few such clusters (see Cudworth & Hanson 1993; Dinescu et al. 1997,1999; Odenkirchen et al. 1997) clearly show disk kinematics. Three major questions confront us. How old is the thick disk? What is its mean metallicity and metallicity spread (i.e., its MDF)? What was its origin?

The Age of the Thick Disk. The most direct means to estimate the age of the thick disk is to derive it from its globular clusters, as discussed in Sects. 8 and 9. Because the thick disk has a very different mean metallicity than the halo, the relative age determinations are complicated by remaining uncertainties in (a) the slope of the M_V *vs.* [Fe/H] relation, discussed in Sect. 3; (b) the fact that many disk clusters only have red horizontal branches, and the differences between M_V(RR) and M_V(RHB) may not be trivial; and (c) the helium mass fraction and the other element-to-iron ratios need careful study. Nonetheless, ignoring point (c), and assuming that the abundance ratios of disk clusters are very similar to those of more metal-poor clusters (which seems to be the case, as Carney 1996 has discussed), one may compare ages following Chaboyer et al. (1996). Using $M_V = 0.15[\mathrm{Fe/H}]+0.98$, the ages of NGC 104 (47 Tuc), NGC 6352, and NGC 6838 (M71) are 16.0 ± 0.6 Gyrs, while 10 "old halo" clusters with known [α/Fe] ratios are 20.4 ± 0.8 Gyrs and the 3 "young halo" clusters are 16.5 ± 1.5 Gyrs. The absolute ages depend strongly on the uncertain zero point of the M_V *vs.* [Fe/H] relation, but the relative ages, which depend on the better-determined slope, suggest that the disk clusters are indeed very old. But they are not, apparently, quite as old as the "old halo" clusters, which, as we have seen, have in common with the disk populations in general a net prograde rotation and a modest metallicity gradient. The thick disk clusters ages are more comparable to the "young halo" clusters, despite their very significant differences in kinematics.

Three clusters do not comprise very large foundation on which to base an estimate for the age of the thick disk. Are they representative? In Fig. 45

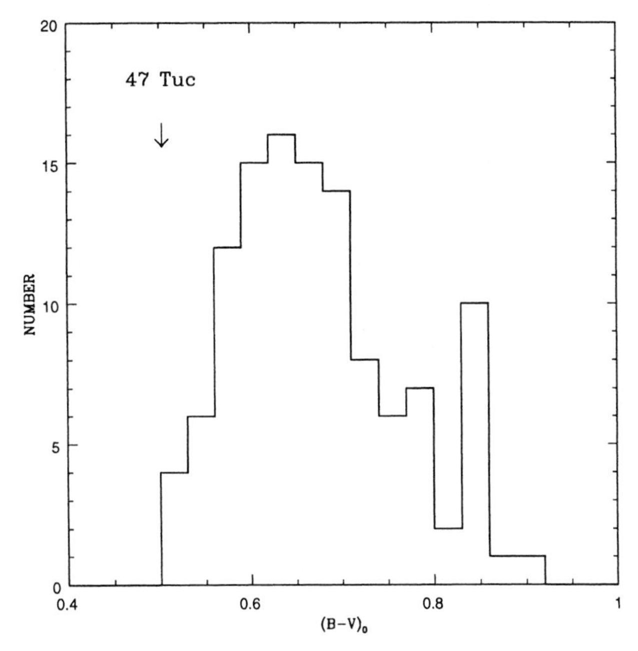

Fig. 45. The distribution of dereddened $B - V$ colors for field stars with [Fe/H] within ± 0.10 dex of the metallicity of 47 Tuc. The *arrow* indicates the color of the cluster's main sequence turn-off

we show the distribution of $(B - V)_0$ values for the stars in the Carney et al. (1994) sample whose metallicities are within 0.1 dex of that estimated for 47 Tuc. The main sequence turn-off color index for the cluster is indicated with an arrow, and it is clear that the field stars do not extend blueward of that, indicating a comparable age.

A second method to compare the age of the thick disk and the halo populations is hinted at in Fig. 18. The short-period binary systems all have circular orbits due to tidal circularization, the effects of which are a strong power of the ratio of the stars' separations and radii. In the Hyades, binary systems with periods of less than 5.7 days are circular, while longer period systems are eccentric (Mayor & Mermilliod 1984; Burki & Mayor 1986). In the older open cluster M67, the transition between circular and eccentric periods is about 11 days (Mathieu & Mazeh 1988; Mathieu et al. 1990) and thus we have another chronometer available. The exact dependence of the transition period on the cluster age is still debated (see Goldman & Mazeh 1991), but relative ages may be estimated from the transition period that divides the circular from the eccentric binary star orbits. In Fig. 46 the orbits of metal-poor ([Fe/H] ≤ -1.0) and high-velocity ($V \leq -100$ km s^{-1}) halo stars are plotted. The details of the orbits have been published (Latham et al. 1988, 1992) or will be published by Latham and his colleagues (Latham et al. 1999; Goldberg et

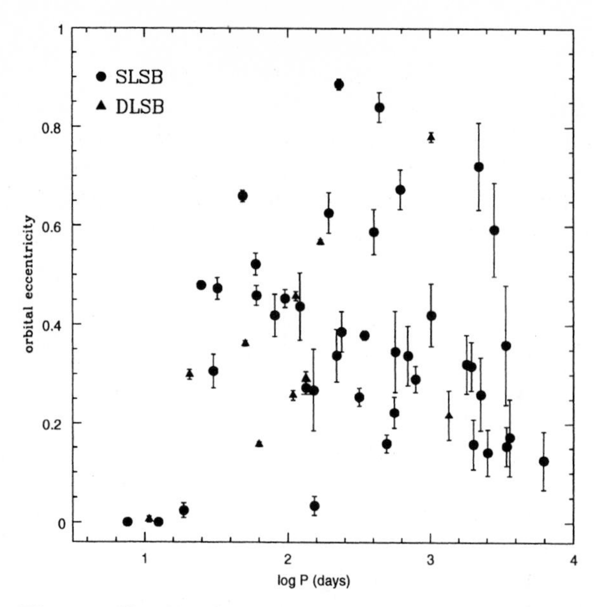

Fig. 46. The distribution of orbital eccentricities against the logarithm of the periods for a sample of halo stars (single- and double-lined spectroscopic binary systems)

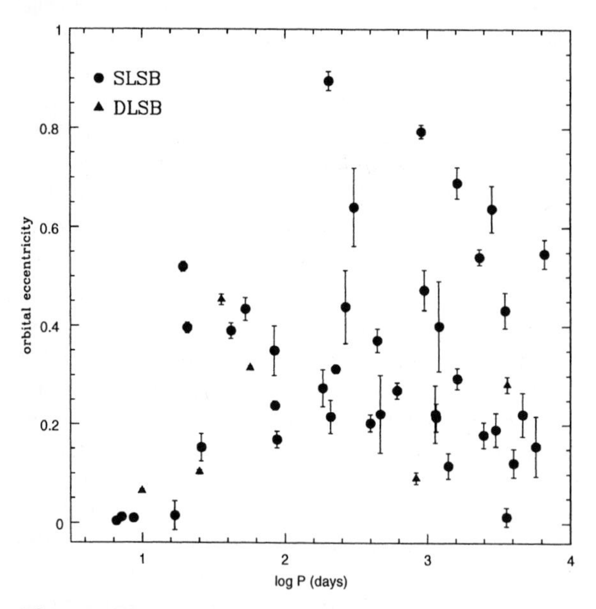

Fig. 47. The distribution of orbital eccentricities against the logarithm of the periods for a sample of thick disk stars

al. 1999), excluding five blue stragglers. The transition period is seen easily at $\log P \approx 1.3$ ($P = 20$ days). For the thick disk subsample, kinematics have been employed to produce a reasonable thick disk-like MDF, with a peak near [m/H] ≈ -0.7, and the sample was then winnowed further by restricting the sample to $-1.0 \leq$ [m/H] ≤ -0.4. The orbits for this sample are shown in Fig. 47 and, again, the transition period appears at $\log P \approx 1.3$ (20 days). To within the precision of this chronometer, the thick disk and metal-poor high-velocity populations have comparable ages.

The Metallicity of the Thick Disk. Figure 36 illustrates that the globular clusters show two distinct MDFs, with the more metal-rich one associated with disk clusters such as 47 Tuc and M71 (see Cudworth & Hanson 1993 for their space motions). The peak in this metal-rich MDF is at [Fe/H] ≈ -0.5. Field stars show similar results. Perhaps the most direct measure is that of Gilmore et al. (1995), who studied a set of stars with $15 \leq V \leq 18$. This *in situ* sample covered distances ranging from 500 to 3000 pc from the disk, and thus enabled them to rather cleanly see the MDF at a position where the "contamination" by the thin disk population was minimized. They found the MDF peak to be at [Fe/H] ≈ -0.7, a bit more metal-poor than the globular clusters. This difference may not be real, however, since the techniques used to estimate [Fe/H] are not the same, and an offset of 0.2 dex is probably within the calibration uncertainties. This is an important question, however, for if the difference is real, then since the globular clusters studied by Zinn (1985) tend to lie closer to the Galactic center, the difference would imply a radial metallicity gradient in the thick disk population. This gradient then would imply that the thick disk evolved slowly, with dissipation, and thus is more likely to be an inherent stage in the evolution of the Galactic disk. The absence of such a metallicity gradient, conversely, would argue against such evolution, and would be mild evidence in favor of an accretion origin for the thick disk.

Carney et al. (1989) studied the MDF of the thick disk using their spectroscopic metallicities for a large number of high proper motion stars in the solar neighborhood. While not an *in situ* sample, they studied the MDF over a range of W velocities, which correlated with the distances the stars reach from the Galactic plane. A characteristic peak in the MDF appeared at [m/H] ≈ -0.5, which they identified with the thick disk. As expected, for the smaller W velocities, this peak was weaker than that of the disk, at [m/H] ≈ 0.0, but dominated at intermediate W velocities, and essentially disappeared for $|W| \geq 100$ km s^{-1}. One should not attach too much significance to the 0.2 dex disagreement with the results of Gilmore et al. (1995) and the agreement with the globular cluster results of Zinn (1985) since the metallicities of Carney et al. (1989) are based on yet another metallicity calibration. Further, use of a sample that is kinematically selected, in this case from proper motion catalogs, introduces some potentially serious systematic biases. The advantage of such a selection is that one identifies very rare high-velocity (i.e., halo) stars much more readily. But this may distort the true kinematic *vs.* metallicity relations. To correct for this bias, Dr. Luis Aguilar has developed a statistical correction procedure that is based, in essence, on the $1/V_{max}$ algorithms of Schmidt (1968, 1975). Basically, this weights the contribution of each star by the ratio of two volumes: that defined by the star's distance and that defined by the distance at which it could still be seen in a magnitude-limited survey. Aguilar has extended the basic idea to include not only magnitude limits, but also the limits in tangential velocity, which is set by $\mu_{lim} \times d$, where d is a star's distance and μ_{lim} is the lower limit to the proper motion catalog. There are other, more subtle effects, such as the fact the sample is limited to the northern hemisphere and the solar peculiar motion and the direction of the Local Standard of Rest also play a role in determining which stars are included in a proper motion catalog. The corrections appear to work, based on Monte Carlo simulations, however, and, more important, on the kinematics derived for metal-rich disk stars. The "velocity ellipsoid" is the set of the U, V, and W velocity dispersions: $[\sigma(U); \sigma(V); \sigma(W)]$ – all in km s^{-1}. Kinematically biased samples will produce unrealistically high velocity dispersions, and this is seen easily in the unweighted data from Carney et al. (1994). For the stars more metal-rich than the thick disk, $[m/H] > -0.5$, the velocity ellipsoid becomes $[65 \pm 3; 35 \pm 2; 31 \pm 2]$, which may be compared to several kinematically unbiased samples: the local G0–K0 results of Delhaye (1965) of [29; 17; 15]; the results of Edvardsson et al. (1993) for stars with $[Fe/H] \geq -0.3$ of $[35 \pm 3; 21 \pm 2; 17 \pm 2]$; and the results of Flynn & Morell (1997) for M stars with $[Fe/H] \geq -0.5$ of $[37 \pm 3; 24 \pm 2; 17 \pm 1]$. With the new weighting algorithm applied to a revised but still preliminary version of the proper motion sample, we find $[29 \pm 2; 18 \pm 1; 16 \pm 1]$, in excellent agreement with the kinematically unbiased samples. (The version is preliminary because we have not yet completed culling subgiants from the sample.) Having established that the corrections for kinematical bias appear to work, we follow the example of Gilmore et al. (1995) and consider the corrected metallicity distribution for stars whose Galactic orbits take them to 1.0 ± 0.2 kpc from the plane, shown in Fig. 48. The peak in the MDF has shifted to $[m/H] \approx -0.7$, in good agreement with Gilmore et al. (1995). There remain signs of the disk population with a peak near -0.1 and the halo, with a weaker peak near -1.7. Or does the thick disk itself extend over such a large range in metallicity?

One interesting point seen in Fig. 48, and in the data of Gilmore et al. (1995) is the low-metallicity extent of the thick disk. In both cases it appears to reach down to metallicities as low as -1.5. This low-metallicity tail has been the subject of significant debate. To summarize, Norris et al. (1985) and Morrison et al. (1990) estimated metallicities using the *DDO* photometric system for a sample of kinematically unbiased stars. They found stars with disk-like kinematics and with metallicities as low as $[Fe/H] = -1.6$. Unfortunately, the *DDO* system utilizes the CN molecular features and so is not a direct

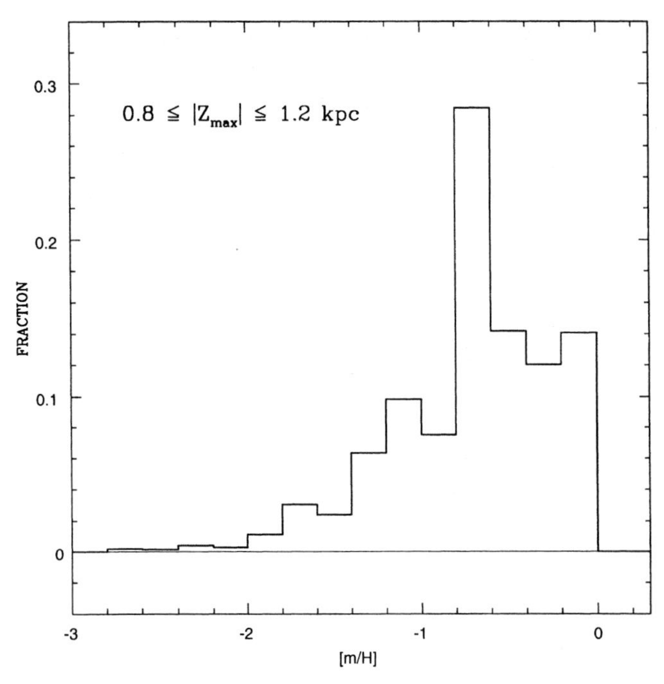

Fig. 48. The weighted metallicity distribution based on data from Carney et al. (1994) for stars whose orbits reach 0.8 to 1.2 kpc from the Galactic plane

measure of [Fe/H]. Recalibrations of the *DDO*–[Fe/H] relation by Twarog & Anthony–Twarog (1994) and by Ryan & Lambert (1995) sharply reduced the significance of the metal-poor tail to the disk or thick disk population, but unless Gilmore et al. (1995) and Carney et al. (1994) have serious problems with their metallicity calibrations, it does appear that the thick disk, at least, has a significant metal-poor tail. Beers & Sommer–Larsen (1995) have also argued in favor of this result. Using a kinematically unbiased sample of metal-poor stars, they used single component models to fit the rotational velocity data, and found that the models underpredict the number of stars with large, disk-like, Galactic rotational velocities. A disk component appears to be necessary, even for $[Fe/H] < -1.5$. The extension to such low metallicities is confirmed from the limited space velocity data for globular clusters. Cudworth & Hanson (1993) found M28 (NGC 6626) to have $\Pi = 3$ km s^{-1}, $\Theta = 167$ km s^{-1}, and $Z = -40$ km s^{-1}. (These velocities are similar in direction to U, V, and W, but are defined at the cluster's position rather than that of the Sun, and Θ refers to the circular orbit velocity at that Galactocentric distance rather than that with respect to the Local Standard of Rest.) M28 has a metallicity of $[Fe/H] = -1.45$ (Zinn 1985). Dinescu et al. (1997) have also found that NGC 6752 ($[Fe/H] = -1.61$) has $\Pi = -20$ km s^{-1}, $\Theta = 199$ km s^{-1}, and $Z = 26$ km s^{-1}.

The Origin of the Thick Disk. The origin of the thick disk is one of the major puzzles in the study of the origin and early evolution of our Galaxy and, by extension, other disk galaxies. Is the thick disk a natural step in the formation of a disk? Or is it the result of a merger event, with the victim being about the size and overall metallicity of the Small Magellanic Cloud? There are at least three separate approaches to answer this question: kinematics *vs.* chemistry; metallicity gradients; and element-to-iron ratios.

Chen (1997) re-analyzed the proper motion data of Majewski's (1992) North Galactic Pole sample and found that the thick disk shows a gradient in its asymmetric drift of -14 ± 5 km s^{-1}/kpc as one moves away from the plane. Such a gradient could have been produced by dissipational settling, but it is not clear if this was due to a tidally-induced merger, or part of the formation of the Galactic disk.

If the thick disk was the result of an accretion event, then we might expect to see a significant change in kinematics, especially in the W velocity component, as a function of metallicity, as we move from the metal-rich domain, where the thin disk presumably dominates, to the intermediate metallicity domain, where the thick disk apparently dominates. A discontinuity in σ(W) around the peak of the thick disk MDF would suggest an accretion event. Figure 49 shows the results of four independent studies. Those of Norris (1987), Strömgren (1987), and Yoss et al. (1987) are based on kinematically unbiased data, while the data from our survey of proper motion stars have been corrected as best as possible for the effects of the kinematic biases. There is generally good agreement, except, perhaps, for the offset to large σ(W) values at high metallicities in the Yoss et al. (1987) sample. The general trend is what one expects from an evolving disk that is dissipating its energy: σ(W) rises as one looks "back in time" (i.e., to lower metallicities). But in the mean, and study by study, it appears that there is rather sudden increase in σ(W) for [Fe/H] ≈ -0.5 and below, roughly the point at which the thick disk is becoming more important (but not, probably) dominant relative to thin disk. Wyse & Gilmore (1995) have also argued that as [Fe/H] declines to about -0.5, σ(W) rises from about 20 km s^{-1} to about 40 km s^{-1}.

The apparent lack of a metallicity gradient in the thick disk would suggest that it has not undergone dissipative processes, as would be expected from a merger event. To be fair, Wyse & Gilmore (1995) noted that the absence of a gradient is not compelling evidence for a merger event since, in essence, it depends upon whether the merger was composed largely of stars or mostly of gas. In the latter case, a merger may still have produced a metallicity gradient as the gas settled into an enlarged and reconstructed disk. There are few data with which to search for a vertical metallicity gradient over a large range in distance from the plane and which employ the same metallicity estimation methods. Gilmore et al. (1995) compared the metallicity distributions from their two fields that sampled $Z \approx 1.0$ and 1.5 kpc, and found no evidence for a gradient. Figure 50 shows the weighted (i.e., corrected for kinematical

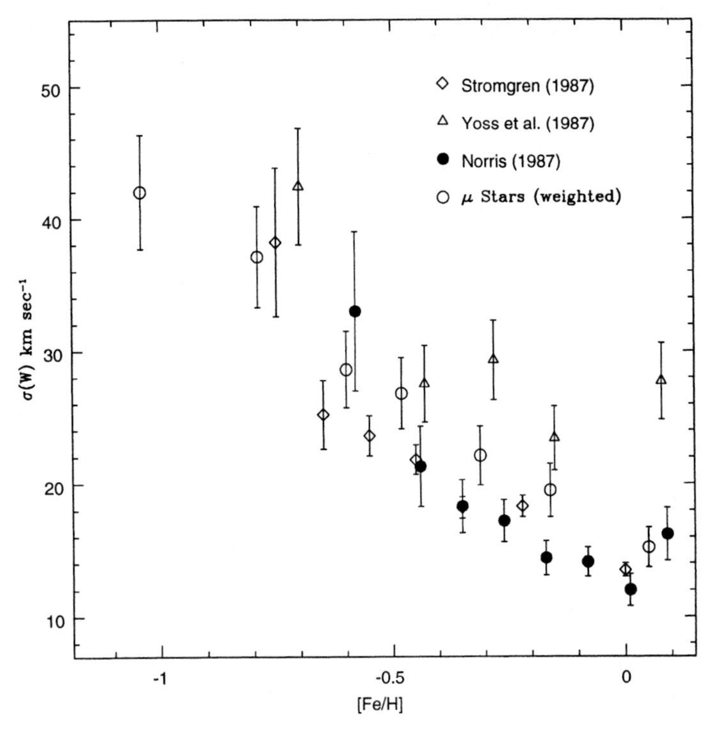

Fig. 49. The variation of the velocity dispersion perpendicular to the Galactic plane *vs.* metallicity from four independent studies

bias) proper motion data for stars whose orbits carry them to 0.6 ± 0.2 kpc from the plane. One sees, as expected, a strong signal from the disk, with a peak at $[m/H] \approx -0.1$, while the thick disk peak is between -0.6 and -0.7. Comparing this to Fig. 48 reveals that it is not clear that *either* the disk or the thick disk show a vertical metallicity gradient in going from 0.6 to 1.0 kpc, *assuming* that the metal-rich stars belong to the thin disk.

A related but perhaps more persuasive argument in favor of an independent origin for the thick disk was advanced by Wyse & Gilmore (1995). They estimated the density of the thick disk stars, those with $[Fe/H] \approx -0.7$, at distances well above the plane, and then extrapolated to estimate how many stars should be found closer to the plane. They then compared this estimate with the number found in the solar neighborhood, and found many more such stars than predicted. The thick disk itself appears to contribute too few stars with such metallicities at mid-plane, and thus a contribution from the disk appears required. Near mid-plane, then, it is the metal-poor tail of the disk population that dominates. Wyse & Gilmore illustrated this nicely using the $\Sigma|W|$ statistic of Freeman (1991). This is, as its name implies, the running sum, ranked by [Fe/H], of the $|W|$ velocities. A constant slope indicates a

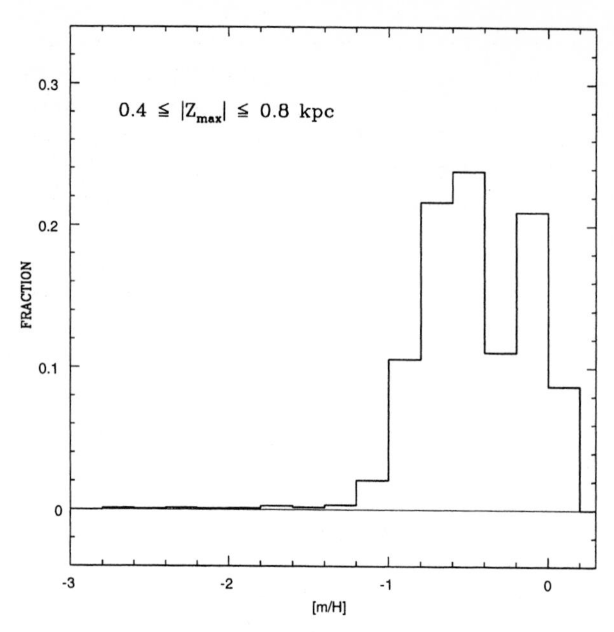

Fig. 50. The weighted metallicity distribution based on data from Carney et al. (1994) for stars whose orbits carry them 0.4 to 0.8 kpc from the plane

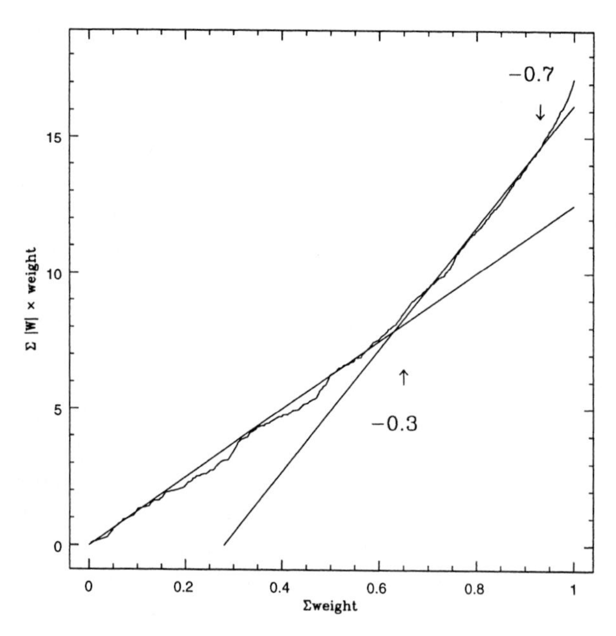

Fig. 51. The running weighted sum of |W|, ordered by [m/H], showing breaks at [m/H] = −0.3 and −0.7

well-behaved single population, but Wyse & Gilmore's (1995) analysis of the kinematically unbiased sample of Edvardsson et al. (1993) showed that at metallicities approaching −0.4, the running sum steepened its slope, signifying the appearance of a second population. In Fig. 51 we show a similar result from our proper motion sample, after correction for the kinematic bias. Here we see the change in slope appearing at [m/H] ≈ −0.3 rather than at −0.4, but it still appears to indicate the appearance of the second, dynamically hotter, thick disk population. The key point here, however, is that Wyse & Gilmore (1995) argued that the thick disk and the thin disk *overlap* in [Fe/H] but are *distinct* in kinematics. This is strong evidence for a merger origin for the thick disk.

Recently, Serge Naumov has completed a study of long-lived stars in the plane of the Galaxy. He used objective prism spectroscopy to identify main sequence stars with effective temperatures cool enough to indicate that the stars' life expectancies exceed the Galaxy's age. This gives equal weights to the thick disk, which as we have seen is ancient, and to the thin disk. High-resolution and low-S/N echelle spectra have been obtained for almost a thousand G and K dwarfs, and radial velocities measured and metallicities estimated following the procedures outlined by Carney et al. (1987). The stars were observed in the "cardinal" directions: $\ell \approx 0^\circ$, 90°, and 180°. Thus the radial velocities alone yield the U or V velocities. Several hundred nearby G and K dwarfs with Strömgren photometry were also observed. Figure 52 shows the distribution of V velocities *vs.* [m/H]. The thick disk is apparent in the distribution of stars with large asymmetric drifts. Note that such stars exist over almost the same range in metallicity as does the thin disk. Another way to consider this is shown in Fig. 53. Here the V velocities are binned for stars with metallicities ranging from solar to −0.5, the nominal metallicity domain of the thin disk. Two velocity components are evident, signifying the separate kinematic properties of the thin and thick disk populations in this metallicity regime. The lower panel shows the resultant "double root residual", described by Beers & Sommer–Larsen (1995). It appears that the thick disk and the thin disk do overlap in metallicity but remain distinct in kinematics, and probably have experienced independent histories.

Finally, in principle the details of the chemical composition may be used to explore the discreteness of the thick disk population relative to the disk. The idea here is that if the thick disk evolved rapidly (as appears to be the case) while the disk evolved slowly, then at the more metal-rich end we may see differences in [α/Fe]. The thick disk, if it evolved rapidly, would retain a stronger signature of SNe II abundance ratios, while the disk would show more contributions from SNe Ia. It should be remembered, however, that similarity would not prove similar histories, but only similar durations of nucleosynthesis timescales. This is a difficult test to make, unfortunately, since [α/Fe] ratios exist for only three thick disk clusters: 47 Tuc, NGC 6352, and M71. Their spectroscopic metallicities are [Fe/H] = −0.73, −0.70, and −0.79 (see the sum-

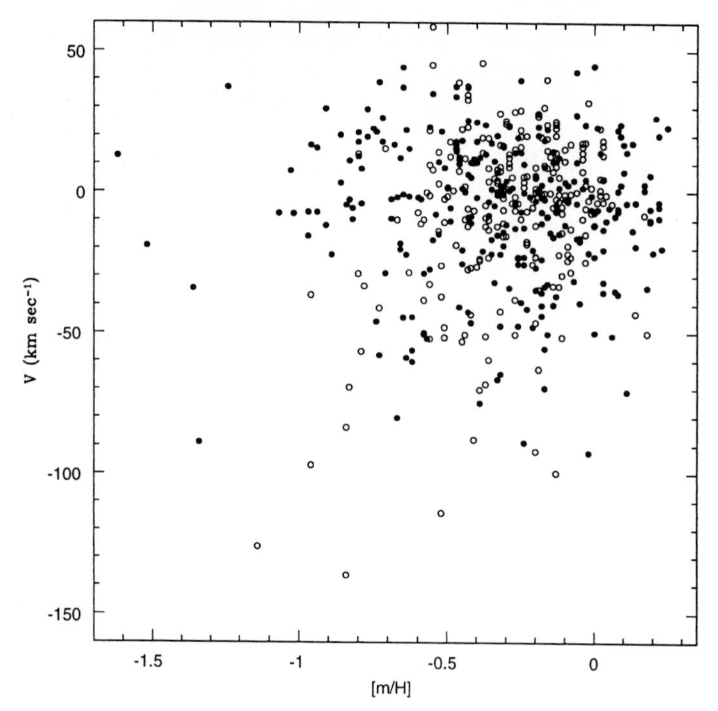

Fig. 52. The distribution of V velocities *vs.* metallicity from the study of Naumov et al. (1999a) of G and K dwarfs in the plane in roughly toward $\ell = 90°$ (*filled circles*) and from a sample of nearby G and K dwarfs with Strömgren photometry (*open circles*)

mary by Carney 1996). Their [α/Fe] ratios are $+0.25 \pm 0.04$, $+0.40 \pm 0.05$, and $+0.46 \pm 0.03$. The unweighted means are then $\langle[\mathrm{Fe/H}]\rangle = -0.74$ and $\langle[\alpha/\mathrm{Fe}]\rangle = +0.37 \pm 0.06$ ($\sigma = 0.11$). The local disk and thick disk star sample of Edvardsson et al. (1993), which as we have seen is probably dominated by the disk rather than the thick disk, even at the lower metallicities, can provide us with a disk comparison sample. Using the 13 stars with $-0.84 \leq [\mathrm{Fe/H}] \leq -0.67$ yields $\langle[\mathrm{Fe/H}]\rangle = -0.74$ and $\langle[\alpha/\mathrm{Fe}]\rangle = +0.18 \pm 0.01 (\sigma = 0.04)$. Taken at face value, this suggests that the thick disk globular clusters do not share the same chemical history as the mid-plane disk stars, but we urgently need more data for both samples, as well as analyses undertaken using similar procedures and atomic line data. We need [α/Fe] data for more metal-rich disk clusters, and, in fact, we also need more data for disk stars closer to the Galactic center. After all, the Edvardsson et al. (1993) sample is confined to the solar neighborhood, while the three globular clusters lie at an average Galactocentric distance of 5.7 kpc. We discuss this further in Sect. 6.

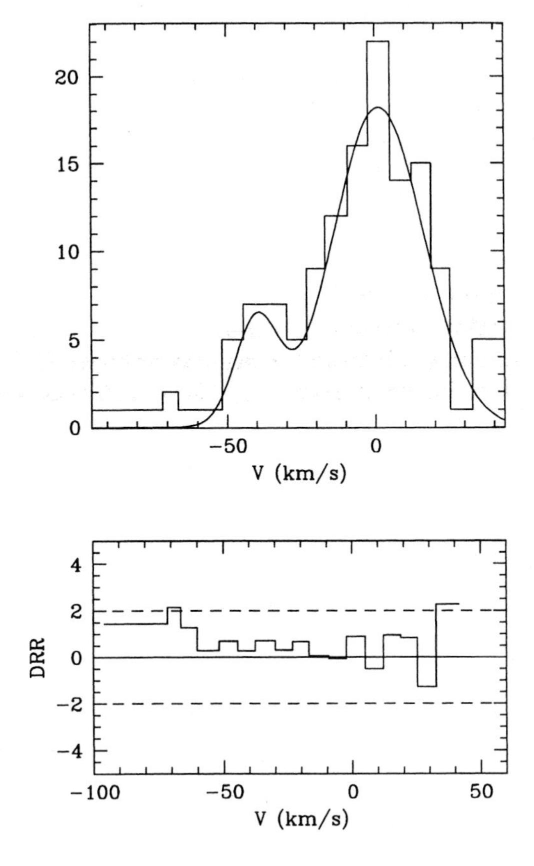

Fig. 53. Stars in Fig. 52 with $-0.5 \leq [m/H] \leq 0.0$ are binned in V velocities, and a double Gaussian distribution is derived. The lower panel shows the "double root residual", following Beers & Sommer–Larsen (1995)

5 Metallicities

We have seen how the heavy element mass fraction, Z, and its finer division to recognize the CNO elements or the "α" elements, are important in the evolution of stars and, consequently, in our interpretation of color–magnitude diagrams. The importance of Z in the estimation of cluster ages is made clear in Fig. 7 and in (29). Mean metallicity is also one of the often-used defining characteristics of stellar populations, and the metallicity distribution function is a major component in our understanding of the evolution of our Galaxy. And finally, insofar as the most metal-poor stars in globular clusters and in the field may retain the signatures of Big Bang nucleosynthesis, their chemical abundances play an important role in cosmology.

5.1 The Metallicity Distribution Function

Figure 34 depicts qualitatively how the distribution of metallicity may evolve in a closed system of stars and gas. This may be appropriate, in principle, to an entire population, such as the Galaxy's halo or bulge, to the proto-Galactic fragments envisioned by SZ, or perhaps even to individual globular clusters that may have undergone some self-enrichment. It is worth attending to some of the basic details of the simpler models, a subject that is reviewed very well by Pagel (1997). The simplest "closed box" models were discussed initially by Schmidt (1963), Searle & Sargent (1972), and Hartwick (1976).

One fundamental assumption in the mathematical modelling is that the gas is well-mixed so that the heavy element mass fraction Z is a simple function of time within the evolution of the system, $Z(t)$. For a total ensemble mass M, then, we have

$$Z(0) = 0 \,, \tag{42}$$

$$M_{\rm gas}(0) = M \,, \tag{43}$$

$$M_{\rm stars}(0) = 0 \,, \tag{44}$$

$$M_{\rm gas}(t) + M_{\rm stars}(t) = M = \text{constant} \,, \tag{45}$$

$$dM_{\rm gas} = -dM_{\rm stars} \,. \tag{46}$$

The initial metallicity need not be zero, of course, but that is the usual assumption for the halo population. The "yield", p, is defined by the increase in heavy element abundances relative to the mass locked up in long-lived stars and stellar remnants:

$$p = M_{\rm gas} \frac{{\rm d}Z}{{\rm d}M_{\rm stars}} = -M_{\rm gas} \frac{{\rm d}Z}{{\rm d}M_{\rm gas}} \,. \tag{47}$$

The metallicity Z then evolves according to:

$$Z = p \ln \frac{M_{\rm gas} + M_{\rm stars}}{M_{\rm gas}} = p \ln \frac{1}{\mu} \,, \tag{48}$$

where μ is the fraction of the total mass remaining in gas. One key result of this is

$$\langle \frac{Z}{p} \rangle = 1 + \frac{\mu \ln(\mu)}{1-\mu} \,, \tag{49}$$

which approaches unity as the fractional gas mass μ approaches zero. Thus the mean metallicity $\langle Z \rangle$ approaches the yield p. This in turn is presumably related to the initial mass function, and since there is little evidence to support a significant variation with metallicity (see Mateo 1993, for example), the yield for the halo, thick disk, and disk populations ought to be similar, and so should their mean metallicities (insofar as such a simple model applies to the Galaxy's true history). The obvious disageement in the mean metallicities can be by invoking a "leaky" rather than a "closed system". If the system can lose mass, presumably in the form of gas, the "effective yield" μ_{eff} may be reduced and so may the mean metallicity. Following Hartwick (1976), Laird et al. (1988), and Pagel (1997), if we assume that the system loses mass at a rate related to the star formation rate, then

$$\frac{\mathrm{d}M}{\mathrm{d}t} = -E = -\frac{\eta}{\alpha\,\psi}\,, \tag{50}$$

where $\mathrm{d}M/\mathrm{d}t$ is the rate of mass expelled, ψ is the star formation rate, η is the proportionality constant, and α is the mass still in stars divided by the mass of stars produced. This proportionality has some physical basis: star formation leads to stars of all masses, and the most massive stars should explode as supernovae. The energetics may be sufficient to expel gas from low-density environments, such as would be expected early in the evolution of the Galactic halo, or in the proto-Galactic fragments of the SZ model. From (50) we obtain

$$\frac{\mathrm{d}M_{\text{gas}}}{\mathrm{d}M_{\text{stars}}} = -\frac{E}{\alpha\,\psi} - 1\,, \tag{51}$$

and with the same definition of the yield p as before, we find

$$M_{\text{gas}}\,\frac{\mathrm{d}Z}{\mathrm{d}M_{\text{gas}}} = -\frac{p}{1+\eta}\,, \tag{52}$$

from which we obtain

$$\frac{\mathrm{d}M_{\text{stars}}}{\mathrm{d}\log(Z/p)} \propto \frac{Z}{p}\,\exp\left[-\frac{Z\,(1+\eta)}{p}\right]\,. \tag{53}$$

If there is no star formation-induced wind, then $\eta = 0$, and we recover the simple case. Pagel (1992, 1997) has argued that the best element in the comparison with observations is oxygen, rather than, say, iron, because oxygen is primarily produced by SNe II, and hence also better satisfies the "instantaneous recycling" assumption. He argues that the solar neighborhood has $\langle\,[\mathrm{O/H}]\,\rangle \approx -0.2$, and if η is assumed to be zero in this case, then $p \approx 0.02$. Thus to explain the halo abundances, η is about 10. This model is what led us (Carney et al. 1990b) to inquire of the fate of the halo's gas, and the conclusion that much of it may have wound up in the reservoir of the Galactic bulge, within which star formation has gone to near completion and, with η again presumed to be near zero due to the large gravitational potential, $\langle Z\rangle$ should now be comparable to that of the solar neighborhood, as found by McWilliam & Rich (1994). This reduced efficiency in the halo caused by mass loss could, of course, be taken to an extreme case. Rather than losing mass in simple proportion to the star formation rate, which produces a metallicity distribution function like that seen in the halo (Laird et al. 1988), the "wind", especially if powerful enough and applied to a system with small enough escape velocity, could completely terminate star formation, leaving behind a terminated metallicity distribution function as in Fig. 54 (Yoshii & Arimoto 1987). This could be the situation within dwarf galaxies or even within individual globular clusters, if they formed over a long enough time for a limited degree of self-enrichment to have occured.

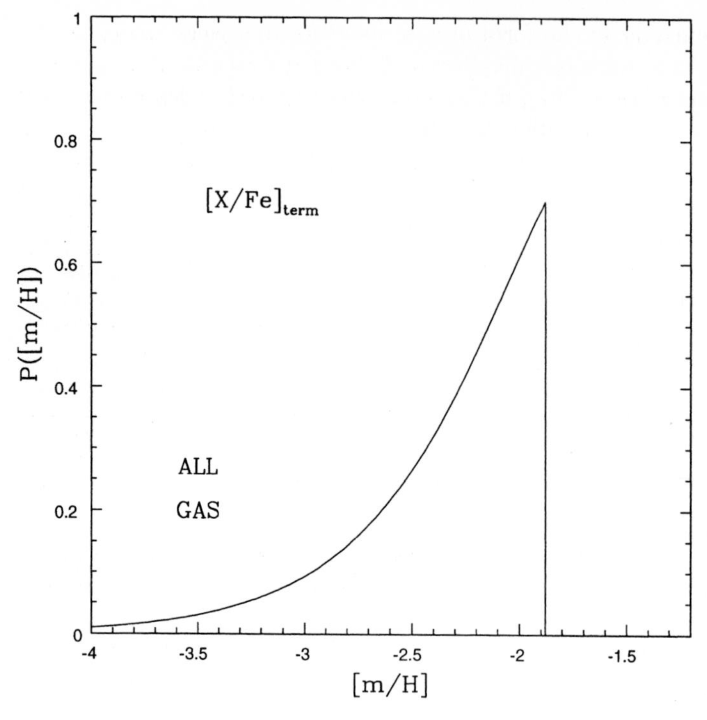

Fig. 54. A model metallicity distribution function terminated abruptly by a vigorous stellar wind. Such a termination will result in a characteristic $[X/Fe]_{term}$

5.2 Measuring Abundances

Because field star and globular cluster abundances are so important, there have been many means devised to either measure them or, at least, infer them. We therefore distinguish between *measurement* of [Fe/H] (or [X/Fe]) and the use of *metallicity indicators* to infer [Fe/H]. Because all metallicity indicators must be tied to some direct measures of [Fe/H], we begin with a discussion of what they require.

High-resolution spectra form the basis for all metallicity measures and the calibration of metallicity indicators. The resolving power must be high enough to avoid line blending, and this means $R \approx 20,000$ or better, along with high signal-to-noise (S/N). The higher the S/N, the weaker lines may be measured and used, and since these are least vulnerable to systematic effects, most high-resolution analyses of field and cluster stars' spectra are done with S/N > 50 or so. Lower S/N spectra may still be useful, especially if there is forehand knowledge of some of the stellar parameters. My own preference is S/N > 100 or better, and 200 or better for the brighter field stars, and $R > 30,000$.

The accurate measurement of equivalent widths is a necessary but hardly sufficient condition to measure [Fe/H] or [X/Fe]. One must know the atomic parameters for the line in question. Usually the statistical weights g are well known, as are the excitation and ionization potentials. But the transition probability f is a major source of uncertainity and often the limitation in the precision of the final results. Accurate laboratory gf values are available for only a small number of atomic absorption lines, although "solar gf values" may be derived for a number of lines by using a model solar atmosphere, accurate equivalent widths, and an adopted solar abundance (from Anders & Grevesse 1989, for example). Finally, one must have a good model atmosphere for the star or stars being studied, including good estimates for the effective temperature, the gravity, the "microturbulence", and a good estimate of the metallicity beforehand or via an iterative process. In principle, the stellar parameters may be derived directly from the spectra, although for cluster stars, the color–magnitude diagram itself can prove a valuable secondary source of such parameters.

The curve of growth is an extremely useful tool for understanding how important these various atomic and stellar parameters are in abundance analyses. Figure 55 was constructed for the λ 6810.27 line of Fe I in a star with $T_{\rm eff} = 6000\,{\rm K}$, and a solar metallicity. The x–axis is the elemental abundance on a scale where $\log n({\rm H}) = 12.00$ and $\log A = \log n({\rm Fe}) - 12$. The solar iron abundance on this scale is $\log n({\rm Fe}) = 7.52$, or $\log A = -4.48$. The y–axis is the logarithm of the equivalent width (in mÅ) of this line, although it is more usual to plot the reduced equivalent width, $\log(W_\lambda/\lambda)$, which is dimensionless. Three curves of growth are shown, and all three reveal the three basic regimes: linear; "flat", and "square root". In the linear regime, increasing the abundance A leads to a linear increase in the equivalent width W_λ. When the abundance is high enough, somewhere around 50 mÅ for this line, saturation sets in. (Absorption lines do not reach zero intensity because the temperature at the depth at which they form produces continuum radiation.) Increasing the abundance increases the optical depth of the line center, but because of the radiative emission, the line cannot become stronger, and the $\log W_\lambda$ *vs.* $\log A$ becomes relatively flat. At even higher abundance levels, the density of atoms in this state becomes large enough for interactions, particularly neutral–neutral van der Waals forces, to broaden the energy levels and produce line shapes with strong damping wings. This allows for increased absorption and the lines again grow, but only approximately with the square root of the abundance. The three separate curves of growth illustrate these effects well.

Consider the two low-gravity models first. The higher microturbulent velocity means that the Doppler effect shifts the line center over a large range in wavelength, so it has a greater "opportunity" to absorb continuum radiation. The line does not therefore saturate as readily, and the "flat part" of the curve of growth occurs at a higher W_λ. But the "square root" part of the curve of

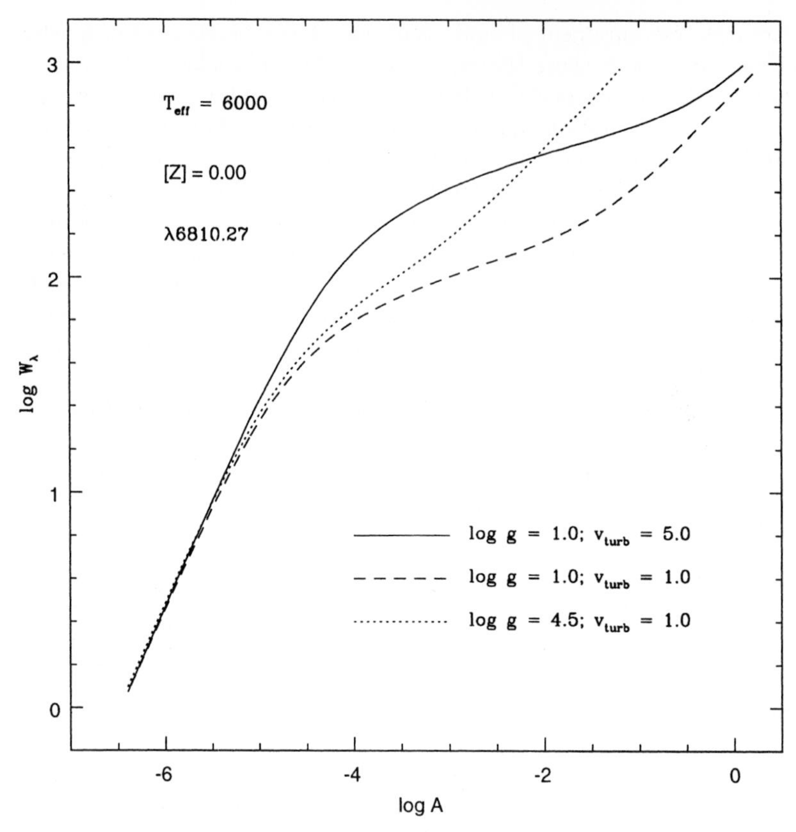

Fig. 55. A model curve of growth for the 6810.27 Å line of Fe I and a model effective temperature of 6000 K

growth is not much affected. To be fair, one does not usually see such a large range in microturbulence: $v_{\rm turb}$ is fairly well correlated with gravity and runs from about 1.0 km s^{-1} for main sequence stars to perhaps 4 or 5 km s^{-1} for the most luminous, lowest-gravity supergiants. Increasing the star's gravity from $\log g = 1.0$ to 4.5 has the effect of sharply narrowing the regime of the "flat part" of the curve of growth. As expected, damping effects occur much more readily. Figure 55 shows why weak lines, and hence high resolution and high S/N, are much preferred: Fe I lines with $\log(W_\lambda/\lambda) < -5.2$ or so are on the linear part of the curve of growth and hence not particularly vulnerable to errors in the adopted stellar gravity or microturbulence. Unfortunately, while the spectrum of iron is rich, for many elements, especially the more interesting ones, lines are few and one has little choice about which of the three regimes of the curve of growth to use. So one should always try to carefully estimate the stellar parameters using all the means at one's disposal. And just to make life a bit more difficult, there are two other processes that may delay saturation and make abundances analyses difficult unless the lines are weak enough and well-measured. Heavy elements often have several stable isotopes, and each of these has a slight shift in the energy levels of the valence electrons, and thus the absorption can occur over a wider range in wavelengths. Barium ($Z = 56$) is an element particularly vulnerable to this effect. In such cases, one must adopt isotopic abundance ratios, divide up the total gf value among them, and compute curves of growth for all of them individually. The sum of their equivalent widths must then be compared with what has been measured for the single, unresolved line. Elements with an odd number of protons in their nuclei have energy levels with hyperfine splitting, again leading to absorption over a wider range in wavelength and a delay in saturation than otherwise expected if one analyzes an absorption line assuming it is single when it is in fact multiple. Manganese ($Z = 25$) is a good example, and Booth et al. (1983, 1984) provide a good summary of the relative contributions from such splitting for a number of frequently measured Mn I lines.

So how does one estimate the stellar parameters?

Using Only the Spectrum. In principle, the stellar spectrum itself is sufficient. Given the basic principle that the analysis of every iron line should yield the same iron abundance, one may follow a fairly straightforward procedure.

- Using only weak lines in one ionization state, on the linear part of the curve of growth, so that microturbulence and gravity are not a serious concern, alter the effective temperature of the model atmosphere until there is no trend in the derived abundances with excitation potential χ of the lines. Trends with χ indicate that the adopted temperature is incorrect.
- Using both weak and strong lines in that same ionization state, alter the microturbulent velocity until there is no trend with reduced equivalent width, $\log(W_\lambda/\lambda)$. Lines on the "flat part" of the curve of growth should yield the same abundances as those on the linear part, after all.
- Using lines from two different ionization states, alter the model's gravity until there is agreement. Because the pressure, hence gravity, sensitivity varies from one ionization state to another, thanks to the Saha equation, one may estimate $\log g$ by altering the values of the models until all ionization states yield the same abundances.

There are two fairly simple rules that apply to analyses of main sequence and red giant stars in the globular clusters, and indeed to any stars where the dominant continuum opacity is from the H^- atom, and where most of the electrons are contributed by the metals one is trying to measure. In this case, the continuum opacity, $\kappa(H^-)$ is proportional to the electron pressure, P_{elec}. The strength of the absorption line is determined by the *ratio* of the line opacity to the continuum opacity, and the line opacity is determined essentially by what fraction of the atoms of a particular element are in the relevant ionization state. The Saha equation tells us that

$$\frac{N_{r+1}}{N_r} \cdot P_{\rm elec} = f(T) \ . \tag{54}$$

In the Sun, most of the iron is singly ionized, so changing the gravity ($P_{\rm elec}$) does not significantly change N(Fe II). Thus the line opacity is not strongly affected by changes in gravity, while the continuum opacity is, and the ratio, therefore, which determines the strength of the Fe II lines, does depend on gravity. For the Fe I lines, however, if N_{r+1} is constant in (54), then at a fixed temperature, $N_r \propto N(\text{Fe I}) \propto P_{\rm elec}$, so the line-to-continuum ratio is independent of gravity. Thus we have two rules relevant in this temperature regime.

- Electron donor element lines arising from a dominant ionization state (i.e., Fe II in the Sun) are pressure sensitive.
- Electron donor element lines arising from the next lower ionization state (i.e., Fe I in the Sun) are not pressure sensitive.

Thus to analyze a star whose temperature is like that of the Sun, or which, like the Sun, has most of its iron singly ionized and whose continuum opacity is dominated by H^-, one uses the Fe I lines to estimate the temperature and microturbulence because they are not very sensitive to gravity, and then one uses the Fe II lines to estimate the gravity.

Making Use of Other Information. Photometry may often be used to estimate a star's temperature, if the reddening is known. Some photometric systems, especially the Strömgren $uvby\beta$ system, may be used to estimate all three stellar parameters as well as reddening. (See Schuster & Nissen 1989a and Edvardsson et al. 1993 for details relevant to main sequence stars, and Anthony–Twarog & Twarog 1994 for details regarding metal-poor red giants.) For stars within globular clusters, one has an extra advantage, especially if the cluster's reddening has already been determined using blue horizontal branch stars and the color–color relation (i.e., Fig. 44). Since the horizontal branch luminosity is known reasonably well, and Fig. 3 shows the range in masses of stars beyond the turn-off is quite small, then

$$\log g = -12.505 + \log\left(\frac{M}{M_\odot}\right) + 4\,\log T_{\rm eff} + 0.4\,(M_V + BC)\ , \tag{55}$$

where BC is the bolometric correction (see Cohen et al. 1978 and Alonso et al. 1999a,b for red giants, and Carney 1983 and Alonso et al. 1995, 1996 for dwarfs).

The key problem in abundance analyses is usually the estimation of the temperature, and there has been considerable discussion in the literature regarding the preferred relations between temperature and various color indices.

Without embarking on an extensive review, much less a conclusion, there are two basic points to keep in mind if a color index is to be employed. First, it should be insensitive to metallicity if at all possible. The Strömgren $b-y$ index is quite suitable, while $B-V$ is less so. For red stars, $b-y$ becomes more problematical since it is far from the flux peak and becomes more vulnerable to not only atomic but also molecular line absorption. Second, a long wavelength baseline is preferred. Near solar temperatures, an error of ± 0.02 mag in $b-y$ results in an error in $T_{\rm eff}$ of about 130 K, while a similar error in $V-K$ leads to an error in $T_{\rm eff}$ of only 30 K. As a rule of thumb, near solar temperatures an error of 100 K leads to an error in [Fe/H] of about 0.1 dex. For this reason, $V-K$ is often preferred as a temperature indicator, and especially for cooler stars. When TiO absorption becomes strong, however, and the V band becomes more heavily blanketed, one often resorts to $J-K$.

The estimation of temperatures, from which color–temperature relations may be derived, has been accomplished via four methods.

1. The most fundamental method is to measure angular diameters of stars. Since $L \propto R^2T^4$, dividing through by the distance squared leads to $\ell \propto \theta^2T^4$. Measuring the angular diameter θ and the *apparent* bolometric flux ℓ thus leads to the effective temperature (Code et al. 1976; Ridgway et al. 1980). Model atmospheres are required to estimate the effects of limb darkening on the angular diameters measured at different wavelengths.

2. The remaining three methods all make more direct use of model atmospheres. One method uses the measured flux distribution *vs.* wavelength, obtained via spectrophotometry, with model atmosphere surface flux distributions. Peterson & Carney (1979) and Carney (1983) have summarized the results for individual stars and the resultant color–temperature relations (see also Carney et al. 1994). This approach, using the slope of the Paschen continuum, works well for stars with temperatures from about 4750 K to about 7000 K, but has not been applied to cooler stars such as red giants. Figure 56 shows a comparison of models with the flux distribution for HD 103095 (see Balachandran & Carney 1996).

3. Related to method **2** is the use of synthetic colors. From the surface flux distributions, if one knows the filter transmissions *vs.* wavelength (and the reflectivity of aluminum and atmospheric extinction), one may compute relative fluxes through the bandpasses that define a color index. Figures 1 and 2 show the basic ideas. There are several good examples, including Edvardsson et al. (1993) for Strömgren photometry and Bell & Gustafsson (1989) for infrared photometry. Synthetic colors form the basis, in fact, for one of the most commonly used color–temperature calibrations for red giants in globular clusters: the relations between $J-K$ and $V-K$ and $T_{\rm eff}$ and BC of Cohen et al. (1978).

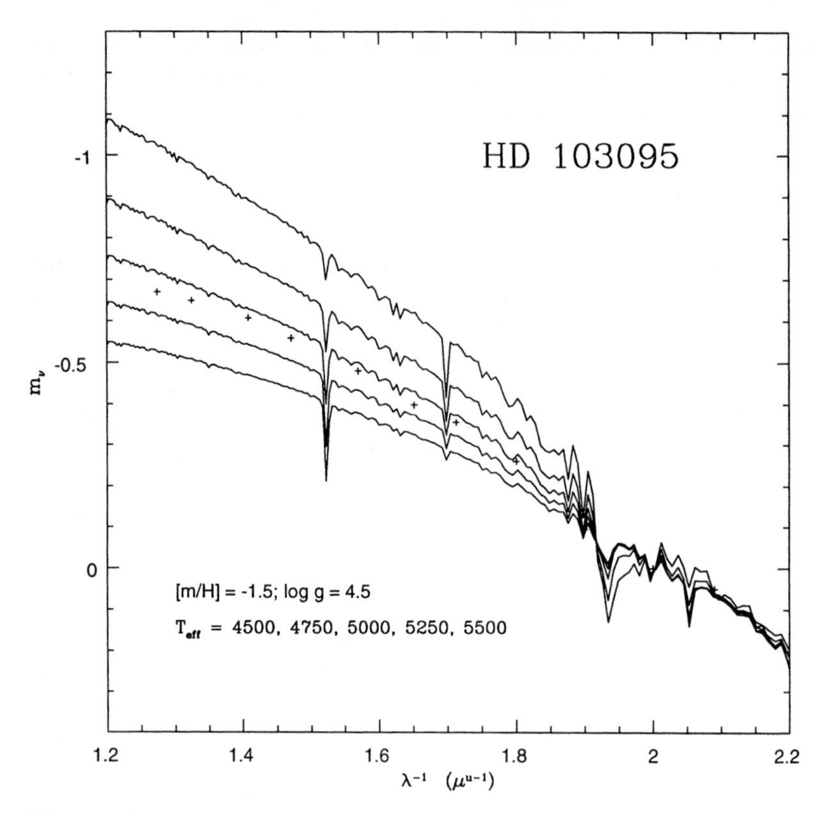

Fig. 56. Spectrophotometry of HD 103095 (*plus signs*) compared to model atmosphere surface flux distributions plotted as a function of inverse wavelength in microns. The star has an effective temperature of about 5050 K

4. Finally, in recent years calibrations involving the infrared flux method (IRFM) have proven very useful. Basically,

$$\frac{F_{\rm bol}}{F_{\lambda,\rm IR}} = \frac{\sigma T_{\rm eff}^4}{F_{\rm model}\,(\lambda({\rm IR})\,;\,T_{\rm eff}\,;\,\log g\,;\,[{\rm Fe/H}])}\,. \tag{56}$$

The reference wavelength chosen is generally selected to be in the infrared, out on the Rayleigh–Jeans tail of the flux distribution, which means K (2.2μ) or L (3.5μ). The gravity and metallicity effects are usually quite small, and so the IRFM is vulnerable primarily to the precision of the photometry (which, however, is necessary in any case to provide a color–temperature relation) and the relation between magnitudes and the energy in erg/s/cm^2/Hz at the top of the Earth's atmosphere. Generally, the temperature obtained for any one star from the IRFM is not as high in precision as desired for an abundance analysis, but the use of many stars yields very well defined color–temperature relations. Good examples of such relations are those of Blackwell et al. (1990)

and Alonso et al. (1996,1999b). The latter works in particular give color–temperature relations for dwarfs and giants for a wide variety of color indices and metallicities, and compare them with prior work.

5.3 Some Results for Globular Clusters

Even with the best observations, including high-resolution and high-S/N, systematic errors can be difficult to control. As noted, a 100 K error in the temperature scale will result in about a 0.1 dex error in [Fe/H]. The *gf* values employed in the analyses must, of course, be free of systematic errors as well, and there have been numerous cases of such effects over the years. For example, some laboratory-based *gf* values may not correctly measure the temperatures of the gas being analyzed, and this will carry all the way through the stellar abundance analyses. That is, a systematic error in *gf vs.* excitation potential will lead to systematically erroneous stellar temperatures and abundances. For this reason, laboratory *gf* values measured in equilibrium conditions, particularly by D. E. Blackwell and his group at Oxford University, are extremely valuable. Fe I *gf* values may be found in Blackwell et al. (1975; 1976; 1979a,b; 1980; 1982a,b; 1986), Blackwell & Shallis (1979a,b) and Andrews et al. (1979). Unfortunately, one cannot attain typical stellar atmospheric temperatures in equilibrium conditions in the laboratory, and so *gf* values measured in this way often refer only to neutral lines, and for only those with modest excitation potentials, $\chi < 2.5$ eV or so. For other lines, or for lines not measured in laboratory conditions because they are too weak, one often must employ other laboratory or observational results. For iron, for example, O'Brian et al. (1991) provide a good summary of Fe I laboratory *gf* values, and their higher quality lines do not appear to show any systematic differences as a function of excitation potential with the results from Blackwell's group (although there is a slight zero point shift of 0.02 dex). For Fe II, a very useful reference is that of Biémont et al. (1991). On the observational side, one may estimate *gf* values by an "inverted analysis". That is, one adopts, for example, a solar elemental abundance, based on, say, Anders & Grevesse (1989). One determines what *gf* value is necessary to produce the measured equivalent width. A subtle systematic error here is in the choice of model solar atmosphere. Does one use the best-determined empirical model, such as that of Holweger & Müller (1974) or does one use the model solar photosphere computed using the same physics as that for the star being analyzed? As Balachandran & Carney (1996) and Carretta & Gratton (1997) point out, the slightly different $T - \tau$ relations result in slightly different spectroscopic temperatures and abundances for the stars being analyzed, the differences amounting to over 0.1 dex.

Carney (1996) has summarized the results of [Fe/H] for individual globular cluster stars, but perhaps it is best to concentrate on the results from three of the most influential programs.

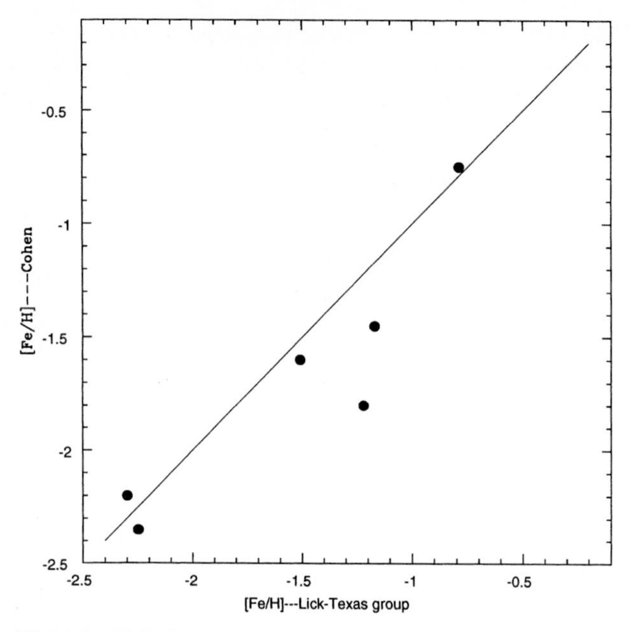

Fig. 57. Globular cluster abundances from the Lick–Texas group compared to those of Cohen

Cohen (1978, 1979, 1980, 1981, 1983; hereafter simply "Cohen") was among the first to utilize new cross-dispersed spectrographs and high-quantum efficiency detectors to study bright red giants in a number of globular clusters. Her results had a very significant impact on the study of globular clusters since they provided the calibrations for a large number of secondary metallicity indicators, examples of which are discussed in the following section.

A joint effort by C. Sneden and R. Kraft and their collaborators (Kraft et al. 1993, 1995, 1997, 1998; Pilachowski et al. 1996; Sneden et al. 1991, 1992, 1994a,b, 1997; hereafter "Lick–Texas") have proven extremely illuminating in the study of element-to-iron ratios in clusters, and in particular the role of mixing, as discussed in Sect. 7. High-S/N, high-resolution spectra were obtained using the 2.7m telescope at McDonald Observatory, the 3.0m Shane Telescope at Lick Observatory, and the 10m Keck I telescope at Mauna Kea. Many stars were studied in each cluster, using a carefully-selected list of *gf* values.

Finally, Carretta & Gratton (1997) undertook a careful re-analysis of older data, and included new data for ten stars in three clusters. The careful re-analysis and establishment of a revised color–temperature calibration enabled them to put data for 24 clusters on a fairly homogeneous scale.

We summarize the comparison of these three programs in Figs. 57, 58, and 59. In general, agreement between all three sets of results is good at the metal-rich and metal-poor ends, but the results from Cohen's work in the

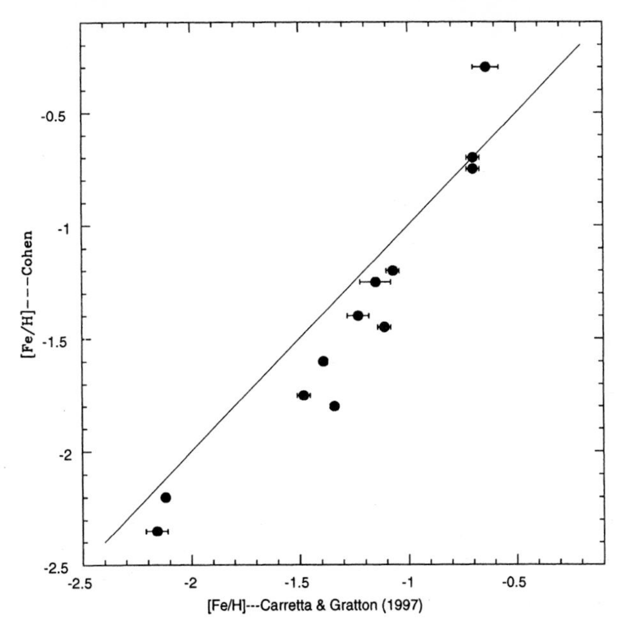

Fig. 58. Globular cluster abundances from Carretta & Gratton (1997) compared to those of Cohen

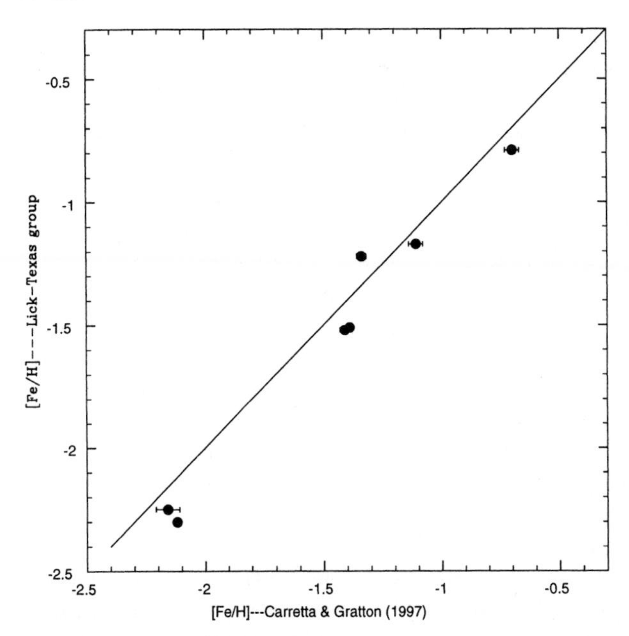

Fig. 59. Globular cluster abundances from the Lick–Texas group compared to those of Carretta & Gratton (1997)

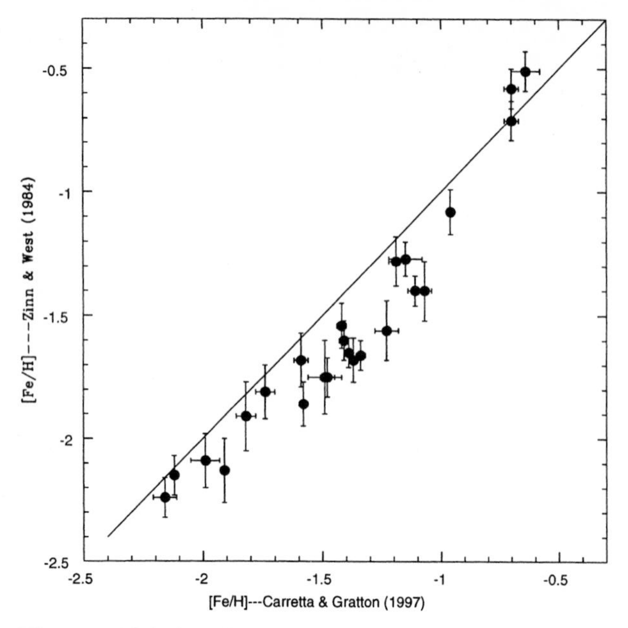

Fig. 60. Globular cluster abundances from Zinn (1985), calibrated using results from Cohen, compared to the abundances from Carretta & Gratton (1997)

intermediate metallicity regime appear to yield lower mean [Fe/H] values than those of the Lick–Texas group and Carretta & Gratton (1997). These latter two groups agree with each other extremely well, and I recommend their use in preference to the older calibrations.

5.4 Metallicity Indicators

Direct measures of [Fe/H] may be used to calibrate a wide variety of metallicity indicators. Zinn & West (1984) presented a thorough analysis and comparison of many such indicators. These provided the basis for the commonly-used scale of cluster abundances summarized by Zinn (1985), which in turn was calibrated using the high-resolution spectroscopic results of Cohen. As Figs. 57 and 58 suggest, this scale may not agree in detail with the modern calibrations, and Fig. 60 confirms the expectation. The two scales appear to agree well at the metal-rich and metal-poor ends of the scale, but not at intermediate metallicities.

In general, metallicity indicators may be grouped into three broad classes. First, there are methods that rely on the strongest lines in stellar spectra, so that one may use lower resolution spectroscopy and reach fainter magnitudes. These apply to individual stars, so that the nominal accuracy of the final result may be improved by studying larger numbers of stars within a cluster. These methods may also be applied to single field stars, thereby linking the study of the field and cluster samples. Further, this is amenable to direct calibration since a large number of field stars are bright enough for measurement of [Fe/H] through high-resolution spectroscopy.

Second, strong lines in stellar spectra may be used, but additional information is needed, such as the location of the star in a cluster color–magnitude diagram because the line strengths change with temperature and gravity and the CMD provides the necessary information. These samples cannot be directly related to field star studies, and calibration using high-resolution spectroscopy is less direct. (Cluster metallicities are determined from luminous red giants, for example, and that metallicity is then used to calibrate the metallicity indicator using other stars in the cluster.)

Finally, one may use a single property of a cluster color–magnitude diagram which is known to have metallicity sensitivity, such as the color of the red giant branch (see Figs. 9 and 10). This is obviously not applicable to field stars and requires [Fe/H] analyses of a large number of clusters to be useful.

Rather than attempt an incomplete review of the large number of metallicity indicators, we review here a single example of each of the three above classes.

ΔS. The ΔS method of estimating [Fe/H] was devised by George Preston (1959), and involves comparing the strengths of the Ca II K line with the Balmer lines near minimum light for RR_{ab} variables (although it has also been extended to the RR_c variables by Kemper 1982). The advantage of minimum light is that RR_{ab} variables have quite similar temperatures at that phase, and since that is also when the stars are coolest, it is when the Ca II lines are strongest. The stellar spectral types are estimated using the hydrogen lines, which are insensitive to metallicity, and the K line, which is obviously sensitive. Then ΔS is defined to be

$$\Delta S = 10\,[\mathrm{Sp(H)} - \mathrm{Sp(K)}]\ . \tag{57}$$

Recently, three new calibrations of ΔS *vs.* [Fe/H] have appeared, based on high-resolution, high-S/N analyses of field RR Lyraes, and the results are very similar. Clementini et al. (1995) found

$$[\mathrm{Fe/H}]_{\mathrm{C}} = (-0.194 \pm 0.011) \cdot \Delta S - (0.08 \pm 0.18)\ . \tag{58}$$

Lambert et al. (1996) found

$$[\mathrm{Fe/H}]_{\mathrm{L}} = (-0.19 \pm 0.01) \cdot \Delta S - (0.15 \pm 0.07)\ . \tag{59}$$

Finally, Fernley & Barnes (1996) obtained

$$[\mathrm{Fe/H}]_{\mathrm{FB}} = -0.195\,\Delta S - 0.16\ . \tag{60}$$

Direct comparison between individual stars observed by each group indicate differences of -0.10 ± 0.03 dex (Lambert et al. minus Clementini et al.;

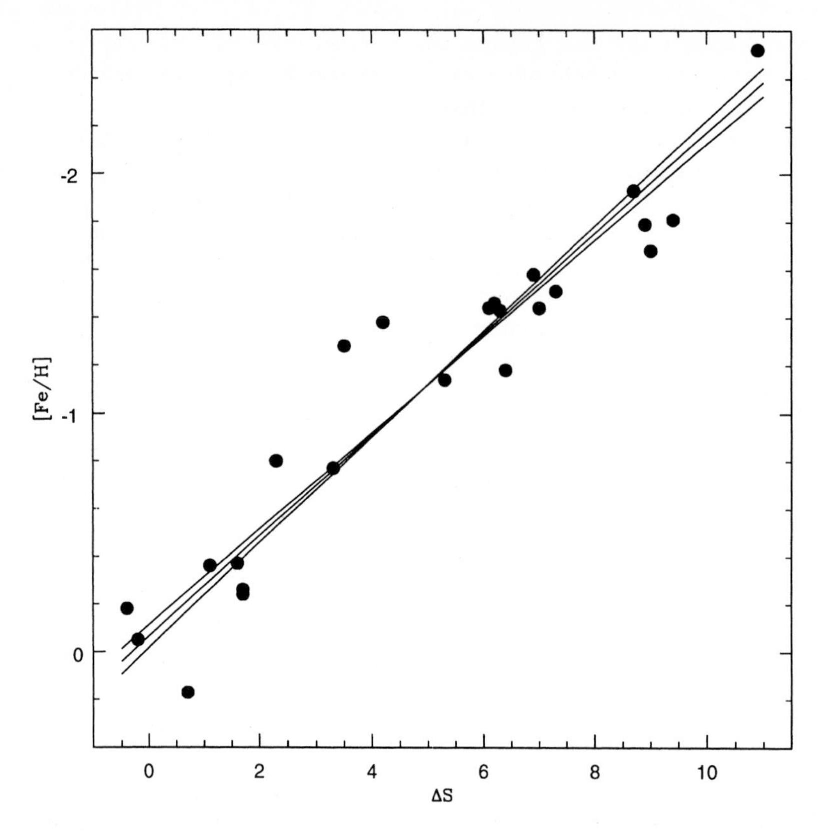

Fig. 61. A recalibration of ΔS *vs.* [Fe/H] using results from Clementini et al. (1995), Lambert et al. (1996), and Fernley & Barnes (1996). The middle line is the result of the bisector analysis

4 stars); $+0.07 \pm 0.04$ (Lambert et al. minus Fernley & Barnes; 6 stars); and $+0.16 \pm 0.03$ (Clementini et al. minus Fernley & Barnes; 3 stars). A consistent re-analysis of all the data using the same set of *gf* values and temperature estimation procedures would be a useful enterprise. We adopt a less time-consuming (and less rigorous) procedure here, however, and simply accept the [Fe/H] values reported, and average the results when more than one is available for a star. The ΔS values were taken from Blanco (1992), and a bisector analysis approach (Isobe et al. 1990) was employed to rederive the relationship. Normal linear least squares fitting, employed in all three of the above studies, is not entirely permissable here because there are errors in both quantities, [Fe/H] and ΔS. We adopt typical errors of 0.10 dex for $\sigma([\mathrm{Fe/H}])$ and 0.5 for $\sigma(\Delta S)$. Figure 61 summarizes the results. The final relation is

$$[\mathrm{Fe/H}] = (-0.211 \pm 0.013) \cdot \Delta S - (0.065 \pm 0.078) . \tag{61}$$

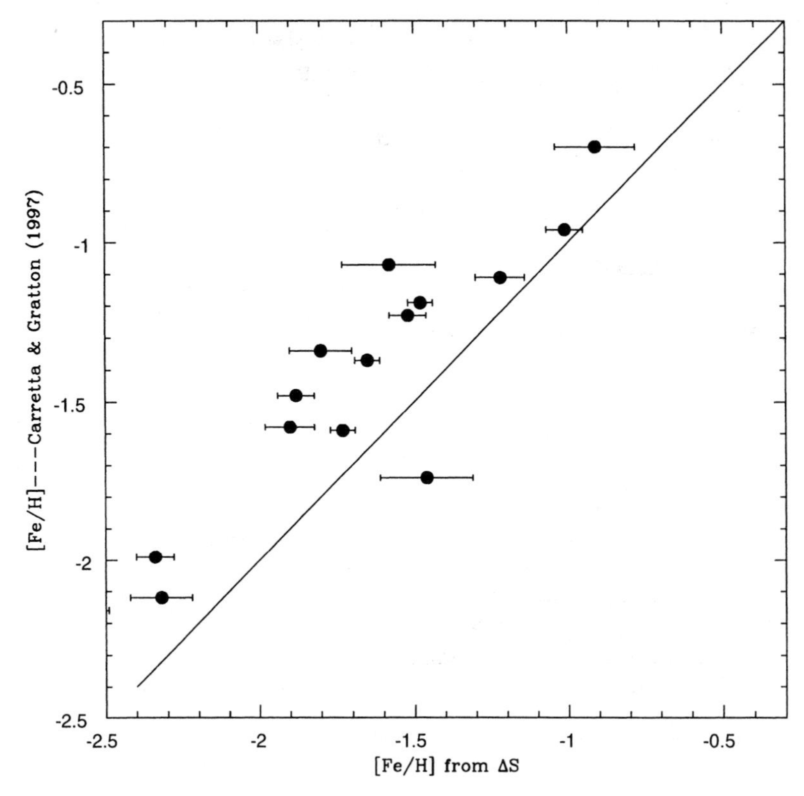

Fig. 62. The recalibrated ΔS *vs.* [Fe/H] relation given in (61) applied to globular clusters and compared with the results of Carretta & Gratton (1997)

How does this calibration compare when applied to globular clusters with those derived from more direct measures? Carretta & Gratton (1997) made such a comparison with the calibration of Clementini et al. (1995). A major advantage of the comparison was that the *gf* values employed in the two studies came from the same sources, reducing the potential systematic errors. Agreement was quite good in general, although no cluster RR Lyraes were studied with $\Delta S \leq 4$. At the metal-poor end, there was a hint that for ΔS values of 10 to 12, the globular clusters may be 0.2 to 0.3 dex more metal-rich than the field stars at the same ΔS. The statistics do not warrant confidence (2 field stars; 4 clusters), but could signify problems of ΔS measurement in distant metal-poor stars, or differences in chemical composition patterns since ΔS measures the strength of a strong calcium line while the Carretta & Gratton (1997) study is a direct measure of [Fe/H]. Using (61) and cluster ΔS measures from Smith (1981, 1984), Smith & Manduca (1983), Costar & Smith (1988), Smith & Butler (1978), Smith & Perkins (1982), and Butler (1975) leads to Fig. 62, which compares the resulting [Fe/H] values to the spectroscopic abundance analyses of clusters' red giants from Carretta & Gratton (1997). There is an offset of roughly 0.2 dex between the two calibrations, especially for lower metallicities, signifying that (61) has produced a systematic offset from the merged samples and the bisector analysis. This is not an artifact of mixing studies with differing sets of *gf* values. If we employ only the results from Clementini et al. (1995), but fit ΔS and [Fe/H] using a bisector analysis, we obtain a result essentially identical to that of (61). The [Fe/H] values for the globular clusters obtained from ΔS change by, at most, 0.03 dex when the Clementini et al. (1995) data are used for the calibration. We conclude that the cluster ΔS values have some systematic errors due to lower S/N levels, that the field and cluster RR Lyraes differ in their [Ca/Fe] values, or that, perhaps, there are some more subtle differences between the field RR Lyraes that calibrate the [Fe/H] *vs.* ΔS relation and those in clusters. Differences in [Ca/Fe] are *not* supported by detailed studies, as described in the following section on the abundances measured in cluster red giants.

As noted, one powerful advantage of the ΔS method is that it may be applied to individual RR Lyrae variables in the field and in clusters. It is therefore a very widely used metallicity indicator. However, ΔS has two drawbacks. First, it measures the calcium abundance, not iron. [Ca/Fe] is not constant among stars of all metallicities (see, for example, Wheeler et al. 1989; Edvardsson et al. 1993; Sect. 6). In principle, as long as [Ca/Fe] varies monotonically with [Fe/H], the calibration of ΔS in terms of [Fe/H] should still work reasonably well, but as discussed in the following section, there are stars and clusters whose [Ca/Fe] values differ from what is expected for their [Fe/H] values. Second, the Ca II K line arises from the ground state, and thus is vulnerable to contamination by interstellar lines. Radial velocities may help separate the stellar and interstellar features, but generally the metal-rich RR Lyraes are of low velocity and therefore correction for the effect can be difficult.

The Calcium Triplet and W'. There is another set of calcium lines that are quite strong in the spectra of red giants: the "triplet" at 8498, 8542, and 8662 Å. The latter two have been used to establish a metallicity indicator of the second class discussed above. In this case, equivalent widths are measured for the latter two, stronger, lines (or really, "pseudo equivalent widths" since the lines are so strong that other features underlie them and the resolutions employed often are insufficient to locate the true continuum). The difficulty faced in using these lines is that they are strong and on the "square root" portion of the curve of growth, and hence quite sensitive to gravity as well as temperature and abundance. The gravity and temperature sensitivities are circumvented somewhat by plotting the combined equivalent width of the two lines *vs.* the V magnitude of the star minus that of the cluster's horizontal branch, $V - V_{HB}$. This magnitude difference is related to both temperature and gravity, of course. Figure 63 shows data from five clusters taken from

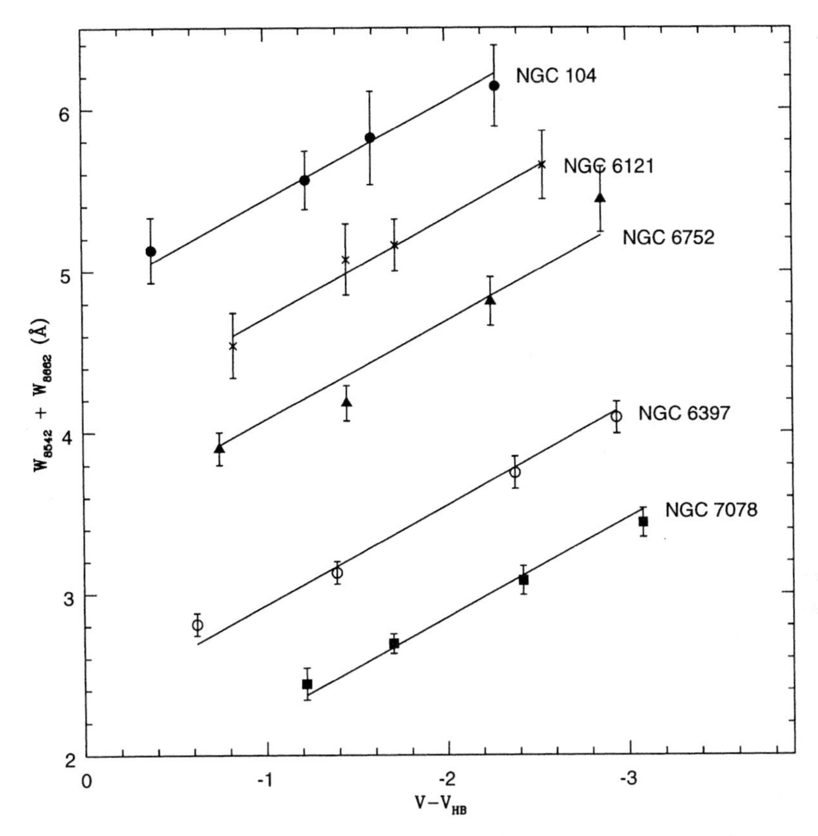

Fig. 63. The sum of the equivalent widths of two of the Ca II infrared triplet plotted against the magnitude difference of individual clusters' stars and the horizontal branch magnitude. Data are taken from Da Costa & Armandroff (1995)

Da Costa & Armandroff (1995). The paper also includes an excellent discussion of how to correct for systematic effects on the measurement of equivalent widths from other studies. The figure displays the basic points. A line of constant slope (−0.62 dex/mag) is fit to the data, and the clusters separate quite nicely. The intercept of the lines defines W':

$$W' = (W_{8542} + W_{8662}) + 0.62\,(V - V_{\mathrm{HB}})\,. \tag{62}$$

The advantages of W' are that it is measured near the flux peak of red giants and that the lines arise from a lower energy level of 1.7 eV, so they do not suffer from interstellar line contamination. The primary drawback is that one cannot compare abundances of clusters with field stars directly because part of the measurement requires the comparative V magnitudes of the stars and the horizontal branches. A less common but potentially serious drawback is that, like ΔS, the measurement is of calcium, not iron. Again, as long as

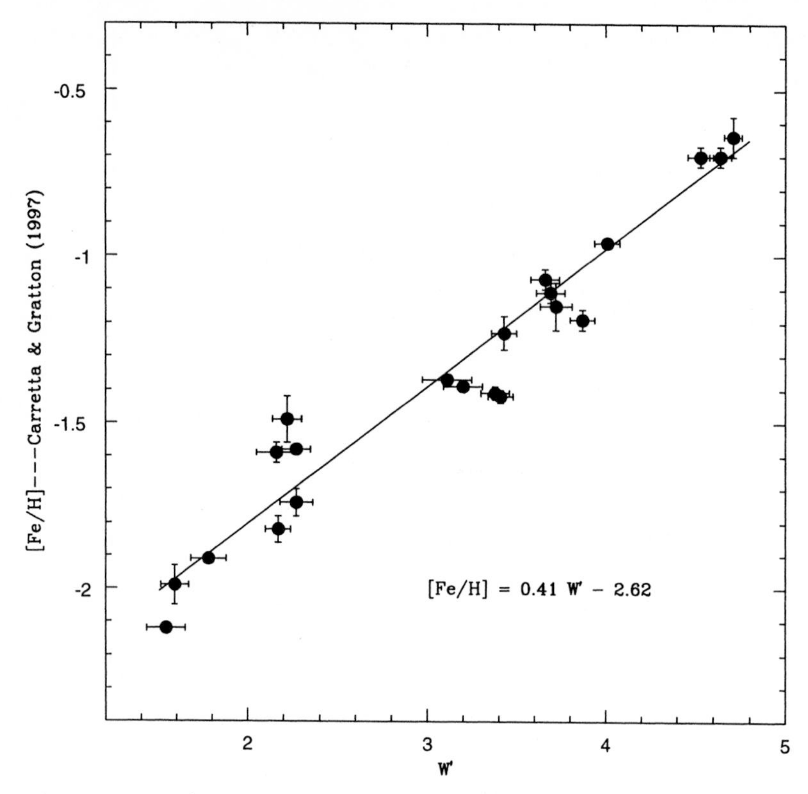

Fig. 64. The W' index *vs.* [Fe/H] taken from Carretta & Gratton (1997)

calcium and iron are monotonically linked, this is unlikely to be a problem, but a cluster with an unusual [Ca/Fe] ratio will yield an unreliable [Fe/H] value.

The calibration of W' in terms of [Fe/H] was established by Da Costa & Armandroff (1995), using the Zinn (1985) secondary calibration of cluster metallicities. As we have seen, this calibration does not compare well with those from the Lick–Texas group or that from Carretta & Gratton (1997) at intermediate metallicities. Since that is the regime where the Da Costa & Armandroff (1995) calibration has an odd change in the slope of W' *vs.* [Fe/H], it is worth undertaking a recalibration, as shown in Fig. 64. The W' data have been taken from Rutledge et al. (1997). The linear fit using the 20 clusters with the necessary data results in a good fit with a correlation coefficient of 0.966:

$$[\mathrm{Fe/H}] = 0.41\, W' - 2.62\,. \tag{63}$$

The Lick–Texas results include only five clusters, but a higher correlation coefficient (0.992):

$$[\mathrm{Fe/H}] = 0.49\,W' - 3.08\,. \tag{64}$$

Giant Branch Slopes and Color. As an example of the third type of metallicity indicator, one that uses properties of the color–magnitude diagram itself, we consider the work of Sarajedini (1994). It has been known for a long time what is evident in Figs. 9 and 10: the position of the red giant branch and its slope are affected by metallicity. The position of the red giant branch at the luminosity of the horizontal branch is affected by three factors. **1.** The Hayashi track is weakly sensitive to mass. A star with a higher mass results in a slightly hotter track, but the effect is small. Thus at a fixed age, metal-rich clusters with very slightly higher turn-off masses will have very slightly hotter tracks, as (24) indicates. **2.** The Hayashi track is quite sensitive to opacity (i.e., metallicity). Higher metallicity/opacity leads to cooler temperatures at fixed luminosities (24). **3.** At higher metallicities the red giant branch shifts to cooler temperatures and redder colors due to line blanketing effects on the $B-V$ and, to a smaller extent, on the $V-I_c$ color.

The slope of the red giant branch is affected by metallicity primarily through its effects on the emergent spectrum: cooler temperatures increase the line blanketing, and hence the slopes are greatest for the clusters containing the coolest, most metal-rich stars. The bolometric corrections are also affected, leading in the most metal-rich cases to red giants branches that seem to turn over and even "droop" in V *vs.* $B-V$ diagrams. One of the most common metallicity indicators that relies on the color of the red giant branch at the luminosity of the horizontal branch is $(B-V)_{0,\mathrm{g}}$, which is discussed by Zinn & West (1984). Sarajedini (1994) chose to use the $V-I_c$ color index since cluster color–magnitude diagrams are making more common use of VI_c rather than BV photometry, and, in the case of red giant stars, VI_c is closer to the flux peak. His calibration of $(V-I_c)_{0,\mathrm{g}}$ in terms of [Fe/H] is shown in Figure 65. The drawbacks of both $(B-V)_{0,\mathrm{g}}$ and $(V-I_c)_{0,\mathrm{g}}$ are that they provide only one measure of metallicity per cluster, the metallicity indicator cannot be measured for field stars, and that either must be corrected for reddening. This last point, however, was turned to an advantage by Sarajedini (1994), since he realized that the metallicity obtained from the reddening-insensitive slope of the giant branch should yield the same results as the reddening-sensitive color of the giant branch. Thus measurement of both may, in principle, lead to estimates of *both* the reddening and the metallicity of a cluster. The slope is measured by the magnitude difference between the horizontal branch and the V magnitude where $(V-I_c)_0 = 1.2$, and is defined as $\Delta V_{1.2}$. Figure 66 shows his calibration of that metallicity indicator. The details of how to exploit these calibrations to determine $E(B-V)$ and [Fe/H] are discussed in his paper, and the method has found widespread use.

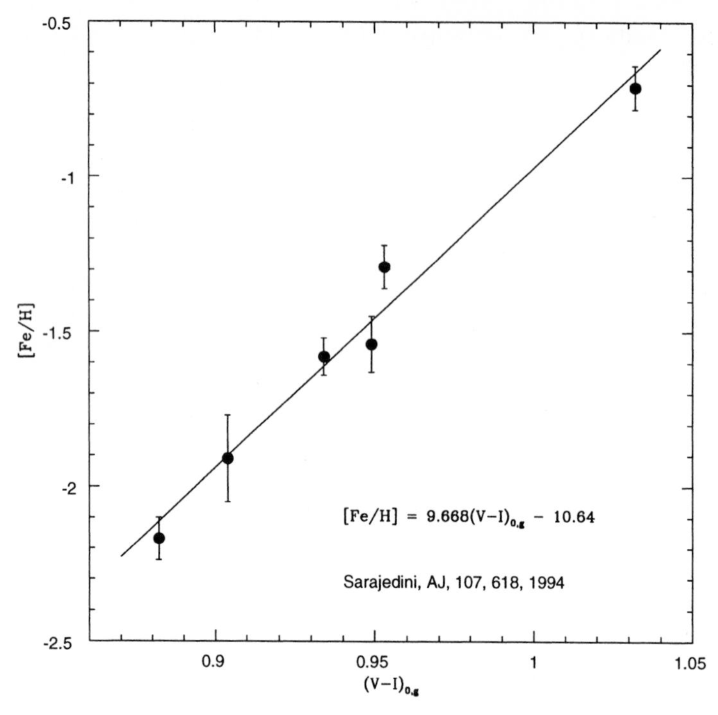

Fig. 65. The dereddened $V - I_c$ color index of the red giant branch at the luminosity of the horizontal branch *vs.* [Fe/H], as given by Sarajedini (1994)

5.5 The Importance of Proper Metallicity Scales

Much of the interest in globular clusters and their field star counterparts arises from their great ages, either inferred from their very low metallicities or derived indirectly or directly (see Sects. 4 and 8). Metallicity is often used as the surrogate for age, so it is very important that we at least get the correct rankings of one set of clusters or fields stars *vs.* another. The lingering disagreement between the RR Lyrae and red giant abundance scales (see Fig. 62) complicates direct comparisons of such samples to interpret the history of chemical enrichment in the Galaxy.

Of more fundamental concern, however, is the effect of the uncertainties in [Fe/H] on deriving accurate relative and absolute ages. Figure 7 makes it clear that both depend substantially on the heavy element mass fraction. Further, as discussed in Sects. 4 and 8, the second parameter is thought to be caused in part, and perhaps in large part, by modest differences in cluster ages. Exploiting this assumption to *derive* relative ages requires accurate metallicity measures since [Fe/H] is the first parameter. A change in the abundance scale, especially in the intermediate metallicity regime suggested by Fig. 57 where the second parameter is most useful, may have serious implications for the estimated age spreads of globular clusters.

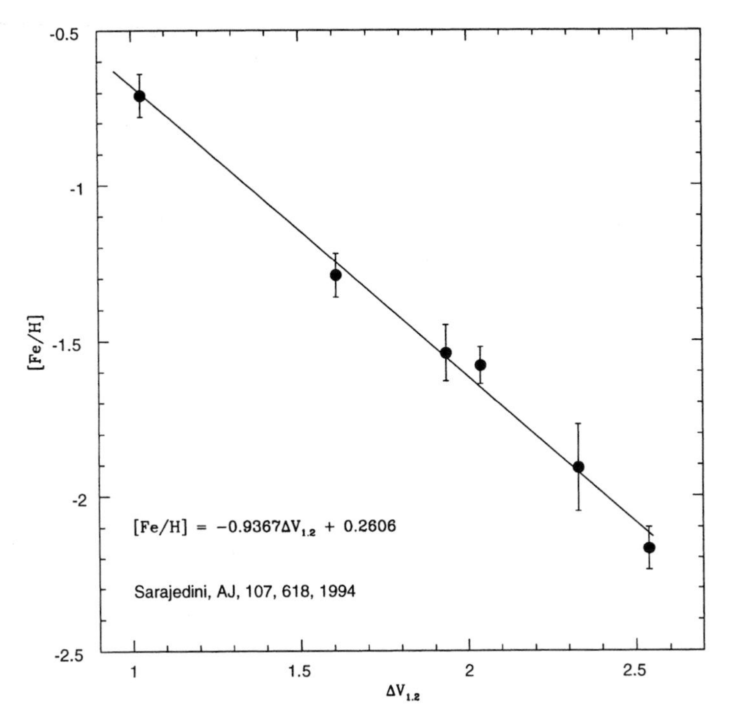

Fig. 66. The metallicity sensitivity of the slope of the red giant branch, as given by Sarajedini (1994)

6 Elemental Abundance Ratios I. Clues to the History of Nucleosynthesis

The measurement or estimation of a star's mean metallicity is crucial to the estimation of stellar distances, cluster ages, and the overall chemical history of our Galaxy. But there are many more fascinating clues if we look more closely at stellar spectra, primarily because there is more than one source for nucleosynthesis products. The Big Bang was the first source, although it produced, almost exclusively, only the lightest elements. We therefore recognize that the abundances of these light elements, especially lithium, beryllium, and boron, in the Galaxy's oldest stars may provide a record of the nucleosynthesis that happened in the first few minutes of the Universe. Supernovae were probably the next nucleosynthesis sources to appear, at least those that arise from the deaths of massive stars. Mass loss from intermediate mass AGB stars may have begun to contribute next to the Galactic stew, and, somewhat later, supernovae that result from the detonation (or deflagration) of massive white dwarfs. Careful study of the Galaxy's oldest stars, both in globular clusters and metal-poor stars in the field, may be used to test our understanding of

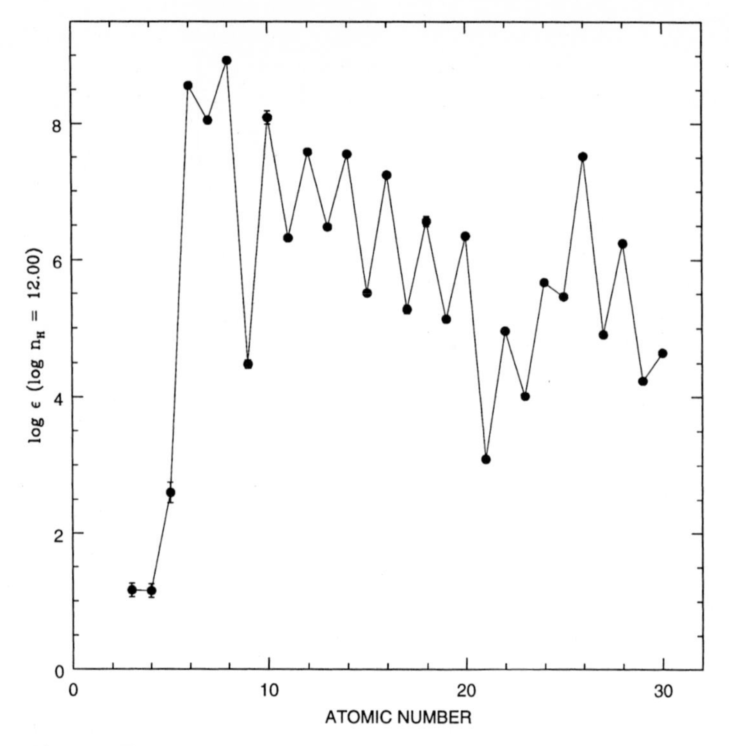

Fig. 67. The relative abundances of elements with $Z = 3$ (lithium) to 30 (zinc) in the Sun, using data from Anders & Grevesse (1989)

these nucleosynthesis sources, and with improved understanding, help us piece together the early chemical history of the Galaxy.

6.1 The "α" Elements

General Trends. Figure 67 shows the relative abundances of elements up to and slightly beyond the "iron peak" ($Z = 26$). This includes elements manufactured during all the stages through "silicon melting" in which nuclear statistical equilibrium is more or less established and the abundance patterns reflect the binding energies per nucleon. Note that there is a clear pattern: even-numbered elements are more abundant, which can be understood in terms of nuclear binding energies. Elements whose nuclei may be thought of as being collections of helium nuclei, also known as α particles, are especially tightly bound and abundant. These include carbon ($Z = 6$), oxygen (8), neon (10), magnesium (12), silicon (14), sulfur (16), argon (18), calcium (20), and titanium (22). (Actually, titanium's most abundant isotope, ^{48}Ti, does not have equal numbers of protons and neutrons.) Of these abundant elements, neon and argon are nearly impossible to measure in the spectra of old, cool stars (although they can be measured via emission lines in planetary nebulae). The two most abundant elements, carbon and oxygen, are, unfortunately, somewhat difficult to measure. Carbon abundances are often derived from the CH molecular features (see Laird 1985), while oxygen is derived from the [O I] $\lambda\lambda$6300, 6363 lines in red giants and the O I triplet near λ7771 in dwarfs and horizontal branch stars (see Tomkin et al. 1992; Carney et al. 1997). In dwarfs, one may also exploit the OH lines in the ultraviolet (see, for example, Bessell et al. 1991; Nissen et al. 1994; Israelian et al. 1998; Boesgaard et al. 1999) or in the infrared (see Balachandran & Carney 1996). Unfortunately, there is not very good agreement among the various methods. The oxygen triplet yields [O/Fe] values that are typically +0.9, while the forbidden lines, applied to both red giants and to some dwarfs (Spite & Spite 1991; Spiesman & Wallerstein 1991; Nissen & Edvardsson 1992; Fulbright & Kraft 1999) yield +0.3. The IR OH lines also yield +0.3, but the UV OH lines have often resulted in much higher [O/Fe] values (see Israelian et al. 1998 and references therein), increasing, in fact, as [Fe/H] drops. Since oxygen is such an important element, this is a serious issue, and one must look carefully to assess which is most likely to be correct. The evidence is probably in favor of the roughly constant value of +0.3 for the metal-poor stars. The two methods that yield high values are the most difficult and most vulnerable to systematic errors. The UV OH lines lie in a region where the continuum is hard to discern due to the sea of weak lines. This is particularly true for the solar spectrum, which is often used to estimate gf values so that the method of synthetic spectra may be applied to other stars (Nissen et al. 1994). The oxygen triplet, on the other hand, arises from a very high energy level (9.14 eV), and so forms at a very different depth than all other lines studied. (Indeed, Tomkin et al. 1992 noted that the high excitation potential atomic carbon lines behaved the same, and produced abnormally high carbon abundances compared to CH lines. They also found that the high-excitation carbon and oxygen atomic lines produced C/O ratios in agreement with other results.) Further, Fulbright & Kraft (1999) have shown that the high [O/Fe] ratios derived by Israelian et al. (1998) for two subgiants are inconsistent with [O/Fe] derived from the [O I] λ6300 line using very high S/N spectra. Finally, it is hard to understand why oxygen alone would have such large enhancements when the other elements such as silicon, magnesium, calcium, and titanium seem to be constant among metal-poor dwarfs (see Fig. 68).

In evolved stars, as we will see in the next section, carbon and oxygen abundances appear to change, even among an individual globular cluster's red giants, due to CNO cycle processing and mixing of those products into the stars' photospheres. Magnesium also appears to be affected in some evolved stars. And sulfur's few lines are far in the red and to date received too little attention (but see François 1988). Attention here will concentrate on oxygen, silicon, calcium, and titanium in consequence, with special care given to oxygen.

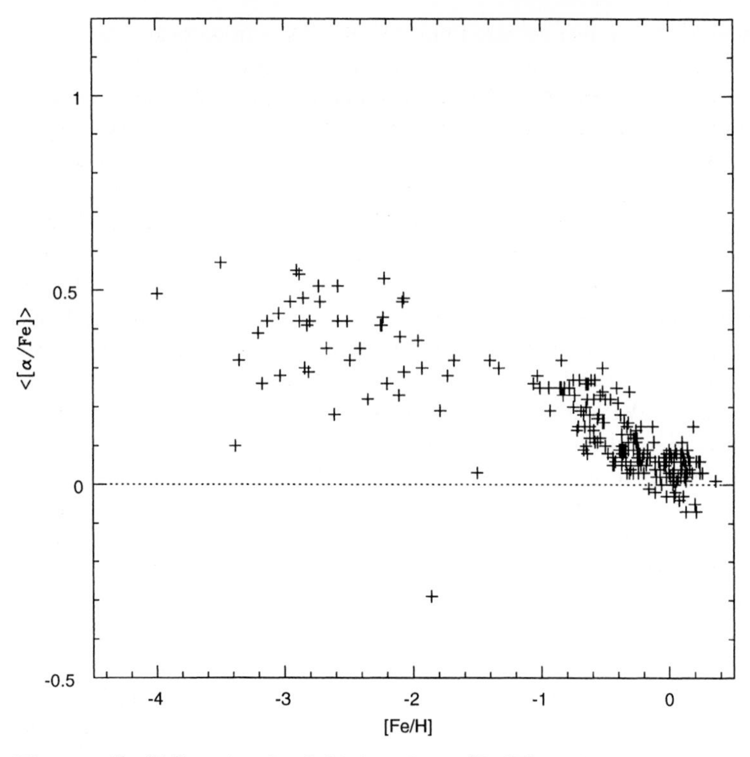

Fig. 68. $\langle[\alpha/\text{Fe}]\rangle$ ratios for field dwarfs *vs.* [Fe/H], with data taken from Gratton & Sneden (1988), Edvardsson et al. (1993), McWilliam et al. (1995), King (1997), and Carney et al. (1997)

Wheeler et al. (1989) have nicely summarized the trend of $\langle[\alpha\ /\text{Fe}]\rangle$, defined by magnesium, silicon, calcium, and titanium, *vs.* [Fe/H]. In Fig. 68 we show similar results for field dwarfs, without the use of magnesium, with data taken from Gratton & Sneden (1988), Edvardsson et al. (1993), McWilliam et al. (1995), King (1997), and Carney et al. (1997). What is immediately clear is that the clouds out of which metal-poor stars formed had been enriched by a different mix of nucleosynthesis sources than have the more metal-rich stars. At the lowest metallicities, it appears (generally) that a single class of source was operating, presumably supernovae of Type II. Assuming for the moment that [Fe/H] behaves as a (non-linear) chronometer, when [Fe/H] approached values near -1.0, a new source of nucleosynthesis began to contribute to the chemical mix. This new source produced or produces more iron relative to the lighter α elements, and is thought to be due to detonating massive white dwarfs (see Nomoto et al. 1997 for a review), thought to be the sources of Type Ia supernovae. As more of these have occurred relative to the Type II supernovae, the overall chemical abundances in the interstellar medium have changed; specifically, the [α/Fe] ratio has declined, as discussed by Matteucci & Greggio (1986).

Wyse & Gilmore (1988) and Smecker–Hane & Wyse (1992) have discussed the possible utility of [α/Fe] as a chronometer. If the models for Type Ia supernovae are a good guide, once there has been a burst of star formation it should take roughly 10^9 years for the first Type Ia supernovae to appear, although timescales ranging from 0.5 to 3 Gyrs have been suggested by Yoshii et al. (1996). Thus the simplest interpretation of Fig. 68 is that the Galaxy took only about that long ($\approx 10^9$ years) to rise from primordial chemical abundances to [Fe/H] ≈ -1. Further, as Wyse & Gilmore (1988) noted, the relatively small scatter at this transitional metallicity suggests that the Galaxy was also quite well mixed. This is a quite plausible interpretation, but one must recall that within the model, the time for the Type Ia to appear is a *duration*, not a fixed time. Two interstellar clouds could begin their star formation processes at very different times, enrich their gas with the ejecta from Type II, then Type Ia supernovae, and each would yield a distribution similar to Fig. 68. The figure does not, in fact, rule out accretion of smaller fragments. If a small galaxy or fragment underwent self-enrichment but star formation was terminated early, when the metallicity was low, it would show only the effects of Type II supernovae, independent of when star formation began in the galaxy/fragment. A small galaxy in which star formation and nucleosynthesis was prolonged would tend to also show the same trend as Fig. 68, and thus its accreted stellar remains would not be easily distinguishable from stars formed within the Galaxy itself. Only small galaxies whose chemical enrichment history were much slower than that of the Galaxy might be recognized by unusual locations in the figure.

Before we turn to the apparent oddities in Fig. 68, let us consider the globular cluster data. In Fig. 69 we add in the globular cluster data, as summarized by Carney (1996), with recent data on Palomar 12 and Ruprecht 106 from Brown et al. (1997), NGC 7006 from Kraft et al. (1998), and NGC 3201 from Gonzalez & Wallerstein (1998). The clusters generally follow the same trend as the field dwarfs. The clusters and field stars have therefore experienced similar chemical histories, whether we are seeing the chemical evolution of a single entity or a multitude of smaller ones.

Some Exceptional Cases. Figures 68 and 69 both show objects with unusual [α/Fe] values for their metallicities. The unusual clusters are Ruprecht 106 and Palomar 12, both studied by Brown et al. (1997). The first-order validity of the use of [α/Fe] as a chronometer is established by the relatively young ages for these clusters. Stetson et al. (1989) argued that Palomar 12 is 25% to 30% younger than other comparable metallicity clusters, while Buonanno et al. (1990, 1993) argued that Ruprecht 106 is 4 to 5 Gyrs younger than globular clusters in general. Unfortunately, the clear signature of Type Ia ejecta in these two clusters does not help us refine the timescale of this chrono-

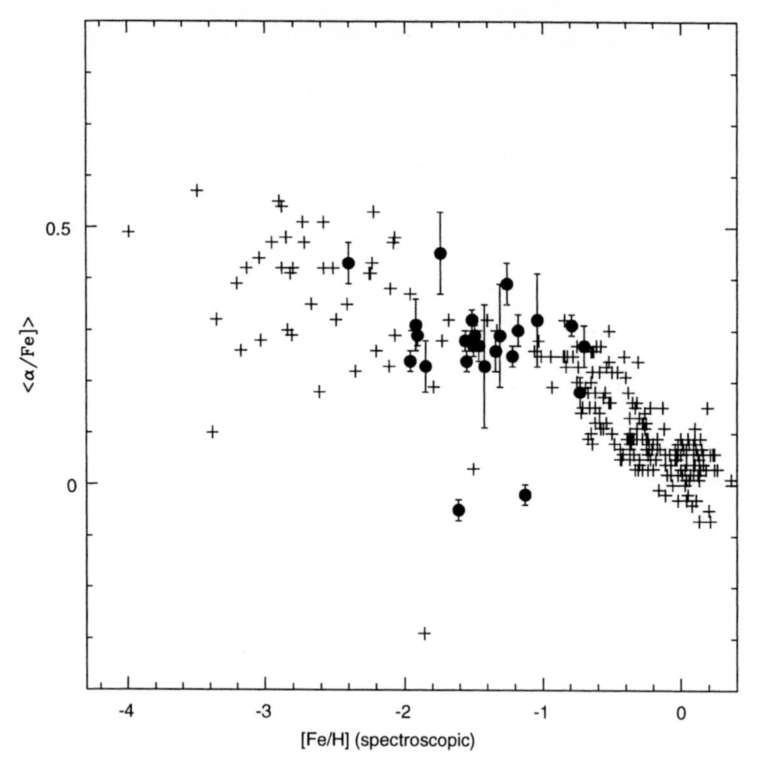

Fig. 69. $\langle[\alpha/\text{Fe}]\rangle$ ratios for field dwarfs (*plus signs*) and globular clusters (*filled circles*) *vs.* [Fe/H]

meter: we can say only that the timescale is probably less than several billion years, assuming, of course, that these clusters formed within our own Galaxy and that their chemical evolution is closely coupled to it. In this case, however, it is hard to understand the unusually low iron abundances for the near-solar $[\alpha/\text{Fe}]$ ratios. The unusual field stars are HD 134439 and HD 134440 (King 1997), a common proper motion pair, and BD+80° 245 (Carney et al. 1997). Signs of unusual ratios had already been seen in some metal-poor stars, based on less complete spectroscopic data. Peterson (1981) had already found that the HD 134439/40 pair was deficient in silicon and calcium relative to other metal-poor stars, while Carney & Latham (1985) drew attention to the pronounced silicon deficiency ([Si/Fe] = −0.37) of BD−6° 855. King (1997) has suggested that BD+3° 740 may be yet another example. What these field stars and clusters have in common is a relatively large distance from the Galactic center. Figure 70 shows $[\alpha/\text{Fe}]$ plotted against Galactocentric distance R_{GC} while Fig. 71 shows the stars from the survey of Carney et al. (1994) plotted according to their V velocities (roughly disk angular momentum) *vs.* computed mean apogalacticon distances. The unusual field stars and cluster are

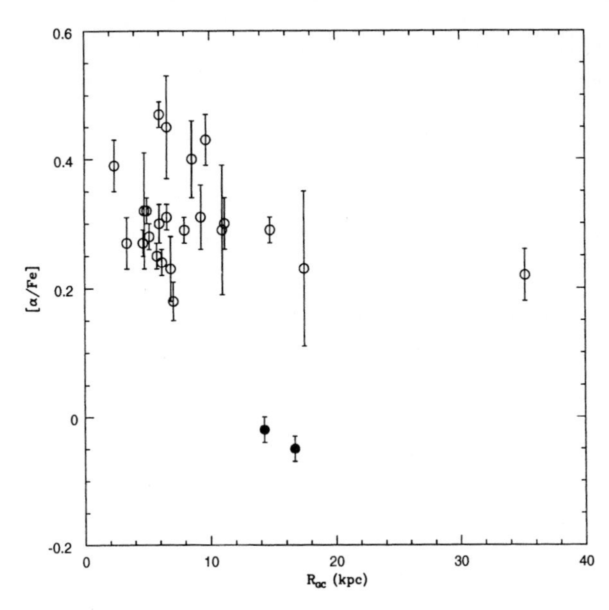

Fig. 70. [α/Fe] ratios for globular clusters plotted against current Galactocentric distance

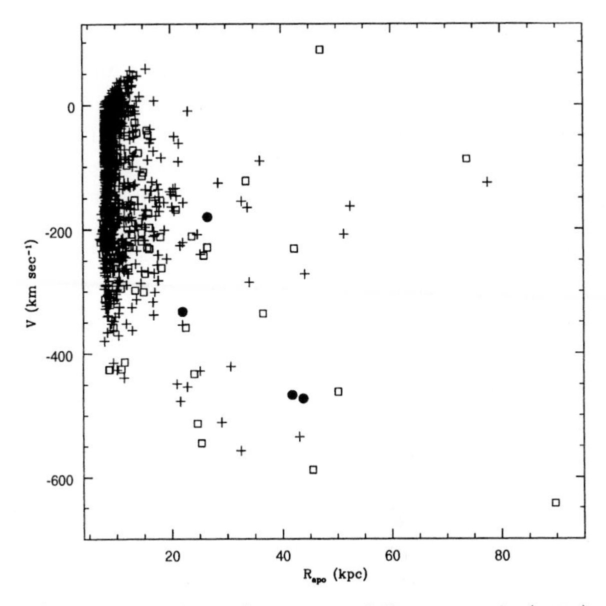

Fig. 71. Stars from the survey of Carney et al. (1994) plotted in V velocity (disk angular momentum) *vs.* apogalacticon distance in kpc. Stars with unusual [α/Fe] are plotted as *filled circles*. *Open squares* are stars with estimated reddening values of $E(B-V) \geq 0.05$ mag

plotted as filled circles. The only remarkable point about the stars and clusters with unusual [α/Fe] values is that they are relatively far from the Galactic center.

The new question is whether the unusual abundances and large apogalacticon distances indicate merger events, with the victims having experienced star formation that was slow enough to permit inclusion of Type Ia ejecta into the star-forming gas, and also produce relatively low metallicities. The *very* low [α/Fe] values could be produced by *very* slow, perhaps episodic, star formation, as Gilmore & Wyse (1998) have suggested. We may avoid invocation of a merger if we postulate that this same chemical history happened within relatively isolated, perhaps distant, pockets in our Galaxy. The problem is illustrated well by the recent abundance analyses published by Nissen & Schuster (1997). They selected stars whose [Fe/H] ratios were known to span the region around the "turn–down" in [α/Fe]. One set of stars was chosen for its disk-like kinematics, with $V > -70$ km s^{-1}. A second was chosen for halo kinematics, with $V < -170$ km s^{-1}. They noted that while some of the halo stars followed the same trend in [α/Fe] as did the disk stars, some of the halo stars were clearly deficient in [α/Fe]. These deficiencies also seemed to be related to the maximum distance these stars reached from the Galactic plane and their maximum radial distances from the Galactic center. This last point is illustrated in Fig. 72. Here we have followed their example and used [Na/Fe] as a surrogate for [α/Fe]. Clearly these data indicate that the chemical history of a star and the gas it formed from is not described simply by [Fe/H]. Variable cloud-to-cloud star formation histories may have been involved, or, as they argued, the data suggest separate chemical histories for the disk stars and at least some of the halo stars. Are these separate histories due to variations within the Galaxy's early star-forming regions or due to accretion of small galaxies with their own histories? Gilmore & Wyse (1998) have argued against the latter hypothesis, at least for the stars studied by Nissen & Schuster (1997), because the halo stars with large $R_{\rm max}$ distances also have very small perigalacticon distances. This is a natural consequence of the selection criteria employed since $V < -170$ km s^{-1} tends to yield stars with very little angular momentum, hence on plunging orbits. Gilmore & Wyse argued that the relatively high metallicities indicate a fairly massive parent galaxy (since metallicity roughly scales with luminosity), but typical low-density dwarf galaxies on such plunging orbits would not survive long enough to allow Type Ia ejecta to significantly enrich the gas. They suggested instead that the variations are more likely to have been created in lower-mass but denser "proto-halo fragments". Do these ideas apply to the other unusual stars discussed already? BD$-6°$ 855 is on a plunging orbit, and may be related to the stars studied by Nissen & Schuster (1997), but both BD$+80°$ 245 and HD 134439 and HD 134440 have significant amounts of angular momentum, so they are not on plunging orbits. All three stars are,

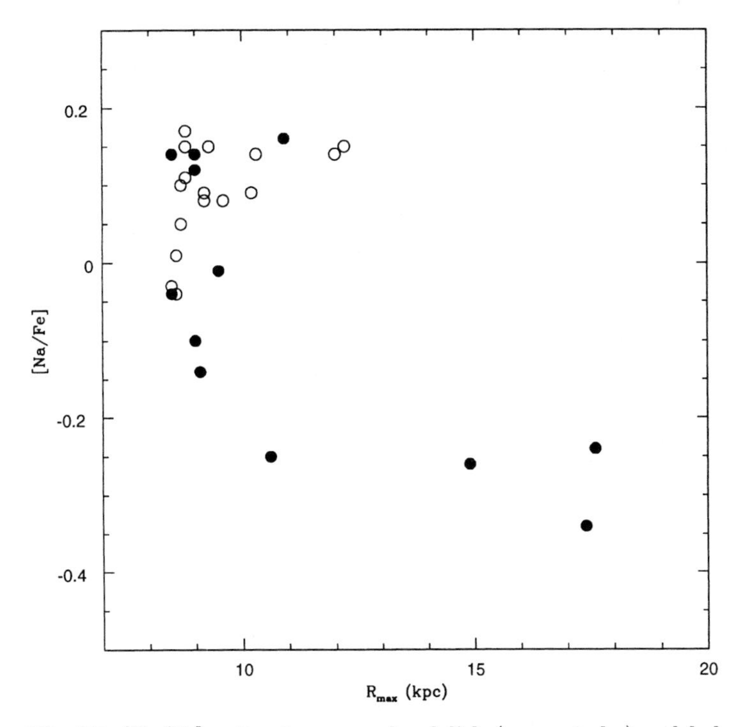

Fig. 72. [Na/Fe] ratios for a sample of disk (*open circles*) and halo (*filled circles*) stars plotted against maximum radial distance from the Galactic center. The data are taken from Nissen & Schuster (1997)

however, on retrograde orbits, so this may also be a clue regarding a possible accretion origin.

Several studies of high velocity (large apogalacticon distance) and retrograde orbit ($V < -220$ km s^{-1}) stars are now underway. With Chris Sneden and Inese Ivans at the University of Texas, we have acquired spectra of roughly two dozen such stars in both hemispheres. Luisa de Almeida at UNC has re-analyzed BD$-6°$ 855, confirming its low [α/Fe] ratio and signs of low values for CPD$-80°$ 349 and CD$-29°$ 2277. Hanson et al. (1998) have obtained detailed abundance ratios for a sample of metal-poor field red giants with well-determined proper motions. They noted that [Na/Fe] ratios tend to divide into two groups for the most metal-poor red giants that are in retrograde rotation. On the other hand, Alex Stephens at the University of Hawaii has undertaken a program of detailed spectroscopic analyses of several dozen stars with extreme velocities. His first sample (Stephens 1999) of 11 stars showed *all* of them to have normal [α/Fe] ratios, despite a very wide range in calculated apogalacticon distances. He has noted that if some of these stars are the remains of "cannibalized" satellite galaxies, their nucleosynthesis history has been extremely similar to that of the Milky Way, contrary to the

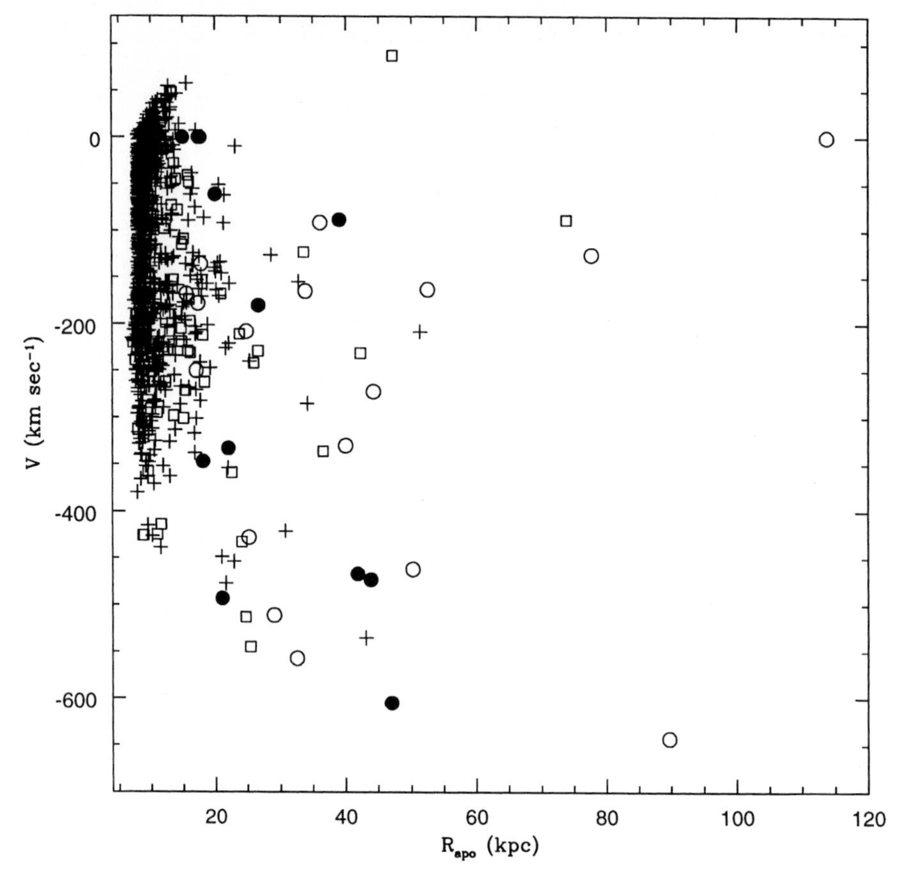

Fig. 73. Stars from the survey of Carney et al. (1994) plotted in V velocity (disk angular momentum) *vs.* apogalacticon distance in kpc. Stars with unusual [α/Fe] are plotted with *filled circles*, based on work of de Almeida (unpublished), Carney & Latham (1985), Nissen & Schuster (1997), King (1997), Carney et al. (1997), Hanson et al. (1998), and Stephens (1999). Stars plotted as *open squares* have uncertain reddening and kinematics. Stars with apogalacticon distances estimated to be greater than 15 kpc but which are known to have normal [α/Fe] values are plotted as *open circles*

few examples cited above. The continuing programs, and others yet to begin, will help clarify how common the apparently separate chemical histories have been. We also seek means by which we can relate the stars with atypical abundance ratios to one another. We illustrate here two possibilities. In Fig. 73 we repeat Fig 71, but adding in this time stars from other programs by using only stars with estimated apogalacticon distances exceeding 15 kpc. The unpublished results of de Almeida are included, as are those of Nissen & Schuster (1997), Hanson et al. (1998), and Stephens (1999). Stars with known normal [α/Fe] are plotted as open circles, while sub-normal [α/Fe] ratios are plotted as filled circles. What is becoming clearer, especially thanks to the work of Stephens, is that most of the stars with extreme apogalacticon distances, thus most likely to have been accreted by the Galaxy, show normal [α/Fe] ratios. So far, the phenomenon appears confined to $R_{\rm apo} < 50$ kpc.

A second way to search for patterns that might signify accretion events is to follow the suggestion of Lynden–Bell & Lynden–Bell (1995) and plot the locations on the sky of the apparent orbital poles of the objects in question. The pole is a single position rather than a complicated set of vectors, and may offer some interesting clues. The method was derived as a means of searching for possible relations in the kinematics of globular clusters and dwarf galaxies ("ghostly streams"), whose tangential velocities are unknown. When all three motions are known, one should also seek shared motions, as, for example, has been done by Chen (1998) and Majewski et al. (1994). However, the "ghostly streams" method is of limited use for local samples since all the stars are very near the Sun and the Galactic plane, so the longitudes of the poles are all near $\pm 90°$. The latitudes of the poles in these cases are defined by the V and W velocity vectors, so we prefer to exploit the data in a slightly different but more physical fashion. The two quantities we expect to be conserved, at least approximately, are the kinetic energy and the angular momentum in a star's orbit. Since all the stars in our program are near the Sun, we may approximate the former by the "rest frame energy" $v_{\rm RF}$ which uses all three local velocity vectors U, V, and W but corrects for the local circular velocity Θ_0, assumed to be 220 km s^{-1},

$$v_{\rm RF} = \left[{\rm U}^2 + ({\rm V} + \Theta_0)^2 + {\rm W}^2\right]^{1/2} . \qquad (65)$$

For the angular momentum, it is common to employ only the V velocity, as in Fig. 39. But in a spherical potential, which may be the case for a dark matter-dominated potential, both the V and the W velocities should be employed, which we have done in Fig. 74. The common proper motion pair HD 134439/40 with a low [α/Fe] ratio shows up at $v_{\rm RF} \approx 400$ km s^{-1}, and angular momentum near -255 km s^{-1}. Two other low [α/Fe] stars that lie close to one another are BD+80° 245 and HD 6755 near (328; -280). HD 108577 is nearby at (312; -234). More analyses must be completed before we are able to confirm what is suggested by these few groupings: specific merger remnants may have been identified.

A Caveat Regarding [α/Fe] *vs.* [Fe/H]. A quite different problem arises in the interpretation of Figs. 68 and 69 if the apparent transition between the enhanced and declining values of [α/Fe] is not caused by age, but by metallicity sensitivity in the formation of SNe Ia. Hachisu et al. (1996; 1999) and Umeda et al. (1998) have explored the SNe Ia model wherein mass transfer of hydrogen-rich material onto a CO white dwarf occurs until it exceeds its Chandrasekhar mass limit and detonates. They find that for a high enough

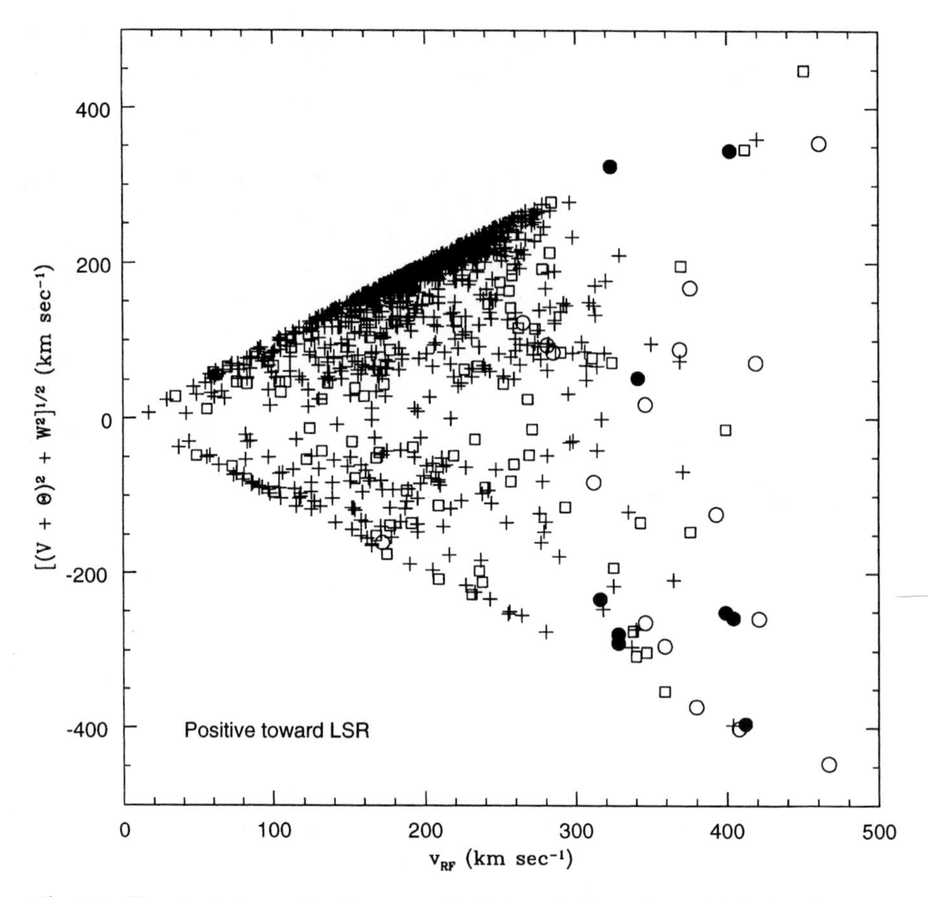

Fig. 74. The "rest frame kinetic energy" of stars in the solar neighborhood compared to their angular momenta, as measured using both the V and W velocities. *Plus signs* are from Carney et al. (1994), as are *open squares* except that such stars have uncertain reddening and kinematics. *Open circles* are stars with apogalacticon distances of 15 kpc or larger, and normal (enhanced) $[\alpha/\mathrm{Fe}]$ ratios. *Filled circles* have similar apogalacticon distances but unusually low $[\alpha/\mathrm{Fe}]$

accretion rate, a strong wind is emitted by the white dwarf, and the hydrogen burns steadily to permit the white dwarf mass to increase. The key to these winds is the opacity due to heavy elements, and so there is a metallicity sensitivity. In fact, Umeda et al. (1998) argued that if $[\mathrm{Fe/H}] < -1$, the wind and the growth in mass of the white dwarf are inhibited, and in consequence a supernova event would not occur. If this model is correct, then the onset of the decline in [O/Fe] and $[\alpha/\mathrm{Fe}]$ values at $[\mathrm{Fe/H}] \approx -1$ is not due to the time required for binary systems to evolve and produce white dwarf detonations, but simply to the need for stellar metallicities to rise to $[\mathrm{Fe/H}] \approx -1$. Kobayashi et al. (1998) discussed the Galactic and cosmic chemical evolution consequences. Of course, it is difficult to reconcile this idea with the observed low [α/Fe] values for a number of stars with [Fe/H] < -2, and clusters with [Fe/H] ≈ -1.6 (Ruprecht 106).

Some Further Caveats on Mean Metallicity Estimations. We have seen that at least some metallicity indicators, such as ΔS or the calcium triplet, actually measure calcium abundances. If there is a monotonic relation between [Ca/Fe] and [Fe/H], calcium line strengths should be good surrogates for the estimation of iron abundances. However, as we have seen, there are some cases where [Ca/Fe] may differ from the majority of clusters. Further, because the "α" elements are significant electron donors, and hence affect the net H^- opacity, the color of the red giant branch may be affected by variations in [α/Fe]. The "simultaneous reddening and metallicity" method discussed above, for example, is prone to such a systematic effect, as discussed by Sarajedini & Layden (1997). Metallicity indicators are very useful tools, but systematic effects like this should always be kept in mind, and, of course, ultimately, there is nothing better than high-dispersion, high-S/N spectroscopic analysis to understand what is going on in stars in the field and in clusters.

[α/Fe] in Thick Disk Stars. In Sect. 4 we argued that the thick disk and the thin disk appear to overlap in metallicity yet are separate in kinematics, suggesting independent histories. Another test of this is to consider the behavior of [α/Fe] *vs.* [Fe/H], as we have done above. Serge Naoumov and I have been working with Jason Prochaska and Art Wolfe to obtain high-resolution ($\approx$ 45,000) and high-S/N ($\approx$ 200–300) spectra of stars whose metallicities are typical of the thick disk *and* whose kinematics are typical of the thick disk. The latter criterion is especially important since it appears that metallicity alone is inadequate to identify thick disk stars, especially in the solar neighborhood. Even stars with [Fe/H] $= -0.7$ in the solar neighborhood are far more likely to belong to the thin disk rather than the thick disk.

We have preliminary results for three thick disk stars, G247-32 ([Fe/H] $= -0.44$; [α/Fe] $= +0.34$), G114-19 ([Fe/H] $= -0.66$; [α/Fe] $= +0.34$), and G92-19 ([Fe/H] $= -0.82$; [α/Fe] $= +0.32$). In Fig. 75 we show these data along with the results from Edvardsson et al. (1993). We also include the results of the three disk globular clusters, 47 Tuc, NGC 6352, and M71, discussed in Sect. 4. At metallicities typical of the disk stars studied by Edvardsson et al. (1993), the three thick disk stars appear to have unusually high [α/Fe] ratios, consistent with the great age (and hence presumably rapid formation) of the thick disk, and also with the concept that the thick disk and the thin disk have had different chemical enrichment histories. The figure tells only part of the story, in fact. The temperature scale employed by Edvardsson et al. (1993) does not agree well with that of Alonso et al. (1996), being almost 200 K too hot for the hotter stars, and only slightly hotter than the

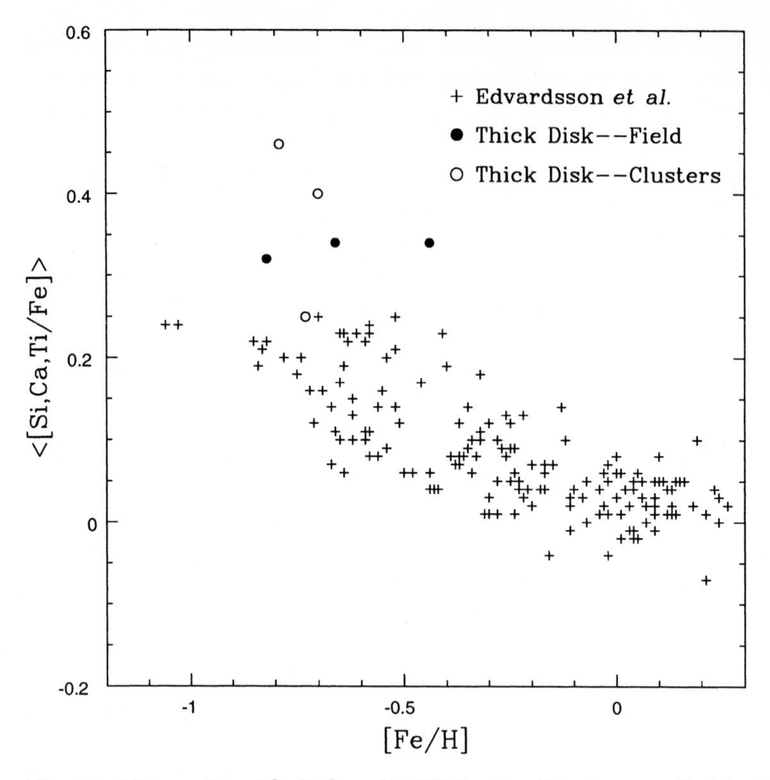

Fig. 75. The results of [α/Fe] *vs.* [Fe/H] for the disk stars studied by Edvardsson et al. (1993) and three stars with similar metallicities but thick disk kinematics

cooler stars. Correction to a similar temperature scale will, in general, shift the Edvardsson et al. (1993) results by, typically, 0.1 dex and toward lower metallicities.

6.2 Neutron Capture Elemental Abundances

As in the case of the "α" elements, other elemental abundance ratios that arise from different nucleosynthesis sources may be employed to explore the general trends in the history of Galactic nucleosynthesis. These may include variations from the mean, due to "pockets" or "proto-Galactic fragments" within the forming Milky Way or to accretion of smaller galaxies. The neutron capture elements are a potentially valuable source of such data. Almost all of the elements beyond the iron peak are formed by neutron capture. These are subdivided into two major cases: the *s–process*, where the neutron flux is low and nuclei have time to β decay before another neutron capture occurs; and the *r–process*, where the neutron flux is very high, and captures continue without β decay until the (n,γ) processes reach a balance with (γ,n) processes. Such a balance requires very high temperatures. The s–process is thought to occur in a number of environments, depending on the reaction responsible for the supply of neutrons. The most common source is thought to be due to α captures on ^{13}C: $^{13}C(\alpha,n)^{16}O$, which can occur at temperatures and densities found in low and intermediate mass asymptotic giant branch stars. The r–process, on the other hand, is *thought* to arise primarily in Type II supernovae. Again, the likely nucleosynthesis sources for the s–process and the r–process probably are stars with rather different masses and evolutionary timescales, and by studying stars formed at the earliest times in the Galaxy's history, we may again hope to (a) see differences in s–process and r–process abundances *vs.* the (time surrogate) metallicity; and (b) exploit these differences and some knowledge of their sources to calibrate the chemical enrichment timescales.

Field Stars. One of the brightest and most metal-poor field stars is HD 122563 (HR 5270). A thorough study of this star's neutron capture elemental abundances by Sneden & Parthasarathy (1983) showed clear signs of both the s–process and especially the r–process. An even more r–process-rich star, HD 110184, was analyzed and discussed by Sneden & Pilachowski (1985). It is not an easy task to distinguish the relative contributions of these two processes since reliance is often made on neutron capture models, tested in comparison with the Sun's abundance pattern, which has contributions from both processes. In fact, we are unable to resolve isotopic abundances of most of the neutron capture elements from our solar or stellar spectra, and therefore we must recognize that almost all of the *elemental* abundances which we can measure have some contributions from both processes. Nonetheless, Sneden & Parthasarathy (1983) have summarized the expected relative contributions of the r–process and s–process to the elemental abundances. Barium, which has relatively strong lines even in metal-poor stars, is dominated by the s–process, but somewhat over 10% of the solar barium abundance is due to r–process contributions. On the other hand, europium is a particularly good indicator of the r–process, with less than 10% of the solar abundance due to the s–process. Figure 76 shows the observed abundance pattern of HD 110184 matched to theoretical s–process (normalized to the observed abundance of barium) and r–process (normalized to the observed abundance of europium). The heaviest elements show that the r–process has played an important role in the chemistry of the star, but an s–process contribution is required to match the abundances of the "light s–process" elements strontium, yttrium, and zirconium. The most extreme r–process contributions have been seen in CS 22892-52 (Sneden et al. 1994b; Cowan et al. 1995). Figure 77 shows the distribution of observed elemental abundances and r–process and s–process abundance patterns, both normalized at neodymium. The s–process contribution is very weak, apparently, and, encouragingly, the model r–process abundance pattern is an excellent match to the observations.

Do we see changes in the relative contributions of the r–process and s–process as a function of time? Within the field stars, Krishnaswamy–Gilroy

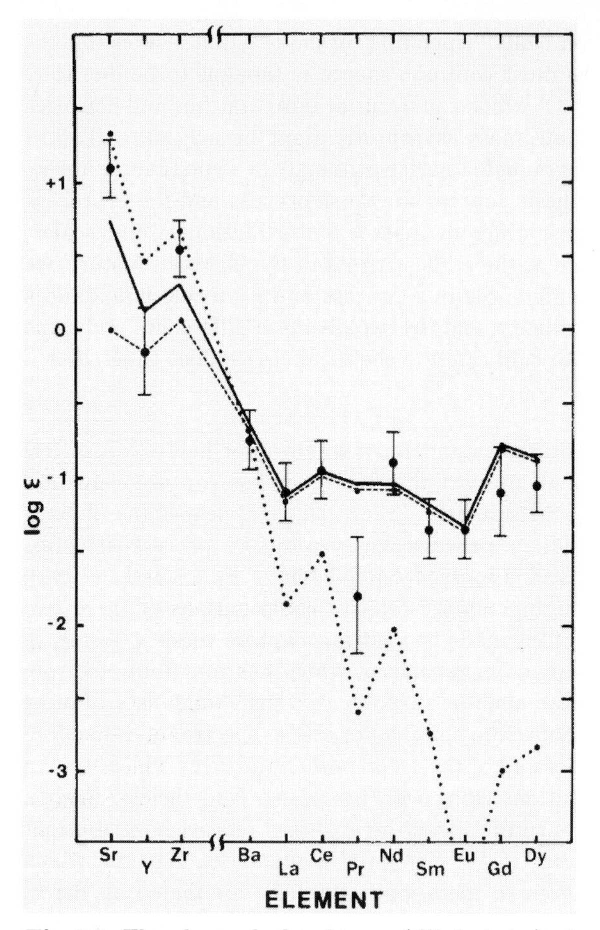

Fig. 76. The observed abundances (*filled circles*) of neutron capture elements in HD 110184 compared to theoretical s–process abundances (*dotted line*), normalized to barium, and theoretical r–process abundances, normalized to europium, taken from Sneden & Pilachowski (1985)

et al. (1988) noted that the slope of the abundances of europium *vs.* barium changed at a metallicity level corresponding to [Fe/H] ≈ -2.5 (see Fig. 78). [Note that François (1996) did not find such a trend, but his data did not extend below [Fe/H] ≈ -2.5.] This change could indicate the point at which the s–process contributions (Ba) become significant. If the s–process site is intermediate mass AGB stars, which should begin to appear about 10^8 years after star formation is initiated, then the timescale to reach these metallicity levels might be as short as a few times 10^8 years. On the other hand, since the abundances of only a few stars signify the trend (note some abundances of europium are only upper limits), it could also indicate a time when the effects of only a few individual supernovae are being seen, as suggested by

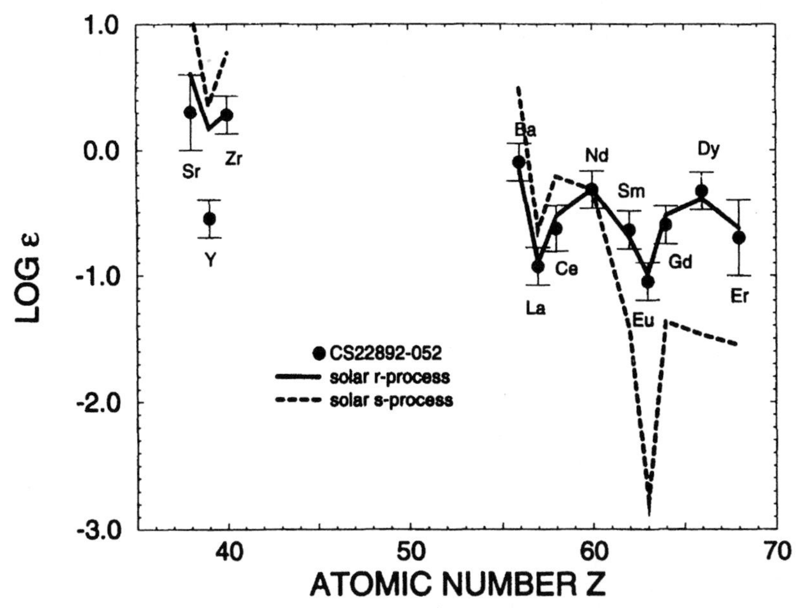

Fig. 77. The abundances of neutron capture elements in CS 22892-52 compared to theoretical s–process and r–process abundances, both normalized at neodymium, taken from Cowan et al. (1995)

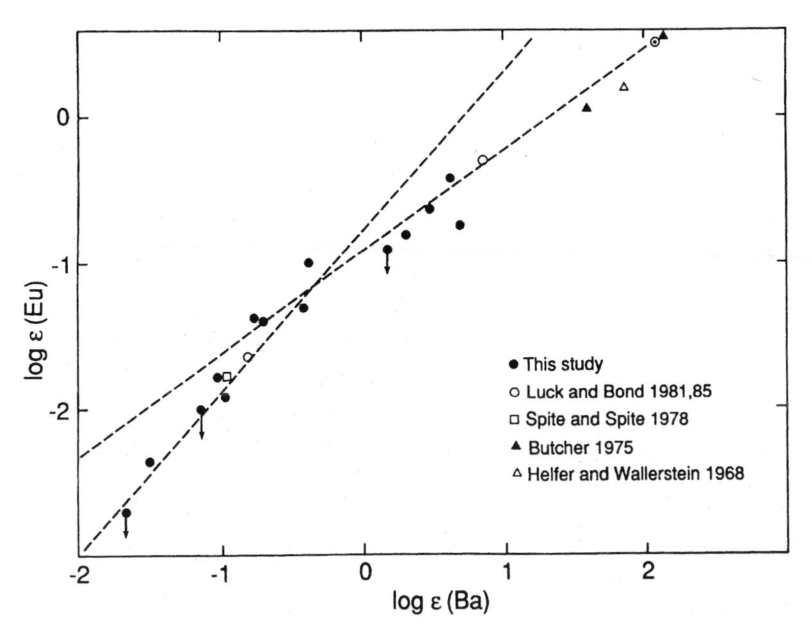

Fig. 78. The abundances of the neutron capture elements barium *vs.* europium for a sample of metal-poor field halo stars, taken from Krishnaswamy–Gilroy et al. (1988)

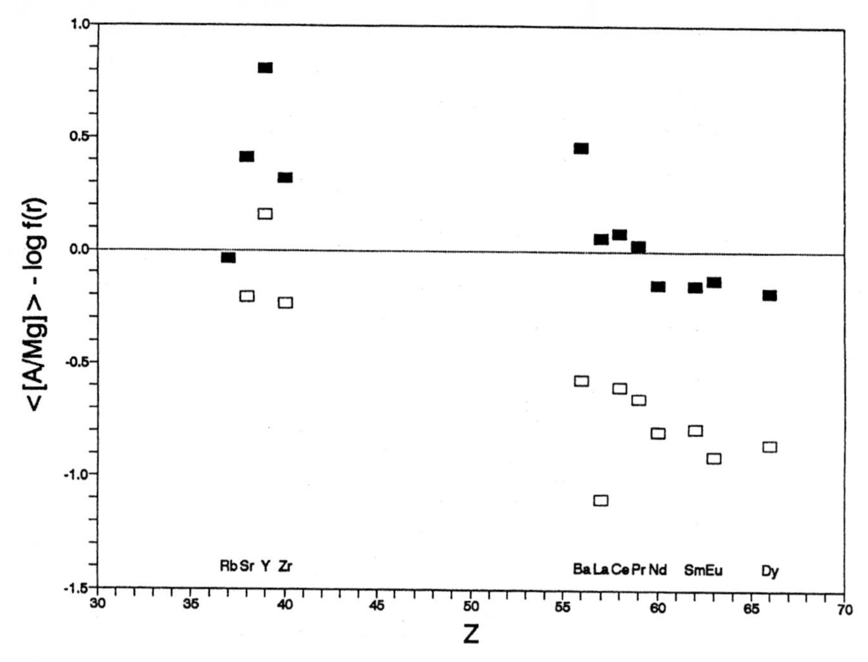

Fig. 79. The abundances of neutron capture elements compared to a theoretical pure r–process contribution, divided into two metallicity domains, taken from Gratton & Sneden (1994). *Filled squares* have [Fe/H] < -2.5 and are nearly pure r–process. *Open squares* have $-2.5 <$ [Fe/H] < -1, and show signs of s–process material

Gratton & Sneden (1991, 1994). A better demonstration of the differences in s–process *vs.* r–process contributions as a function of metallicity is shown in Fig. 79, which is taken from Gratton & Sneden (1994). This adds credence to the idea that for stars with [Fe/H] < -2.5 or so, only the r–process produced the heavier elements, while for more metal-rich stars, s–process contributions became important. Thus as in the case of [α/Fe] *vs.* [Fe/H], it appears that we have an approximate timescale for the estimation of the time required for the Galaxy to achieve a certain metallicity. For [Fe/H] $= -2.5$, the r–process *vs.* s–process abundances indicate a timescale perhaps as short as a few times 10^8 years, while for [α/Fe] *vs.* [Fe/H], it required the SNe Ia timescale (10^9 years?) to achieve [Fe/H] ≈ -1.

Globular Clusters. Such detailed elemental abundance patterns as discussed above for field stars are more difficult to obtain for globular clusters since even the brightest red giants are several magnitudes fainter than the field stars. Nonetheless, we may explore the general trends with the limited data available if we concentrate on the more easily measured elements, such

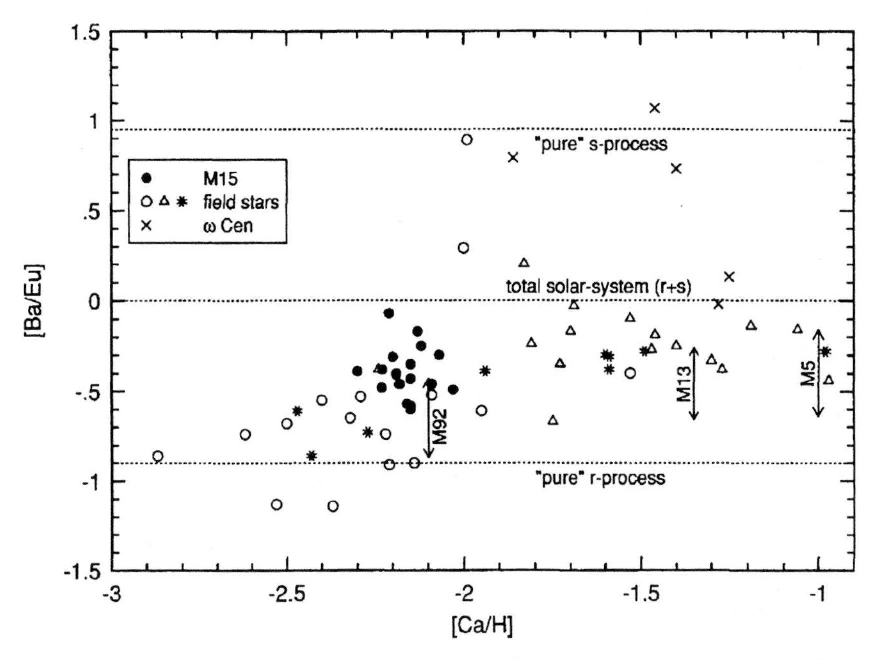

Fig. 80. The relative abundances of barium (dominated by the s–process) and europium (dominated by the r–process) as a function of [Ca/H]. The figure is taken from Sneden et al. (1997). The text discusses the sources of the data

as barium and europium. Figure 80 shows [Ba/Eu] *vs.* [Fe/H] for a sample of field stars and clusters, taken from Sneden et al. (1997).

As they describe, the field stars are represented by open circles, which are data from McWilliam et al. (1995) (but with revised barium abundances), open triangles are from Gratton & Sneden (1991, 1994), and the asterisks are from Shetrone (1996b) (and also with revised barium abundances). The dots are stars within M15, derived by Sneden et al. (1997), while the arrows represent the approximate domains of stars in M92, M13, and M5, with barium abundances taken from Shetrone (1996a) and europium abundances taken from Armosky et al. (1994). The crosses represent stars in ω Cen, with data from Smith et al. (1995). One key point made by Sneden et al. (1997) was that the barium and europium abundances in M15 do not correlate with signs of internal mixing (as discussed in the following section). While ω Cen continues to display its reputation as a chemically unusual globular cluster (with strong signs of s–process material), the other four globular clusters appear to follow the general trend defined by the field stars. The clusters do not extend to low enough metallicities to show convincingly an increased and nearly "pure" r–process signature, but insofar as they behave like the field stars, we have some reason to believe that their chemical enrichment histories and, hence, their chemical enrichment timescales, are the same.

6.3 Lithium

Lithium is one of the few elements manufactured during the Big Bang and is also one of the easiest to measure, thanks to a particularly strong spectral line at 6707 Å. Standard Big Bang nucleosynthesis models (see Boesgaard & Steigman 1985 for a nice review) predict that the lithium abundance is sensitive to η, the ratio of the number of baryons to photons in the early Universe, and hence to its baryonic mass density. For $\eta \approx 10^{-11}$, the predicted value of the logarithm of the number density of lithium, log ε(Li), is 3.3, on a scale where log $\varepsilon(\mathrm{H}) = 12.00$. This declines, reaching a value of slightly under 2.0 at $\eta \approx 3 \times 10^{-10}$, then rises slowly again, reaching a value of about 3.5 at $\eta \approx 2 \times 10^{-9}$.

The results of Spite & Spite (1982) therefore generated a great deal of excitement, for they found that for $T_{\mathrm{eff}} > 5500$ K, metal-poor dwarfs appear to have a roughly constant lithium abundance, log $\varepsilon(\mathrm{Li}) \approx 2$, near the minimum of the Big Bang nucleosynthesis models and therefore an apparently good measure of η. Since then, a great deal of work has been done on this "Spite plateau", as well as the cooler stars, which show declining lithium abundances, presumably due to the increasing depth of the surface convection zone and the transport of lithium to deeper, hotter layers where it may be destroyed by proton capture (see Fig. 81). Spite & Spite (1982) also noted that the apparent lithium abundance of the halo stars was a factor of ten below that of the meteoritic value, the interstellar medium, and that of young disk stars. Two questions arise.

- Do the halo stars have low lithium abundances because their long lives have enabled a greater degree of lithium transport and destruction? In this case, the lithium abundances are valuable probes of stellar mixing.
- Alternatively, has the lithium abundance in the interstellar medium risen over the course of the past 10^{10} years? This preserves the value of the halo stars in the study of Big Bang nucleosynthesis, and forces us to consider sources of lithium production, which is a major challenge given the "fragility" of the lithium nucleus.

If the lithium in metal-poor halo stars was manufactured during the Big Bang, then in the absence of mixing and lithium destruction, all metal-poor stars should have very similar lithium abundances. Thus one "experiment" to test the lithium destruction in halo stars is to determine if the scatter in lithium abundances can be ascribed solely to observational errors or must be due at least in part to real star-to-star differences. Some stellar evolution models in fact predict such a variation in lithium abundances. Stars probably arrive on the main sequence with different degrees of rotation and angular momentum. A magnetically coupled stellar wind may carry away the angular momentum in the outer layers, and the resultant differences in angular momentum as a function of depth may cause turbulent mixing, according to rotating models

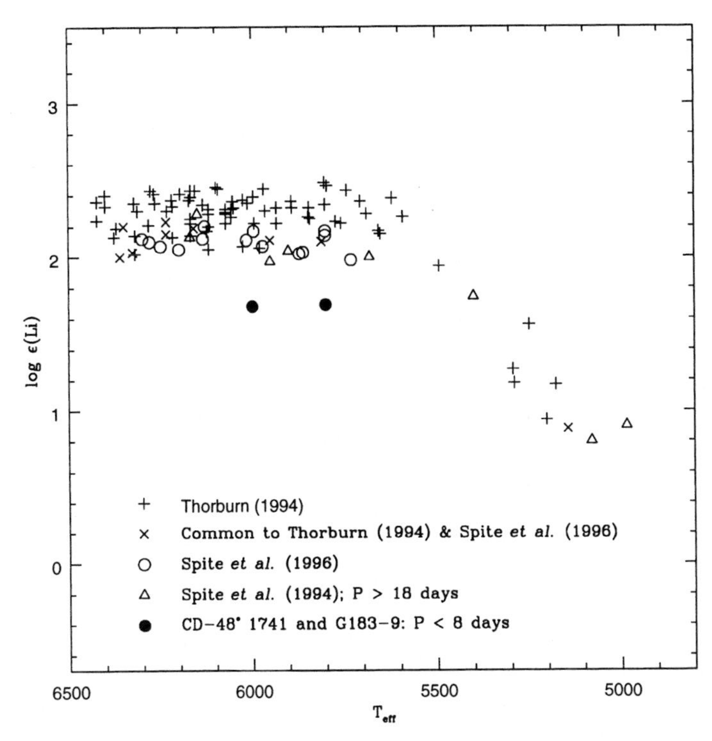

Fig. 81. Lithium abundances for field stars with $[Fe/H] \leq -1.1$, including binary systems studied by Spite et al. (1994) and for G183-9

computed by Pinsonneault et al. (1989, 1990, 1992), and Chaboyer & Demarque (1994). The basis for conducting a test of this specific model is seen in Figs. 18 and 46. Short period binaries experience tidal interactions, which first force co-rotation and then circularize the orbits. As the figures show, systems with periods less than about 20 days should be co-rotating, and the shorter the period, the earlier co-rotation was established. Co-rotation will prevent further angular momentum loss and lithium depletion via the turbulent diffusion mechanism. One final test of the lithium depletion *vs.* lithium production models is the isotopic ratios. The proton capture cross section is much larger for ^{6}Li than for ^{7}Li (Caughlan & Fowler 1988), so it is destroyed at lower temperatures. Hence any mechanism that leads to circulation or settling of photospheric Li to hotter, deeper layers will destroy ^{6}Li much more rapidly. A test of destruction, therefore, is to measure the abundance ratio of these two isotopes: in the zeroeth order, the mere presence of ^{6}Li argues against significant destruction of ^{7}Li.

Lithium in Metal-Poor Binaries. Spite et al. (1994) were the first to recognize the importance of short period halo binaries to test the rotational models of lithium depletion. They observed metal-poor stars with binary orbital solutions obtained by Latham et al. (1988, 1992) and by Lindgren et al. (1987). All the metal-poor binary stars but two have eccentric orbits, suggesting minimal tidal interaction. G65-22, with a period of 18.7 days, probably circularized only relatively recently, being right at the edge of the metal-poor transition period between circular and eccentric orbits. CD$-48°$ 1741, on the other hand, with a period of only 7.6 days, achieved co-rotation and circularization much earlier. To this sample we may now add another, even shorter period binary system: G183-9. Goldberg et al. (1999) have found this star to be double-lined, and an orbital period of 6.2 days. Suchitra Balachandran and I have completed an abundance analysis of this star, using the known mass ratio, model isochrones, and model atmosphere surface fluxes to correct for the secondary's contribution to the continuum. The primary appears to have an effective temperature of 6000 K, [Fe/H] $= -1.17$, and log ε(Li) $\approx$ 1.7. Figure 81 shows the derived lithium abundances for these binary systems and other field stars, using results from Thorburn (1994) and Spite et al. (1996). The figure shows that there may be small differences in the temperature scales employed by Thorburn (1994) and Spite et al. (1996), but that in general, the longer period binary systems show the same abundances as the apparently single stars. However, the two shortest period systems actually indicate *lower* lithium abundances. The inhibition of spin-down of the outer layers of these two stars, and the presumed inhibition of turbulent diffusion, has *not* revealed higher lithium abundances. Instead, it appears that lithium depletion is greater, for reasons as yet unclear.

Tests for Constancy of Lithium Abundances. Thorburn's (1994) extensive study of lithium abundances in metal-poor dwarfs concluded that the observed dispersion along the "Spite plateau" was larger (0.13 dex) than could be attributed to observational errors alone. The small variations would then be due to internal processes, and even if turbulent diffusion is not the cause, some parameter that varies from star to star causes differing degrees of depletion. The logical extension of this argument is that perhaps none of the stars retain Big Bang abundances of lithium in their outer envelopes. As we have discussed already, derived elemental abundances differ from "true" ones due to differences in the *gf* values employed, and the estimation of the stellar temperatures, gravities, and microturbulent velocities. Fortunately (or unfortunately), in the case of lithium, only one line (albeit with four components) is analyzed, and the *gf* values are well determined (Andersen et al. 1984). The line is neutral and hence not very sensitive to gravity, and it is also usually quite weak, and so not very sensitive to microturbulent velocity. But temperature remains the primary variable, and small differences in the adopted temperature calibrations may lead to significant differences in the derived lithium abundances. Spite et al. (1996) made this point clearly in their re-analysis of lithium abundances in some very metal-poor field dwarfs. They were careful

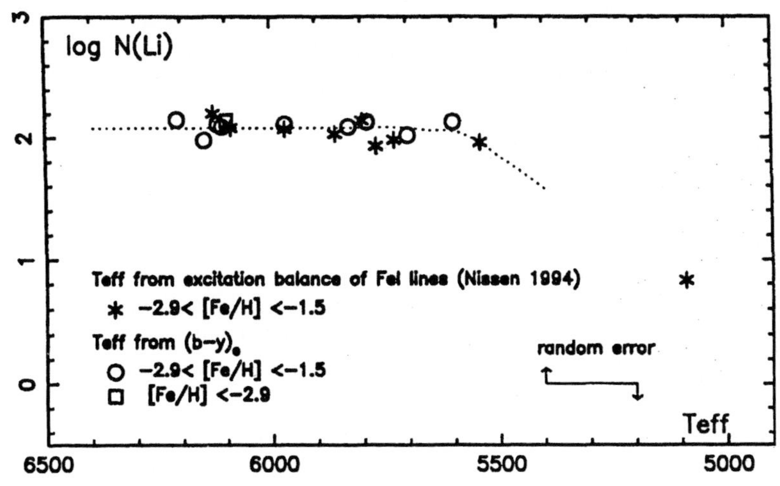

Fig. 82. Lithium abundances for field stars studied by Spite et al. (1996), with careful attention to relative temperatures

to distinguish between temperatures estimated via the Boltzmann equation (using lines of differing excitation potential to derive $T_{\rm eff}$), Strömgren photometry, and from the profiles of the wings of the Hα line. Figure 82 shows one of these comparative analyses. The scatter about the mean value of the "Spite plateau" in this case is completely consistent with observational errors. On the other hand, there are signs that at least some metal-poor stars show considerable larger lithium abundance than that found for the plateau. Are these the stars that have preserved the Big Bang abundances and the rest have become severely depleted (although very uniformly)? Two examples stand out. Among the field stars, King et al. (1996) found that BD+23° 3912 has log ε(Li) $\approx$ 2.5, a factor of at least two higher than that of the "Spite plateau". Not only is the star apparently *abnormal* in its lithium abundances, it is particularly *normal* in all its other elemental abundances. As King et al. (1996) noted, the lack of s–process enhancements is especially important. As discussed below, intermediate mass stars with enhanced lithium abundances always show enhanced s–process abundances, so the lack of such a signature in the spectrum of BD+23° 3912 suggests that the star has not been enriched by a neighbor. Dave Latham and his colleagues have monitored the radial velocity of BD+23° 3912 and see no signs of variability indicative of a companion. The 13 velocities span over 1300 days, and the scatter about the mean is only 0.6 km s^{-1}. BD+23° 3912 itself is clearly a relatively unevolved star, so it has not self-enriched.

The second case comes from spectroscopic studies of lithium abundances in relatively unevolved globular cluster stars, a subject only now becoming possible thanks to efficient spectrographs and large aperture telescopes. Pasquini & Molaro (1996) studied six turn-off and subgiant members of the nearby globular cluster NGC 6397, finding all three turn-off stars to have lithium abundances consistent with the "Spite plateau". The subgiants were depleted, as found in the field subgiant stars by Pilachowski et al. (1993). On the other hand, Boesgaard et al. (1998) studied six subgiants in the metal-poor globular cluster M92 using the Keck I telescope. All are hot enough that, based on similar analyses of metal-poor field subgiants, lithium depletion should be minor (Pilachowski et al. 1993). Despite relatively modest S/N, the spectra appear to reveal that the lithium abundances of these six stars are quite different, despite having identical ages, masses, and chemical composition. This is perhaps the strongest evidence that lithium depletion is real and quite variable in metal-poor stars, and deserves dedicated follow-up efforts, especially using the new large southern hemisphere telescopes to study the nearer globular clusters.

The above two cases are for stars with apparently *excess* lithium. There are also a few cases of field stars that are hot enough to lie on the "Spite plateau" but which are very deficient in lithium: G186-26 (Hobbs et al. 1991); G122-69 and G139-8 (Thorburn 1992); and G66-30 (Spite et al. 1993). As Ryan et al. (1998) have discussed, the stars do not share any other abnormality in their abundance patterns. One way to deplete lithium would be mass transfer, perhaps from a pair of stars so close that the transfer would begin before either star evolved away from the main sequence. The result might be a star with increased mass and decreased lithium, but the star might not be massive enough to be recognizable as a blue straggler. In fact, G66-30 qualifies as a field blue straggler, being slightly bluer than the main sequence turn-off of clusters with a similar metallicity, and is a binary system with a long period (694 days). However, none of the other three stars appear to be binaries: G186-26 (σ = 1.5 km s^{-1}, 43 velocities covering 3219 days); G122-69 (σ = 0.9 km s^{-1}, 10 velocities covering 3358 days); G139-8 (σ = 1.0 km s^{-1}, 18 velocities covering 3189 days), so mass transfer does not seem to be a likely explanation. Without understanding how these stars differ from the stars on the "Spite plateau", one must wonder if they represent a unique and limited phenomenon or whether they are merely extreme cases of lithium depletion in metal-poor dwarfs.

Isotopic Ratios. Smith et al. (1993; 1998) and Hobbs & Thorburn (1994; 1997) detected ^{6}Li in the metal-poor turn-off star HD 84937, and Smith et al. (1998) also detected it in BD+26° 3578. Its abundance is rather high, about 5% that of ^{7}Li. While consistent with the lack of significant destruction, the situation is complicated by three major factors. First, because mixing samples a variety of depths, the presence of ^{6}Li does not preclude *any* mixing, as discussed by Pinsonneault et al. (1999). It does help establish limits, however, to the original lithium abundance, which Pinsonneault et al. (1999) argue is between 0.2 and 0.4 dex above the current "Spite plateau" values. Second, Big Bang nucleosynthesis predicts a much lower value for the ^{6}Li/^{7}Li ratio

(Nollett et al. 1997). If these models are correct, then most of the ^{6}Li was produced prior to the formation of these stars, or perhaps even subsequent to their formation, by the spallation of heavier nuclei (presumably CNO) by cosmic rays. The final problem is that if spallation is the source of the ^{6}Li, we should see higher beryllium abundances in metal-poor stars than is seen. This is discussed at length by Smith et al. (1998), who noted that the ^{6}Li/^{7}Be ratio in these two metal-poor stars is much higher ($\approx$ 60) than is observed in meteorites ($\approx$ 6). The latter ratio is consistent with production via spallation. Their recommendations to the theorists was to refine their predictions of ^{6}Li depletion in stars while their challenge to the observers was, as always, "back to the telescope!"

Lithium Production. There are stars in the Galaxy which have apparently produced lithium in their interiors (see da Silva et al. 1995; de la Reza 1997), probably via "hot bottom burning" and rapid transport to the surface via very deep convection zones (see Boothroyd et al. 1993). The evidence for production in this case is quite simple: they have ten times as much lithium in their atmospheres as is now found in the interstellar medium. Smith & Lambert (1989, 1990) and Plez et al. (1993) have shown that metal-poor AGB stars in the Small Magellanic Cloud are also lithium-rich, although not quite to this degree. Their lithium abundances roughly equal that in the current interstellar medium *of the Galaxy*. Since the Galaxy's young stars are about a factor of 4 times as metal-rich as those in the Small Magellanic Cloud, we might expect that the interstellar medium of the SMC is also metal-deficient. If lithium has been manufactured over time, then the lithium abundance in the interstellar medium of the Small Magellanic Cloud may actually be lower than in the Galaxy, in which case the log ε(Li) values of about 3 found in AGB stars in the Small Magellanic Cloud indicate that the stars are lithium-rich. This lithium production could be part of the reason that the halo and disk stars were born with different amounts of lithium. But it would not, of course, explain the high lithium abundance of BD+23° 3912, nor the variations seen among the six stars in M92.

Recently, three cases indicating some lithium production in low-mass metal-poor stars have been noted. Carney et al. (1998a) have found that the 26-day Cepheid V42 in the globular cluster M5 shows lithium. The lithium abundance is low, lower in fact than the "Spite plateau" mean value, log ε(Li) $\approx$ 1.7. But V42 is apparently a post-AGB star, and it clearly seems to have been able to manufacture lithium in its core. This may be a sign of "cool bottom processing" discussed recently by Wasserburg et al. (1995) and Sackman & Boothroyd (1999). Further, V42 does *not* show any signs of enhanced s–process abundances. Therefore the lack of such enhancements, as noted in BD+23° 3912 by King et al. (1996), may not absolutely be used to rule out lithium production. Lithium is also present but relatively low in abundance in the red giant branch tip variable V2 in NGC 362, log ε(Li) $\approx$ +1.2 (Smith et al. 1999). An even more extreme case is IV-101 in M3, an otherwise normal red giant which has log ε(Li) $\approx +3.0$ (Kraft et al. 1999). We conclude that it certainly would be wise to conduct a more thorough study of low-mass, metal-poor RGB, AGB, and post-AGB stars, especially in globular clusters, to explore the question of whether lithium production by such stars is common and possibly important to the Galactic evolution of lithium abundances in the interstellar medium.

7 Elemental Abundance Ratios II. Internal Processes

Our reliance upon stellar evolution theory is profound. With it, we estimate ages of star clusters and even some field stars, and thereby establish the chronology of the Galaxy's history, and even those of nearby galaxies. Combined with estimates of initial mass functions, we use stellar evolution theory to compute synthetic integrated light spectra for comparison with unresolved galaxies near and far. Many tests have been made of stellar evolution models, using both the Sun and well-studied binary systems and clusters. But our knowledge remains incomplete for there are phenomena that continue to puzzle us and which are likely to prove important in all the applications of theory.

Fortunately, blue stragglers appear to tell us more about the interactions between binary stars in low- and high-density stellar environments, and are probably no longer a challenge to basic stellar evolution theory. But at least three serious challenges still remain: the solar neutrino problem; mixing; and mass loss. The first problem may yet be resolved by neutrino physics, neutrino oscillations in particular, and lie beyond the general scope of these lectures. However, the remaining two phenomena are very important to the evolution of stars in globular clusters, and, indeed, the study of globular cluster stars will prove crucial in improving our understanding of the processes of mixing and mass loss. The latter, for example, is likely responsible for the wide range in envelope masses and resultant range in colors of stars along the horizontal branches in some clusters (see Iben & Rood 1970; Faulkner 1972; Lee et al. 1990, 1994; Catelan 1993). Mass loss therefore no doubt plays a role in the resolution of the "second parameter problem". Mixing, on the other hand, is an excellent test of our understanding of the internal stucture and evolution of stars, as well as the nucleosynthesis processes whose products may be brought up to the stellar photospheres for detailed study.

7.1 Mixing

For over two decades it has been clear that the metal-poor, low-mass stars in globular clusters mix the products of CNO cycle nucleosynthesis into their photospheres to a large degree. Standard theory predicted that no significant contact is made between the convection zone and the hydrogen-burning shell.

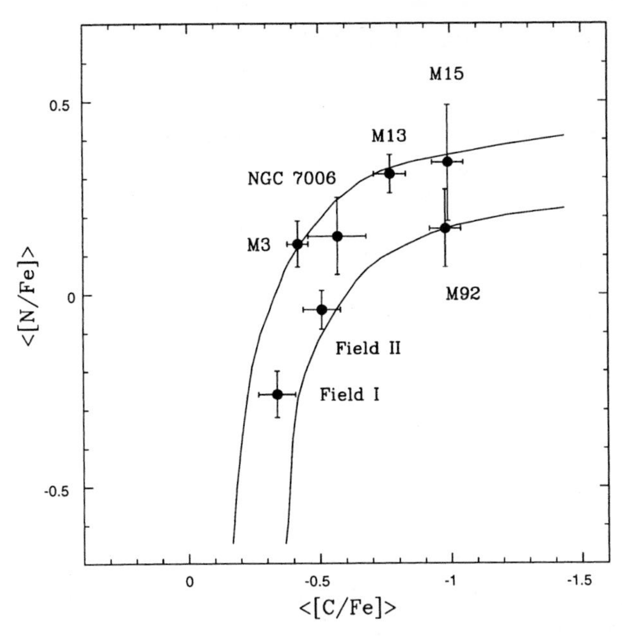

Fig. 83. Carbon and nitrogen abundances for red giants with $-2.0 \leq M_V \leq 0.0$ in metal-poor field stars and globular clusters, following Carbon et al. (1987). Lines represent constant C+N abundances. Cluster data are from Suntzeff (1981: M3 & M13); Carbon et al. (1982: M92); Friel et al. (1982: NGC 7006); and Trefzger et al. (1983: M15). Field giant data are from Kraft et al. (1982). Class II have [Fe/H] ≤ -2.0, class I have $-2.0 <$ [Fe/H] ≤ -1.5.

Nonetheless, signs of CNO cycle products at the stellar surface had been noticed earlier through the decrease in carbon abundances derived from CH band strengths, and the increase in nitrogen abundances derived from CN and NH band strengths. Figure 4 shows how this happens. The hydrogen burning uses ^{12}C as a catalyst, but the slowness of the ^{14}N(p,γ)^{15}O reaction promotes a build-up of ^{14}N at the expense of ^{12}C. In this case the nuclear physics is well understood, but what was not clear was how the material was transported to the surface without a connection to the convection zone. Sweigart & Mengel (1979) offered a possible solution: meridional circulation. They predicted that rotation-driven meridional circulation could connect the hydrogen-burning shell with the convection zone, and that significant mixing would result after the shell crosses the prior point of deepest penetration by the convection zone. (Recall that this also causes a change in the nucleosynthesis rate due to the change in mean molecular weight, and results in the "bump" in the red giant branch luminosity function.) The prediction of this model, then, is that the signs of the mixing should manifest themselves only above a certain luminosity for stars within clusters, and, of course, that the abundances of carbon and nitrogen should be anticorrelated. As Fig. 83 shows, the latter

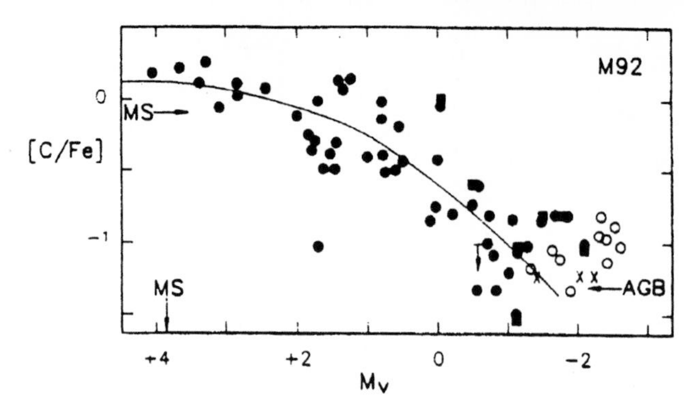

Fig. 84. [C/Fe] values for evolved stars in the metal-poor globular cluster M92, taken from the review article by Kraft (1994), based on work of Suntzeff (1989)

prediction appears to be satisfied. The lines represent loci of roughly constant carbon plus nitrogen abundances, which is expected if only the CN cycle is operating. The field subdwarfs and giants may or may not follow the same general trends seen in the clusters, a point to which we return later. Figure 84, taken from Kraft (1994), shows that the decline in photospheric [C/Fe] begins at a luminosity not too different than the luminosity of the red giant branch bump, so in general meridional circulation appears to be a good beginning to understanding mixing in globular cluster giants. But careful study of the figure reveals that there is a larger spread in [C/Fe] than expected among the higher luminosity stars. This is also clear in the study of M4 by Drake et al. (1994). Even near the horizontal branch luminosity and red giant branch bump, they found spreads in [C/Fe] of over a factor of three.

More recently, it has become clear that the CNO cycle products seen in the photospheres of luminous cluster red giants are not confined to those in the first cycle (CN) of Fig. 4. As elemental abundance analyses have improved due to better spectrographs and detectors, and also in some cases due to use of much larger telescopes such as Keck, it has become clear that almost all the cycles depicted in *both* Figs. 4 and 5 have products being mixed to the surfaces of some stars in some clusters. One of the first signs of this was the observed anticorrelation between sodium and oxygen abundances, shown here (Fig. 85) for the well-studied intermediate-metallicity cluster M13 (Kraft et al. 1997). Sodium abundances rise at first with no apparent change in oxygen, and then as sodium abundances level off after an enhancement of over 0.5 dex, the oxygen abundances decline dramatically. The decline of oxygen can be understood qualitatively on the basis of Fig. 4: the "second" of the CNO cycles is engaged, and with the same bottleneck at the $^{14}N(p,\gamma)^{15}O$ reaction. Oxygen abundances decline while nitrogen rises as the cycles approach equilibrium. The rise in sodium abundances is more complicated, and careful consideration

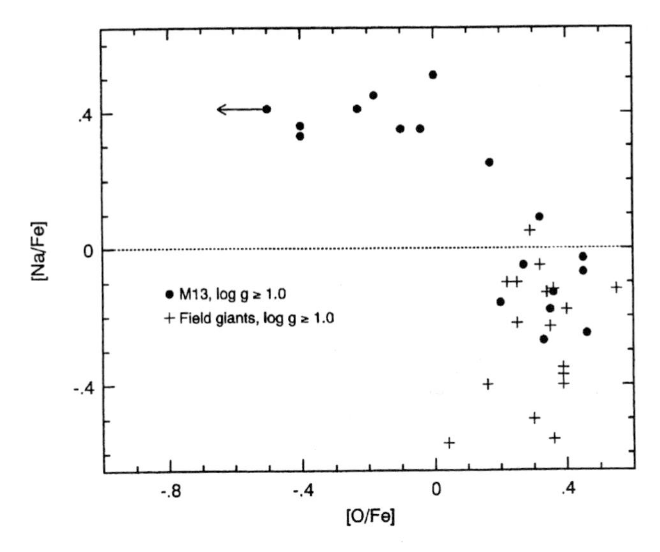

Fig. 85. [Na/Fe] *vs.* [O/Fe] in M13 giants, taken from Kraft et al. (1997)

of the reaction rates of the entire network is required. Langer et al. (1993) have done so, and argued that the general trends can be understood from a nucleosynthesis perspective, as shown in Fig. 86. It shows the expected early anticorrelation between carbon and nitrogen abundances, and the later anticorrelation between oxygen and nitrogren due to the first two CNO cycles of Fig. 4. The decline in ^{16}O is accompanied, even slightly preceded, by a rise in ^{23}Na abundances, as observed.

But it is not quite this simple. While globular clusters in general follow the same sodium–oxygen anticorrelation trend (see Fig. 87), there are very interesting cluster-to-cluster differences. Figure 88 illustrates this point clearly. M13 and M3 have very similar [Fe/H] and [α/Fe] values, and are probably quite similar in age. But their mixing histories are very different. In the case of M3, some red giants retain their initial high [O/Fe] values, even at luminosities near the red giant branch tip. A minority of the (small) sample studied do show oxygen depletions due to mixing. M13, on the other hand, shows the reverse. All but one of the highest luminosity giants ($M_{\rm bol} < -3$) show depleted oxygen abundances, and the depletions are much more extreme than seen in M3. Even at lower luminosities, $M_{\rm bol} \approx -2.2$, oxygen depletions in some M13 red giants have already reached the stellar photospheres. Somehow the depth to which mixing reaches in M13 exceeds that in M3. What variable have we overlooked? Binary companions? Rotation?

Sodium is not the only element in the advanced CNO cycles to show changes in photospheric abundances in globular cluster red giants. Aluminum changes as well, as may be seen in Shetrone's (1996a) study of high-, intermediate-, and low-metallicity field and cluster red giants: see Fig. 89. As before,

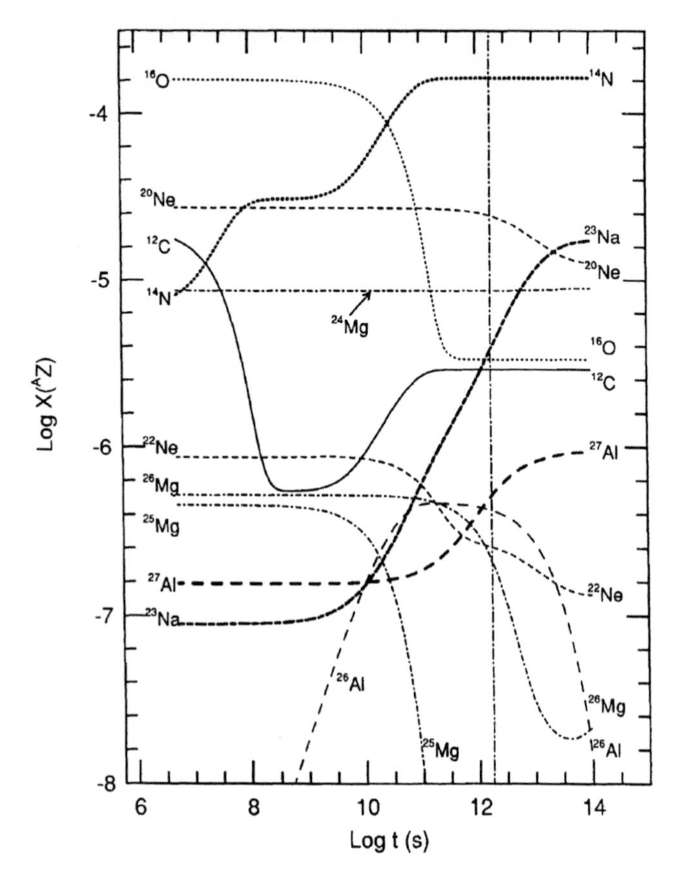

Fig. 86. Predictions of changing abundance ratios *vs.* time due to CNO cycling, taken from Langer et al. (1993). Note the predicted carbon–nitrogen, oxygen–nitrogen, and oxygen–sodium anticorrelations. Note also the predicted constancy of ^{24}Mg

the intermediate- and low-metallicity field stars do not appear to share the same trends as the globular cluster giants. On the other hand, the metal-rich field and cluster stars are similar in that neither show any signs of changing aluminum abundances, suggesting that metallicity does play a role in how deep the mixing may penetrate. Magnesium is also affected, and most importantly, it appears that, unlike the predictions of Langer et al. (1993), it is not the much rarer ^{25}Mg and ^{26}Mg isotopes that are affected, but the much more abundant ^{24}Mg. If the models are correct, such that ^{24}Mg is not affected by nucleosynthesis within red giants and subsequent mixing, the observed variations in elemental magnesium abundances would then require a different mix of isotopic ratios from star to star. The limited data on the metal-poor field dwarfs μ Cas (Tomkin & Lambert 1980) and ν Ind (Lam-

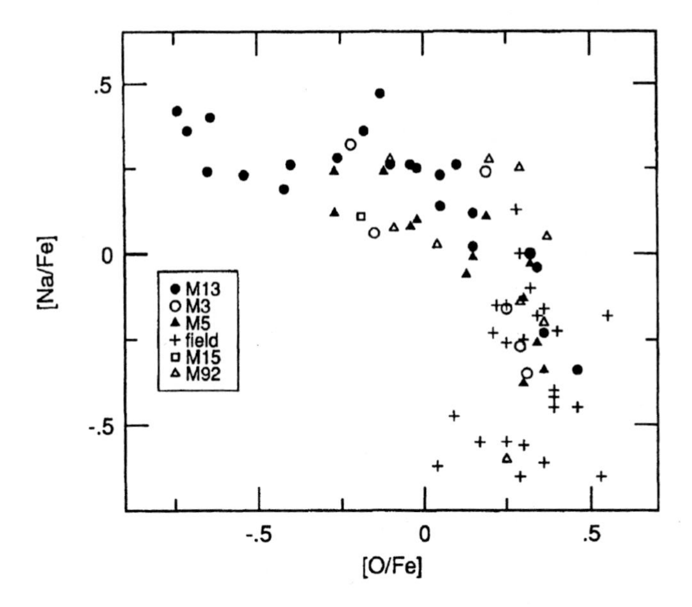

Fig. 87. [Na/Fe] *vs.* [O/Fe] in five intermediate- and low-metallicity globular clusters, plus a sample of metal-poor field red giants, taken from Kraft (1994)

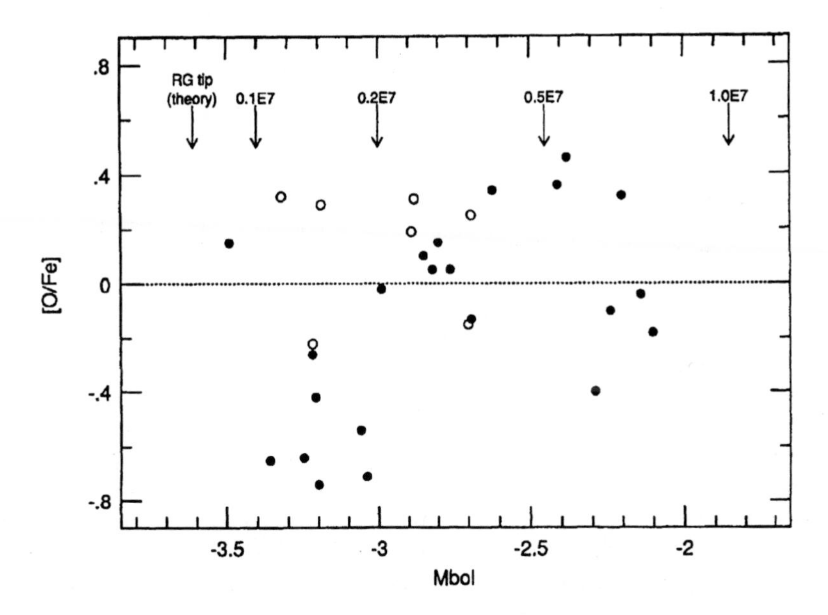

Fig. 88. [O/Fe] *vs.* $M_{\rm bol}$ for red giants in the globular clusters M3 and M13, taken from Kraft et al. (1993). *Open circles* are M3 giants; *filled circles* are M13 giants

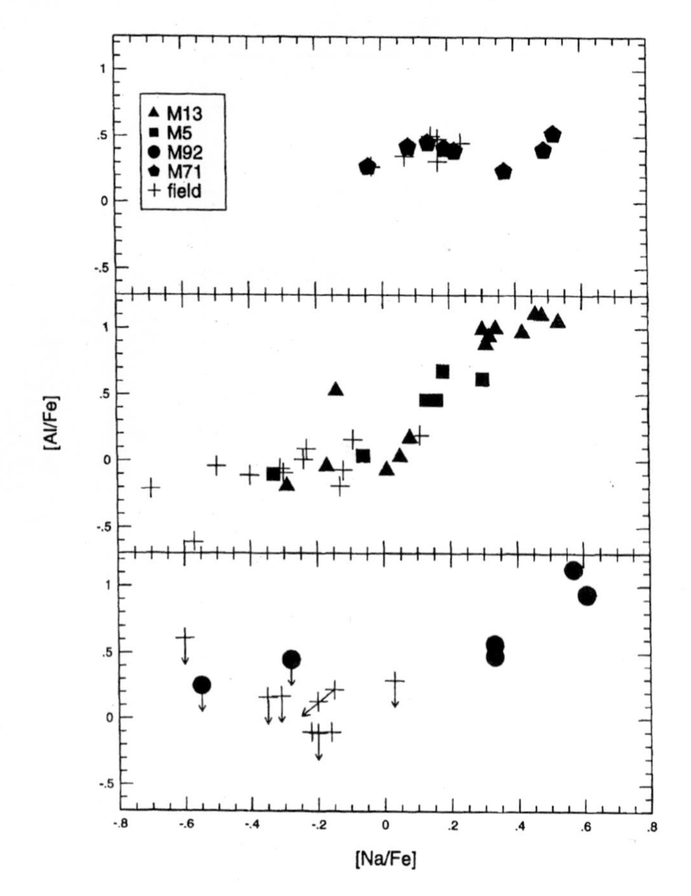

Fig. 89. [Al/Fe] *vs.* [Na/Fe] in a sample of high-, intermediate-, and low-metallicity globular clusters (upper, middle, and lower panel), plus a sample of metal-poor field red giants, taken from Shetrone (1996a)

bert & McWilliam 1986) appear to rule out that possibility. Further, Shetrone (1996b) has shown that in M13, it is the abundances of ^{24}Mg that have been altered, not those of ^{25}Mg or ^{26}Mg, as shown in Fig. 90. It seems clear that the hydrogen burning being "sampled" by the Mg–Al anticorrelation is from a very high-temperature regime indeed. Cavallo et al. (1998) have tried to model the anticorrelation, exploring the effects of the "leakage" from the Ne–Na cycle into the Mg–Al cycle through the ^{23}Na(p,γ)^{24}Mg reaction, and also by varying the ^{24}Mg(p,γ)^{25}Al reaction rate. They find that while aluminum enhancements can be produced, the magnesium abundances remain a serious puzzle. Cavallo et al. (1998) found they could explain the depletion of ^{24}Mg if the proton capture reaction rate was thirty times larger than had been measured. A recent remeasurement of the ^{24}Mg(p,γ)^{25}Al reaction rate by Powell

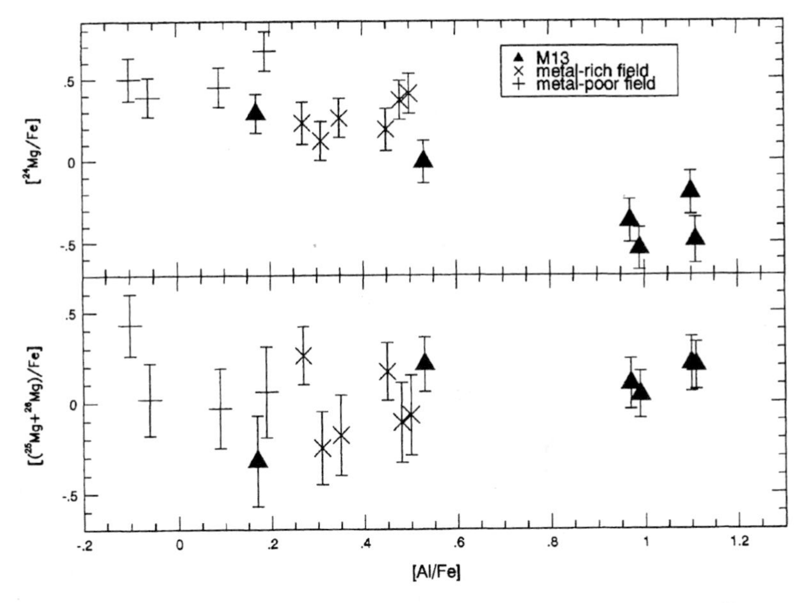

Fig. 90. The abundances of the ^{24}Mg (top panel) and (^{25}Mg + ^{26}Mg) isotopes *vs.* [Al/Fe] in red giants in the field and in the cluster M13, taken from Shetrone (1996b)

(1999) indicates an increase of less than 50%. Apparently, the mixing must reach down into regions of very high temperature.

This deep mixing may be a function of metallicity as well, according to Cavallo et al. (1998). This explains Fig. 91, where the spectroscopic results for [Al/Fe] and [Mg/Fe] obtained by Norris & Da Costa (1995) have been plotted. Different symbols refer to different metallicity domains, and it appears that the stars in ω Cen that have [Fe/H] < -1.3 mix more deeply than do the more metal-rich stars.

Since such deep mixing does apparently occur in some clusters, it is worth considering some of its possible implications. If secondary products of hydrogen burning are reaching the photosphere, so should the primary product – helium. One therefore expects (Langer & Hoffman 1995; Sweigart 1997) that the photospheric helium abundances will be higher in those stars that show signs of deep mixing. This cannot be measured directly, but it may be inferred in the following manner. The increase in helium abundance in the photosphere must be offset by a decrease in the hydrogen abundance, and therefore in the strength of the H^- continuum opacity. All other things being equal, a decrease in the continuum opacity makes absorption lines appear to be stronger. Abundance analyses that do not take this into account will therefore result in higher [Fe/H] values. The effect is not large, but the high-precision results of Kraft et al. (1997) for red giants in M13 may be used

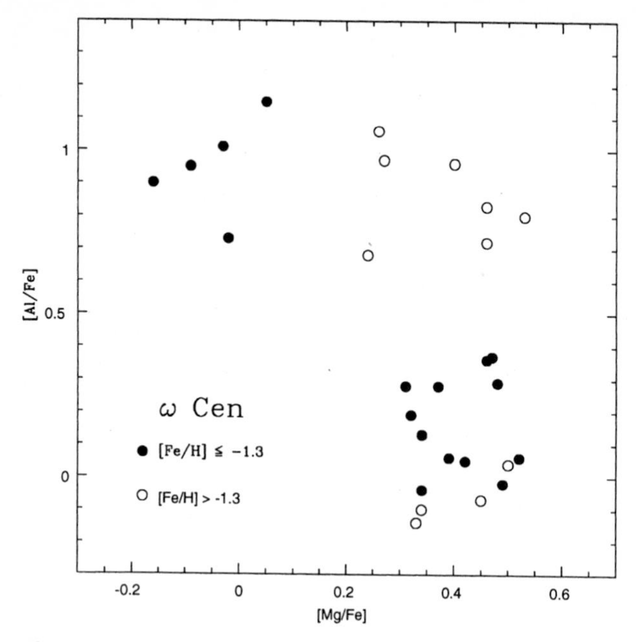

Fig. 91. The abundance ratios [Al/Fe] are compared to [Mg/Fe] in two metallicity domains. The data were taken from Norris & Da Costa (1995)

to test the idea. In Fig. 92 we show their [Fe/H] results plotted against their [O/Fe] results. The effects of variable continuum opacity will not, to first order, affect the [O/Fe] results, so we may distinguish deeply mixed stars (low [O/Fe]) stars from those with little or no deep mixing (high [O/Fe]). There is a very clear trend, represented by a linear fit. (The fit was computed using a least squares linear bisector, but, of course, a linear fit is only a convenient descriptor and may not accurately reflect the proper trend.) The higher [Fe/H] results for the mixed stars can be understood easily if the photospheric hydrogen abundances have been reduced and, presumably, those of helium increased. The magnitude of the effect is hard to estimate without using helium-rich model atmospheres, but a crude first-order estimate is that the helium mass fraction may have increased from $Y = 0.1$ to 0.2 or 0.3 in the photospheres of the highly mixed stars.

7.2 Mixing/Mass Loss/HB Morphology

One of the plausible explanations for the spread in envelope masses and hence in position along the horizontal branch seen in globular clusters, and hence for the second parameter effect, is rotation. Rotation provides extra pressure support for stellar cores, and hence hydrogen burning may be prolonged. If shell hydrogen-burning stars have lengthened lifetimes, this would occur near the red giant branch tip, when the luminosities are high and so, presumably,

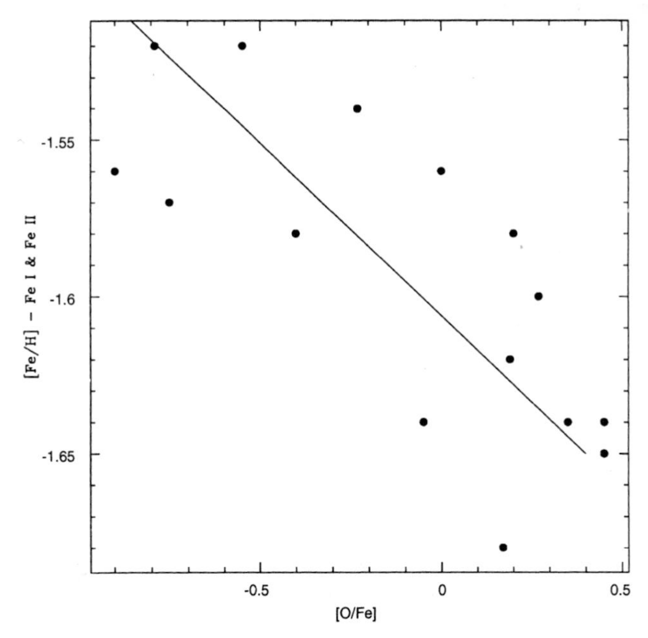

Fig. 92. [Fe/H] *vs.* [O/Fe] in M13 giants, with data taken from Kraft et al. (1997). The higher [Fe/H] values derived for the deeply mixed (low [O/Fe]) stars is consistent with diminished hydrogen and enhanced helium in their photospheres

is the mass loss rate. If higher rotation leads to greater net mass loss, then it should also result in lower envelope masses and bluer zero age horizontal branch positions. Extra rotation would also be expected to alter the degree of mixing, but no detailed models have been computed to predict the magnitude of the effect.

Instead, an empirical approach has been tried, based on the discovery of measurable rotation in blue horizontal branch stars in the field (Peterson et al. (1983) and in clusters (Peterson 1983, 1985a,b). What do we expect to see? At equal [Fe/H] and [α/Fe], a cluster with high rotation might be expected to show signs of deeper mixing, and, thus, more highly depleted oxygen and significantly enhanced aluminum in its most luminous red giants and in its descendent horizontal branch stars. If rotation is also related to mass loss, the bluest horizontal branch stars should be the progeny of the red giants with the highest core rotation rates. Thus a study of blue horizontal branch stars should reveal high rotational velocities and abundance ratios altered significantly by mixing. A comparison between two clusters with similar mean metallicities but differing horizontal branch morphologies might therefore reveal differences in mean rotational velocities and the abundances of CNO cycle elements like oxygen among blue horizontal branch stars. Of course, it may also be the case that all blue horizontal branch stars in *both* clusters

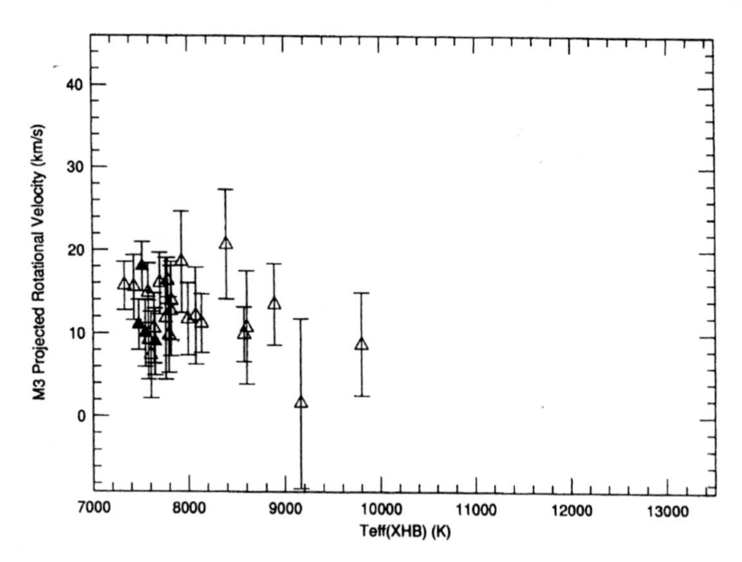

Fig. 93. The distribution of rotational velocities among the blue horizontal branch stars in M3, taken from Peterson et al. (1995, *open symbols*) and Peterson (1985b)

show similar behavior and the differing amounts of rotation simply alter the fraction of stars found on the blue horizontal branch. Peterson *et al.* (1995) have summarized the results of their studies of blue horizontal branch stars in three clusters with very similar metallicities: NGC 288, M13, and M3, the first two having very blue horizontal branches while M3 has stars spread from the blue to the red. The distributions of rotational velocities in M3 and NGC 288 are very similar, suggesting that rotation may not be the cause of the second parameter in this particular pair: see Figs. 93 and 94. On the other hand, the rotational velocities of the blue horizontal branch stars in M13 appear to be bi-modal, with one sample being similar to those of M3 and NGC 288, and the remainder quite high velocity: see Fig. 95. Peterson et al. (1995) also measured the strength of the O I triplet near 7770 Å. Not only were those of M3 and M13 very similar, there were no signs of the extremely low oxygen abundances seen in some of the M13 red giants (see Fig. 85).

At first look, then, none of the (probably naive) predictions have been confirmed. Blue horizontal branches do not necessarily consist of stars rotating rapidly and with signs of deep mixing apparent through lowered oxygen abundances. Of course, one should recall the *caveats* of Peterson et al. (1995). The surfaces of these blue horizontal branch stars were deep within their red giants progenitors. Spin down may have occurred, yet the core might still be rotating rapidly. How well the measured surface rotational velocities reflect the rotation of the core that might enhance mixing and prolong mass loss is not entirely clear. Oxygen abundances may also have been altered by diffu-

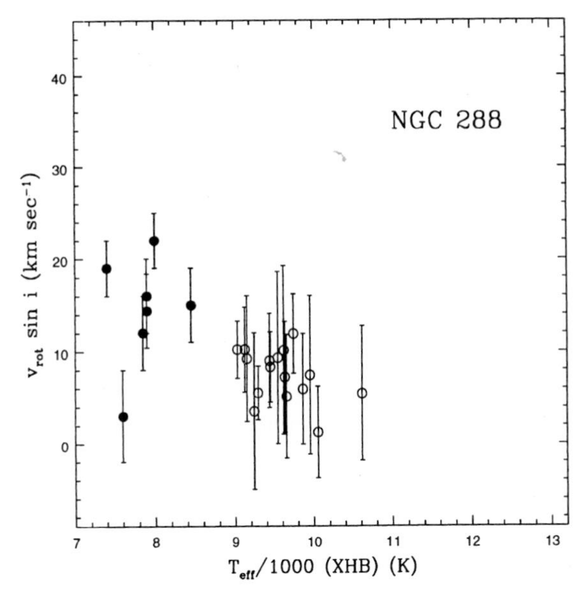

Fig. 94. The distribution of rotational velocities among the blue horizontal branch stars in NGC 288, taken from Peterson et al. (1995, *open symbols*) and Peterson(1983)

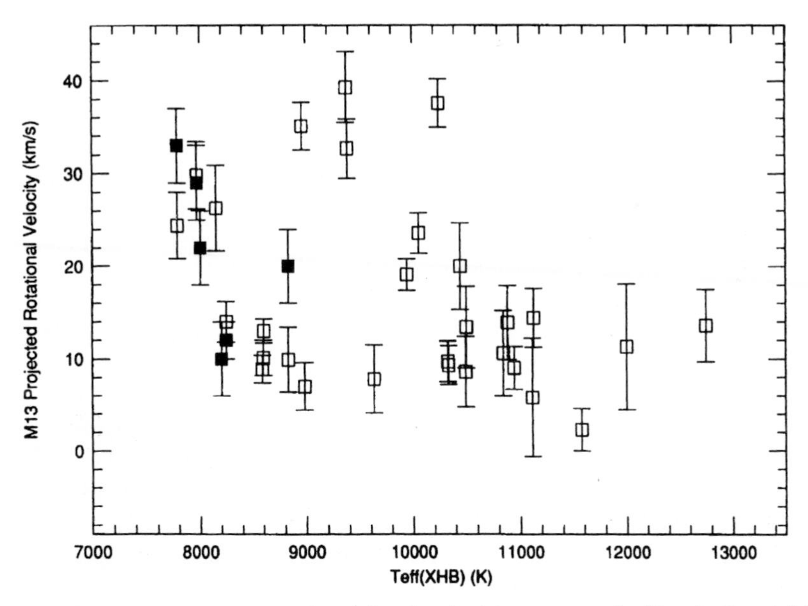

Fig. 95. The distribution of rotational velocities among the blue horizontal branch stars in M13, taken from Peterson et al. (1995, *open symbols*) and Peterson(1983)

sion or “radiative levitation”. And the bluest horizontal branch stars in these clusters were not studied due to their fainter magnitudes, and these may be the stars to study in the quest for the most extreme signs of mixing and its possible relation to mass loss.

7.3 Primordial Abundance Variations Within Clusters

Many globular clusters have masses exceeding 10^5 $M_{\odot}$, and if the star formation process in them was slow enough, we should expect that ejecta from the earliest supernovae might have enriched the gas out of which some stars were still forming. Thus the search for primordial abundance variations is really a search for signs of such a relatively slow star formation process. We study especially those elements likely to have been manufactured in supernovae, including the iron peak and r–process elements, as well as elements manufactured in and expelled by intermediate-mass stars, including material that has undergone extensive CNO cycling and manufactured some of the s–process elements.

Just as there are three broad means by which to estimate metallicities, so there are three general methods by which one may search for abundance variations within clusters. The color–magnitude diagram itself may hold clues. As shown in Figs. 7 and 10, the color and temperature of the red giant branch is sensitive to metallicity, primarily through its dual roles in contributing the electrons that establish the H^- opacity and in contributing line blanketing, especially in the B bandpass. Figure 7 also shows that helium does not strongly affect the red giant branch colors. Even a binary companion to a red giant will not strongly affect the combined magnitude and color as long as the secondary's mass is only slightly smaller than that of the primary. Thus the width in the temperature or color is an excellent means by which a metallicity spread may be determined. Main sequence widths may also be measured, but binary contamination is a more serious problem. The second method is to rely on low- to moderate-resolution spectroscopic metallicity indicators, such as ΔS, if the cluster has many RR Lyrae variables. Finally, high-resolution spectroscopy is especially useful for seeking the smallest variations in abundances, both in the mean and for specific elements. The world being what it is, this most useful method is also the most difficult. To complicate matters for all methods, one must always keep in mind that mixing will alter the primordial abundances of some elements in some, many, or perhaps all red giants within a given cluster. One must approach the question of primordial abundance variations with a clear idea of what effects might be due to mixing, and which, so far as we understand, cannot be caused by mixing.

ω Centauri. The most massive globular cluster in our Galaxy, ω Cen, is one of the most thoroughly studied in the quest for primordial abundance variations, and we use it here as a “case study”. The color distribution of its

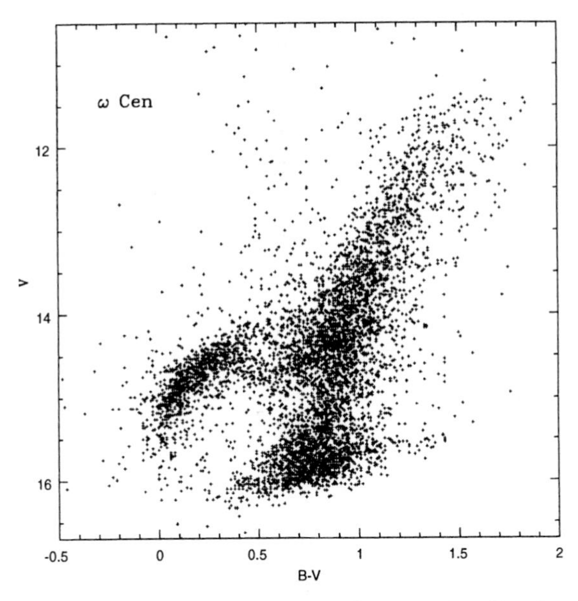

Fig. 96. The color–magnitude diagram of ω Cen, based on the photographic data of Woolley et al. (1961)

red giant branch stars is far wider than observational errors allow. One of the first excellent demonstrations of this was the photographic study by Woolley et al. (1961), with their results shown in Fig. 96. Photographic errors in $B-V$ are about 0.05 mag, far less than the observed width of the red giant branch. The spread was also seen in the infrared color–magnitude diagram by Persson et al. (1980). The implied spread in the heavy element abundances is several tenths dex.

A variant of the spread in color is that provided by Persson et al. (1980). Their infrared photometry included the broadband color $V-K$ as well as a measure of the strong CO bands near 2.3μ. They defined $R(V-K)$ to be a measure of the color of a red giant branch star relative to stars with similar dereddened M_K values in the very metal-poor cluster M92 and the metal-rich cluster M71:

$$R(V-K) = \frac{(V-K)_0 - (V-K)_{0,\mathrm{M92}}}{(V-K)_{0,\mathrm{M71}} - (V-K)_{0,\mathrm{M92}}} \,. \tag{66}$$

As defined, $R(V-K)$ increases as metallicity increases. They defined a similar $R(\mathrm{CO})$ index which increases as the CO band strengths increase. Figure 97 shows the results: as Fig. 96 showed, there is a large spread in the color, as measured by $R(V-K)$, of the red giant branch, ranging from values similar to those found in M92 giants to those in M71 giants, a spread of about 1.6 dex. Since the CNO elements do not contribute significantly to the electron density in red giant atmospheres, $R(V-K)$ is primarily a measure of electron donors,

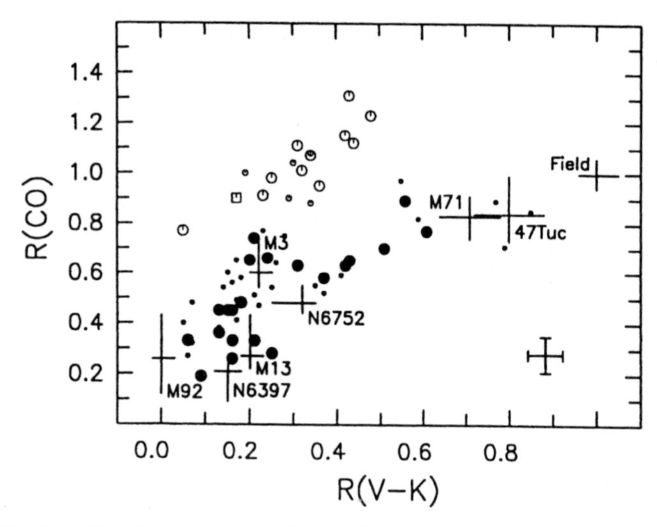

Fig. 97. The distribution of "normalized" $V-K$ colors, which are related to iron-peak abundances, and CO band strengths in red giants in ω Cen. The data are from Persson et al. (1980), and the figure is reproduced from Norris & Da Costa (1995). Large *open* and *filled symbols* refer to CO-strong and CO-weak stars, analyzed by Norris & Da Costa, while smaller symbols refer to the remaining stars in the Persson et al. sample

including iron peak and "α" elements. The distribution almost appears two-pronged, but interpretation of the figure is complicated by the likely presence of mixing, which alters the CO band strength in three ways. First, since the carbon abundance is lower than the oxygen abundance, at least prior to significant mixing, a low iron abundance and hence a low carbon abundance yield a weak CO band strength. Second, as mixing develops and the carbon abundances drop, the band strength weakens further. Finally, as mixing effects become strong, the decline in oxygen abundance may also weaken the band, but keep in mind the predictions in Fig. 86 where the carbon abundances may rise again. Further, the ^{13}C increase may also yield a stronger CO band. Rather than dwell on this figure, then, let us look instead at another moderate resolution metallicity indicator, ΔS. Freeman & Rodgers (1975) and Butler et al. (1978) showed that the RR Lyraes within ω Cen show a wide range in ΔS and, hence, in calcium abundances. Figure 98 shows the results of Butler et al. (1978), and one should recall that the RR Lyraes may not sample the full metallicity spread: the most metal-poor core helium-burning stars may be on the blue horizontal branch, while the most metal-rich ones may be on the red horizontal branch.

High-resolution spectroscopy of ω Cen giants has confirmed the large spread in metallicities (Cohen 1981; Mallia & Pagel 1981; Gratton 1982; François et al. 1988; Paltoglou & Norris 1989; Milone et al. 1992; Brown

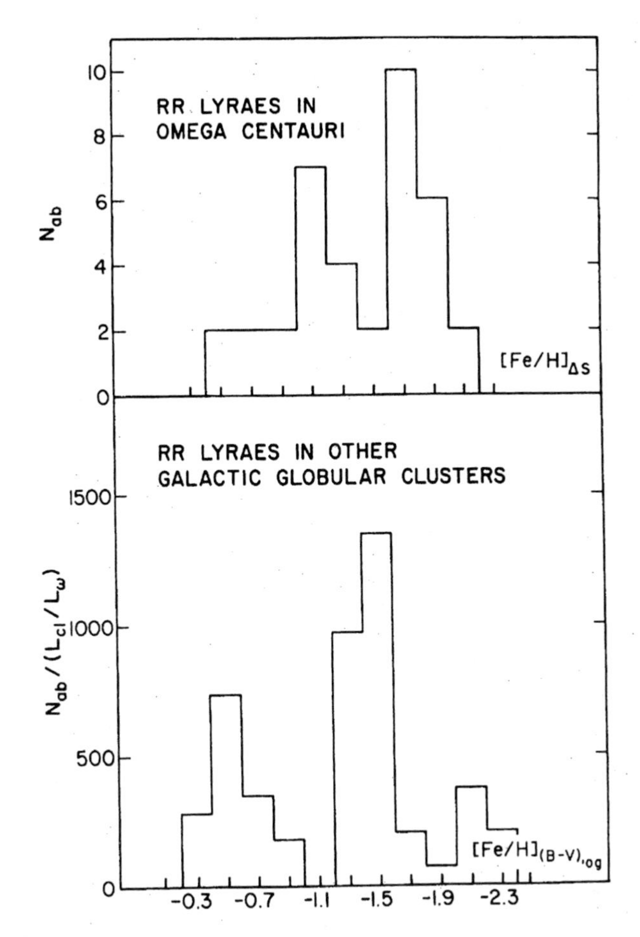

Fig. 98. The distribution of ΔS values for the RR Lyrae variables within ω Cen and in other Galactic globular clusters, taken from Butler et al. (1978)

& Wallerstein 1993; Vanture et al. 1994; Norris & Da Costa 1995; Norris et al. 1996; Smith et al. 1995; Suntzeff & Kraft 1996). The study by Norris & Da Costa (1995) deserves special notice, not only because of the large number of stars studied (40), but also because of the wide range of elements studied. They argued that their observations are consistent with a primordial variation, upon which has been superposed the already-discussed effects of mixing, with the latter somewhat stronger for the more metal-poor stars, as Shetrone (1996a) has also found. Norris & Da Costa (1995) noted further that the s–process elements were strongly enhanced in the more metal-rich stars, suggesting significant contributions from low- to intermediate-mass (1–3 $M_{\odot}$) asymptotic giant branch stars, as shown in Fig. 99. The situation in ω Cen is perhaps not so simple as primordial abundance variations within a single

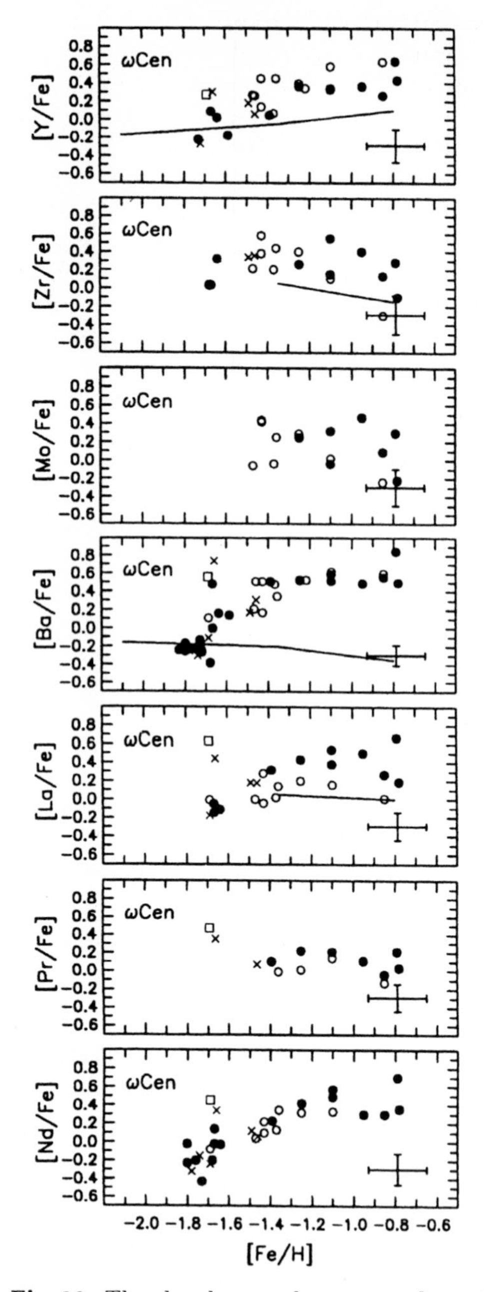

Fig. 99. The abundances of s–process elements relative to [Fe/H] in ω Cen, taken from Norris & Da Costa (1995). Symbols have the same meaning as in Fig. 97, while stars with uncertain CO strength are plotted as *crosses*. The *open square* is a CH star

star-forming protocluster, however. Norris et al. (1997) have combined radial velocity data with high internal precision [Ca/H] measures for almost 400 red giants and found that the dominant ($\approx$ 80%) metal-poor component is more dispersed spatially and is rotating more rapidly than the minority metal-rich component. This is not simply understood in terms of a dissipative origin, and may be more simply explained by a merger event, in which case one might expect abundance variations. If true, the merger was not one simply of two clusters with differing metallicity, and with a small spread in metallicity within each cluster. To explain the large s–process enhancements seen in the metal-rich stars requires star formation to have continued after the merger.

Other Clusters. We explore other clusters using the same general techniques as employed for ω Cen, beginning with the width of the red giant branches. While this is relatively insensitive to mixing effects and binaries, one must keep in mind that differences in reddening among stars in the cluster may lead us to erroneous conclusions. Suntzeff (1993) reviewed the best color–magnitude diagrams available at that time, and found that for the seven clusters with low reddening and sufficiently good photometry, σ([Fe/H]) was always less than 0.10 dex, and, in the case of 47 Tuc, less than 0.04 dex. No other clusters have yet been found that show wide giant branches such as ω Cen, although such claims have occasionally been made for M22, whose reddening is relatively high and probably variable.

Elements that are subject to photospheric variations due to mixing may be employed if one studies stars close to or on the main sequence, where mixing is completely unexpected. Briley et al. (1991) measured CN band strengths in stars in 47 Tuc from near the red giant branch tip down to about one magnitude below the main sequence turn-off, and as Fig. 100 shows, variations are seen at *all* luminosities. This is quite a surprise! As noted above, it is clear that heavier element abundance variations must be very small given the narrowness of the cluster's red giant branch. We are forced to admit that: (a) Briley et al. (1991) had the misfortune to study stars contaminated by binaries or the effects of binary star evolution; or (b) mixing occurs even during the main sequence phase, which is a great challenge to theory; or (c) the cluster's stars have some primordial abundance variations, due perhaps to one of two causes. First, the original protocluster gas cloud may simply not have been well mixed, and therefore polluted by mass loss from low- to intermediate-mass stars from outside the cloud. Second, the pollution may have occurred within the cloud as the protocluster gas was turned into stars. However, intermediate-mass stars have relatively long lives (few $\times 10^8$ years) and such an age spread would be revealed as a broadened main sequence turn-off, which is not seen. The conclusion, that the protocluster was not well mixed, still requires more detailed testing, and certainly we must keep in mind that mixing might actually occur in some main sequence stars for reasons that are not understood.

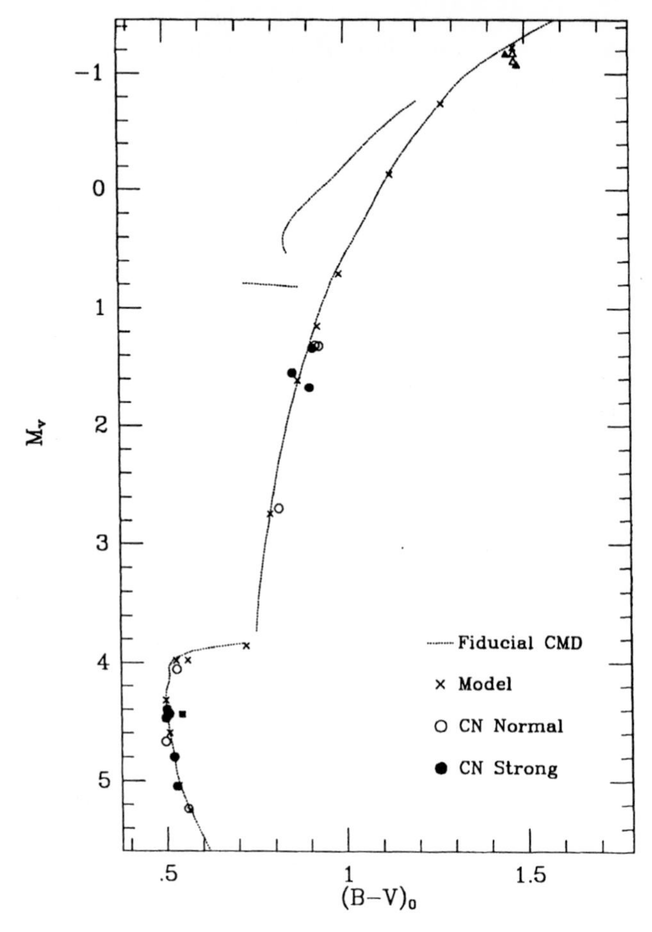

Fig. 100. CN band strengths for stars in 47 Tuc, taken from Briley et al. (1991)

High-resolution spectroscopy has revealed a number of additional surprises, all consistent with some degree of primordial abundance variations within clusters. We begin with evidence that at least some stars within the metal-poor globular cluster M92 have variations in [Fe/H]. Langer et al. (1998) discussed three red giants in the cluster with very similar V magnitudes and $B-V$ color indices, based on CCD photometry. The stars therefore have very similar gravities ($\approx$ 0.05 dex) and temperatures ($\approx$ 20 K), unless an unlikely binary companion has altered the brightness and color of one or more of the stars. This follows up on a suggestion by Langer et al. (1993) that XI-19 is about 0.08 dex more metal-rich than V-45 and XII-8, based on more limited observational data. The more extensive data presented by Langer et al. (1998) show clearly that [Fe/H] appears to be enhanced by

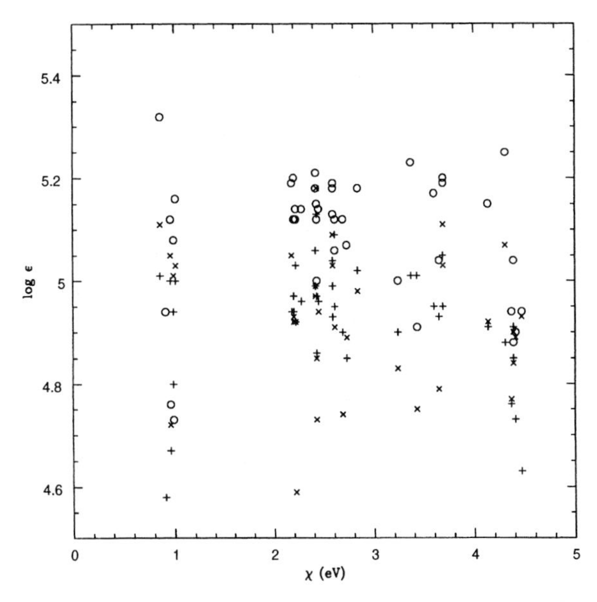

Fig. 101. Abundances derived from Fe I lines by Langer et al. (1998) for three stars in M92 with similar temperatures. *Plus signs* are for V-45, *crosses* for XII-8, and *circles* for XI-19. The lack of trends with excitation potential, χ, indicates the temperatures are well determined, while the offset of the results for XI-19 indicates it has an apparently higher iron abundance

about 0.18 dex in XI-19 relative to the other two stars, based on abundances of individual lines in all three stars plotted as functions of excitation potential and equivalent width (see Fig. 101). The only question, in view of the evidence for mixing of helium into red giant photospheres (Fig. 92), is whether these stars shows signs of different degrees of mixing, which could lead to variations in the derived [Fe/H] values, as discussed above, due to changes in the continuum opacity. Unfortunately, Langer et al. (1998) did not report abundances for the key elements including oxygen, sodium, magnesium, and aluminum. However, Sneden et al. (1991) found the [O/Fe] value for XI-19 (+0.29 dex) to be indistinguishable from those of V-45 (+0.31 dex) and XII-8 (+0.37 dex). It does not appear that we may resort to mixing to explain the difference in derived [Fe/H] value for XI-19. Clearly, further careful studies of this and other clusters would be worthwhile.

Variations in neutron capture elemental abundances may also exist. Sneden et al. (1997) have reported signs of an apparent "bimodality" in the abundances of barium and europium in the metal-poor globular cluster M15 (see Fig. 102). The [Ba/Fe] and [Eu/Fe] ratios are not very sensitive to changes in the atmospheric parameters of the program stars, and neither correlate with the mixing-sensitive and highly-variable [O/Fe] ratio. The abundance differences appear to be real. The ratio of barium to europium appears to be the

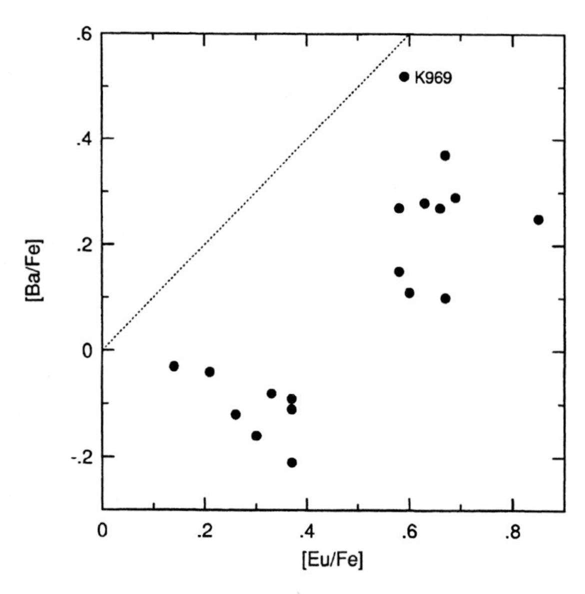

Fig. 102. [Ba/Fe] *vs.* [Eu/Fe] in M15 red giants, taken from Sneden et al. (1997). The *dotted line* indicates the slope if [Ba/Eu] is constant

same for stars in both groups, and consistent with nucleosynthesis involving some s–process and mostly r–process nucleosynthesis, but to the same degree in the two groups (except for K969). Differences in [Eu/Fe] do not correlate with the α elements silicon and calcium thought to be unaffected by mixing, but which are thought to be manufactured in Type II supernovae. M15 therefore presents us with an interesting new puzzle in terms of primordial abundance differences in the heavy neutron capture elements but not in the α elements. Perhaps different mass supernovae were involved prior to and during the formation of the cluster?

Finally, King et al. (1998) have used the same spectra obtained to study the lithium abundances in six subgiants in the metal-poor globular cluster M92 to measure the abundances of some α elements. The greatest surprises are the low mean abundance of magnesium, $\langle[\mathrm{Mg/Fe}]\rangle = -0.15$, and the high mean abundance of sodium, $\langle[\mathrm{Na/Fe}]\rangle = +0.60$. These are very different from unevolved metal-poor field stars, and more similar to more luminous cluster red giants that have experienced significant mixing. This is reminiscent of the CN variations seen in Fig. 100. Do main sequence stars and subgiants, at least in some globular clusters, experience large degrees of mixing? Or are these differences primordial? One additional problem is that the [O/Fe] results for M92 giants obtained by Sneden et al. (1991) show signs of normal [O/Fe] ratios ($\approx$ 0.3 dex) as well as lower values, presumably caused by mixing. If mixing is the explanation for the unusual abundance ratios seen among the M92 subgiants, why is it not seen in all of the red giants? And why are the lithium abundances of the subgiants relatively normal, suggesting little or no mixing? If mixing is not the cause of the unusual sodium and magnesium abundances, yet oxygen abundances are normal in at least some of the red giants, we have an interesting task ahead of us in terms of trying to explain the origin of the primordial abundance pattern. Again, further work, especially new spectroscopic observations, of both red giant and subgiant abundance patterns in this cluster should prove very interesting.

8 Relative Ages of Globular Clusters

The ages of globular clusters are crucial data in our studies of the Galaxy and indeed the Universe because they are one of our best chronometers. Their relative ages reveal the early history of star formation in the Galaxy. How rapidly were the heavy elements synthesized? Did this proceed more rapidly in some parts of the Galaxy than elsewhere? How rapidly did the disk form relative to the halo? Are there populations of clusters with different ages and, hence, origins? What does all this mean for the early history of the Galaxy? The absolute ages, of course, reveal the actual time of the beginnings of star formation and provide a lower limit to the age of the Universe.

The physics and observations employed in estimating relative and absolute ages are the same, of course. However, there are potential systematic effects which may significantly compromise the precision of absolute ages that do not as readily alter the relative ages; hence the focus here on the relative ages. We summarize the procedures and results from two basic methods for relative age determinations, both for globular clusters and field stars. We also compare these results to the [α/Fe] *vs.* [Fe/H] results to check the value of that chronometer, and explore again the utility of the "second parameter" as a relative age indicator. We conclude with a look at one of the key remaining frontiers: the ages of the clusters as a function of Galactocentric distance.

8.1 Relative Ages from Relative Turn-Off Luminosities

It is worth recalling the lessons of Sect. 1 and especially Fig. 3. The main sequence turn-off is the primary age indicator for lower-mass stars. The post-main sequence stars "funnel" into a more or less common red giant branch whose luminosity is not particularly sensitive to mass and, hence, age. The horizontal branch has some age sensitivity but may be easily overwhelmed by metallicity and "second parameter" effects unrelated to age. As ages increase, the turn-offs become fainter and cooler. The essence of the two relative age estimators is that one relies on relative luminosities while the other relies on relative temperatures. Turn-off luminosities also depend on chemical composition (see Fig. 7). One may generally assume that the ratio of the heavy elements to, say, iron, is well-behaved, so that knowledge of [Fe/H] alone leads to Z, the heavy element mass fraction. (We have seen that this is not always true, however.) Under the assumption that the helium abundances are related directly to the heavy element abundances, [Fe/H], in principle, provides us with all the necessary chemical abundance information we need to compute relative ages. Absolute ages still require precise knowledge of the helium mass fraction Y.

Aside from the effects of chemical abundances, it is essential to keep in mind what uncertainties enter into the calculation of Fig. 3, and also into the comparison between the theoretical parameters (luminosity and temperature) and the observed quantities (magnitude and color index). The luminosities arise in the stellar interiors, and so are not especially sensitive to convection (which in turn-off stars is thought to occur only in the outer envelope) or the two outer boundary conditions involved in solving the differential equations given by (1), (2), (3), and (4)/(5). Turning an observed main sequence turn-off magnitude into a cluster *absolute* age estimate therefore requires the bolometric correction, which may be a function of metallicity and temperature, the cluster's true distance, and, of course, detailed knowledge of opacity, energy generation rate, entropy, and second adiabatic exponent. On the other hand, a *relative* age, especially at similar chemical composition, requires primarily only a relative distance estimate since almost everything else remains the same. Even relative ages at different metallicities are primarily sensitive to relative cluster distances. Thus if we have a reliable method to estimate relative cluster distances we may estimate relative cluster ages with some hope of accuracy.

Figure 103 shows the basics of the method in practice. One measures the V magnitude difference between the mean of the RR Lyrae variables on the horizontal branch and the main sequence turn-off, a quantity generally called $\Delta V_{\rm TO}^{\rm HB}$. It is used widely because the difference in *apparent* magnitudes between the horizontal branch and main sequence turn-off is also the difference in *absolute* magnitudes. Further, and this is extremely important for Galaxy-wide studies, it is *independent of reddening.* All that is needed is a reliable relation between M_V(RR) and [Fe/H] and then measurement of $\Delta V_{\rm TO}^{\rm HB}$ suffices for an age estimate.

Considerable work has been dedicated to establishing the relation between M_V(RR) and [Fe/H], as discussed in Sect. 4. The evidence supports a slope of between 0.15 and 0.20 mag/dex. The zero point, which sets the absolute ages scale, remains a serious problem, as discussed in Sect. 9 that follows. Fortunately, Chaboyer et al. (1996) discussed the many uncertainties that affect the transformation of a turn-off absolute visual magnitude into an age estimate, and also utilized a wide range of M_V(RR) *vs.* [Fe/H] relations to derive ages. In Fig. 104 we show their results employing $M_V({\rm RR}) = 0.15[{\rm Fe/H}] + 0.98$, quite close to (40). As long as relatively shallow values for the slope are used, Chaboyer et al. (1996) concluded that:

- The most metal-poor clusters show no sign of differing ages.
- The more metal-rich clusters have a range in ages.

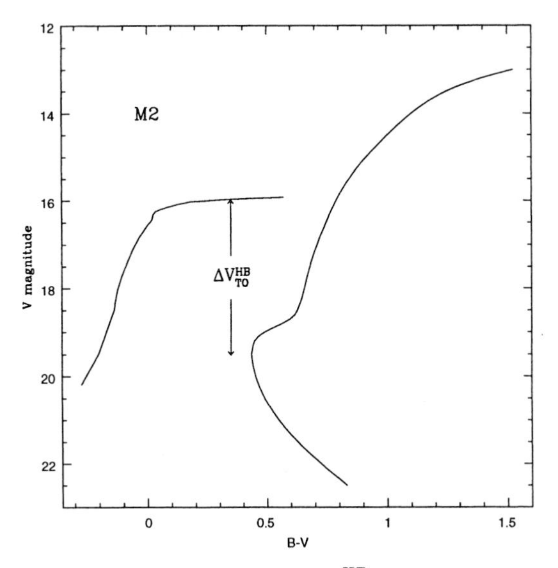

Fig. 103. The definition of $\Delta V_{\mathrm{TO}}^{\mathrm{HB}}$: the difference in V magnitude between the mean level of the horizontal branch as measured using the RR Lyrae variables and that of the main sequence turn-off

- The most metal-poor clusters may be somewhat older than the more metal-rich ones.

We noted at the end of Sect. 4 that at equal metallicities the Oosterhoff II RR Lyraes and horizontal branch luminosities could be about 0.16 mag brighter than those of Oosterhoff I clusters. While based on the comparison of only a single pair of clusters, M2 and M3, it is instructive to investigate the effect on the ages if this applies to all the Oosterhoff II clusters. This assumes that the studies of field RR Lyrae would have somehow selected the RR Lyraes that belong to the population of Oosterhoff I clusters, even at the metal-poor end. The total number of metal-poor field RR Lyraes studied using the Baade–Wesselink method is rather small, and hence this assumption might possibly be correct. In Fig. 104 we therefore show by filled circles the consequences of increasing the zero point of the M_V(RR) *vs.* [Fe/H] relation by 0.16 mag. The ages of these metal-poor clusters drop by several billion years, illustrating the importance of the zero point in deriving relative ages. It is worth re-assessing the idea that the most metal-poor clusters are older than the intermediate-metallicity clusters. Figure 104 suggests that the ages may be comparable, if the Oosterhoff II clusters' horizontal branches are brighter than expected. The situation is unfortunately very complicated and it may be necessary to thoroughly re-analyze the relative ages of the Oosterhoff I and II clusters separatly. At the moment, it appears that: (a) there is little variation in the ages of the most metal-poor (Oosterhoff II) clusters;

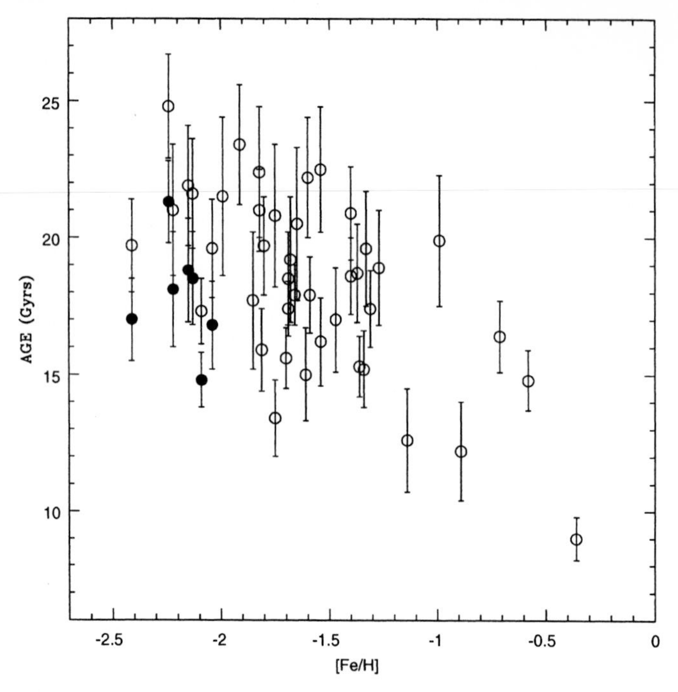

Fig. 104. Ages for globular clusters derived by Chaboyer et al. (1996) from $\Delta V_{\rm TO}^{\rm HB}$ and $M_V(\rm RR) = 0.15[\rm Fe/H] + 0.98$ (*open circles*). *Filled circles* represent the approximate effects of slightly increasing the luminosities of the Oosterhoff II clusters' RR Lyrae variables

(b) the intermediate-metallicity (Oosterhoff I) clusters may be younger than the Oosterhoff II clusters due to differences in origins; and (c) there are age differences among the intermediate-age clusters, but it is difficult to ascertain how much is due to the intermingling of clusters formed outside the Galaxy and accreted by it with clusters that were part of the Galaxy's formation.

A similar analysis has been conducted by Richer et al. (1996). They employed a very similar RR Lyrae luminosity scale, $M_V(\rm RR) = 0.15[\rm Fe/H] + 1.0$, and, of course, similar cluster color–magnitude diagram data. Not surprisingly, their results agree well with those of Chaboyer et al. (1996), and are shown in Fig. 105. There is little age variation among the most metal-poor clusters, a considerable age spread at intermediate metallicities, and a sign of an age–metallicity relation. But again, if the most metal-poor clusters' RR Lyraes are brighter than the adopted $M_V(\rm RR)$-[Fe/H] relation, the derived ages will drop.

Buonanno et al. (1998a) have introduced a modification of the above method for determining ages, but we defer discussion until we take up the issue of $[\alpha/\rm Fe]$ ratios *vs.* ages.

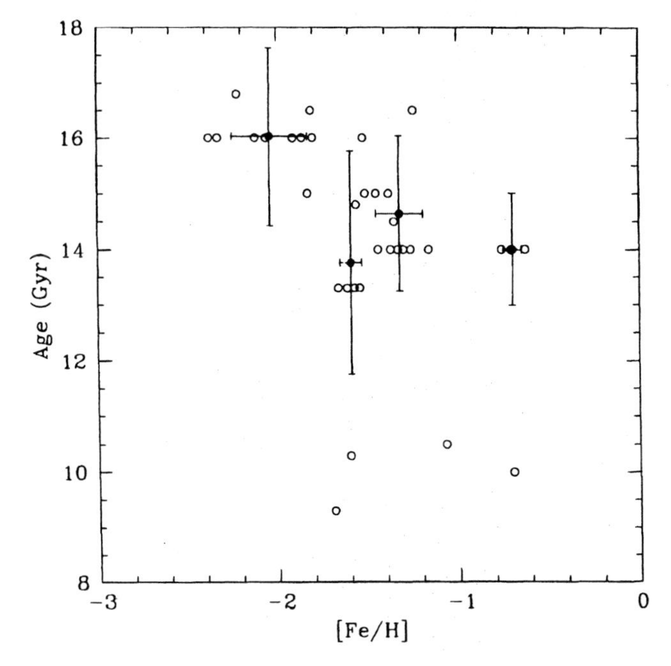

Fig. 105. Ages for globular clusters derived by Richer et al. (1996) from $\Delta V_{\mathrm{TO}}^{\mathrm{HB}}$ and $M_V(\mathrm{RR}) = 0.15[\mathrm{Fe/H}] + 1.0$. Mean points for each metallicity group are plotted as *filled circles*

8.2 The Relative Ages of Field Stars

The range of masses of stars from the main sequence turn-off to the tip of the red giant branch is small in a coeval population. Considering Fig. 3, the decline in T_{eff} at roughly constant luminosity of subgiants means that surface gravities are dropping as T_{eff} declines. The Strömgren photometric system was designed to be sensitive to temperature (the $b-y$ color index) as well as metallicity (the m_1 index) and gravity (the c_1 index). The latter works because at temperatures lower than about 9500 K, the hydrogen is mostly neutral, and the Saha equation says that the electron pressure, which is related to surface gravity, is related to the degree of ionization. Hence the strength of the Balmer jump, the discontinuity in bound-free continuum opacity at $\lambda 3640$, is sensitive to surface gravity. Figure 106 illustrates the principle, using model isochrones and synthetic colors computed by VandenBerg & Bell (1985). If a field star's photometry enables its metallicity to be well determined, its c_1 index compared with its $b-y$ color index may be used to estimate its age, at least for the hotter stars. The method has been employed by Schuster & Nissen (1989b), Nissen & Schuster (1991), and Marquez & Schuster (1994). Results from the latter study are shown in Fig. 107. In agreement with the above results for globular clusters, the field stars' data indicate that the most

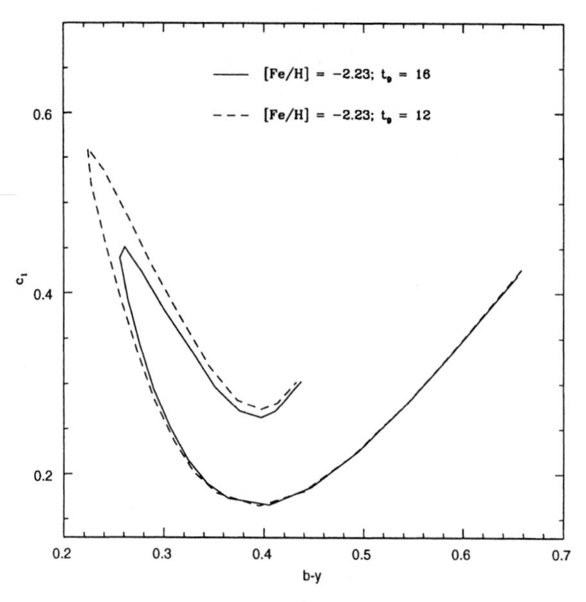

Fig. 106. The Strömgren photometry system $b - y$ and c_1 indices computed by VandenBerg & Bell (1985) for [Fe/H] =−2.23 and ages of 12 and 16 billion years

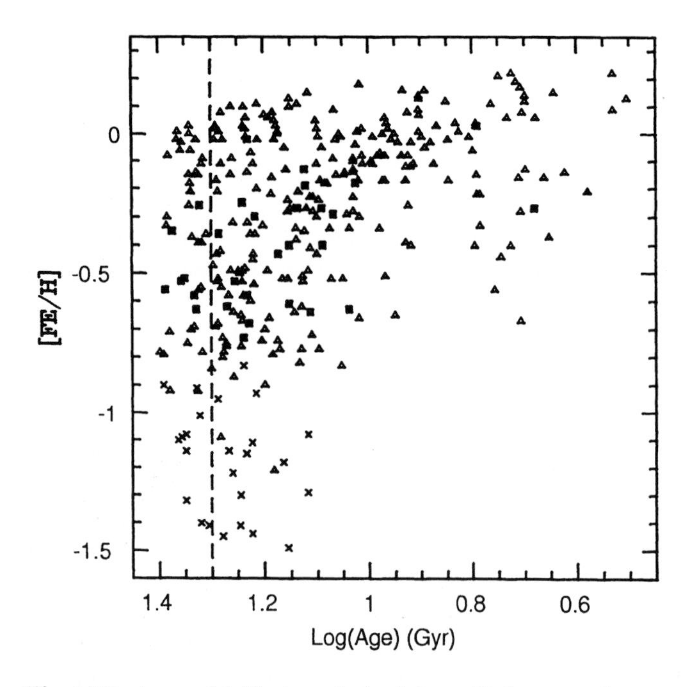

Fig. 107. Ages of field stars derived from Strömgren photometry and model isochrones, taken from Marquez & Schuster (1994)

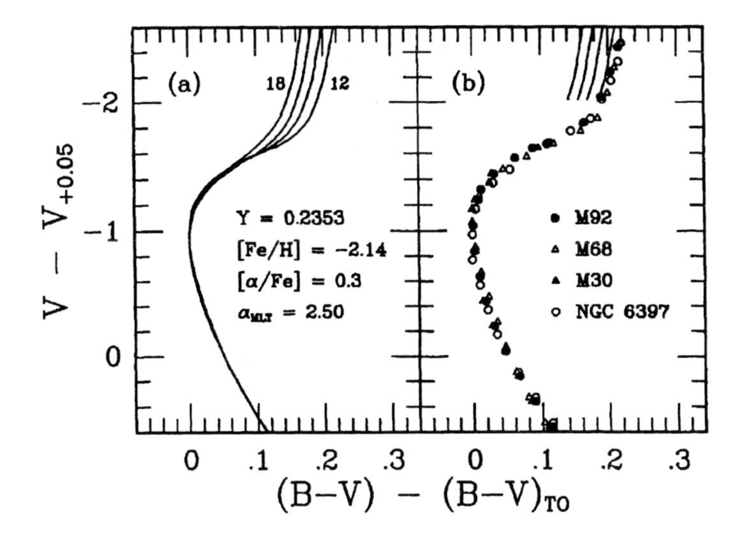

Fig. 108. Left: Metal-poor isochrones with different ages shifted in color to line up turn-off colors, and shifted in magnitude to match at a color 0.05 mag redder than the turn-off. The age sensitivity is revealed in the different colors of the red giant branch. Right: Mean loci for four metal-poor clusters shifted in such a fashion. No age differences are seen. Figure taken from Stetson et al. (1996)

metal-poor stars all have the same ages, at least to within the precision of the method, whereas the intermediate-metallicity stars show a range in ages. There is no sign of an age–metallicity relationship, but the sensitivity of the method to differences in ages declines as the metallicity rises since the atomic line absorption strengthens to the point that measurement of the Balmer jump becomes more difficult. Mixing stars with different origins, as in the case of clusters, remains a concern as well (as seen in Fig. 107).

8.3 Relative Cluster Ages Derived from Color Differences

The independence of $\Delta V_{\mathrm{TO}}^{\mathrm{HB}}$ in terms of reddening makes it a powerful tool, but it has a drawback. The estimation of the magnitude of the turn-off is difficult, and often introduces an uncertainty of ± 0.1 mag into the turn-off luminosity and almost a 10% uncertainty into the derived age. Some of this uncertainty is alleviated by the revision to the method introduced by Buonanno et al. (1998a), and discussed later. However, an alternative approach may often be used, and which is also independent of reddening. The basic idea was introduced by VandenBerg et al. (1990), and is illustrated in the left side of Fig. 108. Model isochrones, or color–magnitude diagrams of clusters, all with similar metallicity, are matched by first shifting the colors so that all have the same main sequence turn-off color. A vertical shift in magnitudes is then made until the main sequences match at a color 0.05 mag *redder* than the main sequence turn-off. The main sequences now more or less overlap, certainly near the turn-off, and the age sensitivity is revealed in the colors of the respective red giant branches. In essence, the method is measuring the difference in color between the main sequence turn-off, which is sensitive to age, and the color of the red giant branch, which is sensitive to metallicity but not age (see Figs. 3, 7, and 9), but transfers the color sensitivity from the turn-off to the giant branch. The method is most useful in comparing relative ages of clusters with similar metallicities. This also requires similar [α/Fe] since Fig. 12 indicates that the turn-off *vs.* red giant branch color difference is sensitive to that ratio. Because colors are affected primarily by surface temperature, and at constant luminosity are affected primarily by stellar radius, any use of colors, relative or absolute, is vulnerable to the two outer boundary conditions employed in the solutions of the equations of stellar structure. Because the outer layers of low mass stars such as are found on globular cluster main sequences and red giant branches are convective, the turn-off *vs.* giant branch color differences are sensitive to the model of convection employed (see Fig. 11). In summary, the method is not advised for the estimation of absolute ages, or even relative ages over a wide range of metallicities. But it is a powerful tool for the estimation of relative ages since it has somewhat greater sensitivity than $\Delta V_{\rm TO}^{\rm HB}$.

Figure 108, taken from Stetson et al. (1996) confirms with improved precision the results from $\Delta V_{\rm TO}^{\rm HB}$ that the most metal-poor clusters have very similar ages, at least to within a precision of about one billion years, including M68 (NGC 4590), M92 (NGC 6341), NGC 6397, and M30 (NGC 7099). Similar ages have also been derived for NGC 5053 by Rey et al. (1998), and for NGC 2419 by Harris et al. (1997). At least at these lowest metallicities, there is no evidence for *any* differences in ages among the globular clusters. (Recall the discussion of the second parameter in Sect. 2. We now see that redder than average horizontal branch colors for NGC 2419, NGC 5053, and NGC 5466 are probably *not* caused by younger-than-average ages.)

At higher metallicities, age differences become detectable. Fig. 109 shows the *differences* derived by VandenBerg et al. (1990) from CMD data then available. The metallicities employed in the figure are taken from Zinn (1985) and the caveats about their accuracy discussed previously (Sect. 5) should be kept in mind. Note that NGC 6254, classified as "metal-rich" by VandenBerg et al. should be classified as an intermediate-metallicity cluster. The figure was generated *assuming that each of the three metallicity groups have the same mean age.* This may or may not be true, as we have seen. But the figure does illustrate some possible age differences among the more metal-rich clusters. The extrema are represented by NGC 288 (+2.5 Gyrs) and Palomar 12 (−3.5) Gyrs. Based on this limited set of data, it does appear that while the most metal-poor clusters have comparable ages, age spreads begin to show up among the more metal-rich systems, as noted earlier.

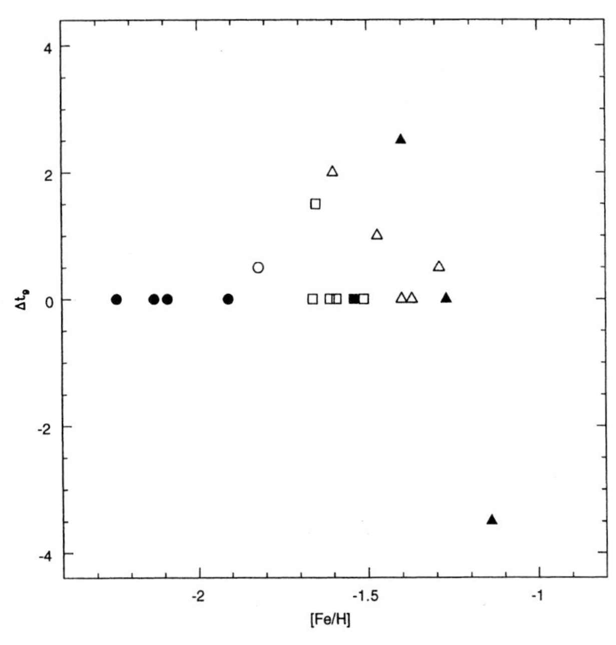

Fig. 109. The relative ages of globular clusters divided into three metallicity regimes by VandenBerg et al. (1990). *Circles* are metal-poor; *squares* are intermediate metallicity; *triangles* are relatively metal-rich. *Filled symbols* represent clusters with "high quality" data while *open symbols* represent clusters with "good quality" data

8.4 Cluster Ages vs. [O/Fe] and [α/Fe]

We have seen that chemistry may also be a useful chronometer, if the transition from enhanced values of [α/Fe] to values nearer solar seen in Figs. 68 and 69 is caused by the transition from a dominance by Type II supernovae to an increasing contribution by Type Ia supernova. (We continue to assume that the transition is not caused by the metallicity effects on the pre-supernova system, as suggested by Hachisu et al. 1996, 1999 and by Umeda et al. 1998.) It is worth making the direct comparison as well between [α/Fe] and cluster ages, following Carney (1996). We have added to that discussion newer results for Ruprecht 106 and Palomar 12 (Brown et al. 1997), NGC 7006 (Kraft et al. 1998), and NGC 3201 (Gonzalez & Wallerstein 1998). Ages were taken from Chaboyer et al. (1996), but *unlike* Fig. 104, we have used a brighter RR Lyrae luminosity *vs.* metallicity calibration, $M_V(\mathrm{RR}) = 0.15[\mathrm{Fe/H}] + 0.725$. This results in considerably younger ages than those in Fig. 104 and also a slightly smaller spread in ages. The choice of distance scales is discussed in the following section. We have used different symbols for clusters whose horizontal branches suggest they are "old halo" or "young halo" according to Da Costa & Armandroff (1995), whose kinematics indicate they belong to a "disk" population, and the special cases of Ruprecht 106 and Palomar 12,

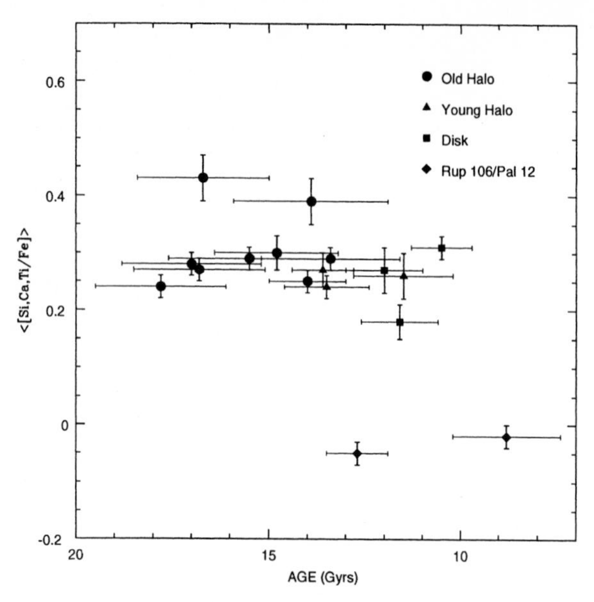

Fig. 110. The relative ages of globular clusters derived from $M_V(\mathrm{RR}) = 0.15[\mathrm{Fe/H}] + 0.725$ (Chaboyer et al. 1996) compared with $\langle[\alpha/\mathrm{Fe}]\rangle$ obtained using silicon, calcium, and titanium

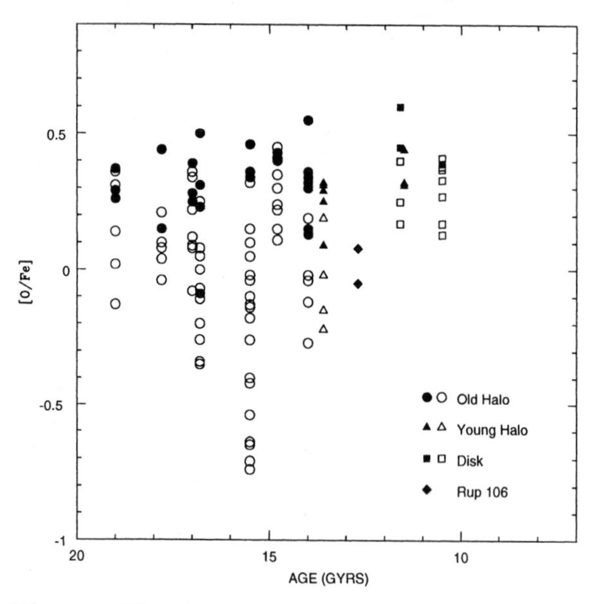

Fig. 111. The relative ages of globular clusters derived from $M_V(\mathrm{RR}) = 0.15[\mathrm{Fe/H}] + 0.725$ (Chaboyer et al. 1996) compared with [O/Fe]. *Filled symbols* represent stars whose surface abundances have probably not been altered significantly by internal mixing

Table 3. Cluster ages *vs.* [α/Fe]. t_9 values are from Chaboyer et al. (1996), assuming $M_V(\mathrm{RR}) = 0.15[\mathrm{Fe/H}] + 0.725$. The Δt_9 values are from Buonanno et al. (1998a)

Cluster	$[\mathrm{Fe/H}]_{\mathrm{Zinn}}$	[α/Fe]	σ([α/Fe])	t_9	σ	Δt_9	σ
			Old Halo				
NGC 2298	−1.81	0.29	0.02	13.4	1.8	+2.0	2.0
NGC 5904	−1.40	0.25	0.02	14.0	1.0	−0.1	1.6
NGC 6121	−1.28	0.30	0.03	14.8	1.6	+0.9	0.7
NGC 6205	−1.65	0.29	0.02	15.5	2.1	−0.7	2.0
NGC 6254	−1.60	0.27	0.02	16.8	1.7	+0.2	2.0
NGC 6397	−1.91	0.24	0.02	17.8	1.7	+1.4	2.0
NGC 6723	−1.09	0.39	0.04	13.9	2.0	....	....
NGC 6752	−1.54	0.28	0.02	17.0	1.8	−0.2	2.0
NGC 7078	−2.15	0.43	0.04	16.7	1.7	+0.7	1.5
			Young Halo				
NGC 3201	−1.56	0.26	0.04	11.5	1.3	−0.5	1.7
NGC 5272	−1.66	0.27	0.03	13.6	0.8	−0.7	1.6
NGC 7006	−1.59	0.24	0.02	13.5	1.1	....	....
			Disk				
NGC 104	−0.71	0.18	0.03	11.6	1.0	−0.3	1.6
NGC 6352	−0.63	0.27	0.04	12.0	1.0	+0.3	1.6
NGC 6838	−0.58	0.31	0.02	10.5	0.8	−0.4	1.8
			Others				
Rup 106	−1.61	−0.05	0.02	12.7	0.8	....	....
Pal 12	−1.13	−0.02	0.02	8.8	1.4	....	....

discussed previously as having unusual [α/Fe] ratios. The results for the mean abundances of silicon, calcium and titanium are shown in Fig. 110 while those for oxygen are shown in Fig. 111. In Fig. 111, filled symbols refer to stars with [Na/Fe] < 0, and presumably little evidence of internal mixing at their surfaces (see Fig. 87). Table 3 lists the data for Fig. 110 and the clusters that are used in Fig. 111. (Detailed [O/Fe] for individual cluster stars may be obtained from the author.)

Figures 110 and 111 indicate that with the exception of Ruprecht 106 and Palomar 12, all the clusters have similar [α/Fe] and [O/Fe] ratios, despite the fairly wide range in ages. The uncertainties in the age estimates make

Table 4. Mean differences in age and abundances derived from Chaboyer et al. (1996) and $M_V(\mathrm{RR}) = 0.15[\mathrm{Fe/H}] + 0.725$

Class	Number	$\langle t_9 \rangle$	σ	σ_μ	$\langle[\mathrm{Fe/H}]\rangle$	$\langle[\alpha/\mathrm{Fe}]\rangle$	σ	σ_μ
Old Halo	9	15.5	1.6	0.5	−1.60	0.30	0.06	0.02
Young Halo	3	12.9	1.2	0.7	−1.60	0.26	0.02	0.01
Disk	3	11.4	0.8	0.4	−0.64	0.25	0.07	0.04

the differences appear to be larger than they are in reality, of course. Nonetheless, age differences of several Gyrs appear to exist, yet $[\alpha/\mathrm{Fe}]$ values are identical among clusters with apparently different ages, whether derived from the M_V(RR) *vs.* [Fe/H] relation and turn-off luminosities, or generally from horizontal branch morphology. Table 4 summarizes the means according to the three broad categories in Table 3. We may draw two interesting conclusions from Table 4, assuming that the ages are approximately correct.

- The three broad classes of clusters appear to span a range of almost 4 Gyrs.
- Their $\langle[\alpha/\mathrm{Fe}]\rangle$ values are essentially identical.

One of the following three ideas may explain these observations.

- Either the timescale for SNe Ia is near the long end of the models, about 3 Gyrs, or
- SNe Ia do not appear until $[\mathrm{Fe/H}] \approx -0.7$, or
- some of the general classes of globular clusters have not interacted chemically, implying independent origins.

We have already commented that the low metallicity of Ruprecht 106 (and of some field stars) make the second option somewhat less likely. Nonetheless, Ruprecht 106 and Palomar 12 complicate simple interpretations. If Palomar 12 is as much younger than the other clusters as it seems, perhaps it does argue for some validity to the SNe Ia enrichment model, but the timescale may be very long indeed. It is also prudent to recall that Richer et al. (1996) assign Ruprecht 106 a much younger age than do Chaboyer et al. (1996), one closer to that of Palomar 12.

Some Important Caveats. The above conclusions depend strongly on the relative ages obtained for the globular clusters.

One minor concern is the issue discussed previously regarding the difference between Oosterhof I and Oosterhoff II horizontal branch luminosities: a linear relation between M_V(RR) and [Fe/H] may be inappropriate over the full range of [Fe/H]. This is probably not a major concern here since, alas,

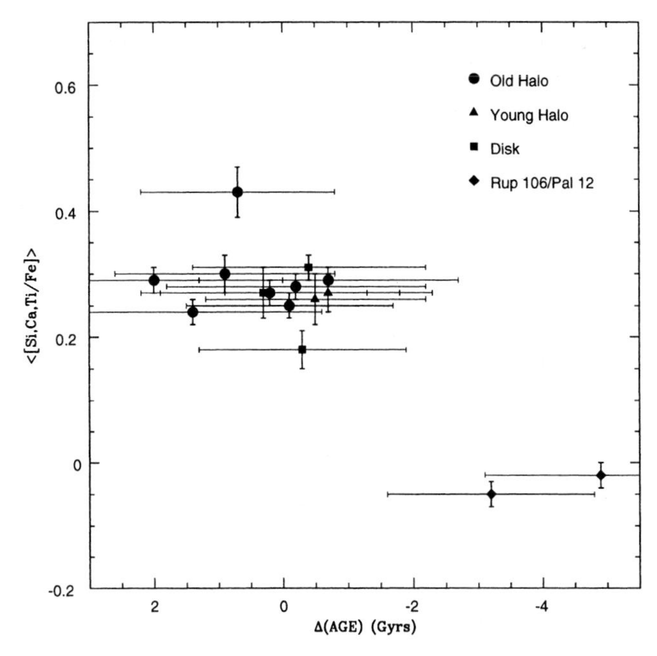

Fig. 112. The difference in age of globular clusters derived by Buonanno et al. (1998a) compared with $\langle[\alpha/\mathrm{Fe}]\rangle$ obtained using silicon, calcium, and titanium

too few Oosterhoff II clusters have well-determined [α/Fe] and [O/Fe] abundance ratios. Only M15 (NGC 7078) and, possibly, NGC 6397 above would be affected. Removing them from the results given in Table 4 does not change the results significantly.

A more serious concern follows from the new relative age estimations from Buonanno et al. (1998a). They have modified the $\Delta V_{\mathrm{TO}}^{\mathrm{HB}}$ technique so that a more readily measurable turn-off parameter is employed. The turn-off itself is nearly vertical for old clusters, and small uncertainties in $M_V(\mathrm{TO})$ result in relatively large uncertainties in age. Buonanno et al. (1998a) have cleverly exploited the reddening insensitivity of $\Delta V_{\mathrm{TO}}^{\mathrm{HB}}$ but avoided the turn-off point itself. Instead, they use the magnitude of the main sequence 0.05 mag redward of the turn-off color. This still provides good age sensitivity. Unfortunately, as they discuss, the relative ages they derive differ somewhat from those obtained by Richer et al. (1996) and quite seriously for clusters with blue horizontal branches from those derived by Chaboyer et al. (1996). The latter difference illustrates how difficult it may be to estimate the horizontal branch mean magnitude when few stars populate the instability strip. In the final two columns of Table 3 we list the age differences for clusters obtained by Buonanno et al. (1998a), and in Fig. 112 we repeat Fig. 110 but use their age differences rather than those of Chaboyer et al. (1996). In this case we reach quite different conclusions than we did above.

- There is a small range in ages among clusters, with at most only modest differences between the "old halo" and "young halo" clusters (perhaps only 10^9 years).
- As before, of course, the $\langle[\alpha/\mathrm{Fe}]\rangle$ values are nearly identical, except for Ruprecht 106 and Palomar 12.
- The timescale for SNe Ia may be as small as 10^9 years, and the unusual abundances of Ruprecht 106 and Palomar 12 may be explained by their much younger ages.

There is not enough information in Table 3 to confirm or reject the reality of the age differences between the Oosterhoff I and II clusters since there are only six of the former (NGC 3201, NGC 5272, NGC 5904, NGC 6121, NGC 6723, and NGC 7006) and only one of the latter (NGC 7078). The age differences are, however, *consistent* with NGC 7078 being older than the other clusters.

The true relative ages of the globular clusters remains a vital but as yet not satisfactorily resolved problem.

8.5 The Second Parameter vs. Age

In the previous section discussing relative ages of clusters, we again raised the issue of the "second parameter", although under the "old halo" and "young halo" characterizations, finding that the relative ages derived by Chaboyer et al. (1996) suggested an age difference of over 2 Gyrs while the difference is 1 Gyrs according to Buonanno et al. (1998a). The importance of age as a possible explanation for the second parameter effect extends beyond the simple existence of a range in globular cluster ages. Perhaps more important in our understanding of the formation of the Galaxy is that the second parameter does have a Galactocentric dependence, as first pointed out by Searle & Zinn (1978), and illustrated in Figs. 21 and 24. If age is the cause of this "global" second parameter, then the outer halo clusters are, on average, younger than those closer to the Galactic center.

In Sect. 3 alternative causes of the second parameter were discussed, including chemical composition, cluster density, and stellar rotational velocities. These effects may be responsible for the wide range in envelope/total masses of horizontal branch stars within individual clusters, while age is very unlikely to be the cause, given the very tight main sequence turn-offs observed in well-studied clusters. Nonetheless, it is relatively easy to imagine a difference in star formation timescales from the dense central regions of the Galaxy to its lower density outer regions and much harder to understand why there would be global variations in composition, cluster densities, or rotational velocities. Indeed, Fig. 24 shows no such global trends in these properties.

Rather than review the entire literature on the second parameter effect, let us concentrate on a carefully-chosen pair of globular clusters with very similar metallicities and other properties, and review the tests of the age hypothesis by

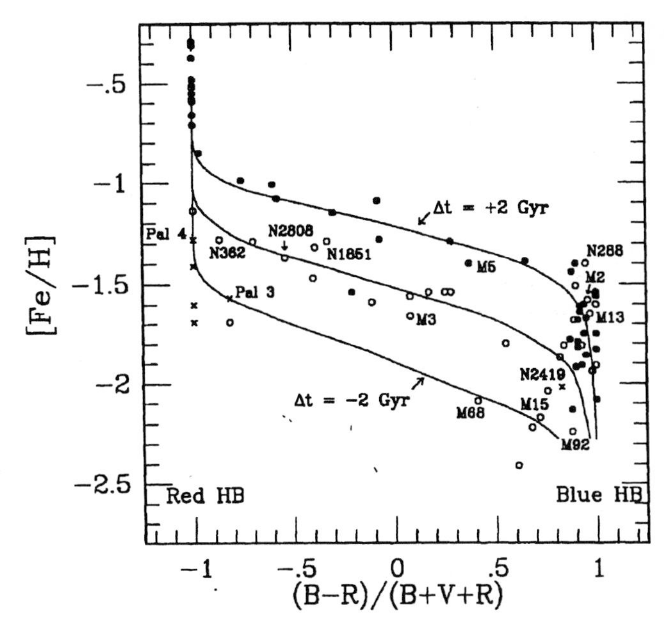

Fig. 113. The horizontal branch morphologies of globular clusters compared to [Fe/H] as a function of distance from the Galactic center. *Dots* have $R_{GC} < 8$ kpc; *circles* have $8 < R_{GC} < 40$ kpc, and *crosses* have $R_{GC} > 40$ kpc. The figure is taken from Stetson et al. (1996), using theoretical isochrones from Lee et al. (1994)

comparing relative ages. There are a number of good candidate pairs. Johnson & Bolte (1998) undertook such a comparison for M3 (NGC 5272) and M13 (NGC 6205). Their metallicities and [α/Fe] values are quite similar: [Fe/H] = −1.34 and −1.39 according to Carretta & Gratton (1997), and [Fe/H] = −1.46 and −1.49; [α/Fe] = +0.27 and +0.30, according to Kraft et al. (1993, 1995). Nonetheless, this pair has proven to be difficult. The data and the analyses of the type carried out by VandenBerg et al. (1990) could not be reconciled with a simple age difference, and Johnson & Bolte (1998) suggested a difference, instead, in helium abundances, with $\Delta Y \approx 0.05$. This is a very large difference, and hard to understand in terms of simple nucleosynthesis and chemical enrichment, where we expect helium abundances to scale with heavy element abundances. New data and a re-analysis for this pair are encouraged, but even if helium differences prove to be the answer here, it seems unlikely that helium can explain the second parameter effect in general, as Lee et al. (1994) discussed in terms of its effects on the pulsational period of RR Lyraes. (M13's two RR Lyraes are probably evolved and so cannot be used readily in such a comparison with M3.)

The more commonly used second parameter pair is NGC 288/NGC 362, which were discussed in Sect. 2. Figure 113 shows why these two clusters are so interesting: despite similar metallicities, their horizontal branch morphologies differ greatly. A very wide range of age differences have been derived: 2.3 Gyrs (Chaboyer et al. 1996); 2.5 Gyrs (Richer et al. 1996); 0 Gyrs (Salaris & Weiss 1997); 3.5 Gyrs (Sandquist et al. 1996); 0 Gyrs (Stetson et al. 1996); and 2 Gyrs (Sarajedini et al. 1997). Since we have already discussed some of the prior results, let us concentrate on the latter two since they demonstrate the problems involved in determining precise age differences. If age is the second parameter, then NGC 288, with its much bluer horizontal branch, should be several Gyrs older than NGC 362. Stetson et al. (1996) evaluated the relative ages using the two general methods discussed already, but in a rather novel manner. They were forced into this for the simple reason that since the two clusters represent *extremes* of the second parameter phenomenon, NGC 288 has an almost entirely blue horizontal branch, while that of NGC 362 is almost entirely red, with no RR Lyrae variables in either cluster. This makes use of $\Delta V_{\rm TO}^{\rm HB}$ very difficult. Stetson et al. instead tried to match the upper main sequence, subgiant, and lower red giant branches of these two clusters, along with those of NGC 1851, which has a very similar metallicity but includes RR Lyrae variables and a somewhat bimodal horizontal branch, so that both the blue and red sides are well populated. The emphasis on the fit was placed on matching the subgiant branches, and the results are shown in part (a) of Fig. 114. It appears that the clusters might have comparable ages. Further, using the differences in distance moduli required for these fits, Stetson et al. (1996) then compared the V magnitude level of the lower red horizontal branch (i.e., the zero-age horizontal branch) of NGC 362 to that of NGC 1851, as shown in Fig. 115. The match is good. They also showed the approximate distribution of stars in NGC 288 compared to that of NGC 1851, corrected for differential distance moduli, and as also shown in Fig. 115. Again, the agreement is quite good, suggesting all the data are consistent with no age difference despite remarkably different horizontal branch morphologies. To illustrate how difficult such comparisons are, however, consider part (b) of Fig. 114, which was derived using the same data employed by Stetson et al. (1996), but which started from a more fundamental assumption: that the *lower main sequences* of the clusters must match if the chemical compositions are the same. It then becomes clear that NGC 362 must be younger than NGC 288. We also reconsider Fig. 115. Walker (1998) has published mean magnitudes and colors for the RR Lyraes in NGC 1851, and it appears that the distance to NGC 288 has been mis-estimated. According to Fig. 115, the blue and red sides of the instability strip differ by about 0.4 mag in V, but according to Walker (1998), they appear to differ by less than 0.2 mag. The relative distances derived by Sarajedini et al. (1997) provide a better fit, and we conclude that, assuming the abundances of the two clusters are well-determined and that all the photometry is accurate, the two clusters do indeed have different ages at a level of about 2 Gyrs. Age appears to be (part of) the explanation of the second parameter phenomenon for this pair of clusters.

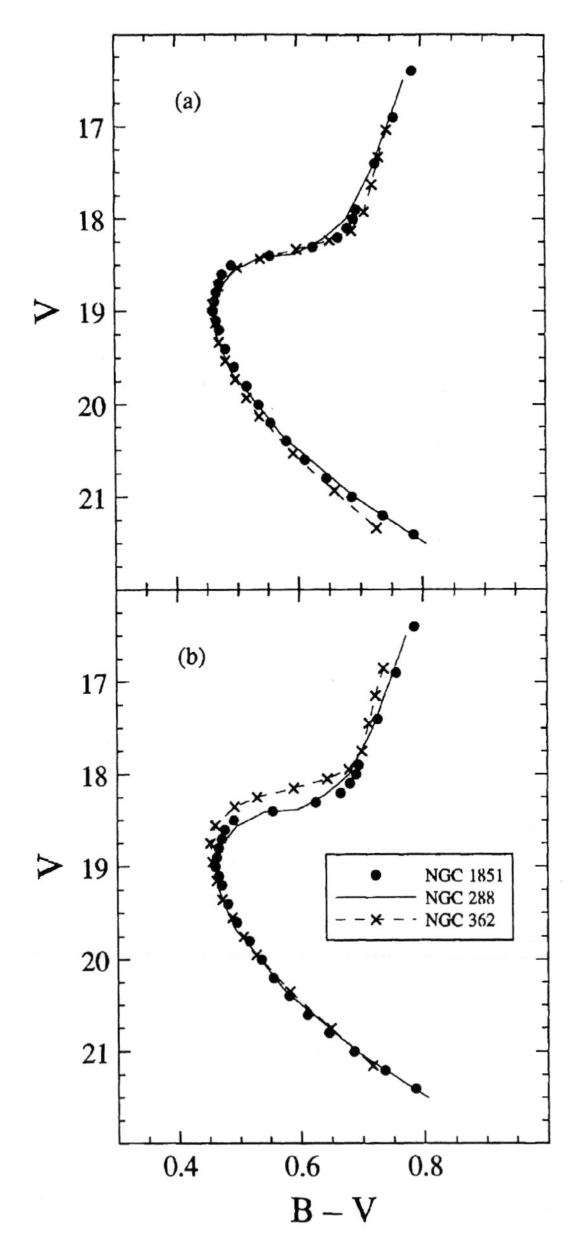

Fig. 114. The relative ages of the globular clusters NGC 288, NGC 362, and NGC 1851 obtained from photometry of the main sequence and differential color and magnitude shifts, taken from Sarajedini et al. (1997). (**a**) represents the fit of Stetson et al. (1996) obtained by matching the subgiant branch. (**b**) represents the fit from the lower main sequence obtained by Sarajedini et al. (1997)

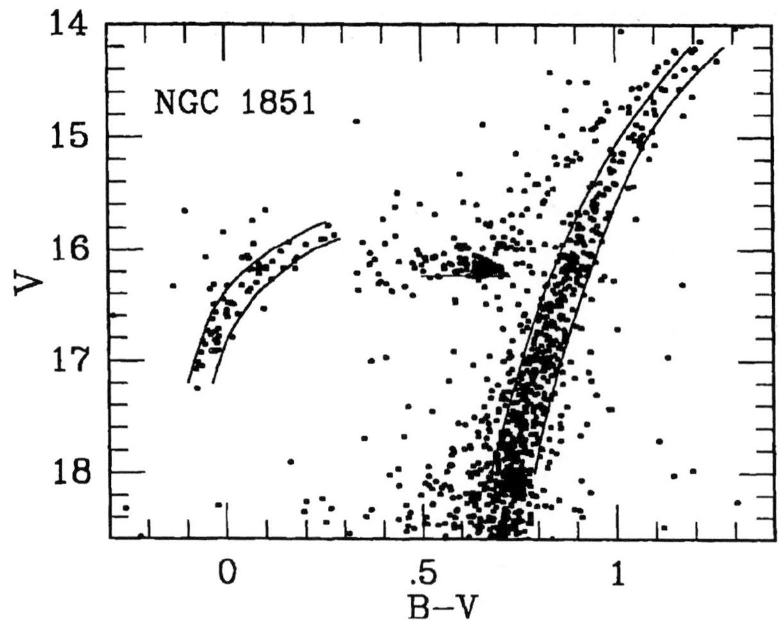

Fig. 115. The resultant overlap on the horizontal branch of NGC 1851 of NGC 288 (blue side) and NGC 362 (red side) using the differential shifts in Fig. 113, taken from Stetson et al. (1996)

8.6 Cluster Ages vs. Galactocentric Distances

Whether or not we find evidence for an age spread among globular clusters close to the Sun, our understanding of the formation history of our Galaxy is very incomplete unless we can also determine the size of any age differences between clusters very near to and very far from the Galactic center. The comparisons are crucial for several reasons. First, the differences in ages between the most metal-poor and most metal-rich clusters in the high-density central regions will reveal how important density is in the Galaxy's history of star formation, and how quickly the system and possibly the bulge population was formed. We require good color–magnitude diagrams in crowded and reddened regions, as well as reliable [Fe/H] and [α/Fe] values. Similarly, age differences between the most metal-poor and most metal-rich clusters in the outer halo will reveal not only how rapidly cluster formation proceeded in that part of the Galaxy, but also if there were offsets in the epochs of star formation compared to the inner Galaxy. And of course, we cannot forget that one of the primary interests in globular cluster ages is cosmological: we won't know the age of the oldest cluster(s) until we have studied them throughout the Galaxy.

The Most Metal-poor Clusters Hither and Thither. We have noted already that the most metal-poor clusters studied to date all appear to have

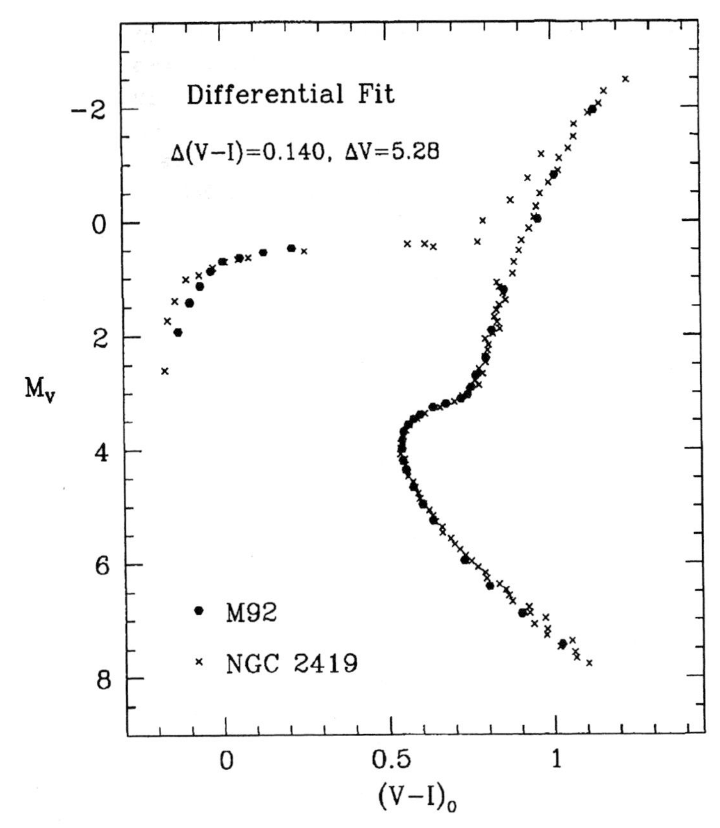

Fig. 116. A comparison of the color–magnitude diagrams of the metal-poor, distant globular cluster NGC 2419 (90 kpc) and that of M92 (9 kpc), taken from Harris et al. (1997)

the same ages, including M68 (NGC 4590), M92 (NGC 6341), NGC 6397, and M30 (NGC 7099) (Stetson et al. 1996), as well as NGC 5053 (Rey et al. 1998). Similarly, the most metal-poor field stars appear to have similar ages (Marquez & Schuster 1994). Figure 116 shows the results of a very careful study of the very distant cluster NGC 2419 (R_{GC} = 90 kpc) by Harris et al. (1997) with data obtained using *HST*. Despite the very large Galactocentric distance, it appears that NGC 2419 formed when the other clusters also formed, despite their much smaller Galactocentric distances (ranging from 6 to 16 kpc, with an average of 9 kpc).

This equality in ages among the most metal-poor clusters appears to extend even beyond the Milky Way. Grillmair et al. (1998) obtained deep *HST* color–magnitude diagram data for the metal-poor Draco dwarf spheroidal galaxy, which lies about 100 kpc from the Milky Way. They noted that the galaxy appears to be older than M68 (NGC 4590) and M92 (NGC 6341) by 1.6 ± 2.5 Gyrs, or, in other words, the same age to within the errors. Likewise, the *HST* data obtained by Buonanno et al. (1998b) of the four most metal-poor clusters in the Fornax dwarf galaxy indicate ages that are identical to each other and to M68 and M92 to within about 10^9 years. This suggests that the "trigger" for the commencement of star formation was a truly global one, extending beyond our Galaxy. But it will still be interesting to determine the relative ages of these clusters and the most metal-poor ones in our Galaxy's central regions.

Does this equality of ages apply for more metal-rich clusters? The initial trigger for star formation may have been global, but did the pace of star formation proceed at different rates in different parts of the Galaxy? Sarajedini (1997) has obtained ground-based photometry of the intermediate metallicity ($[Fe/H] \approx -1.6$) cluster Palomar 14, which lies about 60 kpc from the Galactic center. The cluster shows the second parameter effect, with an unusually red horizontal branch, and it appears that the main sequence turn-off is consistent with an age younger by 3 to 4 Gyrs than other clusters of the same metallicity but closer to the Galactic center. This trend continues for the intermediate metallicity clusters Palomar 3, Palomar 4, and Eridanus, all studied by Stetson et al. (1999). With metallicities [on the Zinn & West (1984) scale] estimated to be -1.57, -1.28, and -1.42, these clusters lie at Galactocentric distances of roughly 80, 100, and 85 kpc, and all have quite red horizontal branches. The cause of the red horizontal branch (second parameter effect) in these clusters is consistent with age effects since Stetson et al. (1999) find all three to be somewhat younger than the similar metallicity clusters M3 and M5. Thus these initial studies of distant intermediate-metallicity clusters suggests that the formation of globular clusters proceeded more slowly in the outer halo, or was more fragmentary as in the Searle & Zinn (1978) scenario. An interesting additional puzzle, however, is the recent comparison between NGC 6229, which lies about 30 kpc from the Galactic center, with M5, which is much closer but whose motions indicate an origin in the outer halo (Cudworth & Hansen 1993). NGC 6229 has a very blue horizontal branch relative to M5, but despite apparently similar metallicities, it does not appear to be significantly older (Borissova et al. 1999). More clusters are under study by Stetson *et al.* Further, they stress the importance of correct metallicities, especially $[\alpha/Fe]$. Errors of only a few tenths dex would alter their conclusions, and the brightest stars in these clusters are very challenging targets for traditional high-resolution stellar spectroscopy.

The Ages of the Inner Galaxy Clusters. Most of the mass of the Galactic halo lies within 5 kpc of the center, but due to crowding and interstellar reddening it has not been thoroughly studied yet.

There is some evidence that the inner halo clusters and field stars are older than the more distant ones. In the case of the field stars, the evidence is somewhat circumstantial, relying on the effects of age on the horizontal branch mor-

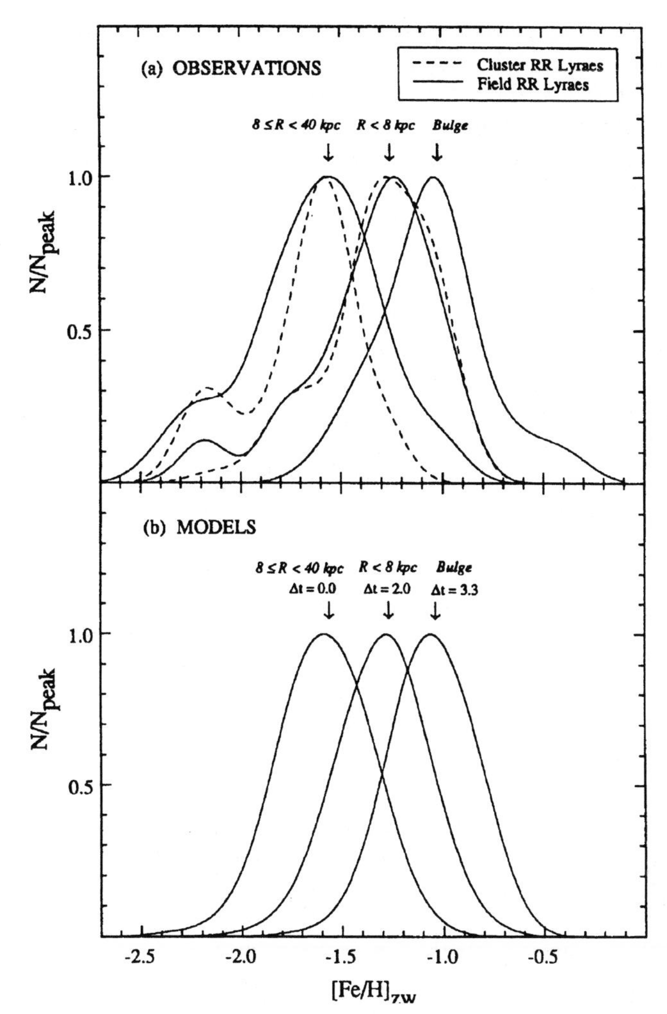

Fig. 117. Model metallicity distributions computed from horizontal branch theory under the assumption that age is the second parameter and compared with observed distributions (Taken from Lee 1992)

phology. The critical observations were made by Walker & Terndrup (1991), who found that spectroscopic abundances (using ΔS) of 59 RR Lyraes in Baade's Window revealed a peak in the metallicity distribution at [Fe/H] ≈ -1.0 rather than -1.5 at the solar Galactocentric distance and beyond. Increased ages may enable more metal-rich stars to populate the instability strip, and Lee (1992) has modelled this effect in his analysis of the Walker & Terndrup (1991) observations. Figure 117 shows his results, arranged by Galactocentric distance. To produce the field and cluster horizontal branch morphologies compared to the observed metallicity distributions, Lee finds that the outer halo is the youngest population, that the clusters and field stars just inside the Solar circle are about 2 Gyrs older, and that the innermost RR Lyraes may be older still. At first glance, this seems to contradict the observations that the most metal-poor clusters all have similar ages, especially since NGC 6397 and M30 have Galactocentric distances of only about 6 kpc. But the comparable ages may be reconciled with differing mean ages if star formation proceeded extremely rapidly in the inner Galaxy and more slowly in the outer Galaxy. In other words, *the key to the puzzle lies in the relative ages of the most metal-rich clusters.* Lee's results predict those in the inner Galaxy are much older than those in the outer halo.

An answer to this puzzle may not be long in coming. Several groups have considered the question of a difference in ages between the inner halo clusters and those at and beyond the solar circle. Richer et al. (1996), Buonanno et al. (1998a), and Salaris & Weiss (1997, 1998) agree that there is as yet no evidence to support a difference in mean ages *vs.* Galactocentric distance. The uncertainties are still too large and the distances sampled too limited, being restricted to $R_{GC} > 4$ kpc, to prove or disprove Lee's conclusions. The differences are certainly not as large as several Gyrs down to this distance, but it is the metal-rich clusters that will provide the key data. Further, *HST* results using both WFPC2 and NICMOS should be available for more inner halo clusters in the near future. A group led by Bob Zinn has obtained WF/PC and WFPC2 data for seven inner halo clusters, four of which are very metal-poor and three of which are very metal-rich. Two other groups, led by Laura Fullton and Sergio Ortolani, have been acquiring NICMOS data for clusters very near the Galactic center, where crowding and differential reddening effects are especially severe. Of course, accurate relative ages still require accurate [Fe/H] and [α/Fe] values. The higher the metallicity, the greater the effects on ages, and hence higher precision is required. The [α/Fe] results are, of course, vital for both the age estimation and the tests of relative ages under the assumption that the SNe II *vs.* SNe Ia enrichment model is valid. Have the most metal-rich inner halo clusters been enriched only by SNe II events, implying a rapid chemical enrichment process? And for the most metal-rich clusters, we must also have a means of estimating the helium abundances since they are likely to be higher than for the most metal-poor clusters. Failure to allow for higher helium abundances may lead to overestimates of cluster ages. How much should we worry? Fullton (1995) and Buonanno et al. (1995) measured the R values (cf. Sect. 1) of NGC 5927 and Terzan 7, both metal-rich clusters. (NGC 5927 lies about 4 kpc from the Galactic center while Terzan 7 belongs to the Sagittarius dwarf galaxy.) In both cases, the R values suggest high helium abundances, $Y \approx 0.28$ to 0.30, and consistent with the value estimated for the Bulge population by Minniti (1995).

To illustrate both the difficulties and the promise ahead, consider Fig. 118. This is a high-resolution ($\lambda/\Delta\lambda$ = 40,000) echelle spectrum of a bright red

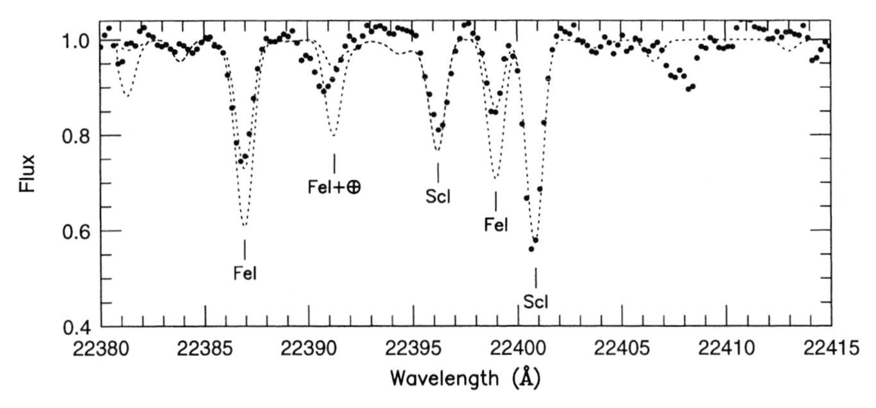

Fig. 118. An echelle spectrum taken near 2.4μm of a heavily reddened giant in Liller 1. The *dotted lines* are synthetic spectra with [Fe/H] = −0.8 and +0.25

giant in the highly extinguished ($A_V = 10$) but nominally metal-rich globular cluster Liller 1, obtained by Suchitra Balachandran, Laura Fullton, John Laird, and myself using the CSHELL instrument on NASA's Infrared Telescope Facility. The star has a V magnitude of 22, but at K, it has a magnitude of 9.3. It is a radial velocity member, so the spectrum illustrates that we will be able to study the chemistry of even the innermost clusters. The interesting initial result is that the spectrum is best modelled by [Fe/H] ≈ -0.8, about that of M71, and almost a factor of ten (1 dex) more metal-poor than integrated spectra (Armandroff & Zinn 1988) or the slope of the giant branch (Frogel et al. 1995) suggest. This is not our final result, but is shown rather to demonstrate what is possible, and to offer the reader the expectation that within a few years we will be able to estimate relative ages and chemical abundances for globular clusters throughout the Galaxy.

8.7 Summary

The most metal-poor clusters and field stars have an undetectable difference in ages, consistent with the r–process dominance in the most metal-poor field stars. At intermediate and higher metallicities, clusters appear to show some differences in ages, the extreme cases being Ruprecht 106 and Palomar 12, and less extreme cases being NGC 288 *vs.* NGC 362. The outermost clusters with intermediate metallicities appear to be younger than those in the inner Galaxy, indicating a slower rate of star formation and suggesting that the second parameter effect in the outer halo may indeed be caused by age. It is not yet clear whether local very metal-poor and intermediate-metallicity clusters have essentially the same ages or differ. The results for the Oosterhoff I and Oosterhoff II clusters M3 and M2 suggest age differences may be present, but this may result from a difference in the origins of these clusters rather than reveal the speed of formation and chemical evolution of the Galaxy independently of mergers or accretion events. The use of [O/Fe] and [α/Fe] ratios and the assumption that the differences reflect differing levels of contributions from SNe II and SNe Ia ejecta suggest the clusters do not differ in age, in general, by more than 1 to 2 Gyrs. However, since [α/Fe] measures the *duration* of star formation rather than date its beginnings, this conclusion is not a compelling one. The ages of the innermost globular clusters have yet to be determined, but there are hints that star formation began and ended earlier in the Galaxy's densest regions.

9 Absolute Ages of Globular Clusters

9.1 Introduction

The absolute ages of the oldest globular clusters is one of the very few means we have to estimate the age of our Milky Way Galaxy and of the Universe itself, and hence considerable investments of theory and observing time have been made to try to refine the precision to the level of 5% to 10%. While the observations and the refinements to the input physics have been remarkable over the past twenty years or so, the goal has remained elusive, and, as we discuss here, a pessimistic but probably honest assessment of the uncertainty is that it remains closer to 30%.

On a secondary level, the absolute ages are important because the *magnitude* of the relative ages depends on the absolute age. Greater absolute ages result in magnifying age spreads, largely due to the "compacting" of the main sequence turn-offs at higher age. Consider Fig. 3, for example. The three turn-offs shown are spread out about equally in stellar mass, luminosity, and temperature. But the first pair of stars differ in turn-off age by 5 Gyrs, while the second pair differ by almost 10 Gyrs.

Sect. 1 discussed the basic physics that is required to derive stellar model isochrones which may be compared with globular cluster color–magnitude diagrams. Clayton's relation (21) contains the heart of the issue: we need to know the mean molecular weight (i.e., the chemical composition) and the opacity to determine the luminosity and hence the lifetime. Figures 6, 7, 9, and 10 illustrate how necessary it is for us to know the helium and heavy element mass fractions. Looking more closely, we see as well that the heavy element mass fraction must be divided into at least two different parts: iron-peak elements and "alpha" elements. The latter are important because some of them affect the overall opacity (and mean molecular weight) and are especially important sources of low-temperature opacity (see Fig. 12). Oxygen also plays an important role because it is part of the CNO cycle, which becomes important, and then dominant, for low-mass stars near hydrogen core exhaustion and the turn-off in the evolutionary tracks. The nuclear energy generation from the proton–proton chains seems to be well understood, despite the continuing discrepancies between observed solar and predicted neutrino fluxes. But

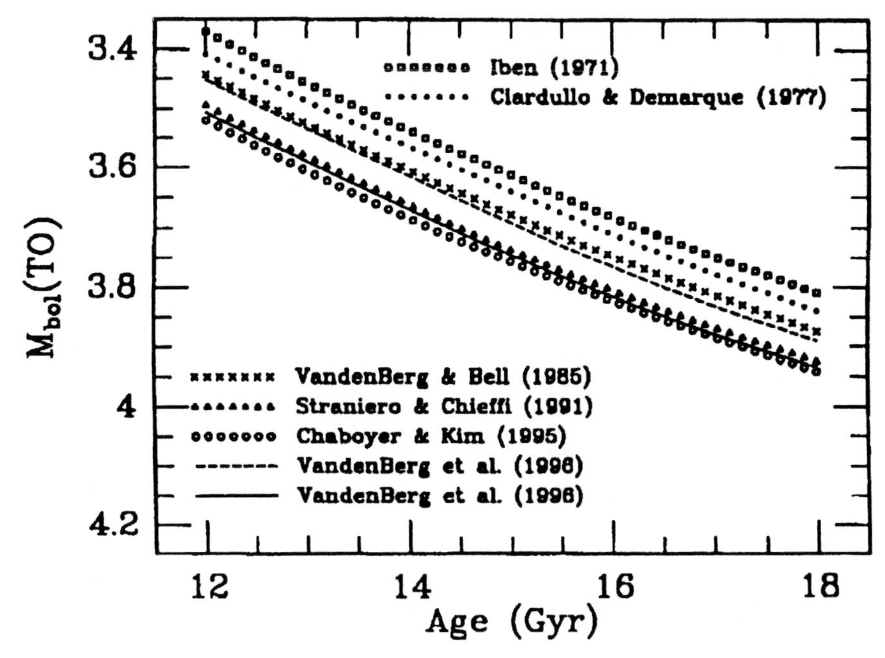

Fig. 119. Turn-off bolometric magnitudes *vs.* age in Gyrs for $Y = 0.20$ and $Z = 0.0001$ from a variety of calculations. Figure taken from VBS96

the CNO cycles are a little more problematic, and we might guess that the greatest concern would be the slowest reaction: $^{14}N(p,\gamma)^{15}O$.

9.2 Sources of Continuing Uncertainties: Theory

Input Physics. There are two excellent recent sources to help understand the improvements that have been made to stellar evolution modelling over the years, and problems that still require solutions: the review article by VandenBerg et al. (1996; hereafter VBS96) and the Monte Carlo study by Chaboyer et al. (1998; hereafter CDKK98).

Figure 119 shows a "history" of the calculated absolute bolometric magnitudes of main sequence turn-offs plotted against age. There has been a fairly steady decrease in the age obtained from a measured turn-off luminosity as the physics has been improved. What is comforting is that the left-most/bottom three loci were all computed with the same input physics, including reaction rates and opacities (including the effects of Coulomb interactions), and obtained essentially identical results. The *computational* methods appear to be robust. Further, much of the input physics now appears to be well understood. The differences between the crosses and dashed lines in Fig. 119 are caused by use of the older Los Alamos Astrophysical Opacity Library (Huebner et al. 1977) and the revised "OPAL" opacities (Rogers & Iglesias

1992), and are obviously small. The new low-temperature opacities provide a good match between Cepheid pulsation predictions and observations (Rogers & Iglesias 1994). The biggest change in recent years was the inclusion of Coulomb interactions in the equation of state (see, for example, Chaboyer & Kim 1995), but unless another such systematic effect is identified, it appears that much of the input physics is under reasonable control. Indeed, the major issues that remain depend more on observational uncertainties than a lack of understanding of physics.

Overlooked physics might also be involved in resolving the solar neutrino problem, as discussed by VBS96. If the problem is in the physics, that electron neutrinos oscillate into either μ or τ neutrinos, then no changes are required in current stellar models. If the solution is "less conventional", involving perhaps weakly interacting massive particles, stellar ages may yet be overestimated. The reader should consult VBS96 for details.

Other examples of overlooked or ignored physics are discussed by VBS96, and include the effects of helium diffusion, stellar/core rotation, and mass loss. Diffusion's potential importance is considerable, as recognized first for globular cluster ages by Noerdlinger & Arigo (1980). If helium diffuses into the hydrogen-burning core, it displaces hydrogen fuel, shortening the lifetime, as well as increasing the mean molecular weight, hence increasing the luminosity and also shortening the lifetime (see (21)). The original estimate of a major effect on cluster ages ($> 20\%$) has not been confirmed. Proffitt & VandenBerg (1991), for example, showed the effect to be relevant at the 10% level, or less. Models computed without diffusion proved better matches to observed color–magnitude diagrams than those computed with diffusion. While diffusion's role in age estimation is due to its effects in the hydrogen-burning core, diffusion as a process, and the helium diffusion coefficient, may be studied as well using stellar photospheres, where lithium may be used as a helium surrogate. Isochrones computed without diffusion by Pinsonneault et al. (1992) of lithium abundances *vs.* $T_{\rm eff}$ for metal-poor stars are a better match to the "Spite plateau" than those with diffusion. And, as discussed in Sect. 6, the apparent constancy of lithium, the presence of ^{6}Li, and the non-enhanced values seen in tidally-locked binaries all suggest that diffusion is not a major process.

Rotation (and any other process such as magnetic fields) provides nonthermal pressure support, and as such may prolong stellar lifetimes, in addition to playing a role in stellar mixing. As was discussed in Sect. 7, the role of rotation is plausible, but the oxygen line strengths and rotational velocities of blue horizontal branch stars are not entirely consistent with rotation being the primary cause of the second parameter. Further, rotation's effect on the estimated ages of globular clusters is probably very small, according to Deliyannis et al. (1989).

The importance of mass loss in producing blue stragglers was discussed in Sect. 2. If the mass loss was significant, Willson et al. (1987) suggested that globular cluster age estimates would be decreased dramatically. The importance of mass loss, however, is probably insignificant. The hypothesized mass loss is driven by pulsational instabilities, but the instability strip defined by the SX Phy variables is very much hotter ($\approx$ 700K) than that of main sequence turn-off stars. Further, we would expect lithium to be depleted in the stars' photospheres (because hotter, deeper layers are now exposed), yet field metal-poor dwarfs with colors like those of globular cluster turn-off stars show lithium in the expected amounts, and lithium is at normal abundance levels in the turn-off stars in NGC 6397 (Pasquini & Molaro 1996).

9.3 Sources of Continuing Uncertainties: Observations

Transformation of Observables into Theoretical Quantities. Stellar model calculations yield luminosities and effective temperatures. Observations provide magnitudes and colors. Transformation in either direction is crucial but difficult, particularly in the color–temperature relations, as discussed in Sect. 5. This is one of the reasons that globular cluster absolute ages are not generally estimated on the basis of their colors. A second good reason is that the model temperatures depend strongly on the adopted ratio of the convective mixing length to pressure scale height (and, indeed, upon the general validity of mixing length theory), as shown in Fig. 11. Since the calculations solve for stellar radius, which depends strongly on this ratio (and on the two outer boundary conditions to solve the four differential equations described in Sect. 1), the stellar temperature is both less certain and less "fundamental" than the stellar luminosity. Thus absolute ages rely solely on luminosities of stars at or near the main sequence turn-off.

Nonetheless, a magnitude is related to a luminosity through the bolometric correction, and even a 0.1 mag error in that quantity may result in an error in the derived age by 10^9 years or more. Consider the isochrones of Straniero & Chieffi (1991), for example. Their bolometric corrections were taken from the calculations of Bell & Gustafsson (1978), and VandenBerg & Bell (1985), which were based on model atmosphere surface flux distributions. Straniero & Chieffi (1991) made a zero point shift to allow for a solar bolometric correction $BC_{\odot}$ of -0.12 mag, rather than that adopted by Bell & Gustafsson (1978), -0.07 mag. A metal-poor star with [Fe/H] ≈ -2 and a color typical of globular cluster main sequence turn-offs has $T_{\rm eff} \approx 6250$ K, based on the temperature scale of Carney (1983) and Alonso et al. (1996). The bolometric correction calculated by VandenBerg & Bell (1985) for such a star is about -0.11 mag. However, bolometric corrections may also be measured empirically by integrating the fluxes obtained through broadband *UBVRIJHK* filters, and using model atmospheres to add in the small amount of flux shortward of U and longward of K. Carney & Aaronson (1979) made the first such measurements for metal-poor stars, Carney (1983) added new data, and Alonso et al. (1995) have provided a more refined analysis and a larger sample. All three studies indicate the bolometric correction for a star with [Fe/H] ≈ -2 and $T_{\rm eff}$ = 6250 K is about -0.20 mag. The difference between what has been used

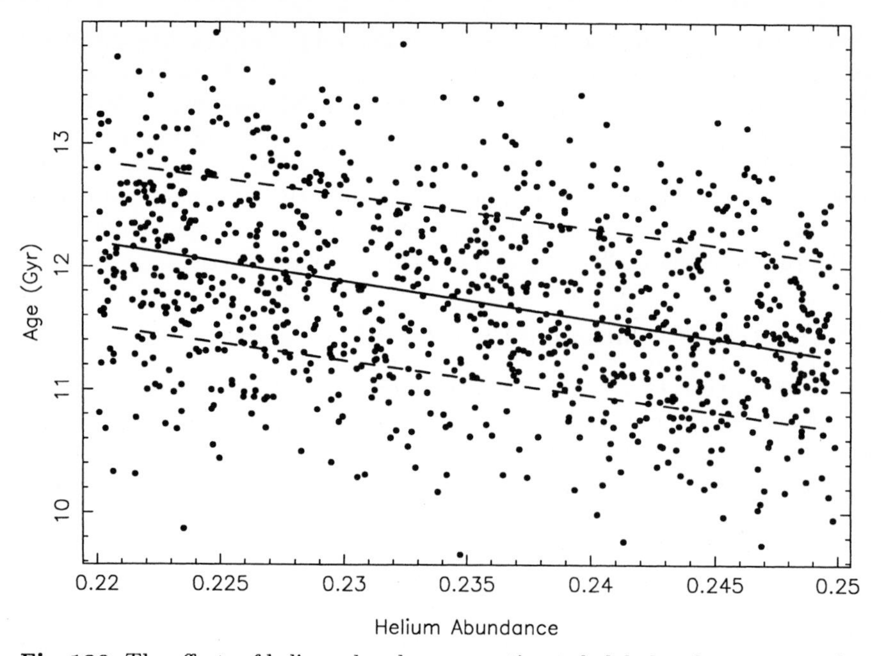

Fig. 120. The effects of helium abundance on estimated globular cluster ages, taken from CDKK98. The *dashed lines* represent 1σ uncertainties. The median and mean metallicities for the clusters being evaluated are [Fe/H] = −1.9

in the calculation of isochrones is large enough to be significant. Use of the empirical bolometric corrections will make the *isochrone* turn-offs brighter, and hence, for a fixed distance scale (see below) and *observed* M_V(TO) value, the derived ages will be larger.

Key Observational Problems. CDKK98 identified the seven most important variables in the estimation of globular cluster ages using a Monte Carlo approach. A large number of isochrones were computed with variations in many key parameters, and compared with the ages derived from turn-off luminosities. Low-temperature opacities, the helium diffusion coefficient, and the $^{14}N(p,\gamma)^{15}O$ reaction rate were found to be of minor importance, but over the plausible ranges of their values, the age estimates varied by ±0.2, ±0.3, and ±0.3 Gyrs (for a typical cluster age of about 12 Gyrs). The four most important variables are considered below, in order of increasing importance.

Helium Abundance. Figure 120 shows the trend of derived age *vs.* Y, the helium mass fraction, along with $\pm\ 1\sigma$ values (the dashed lines). Four methods are available to estimate Y; none of them, alas, are direct.

The first method is theoretical. Models for Big Bang nucleosynthesis (Boesgaard & Steigman 1985) predict that the primordial helium mass fraction, Y_p, which should be present in the most metal-poor stars, is

$$Y_p = 0.230 + 0.011 \ln \eta_{10} + 0.013 (N_\nu - 3) + 0.014 (\tau_n - 10.6) , \qquad (67)$$

where η_{10} is the baryon mass fraction η times 10^{10}, N_ν is the number of neutrino families, and τ_n is the half-life of the neutron in minutes. All these variables are given in (67) at their probable values, so Y_p should be near 0.230.

The value of Y_p may also be estimated by measuring helium abundances in H II regions with a wide range of heavy element abundances, and then extrapolating to zero metallicity, as first proposed by Peimbert & Torres–Peimbert (1974, 1976). Recent measurements of the slope, $\Delta Y/\Delta Z$, and extrapolation to zero metallicity lead to primordial helium mass fractions of slightly under 0.23 (Pagel et al. 1992; Pagel & Kazlauskas 1992). We need not be concerned with the extrapolation to $Z = 0$ for the estimation of Y for globular clusters, of course, but only for helium abundances derived in extragalactic H II regions where the oxygen abundances are very low provide Y values for globular clusters with comparable oxygen abundances. The observations and analysis of H II regions in the blue compact dwarf galaxy I Zw 18 yield $Y = 0.226$ for $[\mathrm{O/H}] = -1.73$ (Pagel et al. 1992), which is only slightly more oxygen-rich than M92, with $[\mathrm{O/H}] = -1.94$ (Sneden et al. 1991).

A third approach is the much less direct "R value" (Iben 1968; Buzzoni et al. 1983) discussed in Sect. 1. In the clusters studied by them, R values (the ratio of the number of horizontal branch stars divided by the number of first ascent red giant branch stars brighter than the horizontal branch) of about 1.4 were derived, indicating $Y = 0.23$ according to (31). Some additional observations and new analyses by Caputo et al. (1987) indicated $Y = 0.24 \pm 0.01$. The quoted uncertainty does not include any systematic error estimates.

Finally, in principle we may exploit the helium sensitivity of the mass luminosity relation depicted in Fig. 6. A metal-poor double-lined eclipsing binary system with a well-determined distance could in principle be used to estimate the helium abundance, and using the same stellar evolution models that are employed in the calculation of main sequence turn-off luminosities. Despite diligent searches over several decades, such stars have proven elusive. However, an interesting object has been identified by Dave Latham and his collaborators (see Goldberg et al. 1999). G24-18p and G24-18f are a common proper motion pair, separated by about 2 arcsec. G24-18p turns out to be an eclipsing binary (Cutispoto et al. 1997). It also is a double-lined system, and the stellar masses are already well determined. Despite the very short period (0.52 day), and consequent large degree of line broadening, we may estimate the metallicity of the system by studying G24-18f, which is a sharp-lined star with no sign of a close companion. Preliminary work by Carney & Lee (1999) indicates $[\mathrm{Fe/H}] \approx -0.8$. Despite the fact that the star is bright and in the *Hipparcos* catalog (HIC 101236), the light from the two stars complicated the trigonometric parallax measurement, so it is not sufficiently well-defined to

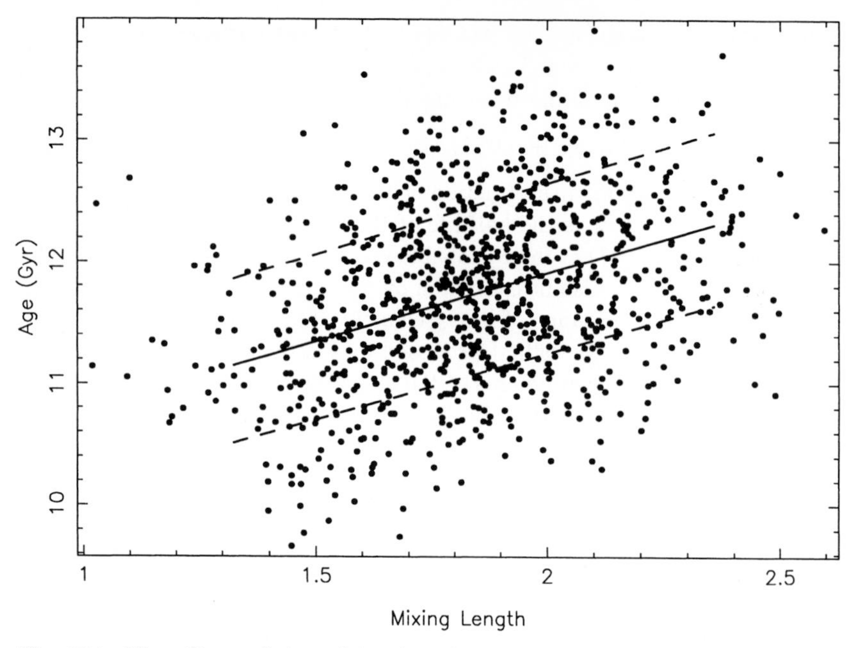

Fig. 121. The effects of the mixing length parameter, α_p, on estimated globular cluster ages, taken from CDKK98

determine the luminosity to the requisite precision [$\pi = 22.53$ mas; $\sigma(\pi) = 5.13$ mas]. Systematic errors aside, $\sigma(M_V) = 0.49$ mag. More work on this star, and continued searches for more metal-poor double-lined eclipsing binaries, are very desirable.

In summary, it appears that the helium mass fraction for metal-poor stars is reasonably well-determined, and that $Y = 0.23$.

The Mixing Length Parameter. Figure 121 shows the sensitivity of the ages of metal-poor clusters to changes in the mixing length parameter α_p ($= \ell/H_p$, where ℓ is the mixing length and H_p is the pressure scale height). This sensitivity was a surprise, since Fig. 11 indicates that, as expected, changing α_p affects the radius of a stellar model but not its luminosity. But model evolutionary tracks of single stars do not map one-to-one into model isochrones, and the results of Chaboyer (1995) and CDKK98 indicate that α_p is a parameter of some significance, even when only the turn-off luminosity is being used to estimate ages. Since α_p is a "free parameter", it is hard to judge precisely how well a set of isochrones has chosen the proper value for α_p. Most models involve a calculation of the evolution of the Sun, adjusting α_p so that the model produces the correct solar temperature, radius, and luminosity for the known solar age and chemical composition.

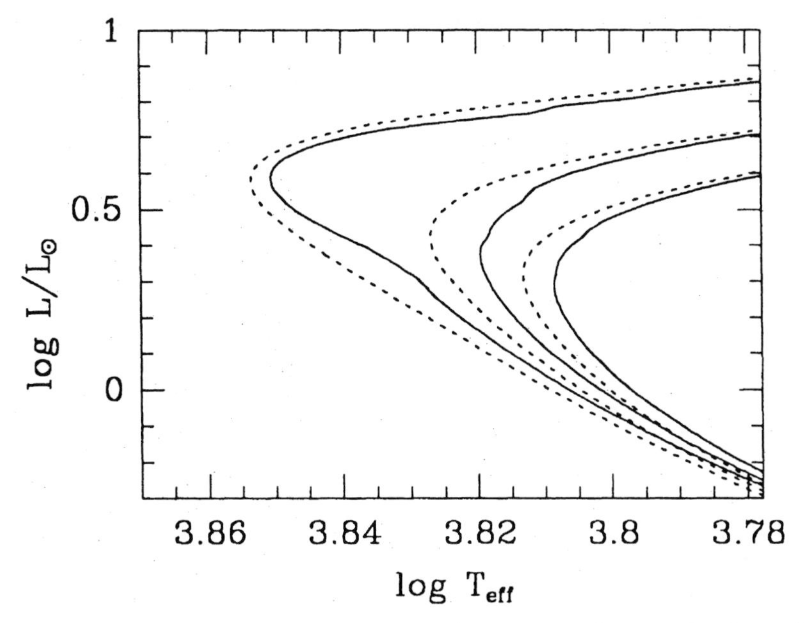

Fig. 122. A comparison of model isochrones computed using the "overadiabatic convection" method of Canuto & Mazzitelli (*solid lines*) and standard mixing length theory (*dotted lines*) for $Z = 0.0001$, $Y = 0.23$, and ages of 10, 14, and 18 Gyrs, taken from Mazzitelli et al. (1995)

Another way to ponder this issue is to explore alternative formulations of convective energy transport. Mazzitelli et al. (1995) have compared mixing length theory with their own theory (Canuto & Mazzitelli 1991, 1992), and Fig. 122 shows the comparisons. Even for the older ages, the two theories predict slightly different turn-off luminosities. There is no clear way to judge which theory, if either, is the better choice to estimate cluster ages, but one may at least invoke the necessary (but not sufficient) condition that the model isochrones provide good matches to the entire color–magnitude diagram of a cluster, from the cooler, fainter, more convective lower main sequence stars to the turn-off region to the cooler, brighter, and also very convective red giant branch. Mixing length theory, properly calibrated, appears to work quite well, as may be seen in the fit to M92 performed by Harris et al. (1997; Fig. 123).

Chemistry: [α/Fe]. Figure 124 shows the sensitivity of metal-poor cluster ages derived by CDKK98 as a function of [α/Fe]. Fortunately, [α/Fe] values are well determined for many globular clusters, and more data will become available in the near future. Excluding oxygen, Fig. 110 shows that [α/Fe] $\approx$ 0.3 is a reasonable value. Oxygen should not be neglected, of course, due to its numerical dominance of heavy element abundances. Although its photospheric

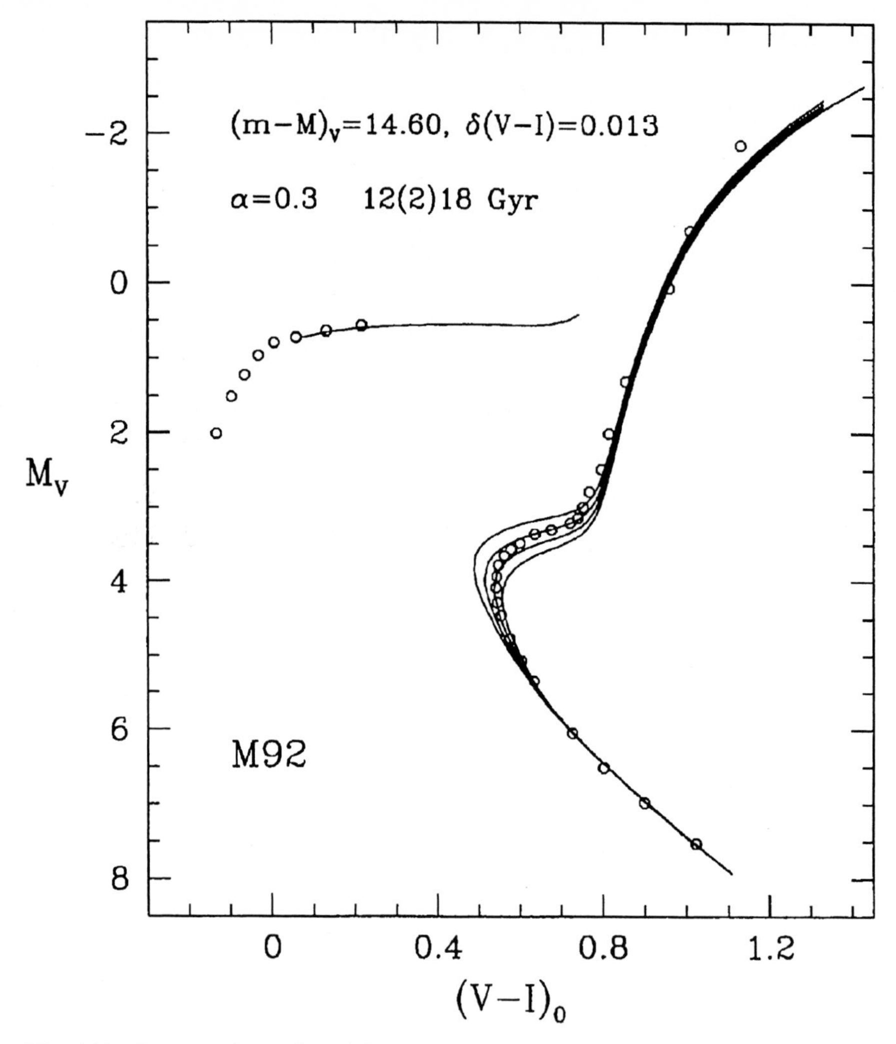

Fig. 123. A comparison of model isochrones computed using mixing length theory with the color–magnitude diagram of M92, taken from Harris et al. (1997)

abundances are vulnerable to mixing effects, Fig. 111 shows that globular cluster giants with minimal contamination from mixing have [O/Fe] $\approx$ 0.3 to 0.4. The reader should remember, however, the continuing debate over [O/Fe] for the most metal-poor stars, discussed in Sect. 6. My own belief is that [O/Fe] $\approx$ 0.3 to 0.4, and, therefore, that the ages estimated by CDKK98 must therefore be revised upward by about 0.5 Gyrs or more to allow for their overestimation of [α/Fe].

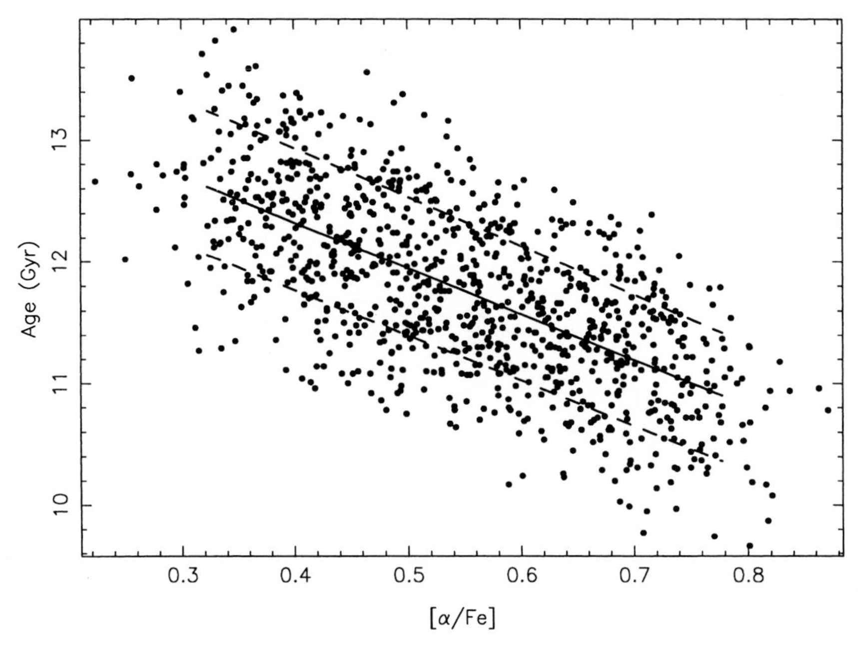

Fig. 124. The effects of [α/Fe] on estimated globular cluster ages, taken from CDKK98

The Distance Scale. The distance scale turns out to be the greatest source of uncertainty in the derivation of absolute cluster ages. Figure 125 shows the results from CDKK98 under different assumptions about the luminosity of the horizontal branch, measured in the regime of the RR Lyrae variables, and at a fiducial metallicity of [Fe/H] = -1.9. Differences of only 0.2 mag in M_V(RR) can result in differences in absolute ages of 3 Gyrs or more (the ages are older than those estimated by CDKK98 if the RR Lyrae luminosities are fainter than they assumed).

Of course, the distance scale of the globular clusters, and of their RR Lyrae variables, extends beyond the age problem: these are standard candles that may be used to estimate the distances to nearby galaxies that help provide the zero point of numerous extragalactic distance indicators. One should keep in mind that if we are convinced that the RR Lyraes are fainter than values we have employed in the past, then the nearby galaxies will be closer than estimated previously, and the extragalactic distance scale will shorten, reducing the Hubble constant and the expansion age of the Universe while simultaneously making the globular clusters older than determined previously. For these reasons the globular cluster distance scale has received very considerable observational and theoretical attention, and a study of the literature suggests to me something of a bias toward trying to reconcile the two ages. Most astronomers hope for such agreement, but we should try to avoid biases in which distance

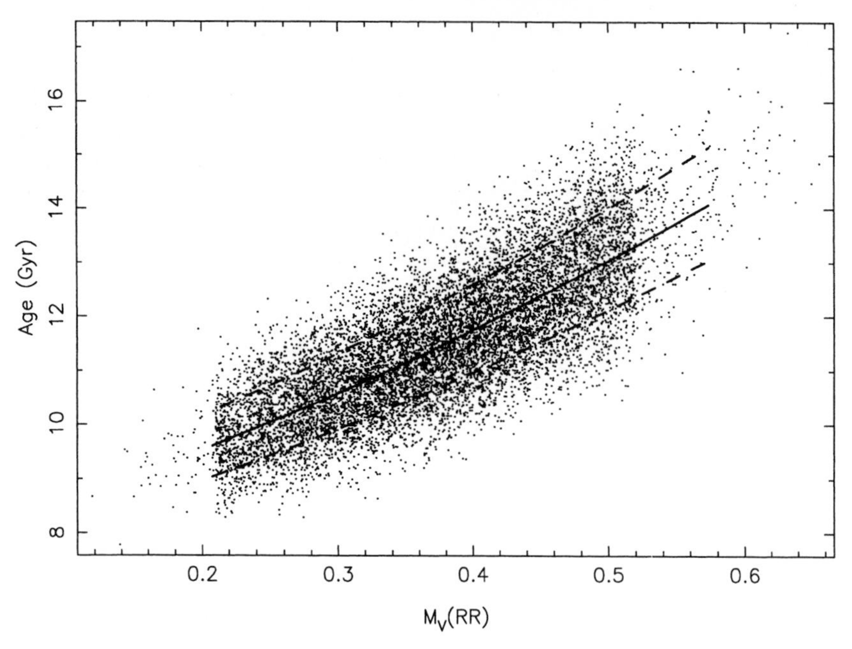

Fig. 125. The effects of the RR Lyrae luminosity, taken at [Fe/H] = −1.9, on estimated globular cluster ages, taken from CDKK98

scale we choose to believe. *There is no reason to let Universal expansion age estimates dictate which globular cluster age scale or distance scale is "most likely" to be correct.* We should simply explore all the procedures, identify the most reliable, and determine what new research efforts are necessary to reduce the uncertainties and disagreements. And let the cluster ages fall where they may. We might even learn something of cosmological significance if we keep our biases in check.

But enough editorializing. Let us review the current situation, which is far from being resolved. To make the comparisons, let us adopt the luminosity of the RR Lyrae variables, estimated either directly or indirectly, as our metric. Faint RR Lyraes favor short distance scales and greater ages (because the turn-off luminosities are consequently fainter). Because there is a mild metallicity sensitivity involved, we must adopt a slope for the M_V(RR) *vs.* [Fe/H] relation, which was discussed extensively in Sect. 3. Then we must adopt a "fiducial" metallicity at which to make the comparisons. For the latter, let us employ [Fe/H] = −1.9, as used by CDKK98. This is convenient since we'll be able to readily see the effects of M_V(RR) on cluster ages using Fig. 125. It also makes sense to use a low metallicity as our reference point since the key issue is that age of the oldest globular clusters, which are likely to be the most metal-poor. For the slope, we adopt 0.20 mag/dex, which is slightly steeper than that derived in the Baade–Wesselink analyses, but consistent with the value obtained by Fernley et al. (1998b) when allowances are made for errors in both the derived M_V values as well as [Fe/H].

Because the zero point is such a fundamental parameter, it is unwise to place too much credence on results that employ free parameters. An example is the Baade–Wesselink method itself. It relies heavily on the zero points of the relations between colors and temperatures, and on the "p factor" that converts observed radial velocities into pulsational velocities. As long as the "p factor" is the same for most RR Lyraes, and the *slopes* of the relations between colors and temperatures are well understood, the derived *relative* distances are accurate, as discussed in Sect. 3. But the derived *absolute* distances retain significant systematic uncertainties. And this is why Carney et al. (1992) employed other means to establish the zero point of the M_V(RR) *vs.* [Fe/H] relation, including trigonometric and statistical parallax results. In fact, Fernley (1994) argued for a change in the value of the "p factor" to obtain better agreement with other distance indicators, but this is a somewhat *ad hoc* procedure. Fortunately, those other zero points actually agree very well with the results from the Baade–Wesselink analyses, but nonetheless, we employ here *only* methods that rely on parallaxes as the basis for the distance estimations.

Two types of parallaxes are available: *trigonometric* and *statistical.* Next generation spacecraft with interferometric capabilities are likely to measure the trigonometric parallaxes for globular clusters directly and to high precision, but at the moment we must employ intermediaries in the estimation of cluster distances, such as main sequence dwarfs and white dwarfs.

Trigonometric parallaxes
Before summarizing the recent results, one must recall that trigonometric parallaxes, like everything else, are sensitive to two types of errors, internal and systematic. The internal or measurement errors are easy to deal with. We are seeking cluster distance moduli, $m - M$, which depend on $\sigma(V)$ and $\sigma(M_V)$. The former, as measured by the $\langle V \rangle$ value for the horizontal branch in the instability strip, are usually small, as long as there are a sufficient number of stars observed. The latter is tied directly to the uncertainties in the parallax measurement:

$$\sigma(M_V) = 2.17 \frac{\sigma_\pi}{\pi} , \qquad (68)$$

where π is the parallax. A potential systematic error may arise for some clusters with only very red or very blue horizontal branches, where it is difficult to establish the magnitude level of the horizontal branch within the instability strip. The comparison between NGC 288, NGC 362, and NGC 1851 discussed in Sect. 8 is a good example of this potential problem, which we avoid here by relying only on clusters with well-populated horizontal branches. A more serious systematic error arises, however, in the transformation of parallaxes into M_V values, due to the non-linear relationship between the two.

Although measurement errors may admit negative values for π, such values have no physical meaning. Corrections for these systematic effects tend to brighten the derived M_V values, depending, again, on σ_π/π, as discussed by Lutz & Kelker (1973), and, later on, by Hanson (1979) and Koen (1992). The analysis by Lutz & Kelker (1973), for example, showed that, depending on the density distribution of stars in a large sample, the systematic corrections reach about 0.1 mag when σ_π/π reaches 0.1. At this level, the random errors have already exceeded 0.2 mag in M_V. Where do we "draw the line" in our use of stars with trignometric parallaxes? The choices result in fewer stars with better-determined parallaxes or more stars with increasing vulnerability to both random and systematic errors.

We begin with a very conservative approach: fewer stars with better parallaxes. Consideration of Fig. 125 suggests that our final M_V(RR) value should have a precision of ±0.10 mag or better if we desire our age estimates to be credible at the 10% level. This was essentially the approach adopted by CDKK98, who selected stars only if $\sigma_\pi/\pi < 0.10$, as determined using the final results from the *Hipparcos* mission (Perryman et al. 1997). They also chose to avoid corrections for significantly differing metallicities between the field dwarfs and the globular clusters to which they were applying main sequence fitting. This removes or at least minimizes model dependencies. Further, they restricted application of their results to only those globular clusters with similar metallicities: $-1.8 <$ [Fe/H] < -1.1. The clusters selected also had to have exceptionally well-defined main sequences to minimize additional errors from the main sequence fitting process itself. They were especially careful with the field stars and clusters in terms of metallicity. The discrepancies already noted between spectroscopic metallicities and the Zinn & West (1984) metallicity scale (see Sect. 5) were avoided by eliminating NGC 288 and NGC 362. The selected clusters all had to have little or no reddening as well. Finally, CDKK98 recognized the disappointing fact that two of the field dwarfs with the best-determined parallaxes, the common proper motion pair HD 134439/40, have an unusal [α/Fe] ratio (King 1997), and were therefore excluded from the calibration. Their final result for M_V(RR) at [Fe/H] $= -1.9$ was $+0.39 \pm 0.08$ mag, with additional systematic uncertainties of the types discussed above. *But let us adhere to even more stringent criteria.* Of the three clusters employed by CDKK98, we eliminate two, M13 and NGC 6752 because they have almost entirely blue horizontal branches. NGC 6752 has no RR Lyrae variables, and the two in M13 are almost certainly highly evolved (Jones et al. 1992). This leaves only M5, with its color–magnitude diagram taken from Richer & Fahlman (1987) and an adopted reddening $E(B-V)$ of 0.03 mag (Zinn 1985). The list of acceptable field dwarfs also must be winnowed, and rather mercilessly, to obtain the most conservative estimate. First, we rely only on field stars that have not undergone much evolution, minimizing possible differences in ages between the field star(s) and M5. Given that age spreads appear to be more common at the metallicity of M5 ([Fe/H] $= -1.17$, Sneden et al. 1992), as discussed in Sect. 8, this seems prudent. We therefore eliminate the two bluest stars, CPD−80° 349 and HD 193901. Further, CPD−80° 349 has been found to be deficient in its [α/Fe] ratio, like HD 134439/40, as we have noted already. In fact, it is wise to eliminate all field stars for which [α/Fe] is as yet unknown. This removes HD 145417 and HD 120559. This leaves only HD 25329, HD 103095, and HD 126681. HD 25329 is 0.6 dex more metal-deficient than M5, and so we eliminate it as well. And, finally, given the stress we place on accuracy, we eliminate HD 126681, whose abundances are normal (Tomkin et al. 1992), but whose value of $\sigma[M_V(\mathrm{RR})]$ exceeds 0.15 mag, independent of systematic effects. While we are left with only one star, HD 103095, and all the uncertainties that are associated with a single star, it does appear to be an excellent match for M5, as discussed in some detail by Jones et al. (1988). Balachandran & Carney (1996) found [Fe/H] $= -1.22 \pm 0.04$, and normal [α/Fe] values. The star appears to be single [our group has obtained 308 radial velocities covering over 17.6 years, and $\sigma(v_{\mathrm{rad}})$ is only 0.5 km/sec]. At a distance of only slightly more than 9 pc, HD 103095 is unreddened, and $M_V = 6.62 \pm 0.02$ mag. This value is 0.17 mag brighter than employed by Jones et al. (1988) in their estimation of $M_V(\mathrm{RR})$ in M5, and their result therefore changes by 0.17 mag. Applied to M5, then, $M_V(\mathrm{RR}) = 0.69 \pm 0.12$ mag at [Fe/H] $= -1.17$, and 0.54 ± 0.12 mag at [Fe/H] $= -1.9$. It is an interesting question why the trigonometric parallax of this star changed so much. The ground-based parallaxes of Jenkins (1963), Beardsley et al. (1974) and Heintz (1984) indicated $\pi = 0.117 \pm 0.003$, while the *Hipparcos* result is 0.109 ± 0.001. The differences are rather larger than the error estimates predict, and this exercise shows that *even for the best-studied metal-poor dwarf*, changes at a significant level in M_V are likely, as well as potential problems that may arise from the use of only one star.

In Table 5 we summarize the above results, along with others discussed below. Reid (1997) applied a limited number of parallaxes from *Hipparcos* to several clusters with very well-observed color–magnitude diagrams, as did Gratton et al. (1997) and Pont et al. (1998). As Table 5 shows, these analyses result in $M_V(\mathrm{RR})$ values that are about 0.15 mag brighter than that obtained by the most conservative approach, but agree with those of CDKK98. Note that for Reid's (1997) estimate, he derived results in two metallicity domains. We have taken an average of both of his estimates, as applied to [Fe/H] $=$ -1.9. The range in the possible results from trigonometric parallaxes of metal-poor main sequence dwarfs can be best judged by comparing the conservative one star/one cluster analysis above, the somewhat less conservative analysis of CDKK98, and the results of Pont et al. (1998). They employed as many stars as possible, although they were very careful to select only metal-poor stars using metallicities estimated from CORAVEL correlation dips. Further, they also included subgiants, which fit quite well onto cluster subgiant sequences.

Another trignometric parallax option is that employed by Renzini et al. (1996). Instead of relying on main sequence fitting, however, they employed

Table 5. Summary of results for the zero point of M_V(RR) at a fiducial metallicity of [Fe/H] = −1.9

Technique	Reference	M_V(RR)
	Trigonometric Parallaxes	
main sequence fit (HD 103095/M5)	text	0.54 ± 0.12
main sequence fit/$\pi_{\rm trig}$	Chaboyer et al. (1998)	0.39 ± 0.08
main sequence fit/$\pi_{\rm trig}$	Reid (1997)	0.28 ± 0.1
main sequence fit/$\pi_{\rm trig}$	Gratton et al. (1997)	0.40 ± 0.05
main sequence fit/$\pi_{\rm trig}$	Pont et al. (1998)	0.37 ± 0.08
$\pi_{\rm trig}$/white dwarfs	Renzini et al. (1996)	0.28 ± 0.2
$\pi_{\rm trig}$/Field HB stars	Gratton (1998)	0.59 ± 0.10
$\pi_{\rm trig}$/HD 17072	Gratton (1998)	0.82 ± 0.15
	Statistical Parallaxes	
Cluster internal dynamics	Rees (1996)	0.55 ± 0.05
Field RR Lyraes (*Hipparcos*)	Fernley et al. (1998a)	0.70 ± 0.15
Field RR Lyraes (*Hipparcos*)	Tsujimoto et al. (1998)	0.63 ± 0.10
Field RR Lyraes (re-analysis)	Popowski & Gould (1998)	0.68 ± 0.12

the white dwarf sequence. This is a very challenging observational problem, given the faintness of the white dwarfs, even in the nearest globular clusters, and the requirement that colors be determined to high accuracy. Figure 126 shows the results applied to NGC 6752, with a cluster modulus of 13.05 mag, and which yields $M_V(\mathrm{RR}) = +0.28 \pm 0.2$ mag. Aside from the uncertainties of the observations and the fitting process, two additional systematic effects might be present. First, as we noted above, NGC 6752 does not have any RR Lyrae variables, and considerable care must be taken to estimate the relative magnitudes that such stars would have compared to the cluster's observed blue horizontal branch stars. Second, the assumption has been made that solar-metallicity white dwarfs and low-metallicity white dwarfs follow the same cooling tracks. This is a much more reasonable assumption than that the main sequences overlap, which as we have seen, do not overlap. If the white dwarf densities are constant, then at fixed temperature/color, $L \propto M^{2/3}$. Thus, for example, if disk white dwarfs have a typical mass of 0.55 $M_\odot$, but halo white dwarfs have a typical mass of 0.50 $M_\odot$, such a fit would introduce a systematic error of only 0.10 mag. In this case, the M_V(RR) results of Renzini et al. (1996) would have to be increased (i.e., dimmed) by 0.10 mag. The post-AGB stars studied by Bond (1997) have luminosities roughly consistent with pre-white dwarfs of about 0.55 $M_\odot$, but Richer et al.

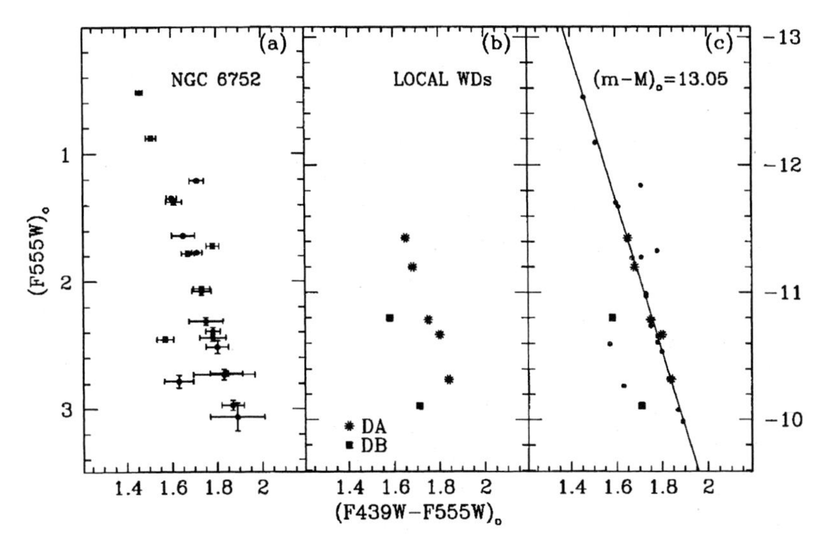

Fig. 126. Fitting of the white dwarf sequence in NGC 6752 to that of field white dwarfs with accurate parallax data, taken from Renzini et al. (1996)

(1997) have suggested the white dwarf sequence in M4 is consistent with M $= 0.51 M_\odot$. This analysis also yields an estimate for M_V(RR) of +0.8 mag at the metallicity of M4, [Fe/H] = −1.19 (Carretta & Gratton 1997). This means M_V(RR) = 0.66 mag at [Fe/H] = −1.9, but a reliable error estimate is hard to obtain because the cluster reddening is both large and abnormal (see Liu & Janes 1990b).

Finally, Gratton (1998) searched the *Hipparcos* database to identify metal-poor field horizontal branch stars (blue, red, and variable) with well-measured trignometric parallaxes. Taking a weighted mean, and upon revising the result to [Fe/H] = −1.9, his results imply M_V(RR) $= 0.59 \pm 0.10$ mag. The results depend fairly strongly on the star with the best parallax, HD 17072. Carney et al. (1998b) have confirmed that the star is a metal-poor ([Fe/H] = −1.17) red horizontal branch star with $[\alpha/\mathrm{Fe}] = +0.3$. If we again take the most conservative approach possible, and use only this best-measured star, M_V(RR) $= 0.82 \pm 0.15$ mag.

Statistical Parallaxes

Statistical parallax is also a powerful tool in the estimation of stellar luminosities, and while it is not as direct a method as trigonometric parallax, and requires much larger samples, such samples exist, in clusters and in the field.

In the case of globular clusters, one compares the dispersions in proper motion within the cluster with those in radial velocity to estimate the cluster distance and, hence, M_V(RR). One must account for the effects of anisotropic velocity vectors, including rotation, and Rees (1996) has presented the latest summary. Taking all the available results together, he estimated M_V(RR) = 0.55±0.05 mag at [Fe/H] = −1.9, a remarkably small error bar. This is a very exciting result, and deserves continued observational efforts and analyses. The current limits are set by the observed proper motion dispersions, and archival and follow-up *HST* observations of many clusters should ultimately provide improved precision and accuracy.

In the field, proper motion and radial velocity data for field RR Lyraes have been analyzed with encouraging results (at least in terms of the error bars). Pre-*Hipparcos* results included those of Barnes & Hawley (1986), Strugnell et al. (1986), and Layden et al. (1996) who found M_V(RR) = 0.62 ± 0.14, 0.64±0.2, and 0.65±0.12 mag, respectively, upon correcting to [Fe/H] = −1.9. (Note that the first result differs slightly from those published. This revised value employs the same reddening scale employed by Strugnell et al.) Fernley et al. (1998a) exploited the proper motions from *Hipparcos*, and found M_V(RR) = 0.70 ± 0.15 mag, essentially the same as the ground-based results. Tsujimoto et al. (1998) have undertaken a similar analysis, and correcting to the same metallicity, found M_V(RR) = 0.63 ± 0.10 mag. A thorough reanalysis of all data by Popowski & Gould (1998) confirms these results, with M_V(RR) = 0.74 ± 0.12 and 0.68 ± 0.12 mag at [Fe/H] = −1.60 and −1.90, respectively. Note that this is also revealed in the analysis of the kinematics of field RR Lyraes by Martin & Morrison (1998), who could obtain agreement with kinematically unbiased samples of metal-poor stars if M_V(RR) lies between 0.60 and 0.70 mag at [Fe/H] = −1.9, but not if M_V(RR) were 0.2 to 0.3 mag brighter.

Summary

Despite all the efforts, involving relatively direct measures of M_V(RR) and employing high-quality data, the spread in the values in Table 5 is quite disappointing, especially considering Fig. 125. We want an uncertainty in M_V(RR) of ±0.10 mag or so. What we find instead are results that are bimodal, with values of about +0.4 mag at [Fe/H] = −1.9 favored by the trignometric parallaxes as applied to clusters using main sequence and white dwarf sequence fitting, and values of about +0.6 mag favored by the trigonometric parallaxes of field horizontal branch stars and statistical parallaxes of field horizontal branch stars and of a modest subset of globular clusters.

What Needs to be Done? Some of the extremely conservative assumptions invoked above may be relaxed once high-resolution, high-S/N spectra are obtained and analyzed for the field dwarfs and horizontal branch stars to verify their [Fe/H] values and see if their [α/Fe] are normal for their metallicities. Obviously more high-quality main sequence photometry is needed for clusters with **(a)** low reddening, **(b)** well populated horizontal branches, and **(c)** spectroscopic [Fe/H] and [α/Fe] values.

One obvious gap in Table 5 is statistical parallaxes of metal-poor main sequence field stars. A very large number of metal-poor stars with good proper motions and radial velocities are known, and such work is underway by myself and my colleagues Dave Latham, John Laird, and Luis Aguilar.

Do the field and cluster horizontal branch and RR Lyrae stars have identical M_V values at equal metallicities? This was suggested by Gratton (1998) as a possible resolution to the dichotomy since, except for the results of Rees (1996), all the methods that measure cluster RR Lyrae luminosities directly through trigonometric parallaxes yield bright values, while all those that measure field RR Lyrae luminosities yield faint values. A clue may be contained in Fig. 87 (taken from Kraft 1994; see also Kraft et al. 1997). The figure suggests that field stars may not mix as deeply or as thoroughly as cluster stars. Figure 92 suggests that deep mixing includes helium. The helium mixing could affect the luminosity of the horizontal branch as well as where on the horizontal branch a star begins burning its core helium. Sweigart (1996) pointed out that the deep mixing would increase the number of stars on the blue side of the horizontal branch and not necessarily the numbers of RR Lyrae variables. To put it more physically, the stars most affected by the mixing would begin core helium burning on the blue horizontal branch. However, many of the RR Lyraes being studied may be *descendents* of such stars. That is certainly a likely explanation for the period shift we have seen between the Oosterhoff I and Oosterhoff II clusters M3 and M2. And it may also explain why the horizontal branch of M13 is so blue compared to that of M3. So if cluster red giants undergo sufficient mixing to produce much bluer horizontal branches than their field counterparts, then the cluster and field RR Lyraes may generally be in different evolutionary states, even at the same metallicity. Field variables, at least those selected in an unbiased fashion, would mostly be near the zero-age horizontal branch. Clusters might more heavily populate the blue side of the horizontal branch and their RR Lyraes would then be more highly evolved and more luminous, as has, perhaps, been suggested by the *Hipparcos* data.

The idea that field and cluster RR Lyraes differ in luminosity at equal metallicity has been tested by Catelan (1998) and by Lee & Carney (unpublished). The basic idea is to exploit the period shift by comparing field and cluster stars at equal temperatures and with similar metallicities. The period shift should reveal a luminosity difference, certainly if it is as large as 0.2 mag in M_V(RR). Catelan (1998) derived a new relation between equilibrium temperature Θ_{eq} for RR Lyraes, [Fe/H], and A_B, using the results of Carney et al. (1992):

$$\Theta_{eq} = (0.868 \pm 0.014) - (0.084 \pm 0.009)A_B - (0.005 \pm 0.003)[Fe/H] \,. \quad (69)$$

He then compared field and cluster RR Lyraes in two metallicity regimes, $-1.50 \leq [Fe/H] \leq -0.95$ and $[Fe/H] \leq -1.85$ using (69) as well as (38) defined by Carney et al. (1992). As one might expect, the additional use of $\log P_0$ in (38) yields smaller scatter in the resultant $\log P$ *vs.* $\log T_{eq}$ plane, but in both

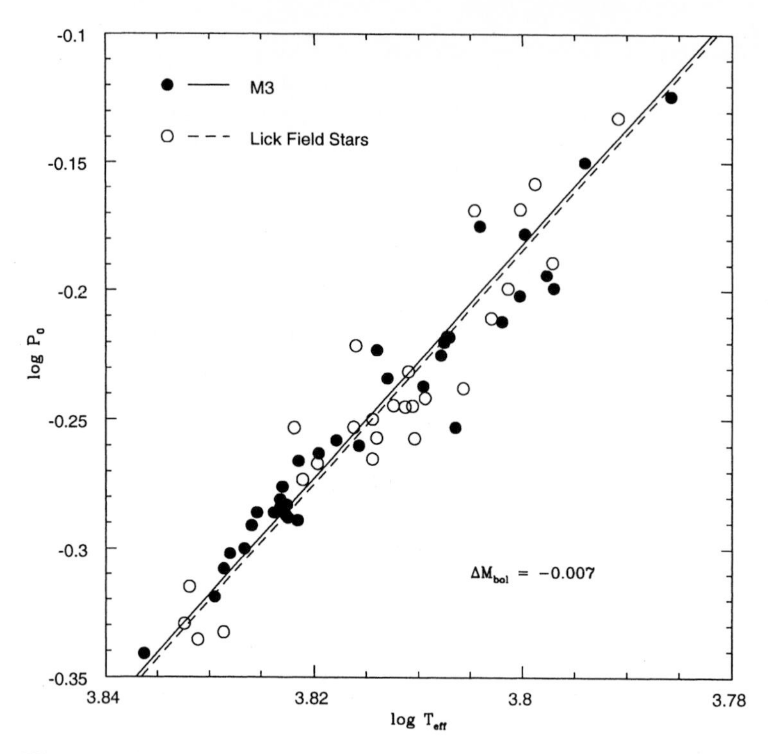

Fig. 127. A comparison of RR Lyraes in M3 with field RR Lyraes with very similar ΔS values. The good agreement in the log P_0 *vs.* $T_{\rm eq}$ plane indicates the luminosities of the two ensembles are the same to within about 0.01 mag

cases, Catelan found no sign of a period shift between the field and cluster variables. We repeat the analysis here, with two variations. First, we compare field and cluster variables in very restricted metallicity domains: ΔS must be the same to within ± 0.5. Equation (61) (and the other calibrations of ΔS *vs.* [Fe/H]) predicts that the [Fe/H] values will be the same to within about 0.1 dex. This is a somewhat tighter comparison than employed by Catelan (1998). Second, we employ a different set of field RR Lyraes. Catelan's variables were those in the *Hipparcos* sample employed by Tsujimoto et al. (1998). Carney et al. (1992) stressed the importance of *unbiased* samples of field RR Lyraes in such analyses, and this has also been discussed in Sect. 3. We therefore again employ the unbiased sample of RR Lyrae variables discovered at Lick Observatory by Kinman et al. (1965, 1966, 1982), and with ΔS values published or unpublished by Butler et al. (1979, 1982), Kinman et al. (1985), and Suntzeff et al. (1991). We compare the values for field RR Lyraes with those in three clusters in the following figures: M3, M68, and M15, with the equilibrium temperatures computed using (38). The field stars' luminosities are expected to agree with those in M3 since the cluster is so rich in RR

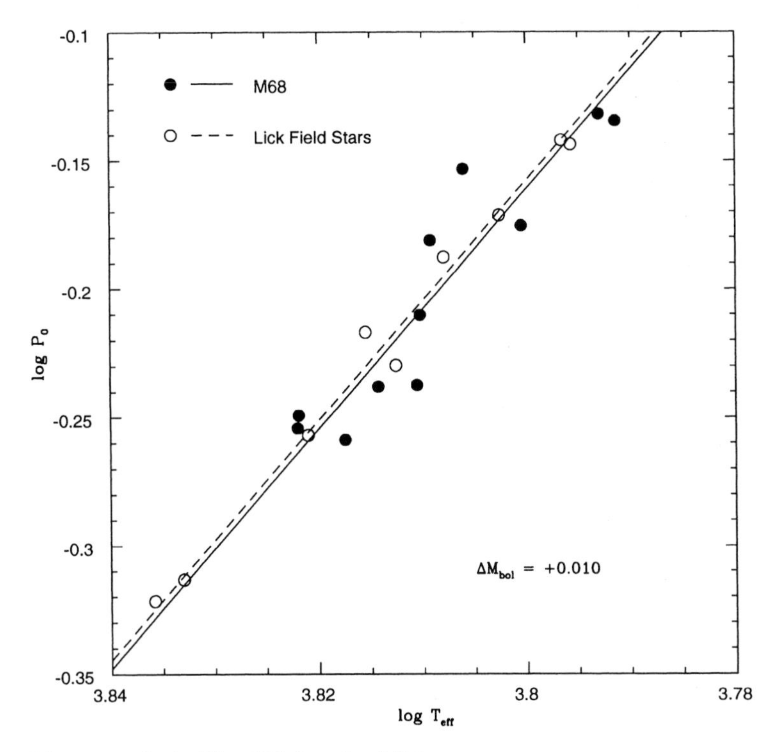

Fig. 128. As in Fig. 127, but for M68

Lyrae variables that they are likely to be near the zero age horizontal branch. Figure 127 shows this is indeed the case. We have already seen that the variables in M2 are brighter than those in M3 (see Fig. 31), and therefore the M2 variables are brighter than those in the field. How about the most metal-poor clusters? Do the field and cluster variables differ in luminosity? Figures 128 and 129 indicate that they are the same. Thus in general main sequence fitting to globular clusters should yield the same values of M_V(RR) and direct determination of field RR Lyraes' luminosities. That they do not therefore remains a very important mystery.

What else might we explore? The period shifts were undertaken assuming that we had successfully matched field and cluster variables at equal metallicities. If sufficient helium has been mixed into the photospheres of the cluster variables but not the field variables, the decrease in the H^- opacity will lead to stronger lines. The ΔS measurement may not be strongly affected, however, since the Ca II lines and hydrogen lines fall in the same wavelength domain and are likely to be affected about equally. But helium could alter the pulsational properties of the stars, as discussed by Catelan (1998). Therefore, it might be interesting to make a thorough comparison of line strengths as a function of

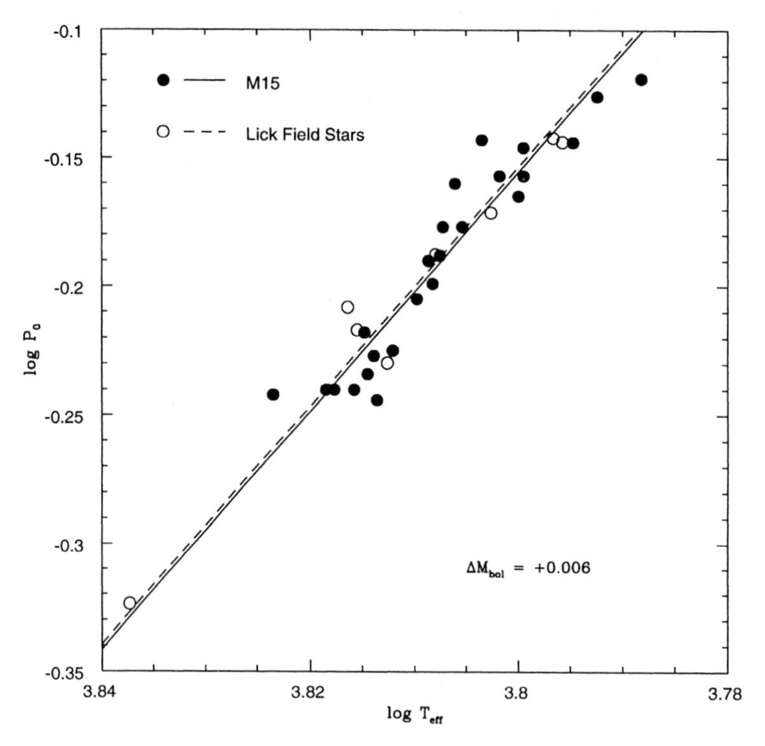

Fig. 129. As in Fig. 127, but for M15

wavelength (so differing degrees of H^- domination of the continuum opacity are sampled) between field and cluster RR Lyraes.

Figure 62 is also a cause for concern. Recall that it compares spectroscopic abundances for globular cluster red giants with those *inferred* from the ΔS metallicity indicator, which was calibrated using spectroscopic abundance analyses of field RR Lyraes. The discrepancy between the cluster red giant and RR Lyrae metallicities may be attributed to one or both of two causes.

- The gf values employed in the two sets of analyses, the cluster red giants and the field RR Lyraes, have systematic errors that are (probably) a function of the excitation potential χ. Thus the abundances of the hotter RR Lyraes differ from those of the cooler red giants. *On the other hand,*
- if the gf values have only minor systematic effects, this may suggest that metallicities derived using ΔS will differ when applied to field and cluster RR Lyraes. The cluster RR Lyraes are interpreted as being more metal-poor than the cluster red giants. This could be due to a continuum opacity effect of the type mentioned.

The first explanation does not appear to be valid. We have seen that Clementini et al. (1995) and Carretta & Gratton (1997) employed the same gf

values in their analyses of field RR Lyrae variables and cluster red giants, but that the discrepancies between the cluster metallicities from ΔS and red giants persists. It would therefore be worthwhile to compute synthetic spectra to assess how ΔS changes as the helium abundance in a stellar photosphere is increased. Figure 62 suggests the differences are largest for the more metal-poor clusters. This is intriguing also since the discrepancies in the derived metallicities seem to be most pronounced for $[\mathrm{Fe/H}]_{\mathrm{spec}} < -1.2$ or so. Recall Fig. 91. Stars more metal-poor than about this metallicity appear to mix more deeply than the more metal-rich stars within ω Cen. If this is true of stars in other clusters, at least those in Fig. 62, then perhaps we have an explanation for the failure of a ΔS calibration obtained from relatively unmixed field stars to work for cluster stars, which may have mixed much more thoroughly. In this case the period shift analyses of Catelan (1998) and those in Figs. 127, 128, and 129 are invalid because the field and cluster variables may well have different chemical compositions and inherently different pulsational properties at equal temperatures.

9.4 Comparison with Results from Cepheid Variables

The RR Lyrae distance scale, as summarized in Table 5, remains indeterminate, with values near $M_V(\mathrm{RR}) = 0.6$ and 0.3 at $[\mathrm{Fe/H}] = -1.9$ both being possible. While a simple average could be taken, it is wisest to understand the systematics that have led to this apparent dichotomy so that we may avoid mixing accurate results with those distorted by some systematic effect(s). The effects of this dichotomy are best illustrated by comparing the ages derived in producing Figs. 104 and 110. In the former case, $M_V(\mathrm{RR}) = 0.20[\mathrm{Fe/H}] + 0.98$ was employed, which, at $[\mathrm{Fe/H}] = -1.9$, yields $M_V(\mathrm{RR}) = 0.60$. In the latter case, the zero point was changed to 0.725, so that at $[\mathrm{Fe/H}] = -1.9$, $M_V(\mathrm{RR}) = 0.35$. In the former case, maximum ages reach about 22 Gyrs, while in the latter case, they are only about 17 Gyrs (and might be less if the M2 *vs.* M3 comparison is correct). It is better to follow through with the two zero points until we have determined which of them is least affected by systematics, or, perhaps more to the point, which is most applicable to the problem at hand, if indeed field and cluster variables differ.

The ages of the oldest globular clusters share a common fundamental reliance on accurate distances with the Universal expansion age since the Hubble constant depends directly on the Cepheid distance scale. It is therefore worth comparing the distances to the Large Magellanic Cloud (LMC) and M31, both of which have distance estimates from Cepheids and from RR Lyraes or globular cluster horizontal branches.

The use of Cepheids to determine extragalactic distances has been reviewed thoroughly by Feast & Walker (1987), and the true distance moduli of the LMC and M31 were estimated to be 18.5 ± 0.1 and 24.25 ± 0.1. We do not discuss distance estimates for the Small Magellanic Cloud because depth effects may be significant and because it is considerably more metal-poor than

Table 6. Summary of distance moduli for the LMC and M31 from Cepheids and RR Lyraes

Technique	LMC	M31
	Cepheids	
Cepheids (pre-*Hipparcos*)	18.5 ± 0.1	24.5 ± 0.1
Cepheids (*Hipparcos*)	18.7 ± 0.1	24.8 ± 0.1
	RR Lyraes	
$M_V(\mathrm{RR}) = 0.20[\mathrm{Fe/H}] + 0.98$	18.4 ± 0.1	24.4 ± 0.1
$M_V(\mathrm{RR}) = 0.20[\mathrm{Fe/H}] + 0.68$	18.7 ± 0.1	24.7 ± 0.1

either M31 or the LMC, and thus any metallicity sensitivity in the Cepheid period–luminosity (P–L) relation would be more significant. (See Caldwell & Coulson 1986 for a discussion.) Using parallaxes from *Hipparcos*, Feast & Catchpole (1997) redetermined the zero point of the P–L relation, finding it 0.2 mag brighter and, hence, that the distance moduli of M31 and the LMC increased by 0.2 mag, as seen in Table 6. Feast et al. (1998) have confirmed the new zero point by combining *Hipparcos* proper motions with values of Oort's constant A derived from radial velocities, essentially a statistical parallax approach. The constant A is related to the Galaxy's rotation at the solar distance: $A = (1/2)[(\Theta_0/R_0 - (\mathrm{d}\Theta/\mathrm{d}R)_{R_0}\cdot)]$ The Cepheid parallaxes are very small, no larger than 3.3 milli-arcseconds (except for Polaris, which is 7.6), but Feast & Catchpole (1997) avoided the systematic errors such as those described above by not computing M_V values, but directly determining the zero point of the P–L relation using their data. Their results are, of course, more vulnerable than most in terms of any very small systematic errors remaining in the *Hipparcos* results, or even systematic errors in the estimation of the parallax errors.

We have estimated distances to the LMC and to M31 using horizontal branches of globular clusters. For the LMC, we combined the globular cluster data summarized by Walker (1992) (who drew attention to the discrepancy between the Cepheid and RR Lyrae distance scales for the LMC) with the new results of Olsen et al. (1998). Cluster metallicities were taken from Olszewski et al. (1991). LMC moduli were computed using the two possible zero points for RR Lyrae luminosities. The formal internal errors are small, with the error of the mean for the LMC distance moduli being only 0.04 mag, much larger than the 0.30 mag difference in the two zero points. For M31, we used the globular cluster data summarized by Fusi Pecci et al. (1996).

Table 6 makes two points clearly. The most striking one is that the RR Lyrae and Cepheid distance scales agree *if*

- The *Hipparcos* parallaxes for field Cepheids are correct and the brighter zero point is appropriate for globular cluster horizontal branches;

  or
- The *Hipparcos* parallaxes are in error so that the pre-*Hipparcos* distance estimates are correct and the fainter RR Lyrae zero point is applicable to globular cluster horizontal branches.

In essence, if the *Hipparcos* parallaxes are correct for the Cepheids as well as the metal-poor field main sequence and horizontal branch stars, then there must be a systematic difference between the luminosities of horizontal branch stars in the field and in globular clusters. This is because the *Hipparcos* results yield consistent distances to the LMC and to M31 from the Cepheids and the RR Lyrae luminosities obtained from main sequence fitting to globular clusters, and applied to globular clusters in the LMC and M31, but the *Hipparcos* results for field horizontal branch stars imply fainter luminosities. If such a difference between field and cluster horizontal branch stars does not exist, the *Hipparcos* results are called into question. One way to choose between these two options would be to derive the distance to the LMC using its own *field* RR Lyraes, of which almost 8000 have been discovered as part of the MACHO project (Alcock et al. 1996). *Does the LMC distance modulus depend on whether field or cluster horizontal branch stars are employed?*

We may take some comfort, at least, from the second, more subtle point revealed by Table 6. The relative distance moduli of the LMC and M31 are the same to within 0.1 mag. Thus if we can resolve the systematic distance scale problem, distances to other galaxies relative to these two should be quite reliable.

9.5 Other Distance Indicators

There are several means by which the RR Lyrae luminosity scale may be tested, as summarized by Huterer et al. (1995). Most have been discussed above, but one additional test is useful. The distance to the Galactic center has been estimated by a variety of techniques, summarized by Reid (1993). There is only one direct method: the measured expansion rates of H_2O masers. Three measurements have been made (Reid et al. 1988a,b; Gwinn et al. 1992), and Reid (1993) concluded that the distance is 7.2 ± 1.3 kpc, which converts, approximately, to $(m-M)_0 = 14.29 \pm 0.39$ mag. Caldwell & Coulson (1987) employed Cepheids to estimate this distance, which was based upon a distance scale in which the distance modulus for the LMC is 18.65 ± 0.07 mag (Caldwell & Coulson 1986), very close to the *Hipparcos*-based results of Feast & Catchpole (1997). They derived a distance modulus for the Galactic center of 14.46 ± 0.18 mag, within the error bars of the maser results. Carney et al. (1995) used the infrared period–luminosity relation for RR Lyraes in Baade's Window and the fainter RR Lyrae luminosity zero point (see Fig. 30) and found $(m - M)_0 = 14.46 \pm 0.10$ mag, identical to the results from the

Cepheids and therefore also consistent with the masers. The zero point of the RR Lyrae luminosities was about 0.70 mag at [Fe/H] = −1.9. The brighter zero point, 0.40 mag, would, of course, change this to 14.76 ± 0.1, which is quite discrepant with the maser results. It appears that a fainter RR Lyrae luminosity applied to *field* RR Lyraes in Baade's Window results in agreement with the *Hipparcos*-based Cepheid distance estimates. As we have seen, the RR Lyraes and Cepheids yield inconsistent distances to the LMC when the fainter zero point is employed, but consistency is achieved for the brighter zero point. But the LMC distance modulus is obtained *via* RR Lyrae luminosities being applied to clusters. This again suggests a difference between field and cluster variables, and underscores the need to compare field and cluster RR Lyraes directly in the LMC.

One useful future approach would be to use primary distance determinations for the LMC. When space-based interferometers capable of positional measurements of a few micro-arcseconds are in operation, the trigonometric parallax of the LMC will be measured directly. But for now, the best method is to exploit the expanding ring of SN 1987A. Panagia et al. (1991) made the first measurement, finding $(m-M)_0 = 18.55 \pm 0.13$ mag. Gould's (1995) subsequent analysis indicated that $(m-M)_0 < 18.37 \pm 0.04$ mag, while Sonneborn et al. (1997) derived 18.43 ± 0.10 mag. Gould & Uza (1998) have produced the most recent analysis, along with a discussion of the remaining uncertainties, and argue that $(m-M)_0 < 18.44 \pm 0.05$ mag. Only the first result, that of Panagia et al. (1991), is marginally consistent with the *Hipparcos*-based Cepheid distance scale. One additional direct measurement technique has become feasible. Guinan et al. (1998) have reported the "first accurate distance determination to the LMC using an eclipsing binary system". After correcting for extinction and the star's approximate position within the LMC, they found $(m-M)_0 = 18.30 \pm 0.07$, consistent with most of the SN 1987A expanding ring results, but not that of the *Hipparcos*-based Cepheid distance scale.

9.6 Additional Age Estimators

Having compared the distance scales that determine globular cluster ages with those from Cepheids, it is now worth comparing the globular cluster ages we have derived with ages obtained by independent methods.

White Dwarf Cooling Ages. A white dwarf is supported by electron degeneracy, and thus in the absence of a companion (and possible mass transfer), should retain a constant radius. The heat from its non-degenerate baryons will be radiated away, rapidly while it is hot, more slowly as it cools. The white dwarf sequence is thus one of constant radius, and because the cooling occurs more slowly at lower temperatures, hence fainter luminosities, a complete census of white dwarfs in the field or in a cluster should find more cool/faint white dwarfs than hot/bright white dwarfs. In other words, the luminosity

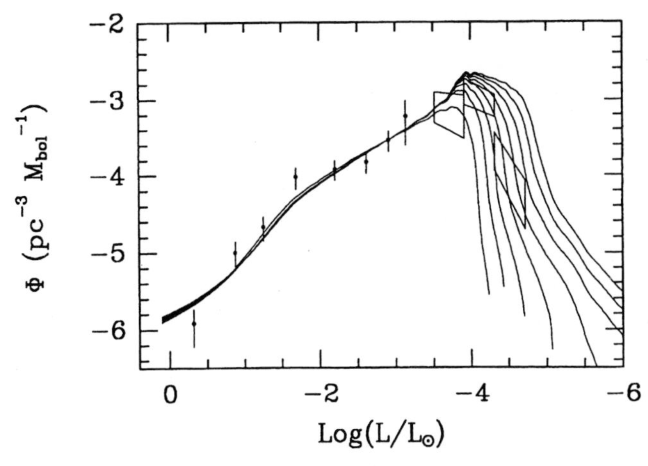

Fig. 130. The white dwarf luminosity function compared to predictions from white dwarf cooling theory for ages of 6 to 13 Gyrs, taken from Wood (1992)

function should increase as the luminosity drops. For a finite age of a stellar population or an individual cluster, there should be a luminosity below which no white dwarfs are found, simply due to the fact that not enough time has elapsed for them to cool to such faint levels. This is the essence of white dwarf cooling ages, first proposed by Schmidt (1959). Winget et al. (1987) used the observed local white dwarf luminosity function and white dwarf cooling theory to estimate the age of, essentially, the Galactic disk, finding it to be around 9 Gyrs. Figure 130 shows the results obtained by Wood (1992) using a very similar analysis, from which he deduced a disk age of between 7.5 and 11 Gyrs. More recent estimates, employing improved opacities and luminosity functions, yield disk ages of 11^{+4}_{-2} Gyrs (Wood et al. 1995). The systematic effects of uncertain physics aside, these ages provide lower limits to the ages of the globular clusters, but are not adequate to distinguish between the smaller and larger ages obtained using the brighter and fainter zero points of the M_V(RR) *vs.* [Fe/H] relation. The age estimates are, however, of great value in understanding the chronology of the Galaxy and answering the question of when star formation began in the disk compared to the halo.

In principle, a white dwarf cooling age estimate could be made for the nearest globular clusters. This would result in an unbiased estimate (in terms of physics) of the timing of initial star formation in the disk and halo. And it would, of course, provide another estimate for the ages of the globular clusters. The problem is daunting observationally, however. Richer et al. (1997) used *HST* to obtain photometry in the nearby globular cluster M4 down to $V \approx 29$. The white dwarf sequence was detected beginning at $V \approx 22$, and was detected down to the limit of their photometry. This established a *minimum age* for M4 of 9 Gyrs.

Radioactive Dating. Aside from the universal expansion age, which is not discussed here, there is one more type of chronometer available: radioactive dating. The ages of rocks, be they terrestrial, lunar, martian, or meteoritic, may be dated by careful analysis of the abundances of stable and radioactive isotopes. Such analyses lead to an estimate of the age of the solar system and the Sun, and, further, to an estimate of the abundances of these stable and radioactive species when the solar system formed. In principle then, one may estimate the "age of the elements" if in addition to these measurements, one knows: (a) the relative *production ratios* for these elements, so that one may estimate for how long the radioactive species have been decaying *prior* to the formation of the solar system, and (b) the *production history* of these species. Both points are vital. For example, if we know the production ratio of, say, europium (which has a stable isotope) and thorium (which does not), and if we measure the ratio of their abundances when the solar system formed, then we can estimate the "average age" of the thorium. The second point is crucial because the age of the Galaxy will depend on whether the europium and thorium were both produced in, for example, a burst when the Galaxy formed, in which case we have measured the age of the Galaxy, or if the elements were formed more steadily, in which case we must make a significant correction to determine the age of the Galaxy because only a smaller fraction of the thorium will have had the full span of time to decay. The subject of nucleocosmochronology has been well reviewed by Cowan et al. (1991), The most comforting result is that the ages derived are broadly consistent with those estimated for the globular clusters using any of the RR Lyrae luminosity scales. The most dissatisfying result is that a wide range of ages are possible, from 10 to 20 Gyrs.

However, radioactive ages of *individual stars* may, in principle, be obtained, circumventing the need to know some details of the Galaxy's star formation history (which depends on either the ages of clusters or white dwarf cooling theory). Butcher (1987), using very high resolution and very high S/N spectra, was able to detect the 4019.13 Å line of thorium and a line of neodymium at 4018.82 Å to measure the ratio of two neutron capture elements: one radioactive (thorium) and the other one stable (neodymium). For a constant rate of formation of these elements, and some assumptions about their production ratios, he derived an age for the Galaxy of about 10 Gyrs. If star formation was higher in the recent past, ages of 11 to 12 Gyrs were possible. Morell et al. (1992) re-analysed Butcher's data, paying special attention to other lines that might overlap or be co-incident with the neodymium and thorium lines, which would obviously lead to systematic errors in their elemental abundance ratios, and confirmed Butcher's general conclusions. Unfortunately, none of the stars studied by Butcher (1987) were particularly metal-poor, so none could be said to have formed very early in the Galaxy's history. They therefore had to make some significant assumptions regarding the Galaxy's star formation history. Recently, however, larger telescopes, and more efficient spectrographs

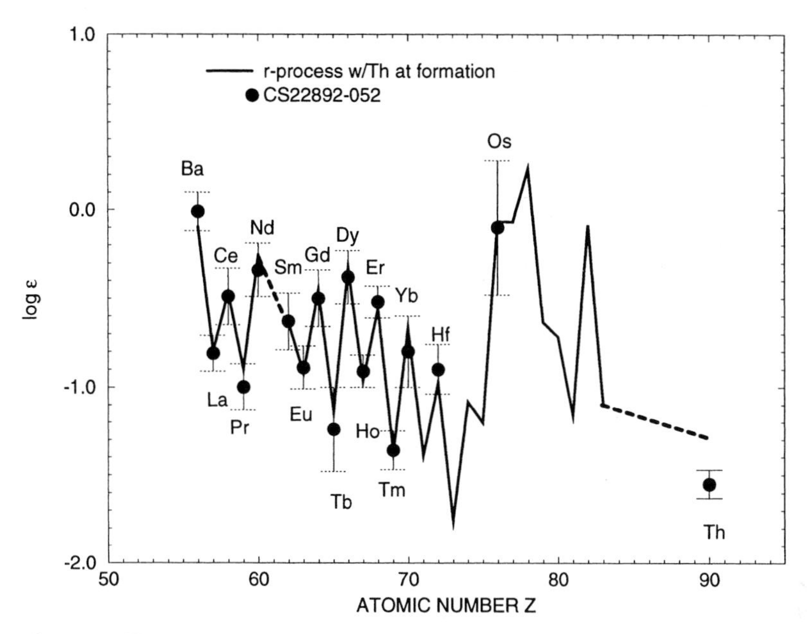

Fig. 131. Neutron capture elemental abundances, including thorium, in CS22892-052 compared to the solar r–process distribution, taken from Cowan et al. (1997). The *dashed line* represents the expected abundance of thorium at the time of the star's formation

and detectors have extended this age-dating method to very metal-poor stars. The most exciting target is CS 22892-052, a star with a very high proportion of r–process nucleosynthesis evident in its spectrum (see Fig. 77). Because the star is so metal-poor, with [Fe/H] $= -3.1$ (Sneden et al. 1994b; McWilliam et al. 1995), a radioactive age estimate will depend much less on the history of star formation. It still depends, of course, on proper understanding of the production ratios of the neutron capture elements, but that seems to be under reasonable control. Recent work using *HST* by Cowan et al. (1996, 1997), for example, has managed to reveal abundances in the "third r–process peak", as shown in Fig. 131. The derived abundances are (remarkably) consistent with the solar system r–process pattern. Sneden et al. (1996) and Cowan et al. (1997) have also measured the thorium abundance in CS 22892-052 and used this to estimate the age of the star and of the Galaxy. Some understanding of the star formation history of the Galaxy is still necessary because that affects the current (and original) solar system abundance patterns. For a disk that is 10.5 Gyrs old and a constant star formation rate, the age of the star and the Galaxy is 18 ± 4 Gyrs. If the disk is younger, say 8 Gyrs, the age estimate decreases by about 1 Gyrs. If the disk has been subject to infall of primordial material throughout its life, on the other hand, the age is reduced by about 0.7 Gyrs. As in the case of the RR Lyrae luminosity scale, it may be adequate to determine the answer on the basis of just one star if it is clearly the best studied, but it is also true that we crave validation through larger samples.

9.7 Summary

It is still too early to have complete confidence in any one answer to the question of how old are the globular clusters. It seems that most of the systematic effects due to physics and computational methods are under control, but Fig. 125 shows clearly where most of the work remains to be done. Table 5 reveals the disappointing and dichotomous suite of results for the M_V(RR) values relevant to Fig. 125. The suggestion that globular cluster horizontal branches may be more luminous than those of field stars explains most, but not all, of the results. (The exception is the statistical parallax work within clusters by Rees 1996.) The suggestion would help resolve the distance estimates to the LMC and M31 using Cepheids and cluster RR Lyraes and the agreement between Cepheids and field RR Lyraes for the distance to the Galactic center. The suggestion fails to explain the lack of a period shift, however, between the Galactic field and cluster RR Lyraes, unless the metallicity scales are systematically affected by photospheric helium enhancements in cluster stars, which would increase line strengths at fixed temperatures. A useful test will be the comparison of apparent RR Lyrae luminosities in globular clusters and of field stars in the LMC. But even if a difference is found, we will still need to understand its cause so that we can understand which clusters are most susceptible to such systematic effects and which are not. And it would not be a bad idea to review again the *Hipparcos* parallaxes and make certain that no systematic effects remain in either the parallaxes or the estimation of their internal errors. Suggestions of small angular scale parallax errors of the order of 0.001 arcsec have been made by Pinsonneault et al. (1998) and Soderblom et al. (1998). Errors of this size will not significantly affect the distances derived to globular clusters *via* main sequence fitting, but may affect the direct distance measures of field horizontal branch stars and Cepheids.

Acknowledgements
I am especially grateful to Lukas Labhardt, Bruno Binggeli, and their colleagues who made the visit to Switzerland so enjoyable personally and professionally. Many colleagues granted permission to use their figures, and Brian Chaboyer is thanked especially for contributing his by email. My fellow lecturers, Bill Harris and Tad Pryor, also taught me much and made me think about many new ideas.

References

1. Aguilar, L., Hut, P., & Ostriker, J. P. 1988, ApJ, 335, 720
2. Alcock, C., Allsman, R. A., Axelrod, T. S., Dennett, D. P., Cook, K. H., Freeman, K. C., Marshall, S. L., Peterson, B. A., Pratt, M. R., Quinn, P. J., Rodgers, A. W., Stubbs, C. W., Sutherland, S., & Welch, D. L. 1996, AJ, 111, 1146
3. Allen, C., & Santillan, A. 1991, Rev. Mex., 22, 255
4. Alonso, A., Arribas, S., & Martinez Roger, C. 1995, A&A, 297, 197
5. Alonso, A., Arribas, S., & Martinez Roger, C. 1996, A&A, 313, 873
6. Alonso, A., Arribas, S., & Martinez Roger, C. 1999a, A&A, 139, 335
7. Alonso, A., Arribas, S., & Martinez Roger, C. 1999b, A&A, 140, 261
8. Anders, R., & Grevesse, N. 1989, Geochim. Cosmochim. Acta, 53, 197
9. Andersen, J., Gustafsson, B., & Lambert, D. L. 1984, A&A, 136, 65
10. Andrews, J. W., Coates, P. B., Blackwell, D. E., Petford, A. D., & Shallis, M. J. 1979, MNRAS, 186, 651
11. Anthony–Twarog, B. J., & Twarog, B. A. 1994, AJ, 107, 1577
12. Armandroff, T. E., & Da Costa, G. S. 1991, AJ, 101, 1329
13. Armandroff, T. E., & Zinn, R. 1988, AJ, 96, 92
14. Armosky, B. J., Sneden, C., Langer, G. E., & Kraft, R. P. 1994, AJ, 108, 1364
15. Arnold, R., & Gilmore, G. 1992, MNRAS, 257, 225
16. Bailyn, C. D. 1995, AAR&A, 33, 133
17. Balachandran, S., & Carney, B. W. 1996, AJ, 111, 946
18. Barnes, T. G., III, & Hawley, S. L. 1986, ApJL, 307, L9
19. Beardsley, W. R., Gatewood, G., & Kamper, K. W. 1974, ApJ, 194, 637
20. Beers, T. C., & Sommer–Larsen, J. 1995, ApJS, 96, 175
21. Bell, R. A., & Gustafsson, B. 1978, A&AS, 34, 229
22. Bell, R. A., & Gustafsson, B. 1989, MNRAS, 236, 653
23. Bessell, M. S., Sutherland, R. S., & Ruan, K., 1991, ApJL, 383, L71
24. Biémont, E., Baudoux, M., Kurucz, R. L., Ansbacher, W., & Pinnington, E. H. 1991, A&A, 249, 539
25. Blackwell, D. E., Booth, A. J., Haddock, D. J., Petford, A. D., & Leggett, S. K. 1986, MNRAS, 220, 549
26. Blackwell, D. E., Ibbetson, P. A., & Petford, A. D. 1975, MNRAS, 171, 195
27. Blackwell, D. E., Ibbetson, P. A., Petford, A. D., & Shallis, M. J. 1979a, MNRAS, 186, 633
28. Blackwell, D. E., Ibbetson, P. A., Petford, A. D., & Willis, R. B. 1976, MNRAS, 177, 219
29. Blackwell, D. E., Petford, A. D., Haddock, D. J., Arribas, S., & Selby, M. J. 1990, A&A, 232, 396
30. Blackwell, D. E., Petford, A. D., & Shallis, M. J. 1979b, MNRAS, 186, 657
31. Blackwell, D. E., Petford, A. D., Shallis, M. J., & Simmons, G. J. 1980, MNRAS, 191, 445
32. Blackwell, D. E., Petford, A. D., Shallis, M. J., & Simmons, G. J. 1982a, MNRAS, 199, 43
33. Blackwell, D. E., Petford, A. D., & Simmons, G. J. 1982b, MNRAS, 201, 595
34. Blackwell, D. E., & Shallis, M. J. 1979a, MNRAS, 186, 669
35. Blackwell, D. E., & Shallis, M. J. 1979b, MNRAS, 186, 673
36. Blanco, V. M. 1992, AJ, 104, 734

37. Blanco, V. M., & Terndrup, D. M. 1989, AJ, 98, 843
38. Boesgaard, A. M., Deliyannis, C. P., Stephens, A., and King, J. R. 1998, ApJ, 493, 206
39. Boesgaard, A. M., King, J. R., Deliyannis, C. P., & Vogt, S. S. 1999, AJ, 117, 492
40. Boesgaard, A. M., & Steigman, G. 1985, AAR&A, 23, 319
41. Boffin, H. M. J., Paulus, G., & Cerf, N. 1992, in Binaries as Tracers of Stellar Formation, ed. A. Duquennoy & M. Mayor (Cambridge: Cambridge Univ. Press), 26
42. Böhm–Vitense, E. 1958, Zs. f. Astrophys., 46, 108
43. Bolte, M. 1992, ApJS, 82, 145
44. Bond, H. E. 1997, in The Extragalactic Distance Scale, ed. M. Livio, M. Donahue, & N. Panagia (Cambridge: Cambridge Univ. Press), 224
45. Booth, A. J., Blackwell, D. E., Petford, A. D., & Shallis, M. J. 1984, MNRAS, 208, 147
46. Booth, A. J., Shallis, M. J., & Wells, M. 1983, MNRAS, 205, 191
47. Boothroyd, A. I., Sackmann, I.–J., & Ahern, S. C. 1993, ApJ, 416, 762
48. Borissova, J., Catelan, M., Ferraro, F. R., Spassova, N., Buonanno, R., Iannicola, G., Richtler, T., & Sweigart, A. V. 1999, A&A, 343, 813
49. Briley, M. M., Hesser, J. E., & Bell, R. A. 1991, ApJ, 373, 482
50. Brown, J. A., & Wallerstein, G. 1993, AJ, 106, 133
51. Brown, J. A., Wallerstein, G., & Zucker, D. 1997, AJ, 114, 180
52. Buonanno, R., Buscema, G., Fusi Pecci, F., Richer, H. B., & Fahlman, G. G. 1990, AJ, 100, 1811
53. Buonanno, R., Corsi, C. E., Bellazzini, M., Ferraro, F. R., & Fusi Pecci, F. 1997, AJ, 113, 706
54. Buonanno, R., Corsi, C. E., Fusi Pecci, F., Richer, H. B., & Fahlman, G. G. 1993, AJ, 105, 184
55. Buonanno, R., Corsi, C. E., Pulone, L., Fusi Pecci, F., & Bellazzini, M. 1998a, A&A, 333, 505
56. Buonanno, R., Corsi, C. E., Pulone, L., Fusi Pecci, F., Richer, H. B., & Fahlman, G. G. 1995, AJ, 109, 663
57. Buonanno, R., Corsi, C. E., Zinn, R., Fusi Pecci, F., Hardy, E., & Suntzeff, N. B. 1998b, ApJL, 501, L33
58. Burki, G., & Mayor, M. 1986, in Instrumentation and Research Programs for Small Telescopes, IAU Sym. No. 118, ed. J. B. Hearnshaw & P. L. Cottrell (Dordrecht: Reidel), 385
59. Buser, R., & Kurucz, R. L. 1978, A&A, 70, 555
60. Buser, R., & Kurucz, R. L. 1992, A&A, 264, 557
61. Butcher, H. R. 1975, ApJ, 199, 710
62. Butcher, H. R. 1987, Nature, 328, 127
63. Butler, D. 1975, ApJ, 200, 68
64. Butler, D., Dickens, R. J., & Epps, E. 1978, ApJ, 225, 148
65. Butler, D., Kemper, E., Kraft, R. P., & Suntzeff, N. B. 1982, AJ, 87, 353
66. Butler, D., Kinman, T. D., & Kraft, R. P. 1979, AJ, 84, 993
67. Buzzoni, A., Fusi Pecci, F., Buonanno, R., & Corsi, C. E. 1983, A&A, 128, 94
68. Caldwell, J. A. R., & Coulson, I.M. 1986, MNRAS, 218, 223
69. Caldwell, J. A. R., & Coulson, I.M. 1987, AJ, 93, 1090
70. Canuto, V. M., & Mazzitelli, I. 1991, ApJ, 370, 295

71. Canuto, V. M., & Mazzitelli, I. 1992, ApJ, 389, 724
72. Caputo, F., Martinez Roger, C., & Paez, E. 1987, A&A, 183, 228
73. Carbon, D. F., Barbuy, B., Kraft, R. P., Friel, E. D., & Suntzeff, N. B. 1987, PASP, 99, 335
74. Carbon, D. F., Langer, G. E., Butler, D., Kraft, R. P., Trefzger, Ch. F., Suntzeff, N. B., Kemper, E., & Romanishin, W. 1982, ApJS, 49, 207
75. Carney, B. W. 1979, ApJ, 233, 211
76. Carney, B. W. 1983, AJ, 88, 623
77. Carney, B. W. 1996, PASP, 108, 900
78. Carney, B. W. 1999, in The Third Stromlo Symposium: The Galactic Halo, ASP Conf. Series 165, ed. B. K. Gibson, T. S. Axelrod, & M. E. Putman (San Francisco: ASP), 230
79. Carney, B. W., & Aaronson, M. 1979, AJ, 84, 867
80. Carney, B. W., Aguilar, L., Latham, D. W., & Laird, J. B. 1990a, AJ, 99, 201
81. Carney, B. W., Fry, A. M., & Gonzalez, G. 1998a, AJ, 116, 2984
82. Carney, B. W., Fulbright, J. P., Terndrup, D. M., Suntzeff, N. B., & Walker, A. R. 1995, AJ, 110, 1674
83. Carney, B. W., Laird, J. B., Latham, D. W., & Aguilar, L. A. 1996, AJ, 112, 668
84. Carney, B. W., Laird, J. B., Latham, D. W., & Kurucz, R. L. 1987, AJ, 94, 1066
85. Carney, B. W., & Latham, D. W. 1985, ApJ, 298, 803
86. Carney, B. W., & Latham, D. W. 1986, AJ, 92, 60
87. Carney, B. W., Latham, D. W., & Laird, J. B. 1989, AJ, 97, 423
88. Carney, B. W., Latham, D. W., & Laird, J. B. 1990b, AJ, 99, 572
89. Carney, B. W., Latham, D. W., Laird, J. B., & Aguilar, L. A. 1994, AJ 107 2240
90. Carney, B. W., Latham, D. W., Laird, J. B., Grant, C. E., & Morse, J. A. 1999, AJ, submitted
91. Carney, B. W., & Lee, J.-W. 1999, in preparation
92. Carney, B. W., Lee, J.-W., & Habgood, M. J. 1998b, AJ, 116, 424
93. Carney, B. W., Storm, J., & Jones, R. V. 1992, ApJ, 386, 663
94. Carney, B. W., Wright, J. S., Sneden, C., Laird, J. B., Aguilar, L. A., & Latham, D. W. 1997, AJ, 114, 363
95. Carretta, E., Cacciari, C., Ferraro, F. R., Fusi Pecci, F., & Tessicini, G. 1998, MNRAS, 298, 1005
96. Carretta, E., & Gratton, R. G. 1997, A&AS, 121, 95
97. Catelan, M. 1993, A&AS, 98, 547
98. Catelan, M. 1998, ApJL, 495, L81
99. Caughlan, G. R., & Fowler, W. A. 1988, Atomic & Nuclear Data Tables, 40, 284
100. Cavallo, R. M., Sweigart, A. V., & Bell, R. A. 1998, ApJ, 492, 575
101. Chaboyer, B. 1995, ApJL, 444, L9
102. Chaboyer, B., & Demarque, P. 1994, ApJ, 433, 510
103. Chaboyer, B., Demarque, P., Kernan, P. J., & Krauss, L. M. 1998, ApJ, 494, 96 (CDKK98)
104. Chaboyer, B., Demarque, P., & Sarajedini, A. 1996, ApJ, 459, 558
105. Chaboyer, B., & Kim, Y.-C. 1995, ApJ, 454, 767
106. Chandrasekhar, S. 1957, An Introduction to the Study of Stellar Structure (New York: Dover)

107. Chen, B. 1997, AJ, 113, 311
108. Chen, B. 1998, ApJL, 495, L1
109. Ciardullo, R. B., & Demarque, P. 1977, Trans. Yale Univ. Obs., 33
110. Clayton, D. D. 1968, Principles of Stellar Evolution and Nucleosynthesis (New York:McGraw–Hill)
111. Clement, C. M., Ferance, S., & Simon, N. R. 1993, ApJ, 412, 183
112. Clementini, G., Carretta, E., Gratton, R., Merighi, R., Mould, J. R., & McCarthy, J. K. 1995, AJ, 110, 2319
113. Code, A. D., Bless, R. C., Davis, J., & Brown, R. H. 1976, ApJ, 203, 417
114. Cohen, J. G. 1978, ApJ, 223, 487
115. Cohen, J. G. 1979, ApJ, 231, 751
116. Cohen, J. G. 1980, ApJ, 241, 981
117. Cohen, J. G. 1981, ApJ, 247, 869
118. Cohen, J. G. 1983, ApJ, 270, 654
119. Cohen, J. G., Frogel, J. A., and Persson, S. E. 1978, ApJ, 222, 165
120. Costar, D., & Smith, H. A. 1988, AJ, 96, 1925
121. Cowan, J. J., Burris, D. L., Sneden, C., McWilliam, A., & Preston, G. W. 1995, ApJL, 439, L51
122. Cowan, J. J., McWilliam, A., Sneden, C., & Burris, D. L. 1997, ApJ, 480, 246
123. Cowan, J. J., Sneden, C., Truran, J. W., & Burris, D. L. 1996, ApJL, 460, L115
124. Cowan, J. J., Thielemann, F.–K., & Truran, J. W. 1991, AAR&A, 29, 447
125. Cox, A. N. 1991, ApJL, 381, L71
126. Cudworth, K. M., & Hanson, R. B. 1993, AJ, 105, 168
127. Cutispoto, G., Kuerster, M., Messina, S., Rodono, M., & Tagliaferri, G. 1997, A&A, 320, 586
128. Da Costa, G. S., & Armandroff, T. E. 1990, AJ, 100, 162
129. Da Costa, G. S., & Armandroff, T. E. 1995, AJ, 109, 2533
130. da Silva, L., de la Reza, R., & Barbuy, B. 1995, ApJL, 448, L41
131. D'Cruz, N. D., Dorman, B., Rood, R. T., & O'Connell, R. W. 1996, ApJ, 466, 359
132. Dearborn, D. S. P., Liebert, J., Aaronson, M., Dahn, C., Harrington, R., Mould, J., & Greenstein, J. L. 1986, ApJ, 300, 314
133. de la Reza, R., Drake, N. A., da Silva, L., Torres, C. A. O., & Martin, E. L. 1997, ApJL, 482, L77
134. Delhaye, J. 1965, in Galactic Structure, ed. A. Blaauw & M. Schmidt (Chicago: Univ. Chicago Press), 61
135. Deliyannis, C. P., Demarque, P., & Pinsonneault, M. H. 1989, ApJL, 347, L73
136. Dickens, R. J., Croke, B. F. W., Cannon, R. D., & Bell, R. A. 1991, Nature, 351, 212
137. Dinescu, D. I., Girard, T. M., van Altena, W. F., Mendez, R. A., & López, C. W. 1997, AJ, 114, 1014
138. Dinescu, D. I., van Altena, W. F., Girard, T. M., & López, C. W. 1999, AJ, 117, 277
139. Doinidas, S. P., & Beers, T. C. 1989, ApJL, 340, L57
140. Drake, J. J., Smith, V. V., & Suntzeff, N.B. 1994, ApJ, 430, 610
141. Eddington, A. S. 1926, The Internal Constitution of the Stars (Cambridge: Univ. Press)
142. Edvardsson, B., Andersen, J., Gustafsson,B., Lambert, D. L., Nissen, P. E., & Tomkin, J. 1993, A&A, 275, 101

143. Eggen, O. J. 1977, ApJ, 215, 812
144. Eggen, O. J. 1978, ApJ, 221, 881
145. Eggen, O. J. 1987, in The Galaxy, ed. G. Gilmore & B. Carswell (Dordrecht: Reidel), 211
146. Eggen, O. J., Lynden-Bell, D., & Sandage, A. R. 1962, ApJ, 136, 748 (ELS)
147. Faulkner, J. 1972, ApJ, 173, 401
148. Feast, M. W. 1997, MNRAS, 284, 761
149. Feast, M. W, & Catchpole, R. M. 1997, MNRAS, 286, L1
150. Feast, M. R., Pont, F. & Whitelock, P. 1998, MNRAS, 298, L43
151. Feast, M. R., & Walker, A. R. 1987, ARA&A, 25, 345
152. Fernley, J. 1994, A&A, 284, L16
153. Fernley, J., & Barnes, T. G. 1996, A&A, 312, 957
154. Fernley, J. A., Barnes, T. G., Skillen, I., Hawley, S. L., Hanley, C. J., Evans, D. W., Solano, E., & Garrido, R. 1998a, A&A, 330, 515
155. Fernley, J., Carney, B. W., Skillen, I., Cacciari, C., & Janes, K. 1998b, MNRAS, 293, L61
156. Ferraro, F. R., Carretta, E., Corsi, C. E., Fusi Pecci, F., Cacciari, C., Buonanno, R., Paltrinieri, B., & Hamilton, D. 1997, A&A, 320, 757
157. Ferraro, F. R., Fusi Pecci, F., Cacciari, C., Corsi, C. E., Buonanno, R., Fahlman, G. G., & Richer, H. B. 1993, AJ, 106, 2324
158. Flynn, C., & Morell, O. 1997, MNRAS, 286, 617
159. François, P. 1988, A&A, 195, 226
160. François, P. 1996, A&A, 313, 229
161. François, P., Spite, M., & Spite, F. 1988, A&A, 191, 267
162. Freeman, K. C. 1991, in Dynamics of Disc Galaxies, ed. B. Sundelius (Göteborg: Göteborgs University), 15
163. Freeman, K. C., & Rodgers, A. W. 1975, ApJL, 201, L71
164. Freedman, W. L., Wilson, C. D., & Madore, B. F. 1991, ApJ, 372, 455
165. Friel, E., Kraft, R. P., Suntzeff, N. B., & Carbon, D. F. 1982, PASP, 94, 873
166. Frogel, J. A., Kuchinski, L. E., & Tiede, G. P. 1995, AJ, 109, 1154
167. Fulbright, J. P., & Kraft, R. P. 1999, AJ, 118, 527
168. Fullton, L. K. 1995, PASP, 108, 545
169. Fusi Pecci, F., Buonanno, R., Cacciari, C., Corsi, C. E., Djorgovski, G., Federici, L., Ferraro, F. R., Parmeggiani, G., & Rich, R. M. 1996, AJ, 112, 1461
170. Fusi Pecci, F., Ferraro, F. R., Bellazzini, M., Djorgovski, S., Piotto, G., & Buonanno, R. 1993, AJ, 105, 1145
171. Fusi Pecci, F., Ferraro, F. R., Crocker, D. A., Rood, R. T., & Buonanno, R. 1990, A&A, 238, 95
172. Gillett, F. C., Jacoby, G. H., Joyce, R. R., Cohen, J. G., Neugebauer, G., Soifer, B. T., Nakajima, T., & Matthews, K. 1988, ApJ, 338, 862
173. Gilmore, G., & Reid, I. N. 1983, MNRAS, 202, 1025
174. Gilmore, G., & Wyse, R. F. G. 1998, AJ, 116, 748
175. Gilmore, G., Wyse, R. F. G., & Jones, J. B. 1995, AJ, 109, 1095
176. Glaspey, J. W., Michaud, G., Moffat, A. F. J., & Demers, S. 1989, ApJ, 339, 926
177. Glaspey, J. W., Pritchet, C. J., & Stetson, P. B. 1994, AJ, 108, 271
178. Gnedin, O. Y., & Ostriker, J. P. 1997, ApJ, 474, 223
179. Goldberg, D., Mazeh, T., Latham D. W., Stefanik, R. P., Carney, B. W., & Laird, J. B. 1999, AJ, submitted

180. Goldman, I., & Mazeh, T. 1991, ApJ, 376, 260
181. Goldsmith, C. G. 1993, in New Perspectives on Stellar Evolution, ed. J. M. Nemec & J. M. Matthews (Cambridge: Cambridge Univ. Press), 358
182. Gonzalez, G., & Wallerstein, G. 1998, AJ, 116, 765
183. Gould, A. 1995, ApJ, 452, 189
184. Gould, A., & Uza, O. 1998, ApJ, 494, 118
185. Gratton, R . G. 1982, A&A, 115, 336
186. Gratton, R . G. 1987a, A&A, 177, 177
187. Gratton, R . G. 1987b, A&A, 179, 181
188. Gratton, R . G. 1998, MNRAS, 296, 739
189. Gratton, R. G., Fusi Pecci, F., Carretta, E., Clementini, G., Corsi, C. E., & Lattanzi, M. 1997, ApJ, 491, 749
190. Gratton, R. G., & Sneden, C. 1988, A&A, 204, 193
191. Gratton, R. G., & Sneden, C. 1991, A&A, 241, 501
192. Gratton, R. G., & Sneden, C. 1994, A&A, 287, 927
193. Green, E. M., Demarque, P., & King, C. R. 1987, The Revised Yale Isochrones and Luminosity Functions (New Haven: Yale Univ. Obs.)
194. Grillmair, C. J., Mould, J. R., Holtzman, J. A., Worthey, G., Ballester, G. E., Burrows, C. J., Clarke, J. T., Crisp, D., Evans, R. W., Gallagher, J. S., III, Griffiths, R. E., Hester, J. J., Hoessel, J. G., Scowen, P. A., Stapelfeldt, K. R., Trauger, J. T., Watson, A. M., & Westphal, J. A. 1998, AJ, 115, 144
195. Guinan, E. F., Fitzpatrick, E. L., DeWarf, L. E., Maloney, F. P., Maurone, P. A., Ribas, I., Pritchard, J. D., Bradstree, D. H., & Giménez, A. 1998, ApJL, 509, L21
196. Gwinn, C. R., Moran, J. M., & Reid, M. J. 1992, ApJ, 393, 149
197. Hachisu, I., Kato, M., & Nomoto, K. 1996, ApJL, 470, L97
198. Hachisu, I., Kato, M., & Nomoto, K. 1999, ApJ, 522, 487
199. Han, Z., Eggleton, P. P., Podsiadlowski, C. A., & Tout, C. A. 1995, MNRAS, 277, 1443
200. Hanson, R. B. 1979, MNRAS, 186, 875
201. Hanson, R. B., Sneden, C., Kraft, R. P., & Fulbright, J. 1998, AJ, 116, 1286
202. Harris, W. E., Bell, R. A., VandenBerg, D. A., Bolte, M., Stetson, P. B., Hesser, J. E., van den Bergh, S., Bond, H. E., Fahlman, G. G., & Richer, H. B. 1997, AJ, 114, 1030
203. Hartwick, F. D. A. 1976, ApJ, 209, 418
204. Hartwick, F. D. A. 1987, in The Galaxy, ed. G. Gilmore & B. Carswell (Dordrecht: Reidel), 281
205. Hazen, M. L., & Nemec, J. M. 1992, AJ, 104, 111
206. Heintz, W. D. 1984, PASP, 96, 557
207. Helfer, H. L., & Wallerstein, G. 1968, ApJS, 16, 1
208. Hobbs, L. M., & Mathieu, R. D. 1991, PASP, 103, 431
209. Hobbs, L. M., & Thorburn, J. A. 1994, ApJL, 428, L25
210. Hobbs, L. M., & Thorburn, J. A. 1997, ApJ, 491, 772
211. Hobbs, L. M., Thorburn, J. A., & Welty, D. E. 1991, ApJL, 373, L47
212. Holweger, H., & Müller, E. A. 1974, Solar Physics, 39, 19
213. Huebner, W. F., Merts, A. L., Magee, N. H., & Argo, M. F. 1977, Los Alamos Sci. Lab. Rept. LA-6760-M
214. Huterer, D., Sasselov, D. D., & Schechter, P. L. 1995, AJ, 110, 2705
215. Hut, P., McMillan, S., Goodman, J., Mateo, M., Phinney, E. S., Pryor, C., Richer, H. B., Verbrunt, F., & Weinberg, M. 1992, PASP, 104, 681

216. Ibata, R. A., Gilmore, G., & Irwin, M. J. 1994, Nature, 370, 194
217. Iben, I., Jr. 1968, Nature, 220, 143
218. Iben, I., Jr. 1971, PASP, 83, 697
219. Iben, I., Jr. 1991, ApJS, 76, 55
220. Iben, I., Jr., & Rood, R. T. 1970, ApJ, 161, 587
221. Isobe, T., Feigelson, E. D., Akritas, M. G., & Babu, G. J. 1990, ApJ, 364, 104
222. Israelian, G., Garcia Lopez, R. J., & Rebolo, R. 1998, ApJ, 507, 805
223. Jacoby, G. H., & Fullton, L. K. 2000, in preparation
224. Jacoby, G. H., Morse, J. A., Fullton, L. K., Kwitter, K. B., & Henry, R. B. C. 1997, AJ, 114, 2611
225. Jenkins, L. F. 1963, General Catalog of Trigonometric Stellar Parallaxes (New Haven: Yale Univ. Obs.)
226. Johnson, J. A., & Bolte, M. 1998, AJ, 115, 693
227. Jones, R. V., Carney, B. W., & Latham, D. W. 1988, ApJ, 332, 206
228. Jones, R. V., Carney, B. W., Latham, D. W., & Kurucz, R. L. 1987, ApJ, 312, 254
229. Jones, R. V., Carney, B. W., Storm, J., & Latham, D. W. 1992, ApJ, 386, 646
230. Jorissen, A., Hennen, O., Mayor, M., Bruch, A., & Sterken, C. 1995, A&A, 301, 707
231. Kemper, E. 1982, AJ, 87, 1395
232. King, J. R. 1997, AJ, 113, 2302
233. King, J. R., Deliyannis, C. P., & Boesgaard, A. M. 1996, AJ, 112, 2839
234. King, J. R., Stephens, A., Boesgaard, A. M., & Deliyannis, C. P. 1998, AJ, 115, 666
235. Kinman, T. D. 1998, private communication
236. Kinman, T. D., Kraft, R. P., Friel, E., & Suntzeff, N. B. 1985, AJ, 90, 95
237. Kinman, T. D., Mahaffey, C. T., & Wirtanen, C. A. 1982, AJ, 87, 314
238. Kinman, T. D., Pier, J. R., Suntzeff, N. B., Harmer, D. L., Valdes, F., Hanson, R. B., Klemola, A. R., & Kraft, R. P. 1996, AJ, 111, 1164
239. Kinman, T. D., Suntzeff, N. B., & Kraft, R. P. 1994, AJ, 108, 1722
240. Kinman, T. D., Wirtanen, C. A., & Janes, K. A. 1965, ApJS, 11, 223
241. Kinman, T. D., Wirtanen, C. A., & Janes, K. A., 1966, ApJS, 13, 379
242. Kinman, T. D., Wong–Swanson, B., Wenz, M., & Harlan, E. A. 1984, AJ, 89, 1200
243. Kippenhahn, R., & Weigert, A. 1990, Stellar Structure and Evolution (Berlin, Heidelberg: Springer)
244. Kobayashi, C., Tsujimoto, T., Nomoto, K., Hachisu, I., & Kato, M. 1998, ApJL, 503, L155
245. Koen, C. 1992, MNRAS, 256, 65
246. Kraft, R. P. 1994, PASP, 106, 553
247. Kraft, R. P., Peterson, R. C., Guhathakurta, P., Sneden, C., Fulbright, J. P., & Langer, G. E. 1999, ApJL, 518, L53
248. Kraft, R. P., Sneden, C., Langer, G. E., & Shetrone, M. D. 1993, AJ, 106, 1490
249. Kraft, R. P., Sneden, C., Langer, G. E., Shetrone, M. D., & Bolte, M. 1995, AJ, 109, 2586
250. Kraft, R. P., Sneden, C., Smith, G. H., Shetrone, M. D., & Fulbright, J. 1998, AJ, 115, 1500
251. Kraft, R. P., Sneden, C., Smith, G. H., Shetrone, M. D., Langer, G. E., & Pilachowski, C. A. 1997, AJ, 113, 279

252. Kraft, R. P., Suntzeff, N. B., Langer, G. E., Carbon, D. F., Trefzger, Ch. F., Friel, E., & Stone, R. P. S. 1982, PASP, 94, 55
253. Krishnaswamy-Gilroy, K., Sneden, C., Pilachowski, C. A., & Cowan, J. J. 1988, ApJ, 327, 298
254. Laird, J. B., 1985, ApJ, 289, 556
255. Laird, J. B., Rupen, M. P., Carney, B. W., & Latham, D. W. 1988, AJ, 96, 1908
256. Lambert, D. L., Heath, J. E., Lemke, M., & Drake, J. 1996, ApJS, 103, 183
257. Lambert, D. L., & McWilliam, A. 1986, ApJ, 304, 436
258. Langer, G. E., Fischer, D., Sneden, C., & Bolte, M. 1998, AJ, 115, 685
259. Langer, G. E., & Hoffman, R. D. 1995, PASP, 107, 1177
260. Langer, G. E., Hoffman, R. D., & Sneden, C. 1993, PASP, 105, 301
261. Latham, D. W., Mazeh, T., Carney, B. W., McCrosky, R. E., Stefanik, R. P., & Davis, R. J. 1988, AJ, 96, 567
262. Latham, D. W., Mazeh, T., Stefanik, R. P., Davis, R. J., Carney, B. W., Krymolowski, Y., Laird, J. B., Torres, G., & Morse, J. A. 1992, AJ, 104, 774
263. Latham, D. W., Stefanik, R. P., Torres, G., Davis, R. J., Mazeh, T., Carney, B. W., Laird, J. B., & Morse, J. A. 1999, AJ, submitted
264. Layden, A. C. 1995, AJ, 110, 2288
265. Layden, A. C., Hanson, R. B., Hawley, S. L., Klemola, A. R., & Hanley, C. J. 1996, AJ, 112, 2110
266. Lee, J.-W., & Carney, B. W. 1999a, AJ, 117, 2868
267. Lee, J.-W., & Carney, B. W. 1999b, AJ, 118, 1373
268. Lee, M.-G., Freedman, W. L., & Madore, B. F. 1993, ApJ, 417, 553
269. Lee, Y.-W. 1992, AJ, 104, 1780
270. Lee, Y.-W., Demarque, P., & Zinn, R. 1990, ApJ, 350, 155
271. Lee, Y.-W., Demarque, P., & Zinn, R. 1994, ApJ, 423, 248
272. Leonard, P. J. T. 1996, in The Origins, Evolution, and Destinies of Binary Stars in Clusters, ASP Conf. Series 90, ed. E. F. Milone & J.-C. Mermilliod (San Francisco: ASP), 385
273. Lindgren, H., Ardeberg, A., & Zuiderwijk, E. 1987, A&A, 188, 39
274. Livio, M. 1993, in Blue Stragglers, ASP Conf. Series 53, ed. R. Saffer (San Francisco: ASP), 3
275. Liu, T., & Janes, K. A. 1990a, ApJ, 354, 273
276. Liu, T., & Janes, K. A. 1990b, ApJ, 360, 561
277. Longmore, A. J., Dixon, R., Skillen, I., Jameson, R. F., & Fernley, J. A. 1990, MNRAS, 247, 684
278. Lynden-Bell, D. A., & Lynden-Bell, R. M. 1995, MNRAS, 275, 429
279. Lub, J. 1977, Ph. D. Thesis, Univ. of Leiden
280. Luck, R. E., & Bond, H. E. 1981, ApJ, 244, 919
281. Luck, R. E., & Bond, H. E. 1985, ApJ, 292, 559
282. Lutz, T. E., & Kelker, D. H. 1973, PASP, 85, 573
283. Madore, B. F., Freedman, W. L., & Sakai, S. 1997, in The Extragalactic Distance Scale, ed. M. Livio, M. Donahue, & N. Panagia (Cambridge: Cambridge Univ. Press), 239
284. Majewski, S. R. 1992, ApJS, 78, 87
285. Majewski, S. R. 1993, AAR&A, 31, 575
286. Majewski, S. R., Munn, J. A., & Hawley, S. L. 1994, ApJL, 427, L37
287. Majewski, S. R., Munn, J. A., & Hawley, S. L. 1996, ApJL, 459, L73

288. Mallia, E. A., & Pagel, B. E. J. 1981, MNRAS, 194, 421
289. Marquez, A., & Schuster, W. J. 1994, A&AS, 108, 341
290. Martin, J. C., & Morrison, H. L. 1998, AJ, 116, 1724
291. Mateo, M. 1993, in The Globular Cluster–Galaxy Connection, ASP Conf. Series 48, ed. G. H. Smith & J. P. Brodie (San Francisco: ASP), 387
292. Mathieu, R. D., Latham, D. W., & Griffin, R. F. 1990, AJ, 100, 1859
293. Mathieu, R. D., & Mazeh, T. 1988, ApJ, 326, 256
294. Mathys, G. 1991, A&A, 245, 457
295. Matteuci, F., & Greggio, L. 1986, A&A, 154, 279
296. Mayor, M., & Mermilliod, J.–C. 1984, in Observational Tests of Stellar Evolution Theory, IAU Sym. No. 105, ed. A. Maeder & A. Renzini (Dordrecht: Reidel), 411
297. Mazzitelli, I., D'Antona, F., & Caloi, V. 1995, A&A, 302, 382
298. McClure, R. D. 1997, PASP, 109, 536
299. McClure, R. D., & Woodsworth, A. W. 1990, ApJ, 352, 709
300. McWilliam, A., Preston, G. W., Sneden, C., & Searle, L. 1995, AJ, 109, 2757
301. McWilliam, A., & Rich, R. M. 1994, ApJS, 91, 749
302. Milone, A., Barbuy, B., Spite, M., & Spite, F. 1992, A&A, 261, 551
303. Minniti, D. 1995, A&A, 300, 109
304. Morell, O., Källander, D., & Butcher, H. R. 1992, A&A, 259, 543
305. Morrison, H. L., Flynn, C., & Freeman, K. C. 1990, AJ, 100, 1191
306. Murali, C., & Weinberg, M. D. 1997, MNRAS, 288, 749
307. Naumov, S., Carney, B. W., Latham, D. W., & Laird, J. B. 1999a, in preparation
308. Nemec, J. M. 1985, AJ, 90, 204
309. Nemec, J. M., & Cohen, J. G. 1989, ApJ, 336, 780
310. Nemec, J. M., Mateo, M., Burke, M., & Olszewski, E. O. 1995, AJ, 110, 1186
311. Nemec, J. M., Nemec, A. F. L., & Lutz, T. E. 1994, AJ, 108, 222
312. Nemec, J. M., Wehlau, A., & Mendes de Oliviera, C. 1988, AJ, 96, 528
313. Nissen, P. E., & Edvardsson, B. 1992, A&A, 261, 255
314. Nissen, P. E., Gustafsson, B., Edvardsson, B., & Gilmore, G. 1994, A&A, 285, 440
315. Nissen, P. E., & Schuster, W. J. 1991, A&A, 251, 457
316. Nissen, P. E., & Schuster, W. J. 1997, A&A, 326, 751
317. Nollett, K. M., Lemoine, M. & Schramm, D. N. 1997, Phys. Rev. C., 56, 1144
318. Noerdlinger, P., & Arigo, R. J. 1980, ApJL, 237, L15
319. Nomoto, K., Iwamoto, K., & Kishimoto, N. 1997, Science, 276, 1378
320. Norris, J. 1986, ApJS, 61, 667
321. Norris, J. 1987, ApJL, 314, L39
322. Norris, J., Bessell, M. S., & Pickles, A. J. 1985, ApJS, 58, 463
323. Norris, J., & Da Costa, G. S. 1995, ApJ, 447, 680
324. Norris, J. E., Freeman, K. C., Mayor, M., & Seitzer, P. 1997, ApJL, 487, L187
325. Norris, J. E., Freeman, K. C., & Mighell, K. J. 1996, ApJ, 462, 241
326. Norris, J., & Ryan, S. 1989, ApJ, 340, 739
327. O'Brian, T. R., Wickliffe, M. W., Lawler, J. E., Whaling, W., & Brault, J. W. 1991, J. Opt. Soc. America, B8, 1185
328. Odenkirchen, M., Brosche,P., Geffert, M., & Tucholke, H.–J. 1997, New Astr., 2, 477
329. Olsen, K. A. G., Hodge, P. W., Mateo, M., Olszewski, E. W., Schommer, R. A., Suntzeff, N. B., & Walker, A. R. 1998, MNRAS, 300, 6650

330. Olszewski, E. W., Schommer, R. A., Suntzeff, N. B., & Harris, H. C. 1991, AJ, 101, 515
331. Oosterhoff, P. Th. 1939, Observatory, 62, 104
332. Pagel, B. E. J. 1992, in The Stellar Populations of Galaxies, IAU Sym. No. 149, ed. B. Barbuy & A. Renzini (Dordrecht: Kluwer), 133
333. Pagel, B. E. J. 1997, Nucleosynthesis and Chemical Evolution of Galaxies (Cambridge: Cambridge Univ. Press)
334. Pagel, B. E. J., & Kazlauskas, A. 1992, MNRAS, 256, 49P
335. Pagel, B. E. J., Simonson, E. A., Terlevich, R. J., & Edmunds, M. G. 1992, MNRAS, 255, 325
336. Paltoglou, G., & Norris, J. E. 1989, ApJ, 336, 185
337. Panagia, N., Gilmozzi, R., Macchetto, F., Adorf, H.–M., & Kirshner, R. P. 1991, ApJL, 380, L23
338. Pasquini, L., & Molaro, P. 1996, A&A, 307, 761
339. Pease, F. G. 1928, PASP, 40, 342
340. Peimbert, M., & Torres–Peimbert, S. 1974, ApJ, 193, 327
341. Peimbert, M., & Torres–Peimbert, S. 1976, ApJ, 203, 581
342. Perryman, M. A. C., Lindegren, L., Kovalesky, J., Høg, E., Bastian, U., Bernacca, P. L., Crézé, M., Donati, F., Grenon, M., Grewing, M., van Leeuwen, F., van der Marel, H., Mignard, F., Murrary, C., Le Poole, R. S., Schrijver, H., Turon, C., Arenou, F., Froeschlé, M., & Petersen, C. S. 1997, A&A, 323, L49
343. Persson, S. E., Frogel, J. A., Cohen, J. G., Aaronson, M., & Matthew, K. 1980, ApJ, 235, 452
344. Peterson, R. C. 1981, ApJ, 244, 989
345. Peterson, R. C. 1983, ApJ, 275, 737
346. Peterson, R. C. 1985a, ApJ, 289, 320
347. Peterson, R. C. 1985b, ApJL, 294, L35
348. Peterson, R. C., & Carney, B. W. 1979, ApJ, 231, 762
349. Peterson, R. C., Rood, R.T., & Crocker, D.A. 1995, ApJ, 453, 214
350. Peterson, R. C., Tarbell, T. D., & Carney, B. W. 1983, ApJ, 265, 972
351. Pilachowski, C. A., Sneden, C., & Booth, J. 1993, ApJ, 407, 699
352. Pilachowski, C. A., Sneden, C., Kraft, R. P., & Langer, G. E. 1996, AJ, 112, 545
353. Pinsonneault, M. H., Deliyannis, C. P., & Demarque, P. 1992, ApJS, 78, 179
354. Pinsonneault, M. H., Kawaler, S. D., & Demarque, P. 1990, ApJS, 74, 501
355. Pinsonneault, M. H., Kawaler, S. D., Sofia, S., & Demarque, P. 1989, ApJ, 338, 424
356. Pinsonneault, M. H., Stauffer, J., Soderblom, D. R., King, J. R., & Hanson, R. B. 1998, ApJ, 504, 170
357. Pinsonneault, M. H., Walker, T. P., & Narayanan, V. K. 1999, ApJ, 527, 180
358. Plez, B., Smith, V. V., & Lambert, D. L. 1993, ApJ, 418, 812
359. Pont, F., Mayor, M., Turon, C., & VandenBerg, D. A. 1998, A&A, 329, 87
360. Popowski, P., & Gould, A. 1998, ApJ, 506, 271
361. Powell, D. C. 1999, Ph.D. thesis, University of North Carolina
362. Preston, G. W. 1959, ApJ, 130, 507
363. Preston, G. W., Beers, T. C., & Shectman, S. A. 1994, AJ, 108, 538
364. Preston, G. W., & Landolt, A. U. 1998, AJ, 115, 2526
365. Pritchet, C. J., & Glaspey, J. W. 1991, ApJ, 373, 105

366. Proffitt, C. P., & VandenBerg, D. A. 1991, ApJS, 77, 473
367. Rees, R. F., Jr. 1996, in Formation of the Galactic Halo...Inside and Out, ASP Conf. Series 92, ed. H. Morrison & A. Sarajedini (San Francisco: ASP), 289
368. Reid, I. N. 1997, AJ, 114, 1611
369. Reid, M. J. 1993, ARA&A, 31, 345
370. Reid, M. J., Gwinn, C. R., Moran, J. M., & Matthews, A. 1988a, BAAS, 20, 1017
371. Reid, M. J., Schneps, M. H., Moran, J. M., Gwinn, C. R., & Genzel, R., Downes, D., & Rönnäng, B. 1988b, ApJ, 330, 809
372. Renzini, A. 1977, Advanced Stages in Stellar Evolution (Geneva: Geneva Obs.)
373. Renzini, A., Bragaglia, A., Ferraro, F. R., Gilmozzi, R., Ortolani, S., Holberg, J. B., Liebert, J., Wesemael, F., & Bohlin, R. C. 1996, ApJL, 465, L23
374. Renzini, A., & Fusi Pecci, F. 1988, AAR&A, 26, 199
375. Rey, S.–C., Lee, Y.–W., Byun, Y.–I., & Chun, M.–S. 1998, AJ, 116, 1775
376. Richer, H. B., & Fahlman, G. G. 1987, ApJ, 316, 189
377. Richer, H. B., Fahlman, G. G., Ibata, R. A., Pryor, C., Bell, R. A., Bolte, M., Bond, H. E., Harris, W. E., Hesser, J. E., Holland, S., Ivanans, N., Mandushev, G., Stetson, P. B., & Wood, M. A. 1997, ApJ, 484, 741
378. Richer, H. B., Harris, W. E., Fahlman, G. G., Bell, R. A., Bond, H. E., Hesser, J. E., Holland, S., Pryor, C., Stetson, P. B., VandenBerg, D. A., & van den Bergh, S. 1996, ApJ, 463, 602
379. Ridgway, S. T., Joyce, R. R., White, N. M., & Wing, R. F. 1980, ApJ, 235, 126
380. Rodgers, A. W., & Paltoglou, G. 1984, ApJL, 283, L5
381. Rogers, F. J., & Iglesias, C. A. 1992, ApJS, 79, 507
382. Rogers, F. J., & Iglesias, C. A. 1994, in The Equation of State in Astrophysics, IAU Coll. No. 147, ed. G. Chabrier & E. Schatzman (Cambridge: Cambridge Univ. Pr.), 16
383. Rutledge, G. A., Hesser, J. E., & Stetson, P. B. 1997, PASP, 109, 907
384. Ryan, S. G., & Lambert, D. L. 1995, AJ, 109, 2068
385. Ryan, S. G., Norris, J. E., & Beers, T. C. 1998, ApJ, 506, 892
386. Sackmann, I.–J., & Boothroyd, A. I. 1999, ApJ, 510, 217
387. Saha, A. 1984, ApJ, 283, 580
388. Saha, A., Monet, D. G. & Seitzer, P. 1986, AJ, 92, 302
389. Saha, A., & Oke, J. B. 1984, ApJ, 285, 688
390. Saio, H., & Wheeler, J. C. 1980, ApJ, 242, 1176
391. Salaris, M., Chieffi, A., & Straniero, O. 1993, ApJ, 414, 580
392. Salaris, M., & Weiss, A. 1997, A&A, 327, 107
393. Salaris, M., & Weiss, A. 1998, A&A, 335, 943
394. Sandage, A. R. 1953, AJ, 58, 61
395. Sandage, A. 1969a, ApJ, 157, 515
396. Sandage, A. 1969b, ApJ, 158, 1115
397. Sandage, A. 1982, ApJ, 252, 553
398. Sandage, A. 1990a, ApJ, 350, 603
399. Sandage, A. 1990b, ApJ, 350, 631
400. Sandage, A., & Fouts, G. 1987, AJ, 93, 74
401. Sandage, A., & Wildey, R. 1967, ApJ, 150, 469
402. Sandquist, E. L., Bolte, M., Stetson, P. B., & Hesser, J. E. 1996, ApJ, 470, 910

403. Sarajedini, A. 1994, AJ, 107, 618
404. Sarajedini, A. 1997, AJ, 113, 682
405. Sarajedini, A., Chaboyer, B., & Demarque, P. 1997, PASP, 109, 1321
406. Sarajedini, A., & Forrester, W. 1995, AJ, 109, 1112
407. Sarajedini, A., & Layden, A. 1995, AJ, 109, 1086
408. Sarajedini, A., & Layden, A. 1997, AJ, 113, 264
409. Saxner, M., & Hammarbäck, G. 1985, A&A, 151, 372
410. Schmidt, M. 1959, ApJ, 129, 243
411. Schmidt, M. 1963, ApJ, 137, 758
412. Schmidt, M. 1968, ApJ, 151, 393
413. Schmidt, M. 1975, ApJ, 202, 22
414. Schuster, W. J., & Allen, C. 1997, A&A, 310, 796
415. Schuster, W. J., & Nissen, P. E. 1989a, A&A, 221, 65
416. Schuster, W. J., & Nissen, P. E. 1989b, A&A, 222, 69
417. Searle, L., & Sargent, W. L. W. 1972, ApJ, 173, 25
418. Searle, L., & Zinn, R. 1978, ApJ, 225, 357 (SZ)
419. Shara, M. M., Saffer, R. A., & Livio, M. 1997, ApJL, 489, L59
420. Shetrone, M. D. 1996a, AJ, 112, 1517
421. Shetrone, M. D. 1996b, AJ, 112, 2639
422. Smecker-Hane, T. A., Stetson, P. B., Hesser, J. E., & Lehnert, M. D. 1994, AJ, 108, 507
423. Smecker-Hane, T. A., & Wyse, R. F. G. 1992, AJ, 103, 1621
424. Smith, H. A. 1981, ApJ, 250, 719
425. Smith, H. A. 1984, ApJ, 281, 148
426. Smith, H. A. 1995, RR Lyrae Stars (Cambridge: Cambridge Univ. Press)
427. Smith, H. A., & Butler, D. 1978, PASP, 90, 671
428. Smith, H. A., & Manduca, A. 1983, AJ, 88, 982
429. Smith, H. A., & Perkins, G. J. 1982, ApJ, 261, 576
430. Smith, V. V., Cunha, K., & Lambert, D. L. 1995, AJ, 110, 2827
431. Smith, V. V., & Lambert, D. L. 1989, ApJL, 345, L75
432. Smith, V. V., & Lambert, D. L. 1990, ApJL, 361, L69
433. Smith, V. V., Lambert, D. L., & Nissen, P. E. 1993, ApJ, 408, 262
434. Smith, V. V., Lambert, D. L., & Nissen, P. E. 1998, ApJ, 506, 405
435. Smith, V. V., Shetrone, M. D., & Keane, M. J. 1999, ApJL, 516, L73
436. Sneden, C., Kraft, R. P., Langer, G. E., Prosser, C. F., & Shetrone, M. D. 1994a, AJ, 107, 1773
437. Sneden, C., Kraft, R. P., Prosser, C. F., & Langer, G. E. 1991, AJ, 102, 2001
438. Sneden, C., Kraft, R. P., Prosser, C. F., & Langer, G. E. 1992, AJ, 104, 2121
439. Sneden, C., Kraft, R. P., Shetrone, M. D., Smith, G. H., Langer, G. E., & Prosser, C. F. 1997, AJ, 114, 1964
440. Sneden, C., McWilliam, A., Preston, G. W., Cowan, J. J., Burris, D. L., & Armosky, B. J. 1996, ApJ, 467, 819
441. Sneden, C., & Parthasarathy, M. 1983, ApJ, 267, 757
442. Sneden, C., & Pilachowski, C. A. 1985, ApJL, 288, L55
443. Sneden, C., Preston, G. W., McWilliam, A., & Searle, L. 1994b, ApJL, 431, L27
444. Soderblom, D. R., King, J. R., Hanson, R. B., Fischer, D., Stauffer, J. R., & Pinsonneault, M. H. 1998, ApJ, 504, 192
445. Sommer-Larsen, J., & Zhen, C. 1990, MNRAS, 242, 10

446. Sonneborn, G., Fransson, C., Lundqvist, P., Cassatella, A., Gilmozzi, R., Kirshner, R. P., Panagia, N., & Wamsteker, W. 1997, ApJ, 477, 848
447. Spiesman, W. J., & Wallerstein, G. 1991, AJ, 102, 1790
448. Spite, M., & Spite, F. 1978, A&A, 67, 23
449. Spite, F., & Spite, M. 1982, A&A, 115, 357
450. Spite, F., & Spite, M. 1991, A&A, 252, 689
451. Spite, M., Francois, P., Nissen, P. E., & Spite, F. 1996, A&A, 307, 172
452. Spite, M., Molaro, P., Francois, P., & Spite, F. 1993, A&A, 271, L1
453. Spite, M., Pasquini, L., & Spite, F. 1994, A&A, 290, 217
454. Stephens, A. 1999, AJ, 117, 1771
455. Stetson, P. B., Bolte, M., Harris, W. E., Hesser, J. E., van den Bergh, S., VandenBerg, D. A., Bell, R. A., Johnson, J. A., Bond, H. E., Fullton, L. K., Fahlman, G. G., & Richer, H. B. 1999, AJ, 117, 247
456. Stetson, P. B., VandenBerg, D. A., & Bolte, M. 1996, PASP, 108, 560
457. Stetson, P. B., VandenBerg, D. A., Bolte, M., Hesser, J. E., & Smith, G. H. 1989, AJ, 97, 1360
458. Storm, J., Carney, B. W., & Latham, D. W. 1994a, A&A, 290, 443
459. Storm, J., Nordström, B., Carney, B. W., & Andersen, J. 1994b, A&A, 291, 121
460. Straniero, O., & Chieffi, A. 1991, ApJS, 76, 525
461. Strömgren, B., 1987, in The Galaxy, ed. G. Gilmore & B. Carswell (Dordrecht: Reidel), 229
462. Strugnell, P., Reid, I. N., & Murray, C. A. 1986, MNRAS, 220, 413
463. Stryker, L. L. 1993, PASP, 105, 1081
464. Suntzeff, N. B. 1981, ApJS, 47, 1
465. Suntzeff, N. B. 1989, in The Abundance Spread within Globular Clusters, ed. G. Cayrel de Strobel, M. Spite, & T. Lloyd Evans (Paris: Obs. de Paris), 71
466. Suntzeff, N. B. 1993, in The Globular Cluster–Galaxy Connection, ASP Conf. Series 48, ed. G. H. Smith & J. Brodie (San Francisco: ASP), 167
467. Suntzeff, N. B. 1998, private communication
468. Suntzeff, N. B., Kinman, T. D., & Kraft, R. P. 1991, ApJ, 367, 528
469. Suntzeff, N. B., & Kraft, R. P. 1996, AJ, 111, 1913
470. Suntzeff, N. B., Kraft, R. P., & Kinman, T. D. 1994, ApJS, 93, 271
471. Sweigart, A. V. 1996, in Third Conference on Faint Blue Stars, ed. A. G. D. Philip, J. Liebert, & R. A. Saffer (Cambridge: Cambridge Univ. Press), 3
472. Sweigart, A. V. 1997, ApJL, 474, L23
473. Sweigart, A. V., & Gross, P. G. 1976, ApJS, 32, 367
474. Sweigart, A. V., & Mengel, J. G. 1979, ApJ, 229, 624
475. Thomas, H.-C. 1967, Zs. f. Astrophys., 67, 420
476. Thorburn, J. A. 1992, ApJL, 399, L83
477. Thorburn, J. A. 1994, ApJ, 421, 318
478. Tomkin, J., & Lambert, D. L. 1980, ApJ, 235, 925
479. Tomkin, J., Lemke, M., Lambert, D. L., & Sneden, C. 1992, AJ, 104, 1568
480. Trefzger, Ch. F., Carbon, D. F., Langer, G. E., Suntzeff, N. B., & Kraft, R. P. 1983, ApJ, 51, 29
481. Trimble, V. 1993, in Blue Stragglers, ASP Conf. Series 53, ed. R. Saffer (San Francisco: ASP), 155
482. Tsujimoto, T., Miyamoto, M., & Yoshii, Y. 1998 ApJL, 492, L79
483. Twarog, B. A., & Anthony-Twarog, B. J. 1994, AJ, 107, 1371

484. Udry, S., Jorissen, A., Mayor, M., & Van Eck, S. 1998a, A&AS, 131, 25
485. Udry, S., Mayor, M., Van Eck, S., Jorissen, A., Prévot, L., Grenier, S., & Lindgren, H. 1998b, A&AS, 131, 43
486. Umeda, H., Nomoto, K., Kobayashi, C., Hachisu, I., & Kato, M. 1999, ApJL, 522, L43
487. Unavane, M., Wyse, R. F. G., & Gilmore, G. 1996, MNRAS, 278, 727
488. van Agt, S. 1973, in Variable Stars in Globular Clusters, ed. J. D. Fernie (Dordrecht: Reidel), 35
489. van Albada, T. S., & Baker, N. 1971, ApJ, 169, 311
490. VandenBerg, D. A. 1983, ApJS, 51, 29
491. VandenBerg, D. A., & Bell, R. A. 1985, ApJS, 58, 561
492. VandenBerg, D. A., Bolte, M., & Stetson, P. B. 1990, AJ, 100, 445
493. VandenBerg, D. A., Bolte, M., & Stetson, P. B. 1996, AAR&A, 34, 461 (VBS96)
494. van den Bergh, S. 1993, MNRAS, 262, 588
495. Vanture, A. D., Wallerstein, G., & Brown, J. A. 1994, PASP, 106, 835
496. Walker, A. R. 1992, ApJL, 390, L81
497. Walker, A. R. 1998, AJ, 116, 220
498. Walker, A. R., & Terndrup, D. M. 1991, ApJ, 378, 119
499. Wasserburg, G. J., Boothroyd, A. I., & Sackmann, I.-J. 1995, ApJL, 447, L37
500. Wheeler, J. C. 1979a, ApJ, 234, 569
501. Wheeler, J. C. 1979b, Comm. Astrophys., 8, 133
502. Wheeler, J. C., Sneden, C., & Truran, J. W., Jr. 1989, ARA&A, 27, 279
503. Willson, L. A., Bowen, G., & Struck-Marcel, C. J. 1987, Comm. Astrophys. C, 12, 7
504. Winget, D. E., Hansen, C. J., Liebert, J., Van Horn, H. M., Fontaine, G., Nather, R. E., Kepler, S. O., & Lamb, D. Q. 1987, ApJL, 315, L77
505. Wood, M. A. 1992, ApJ, 386, 539
506. Wood, M. A., Oswalt, T. D., & Smith, J. A. 1995, BAAS, 27, 1310
507. Woolley, R. v. d. R., Alexander, J. B., Mather, L., & Epps, E. 1961, Royal Obs. Bull., No. 43
508. Wyse, R. F. G., & Gilmore, G. 1988, AJ, 95, 1404
509. Wyse, R. F. G., & Gilmore, G. 1992, AJ, 104, 144
510. Wyse, R. F. G., & Gilmore, G. 1993, in The Globular Cluster–Galaxy Connection, ASP Conf. Series 48, ed. G. H. Smith & J. Brodie (San Francisco: ASP), 727
511. Wyse, R. F. G., & Gilmore, G. 1995, AJ, 110, 2771
512. Yi, S., Demarque, P., & Kim, Y.-C. 1997, ApJ, 482, 677
513. Yoshii, Y. 1982, PASJ, 34, 365
514. Yoshii, Y., & Arimoto, N. 1987, A&A, 188, 13
515. Yoshii, Y., Tsujimoto, T., & Nomoto, K. 1996, ApJ, 462, 266
516. Yoss, K. M., Neese, C. L., & Hartkopf, W. I. 1987, AJ, 94, 1600
517. Zinn, R. 1985, ApJ, 293, 424
518. Zinn, R. 1993, in The Globular Cluster–Galaxy Connection, ASP Conf. Series 48, ed. G. H. Smith & J. Brodie (San Francisco: ASP), 38
519. Zinn, R., & West, M. J. 1984, ApJS, 55, 45

Globular Cluster Systems

William E. Harris

Department of Physics & Astronomy, McMaster University
Hamilton ON L8S 4M1, Canada, **harris@physics.mcmaster.ca**

Abstract. 1. The basic features of the globular cluster system of the Milky Way are summarized: the total population, subdivision of the clusters into the classic metal-poor and metal-rich components, and first ideas on formation models. The distance to the Galactic center is derived from the spatial distribution of the inner bulge clusters, giving $R_0 = (8.1 \pm 1.0)$ kpc.

2. The calibration of the fundamental distance scale for globular clusters is reviewed. Different ways to estimate the zero point and metallicity dependence of the RR Lyrae stars include statistical parallax and Baade–Wesselink measurements of field RR Lyraes, astrometric parallaxes, white dwarf cluster sequences, and field subdwarfs and main sequence fitting. The results are compared with other distance measurements to the Large Magellanic Cloud and M31.

3. Radial velocities of the Milky Way clusters are used to derive the kinematics of various subsamples of the clusters (the mean rotation speed about the Galactic center, and the line-of-sight velocity dispersion). The inner metal-rich clusters behave kinematically and spatially like a flattened rotating bulge population, while the outer metal-rich clusters resemble a thick-disk population more closely. The metal-poor clusters have a significant prograde rotation (80 to 100 km s^{-1}) in the inner halo and bulge, declining smoothly to near-zero for $R \gtrsim R_0$. No identifiable subgroups are found with significant retrograde motion.

4. The radial velocities of the globular clusters are used along with the spherically symmetric collisionless Boltzmann equation to derive the mass profile of the Milky Way halo. The total mass of the Galaxy is near $\simeq 8 \times 10^{11}$ $M_\odot$ for $r \lesssim 100$ kpc. Extensions to still larger radii with the same formalism are extremely uncertain because of the small numbers of outermost satellites, and the possible correlations of their motions in orbital families.

5. The luminosity functions (GCLF) of the Milky Way and M31 globular clusters are defined and analyzed. We search for possible trends with cluster metallicity or radius, and investigate different analytic fitting functions such as the Gaussian and power-law forms.

6. The global properties of GCSs in other galaxies are reviewed. Measureable distributions include the total cluster population (quantified as the specific frequency S_N), the metallicity distribution function (MDF), the luminosity and space distributions, and the radial velocity distribution.

7. The GCLF is evaluated as a standard candle for distance determination. For giant E galaxies, the GCLF turnover has a mean luminosity of $M_V = -7.33$ on a distance scale where Virgo has a distance modulus of 31.0 and Fornax is at 31.3, with galaxy-to-galaxy scatter $\sigma(M_V) = 0.15$ mag. Applying this calibration to more remote galaxies yields a Hubble constant $H_0 = (74 \pm 9)$ km s^{-1} Mpc^{-1}.

8. The observational constraints on globular cluster formation models are summarized. The appropriate host environments for the formation of $\sim 10^5$ to 10^6 $M_\odot$ clusters are suggested to be kiloparsec–sized gas clouds (Searle/Zinn fragments or supergiant molecular clouds) of 10^8 to 10^9 $M_\odot$. A model for the growth of protocluster clouds by collisional agglomeration is presented, and matched with observed mass distribution functions. The issue of globular cluster formation efficiency in different galaxies is discussed (the "specific frequency problem").

9. Other influences on galaxy formation are discussed, including mergers, accretions, and starbursts. Mergers of disk galaxies almost certainly produce elliptical galaxies of *low* S_N, while the high-S_N ellipticals are more likely to have been produced through *in situ* formation. Starburst dwarfs and large active galaxies in which current globular cluster formation is taking place are compared with the key elements of the formation model.

10. (Appendix) Some basic principles of photometric methods are gathered together and summarized: the fundamental signal-to-noise formula, objective star finding, aperture photometry, PSF fitting, artificial-star testing, detection completeness, and photometric uncertainty. Lastly, we raise the essential issues in photometry of nonstellar objects, including image moment analysis, total magnitudes, and object classification techniques.

Introduction

When the storm rages and the state is threatened by shipwreck, we can do nothing more noble than to lower the anchor of our peaceful studies into the ground of eternity.

Johannes Kepler

Why do we do astronomy? It is a difficult, frustrating, and often perverse business, and one which is sometimes costly for society to support. Moreover, if we are genuinely serious about wanting to probe Nature, we might well employ ourselves better in other disciplines like physics, chemistry, or biology, where at least we can exert experimental controls over the things we are studying, and where progress is usually less ambiguous.

Kepler seems to have understood why. It is often said that we pursue astronomy because of our inborn curiosity and the need to understand our place and origins. True enough, but there is something more. Exploring the universe is a unique adventure of profound beauty and exhilaration, lifting us far beyond our normal self-centered concerns to a degree that no other field can do quite as powerfully. Every human generation has found that the world beyond the Earth is a vast and astonishing place.

In these chapters, we will be taking an all-too-brief tour through just one small area of modern astronomy – one which has roots extending back more than a century, but which has re-invented itself again and again with the advance of both astrophysics and observational technology. It is also one which draws intricate and sometimes surprising connections among stellar populations, star formation, the earliest history of the galaxies, the distance scale, and cosmology.

Fig. 1. *Left panel:* Palomar 2, a globular cluster about 25 kiloparsecs from the Sun in the outer halo of the Milky Way. The picture shown is an I-band image taken with the Canada-France-Hawaii Telescope (see Harris et al. 1997c). The field size is 4.6 arcmin (or 33 parsecs) across and the image resolution ("seeing") is 0.5 arcsec. *Right panel:* A globular cluster in the halo of the giant elliptical galaxy NGC 5128, 3900 kiloparsecs from the Milky Way. The picture shown is an I-band image taken with the *Hubble Space Telescope* (see G. Harris et al. 1998); the field size is 0.3 arcmin (or 340 parsecs) across, and the resolution is 0.1 arcsec. This is the most distant globular cluster for which a color–magnitude diagram has been obtained

The sections to follow are organized in the same way as the lectures given at the 1998 Saas–Fee Advanced Course held in Les Diablerets. Each one represents a well defined theme which could in principle stand on its own, but all of them link together to build up an overview of what we currently know about globular cluster systems in galaxies. It will (I hope) be true, as in any active field, that much of the material will already be superseded by newer insights by the time it is in print. Wherever possible, I have tried to preserve in the text the conversational style of the lectures, in which lively interchanges among the speakers and audience were possible. The literature survey for this paper carries up to the early part of 1999.

A *globular cluster system* (GCS) is the collection of all globular star clusters within one galaxy, viewed as a subpopulation of that galaxy's stars. The essential questions addressed by each section of this review are, in sequence:

- What are the size and structure of the Milky Way globular cluster system, and what are its definable subpopulations?

- What should we use as the fundamental Population II distance scale?
- What are the kinematical characteristics of the Milky Way GCS? Do its subpopulations show traces of different formation epochs?
- How can the velocity distribution of the clusters be used to derive a mass profile for the Milky Way halo?
- What is the luminosity ($\equiv$ mass) distribution function for the Milky Way GCS? Are there detectable trends with subpopulation or galactocentric distance?
- What are the overall characteristics of GCSs in other galaxies – total numbers, metallicity distributions, correlations with parent galaxy type?
- How can the luminosity distribution function (GCLF) be used as a "standard candle" for estimation of the Hubble constant?
- Do we have a basic understanding of how globular clusters formed within protogalaxies in the early universe?
- How do we see globular cluster populations changing today, due to such phenomena as mergers, tidal encounters, and starbursts?

The study of globular cluster systems is a genuine hybrid subject mixing elements of star clusters, stellar populations, and the structure and history of all types of galaxies. Over the past two decades, it has grown rapidly along with the spectacular advances in imaging technology. The first review article in the subject (Harris & Racine 1979) spent its time almost entirely on the globular clusters in Local Group galaxies and only briefly discussed the little we knew about a few Virgo ellipticals. Other reviews (Harris 1988a,b, 1991, 1993, 1995, 1996b, 1998, 1999) demonstrate the growth of the subject into one which can put a remarkable variety of constraints on issues in galaxy formation and evolution. Students of this subject will also want to read the recent book *Globular Cluster Systems* by Ashman & Zepf (1998), which gives another comprehensive overview in a different style and with different emphases on certain topics.

I have kept abbreviations and acronyms in the text to a minimum. Here is a list of the ones used frequently:

CMD: color–magnitude diagram
GCS: globular cluster system; the collection of all globular clusters in a given galaxy
GCLF: globular cluster luminosity function, conventionally defined as the number of globular clusters per unit *magnitude* interval $\phi(M_V)$
LDF: luminosity distribution function, conventionally defined as the number of globular clusters per unit *luminosity*, dN/dL. The LDF and GCLF are related through $\phi \sim L(dN/dL)$
MDF: metallicity distribution function, usually defined as the number of clusters (or stars) per [Fe/H] interval
MPC: "metal-poor component"; the low-metallicity part of the MDF
MRC: "metal-rich component"; the high-metallicity part of the MDF

ZAMS: zero-age main sequence; the locus of unevolved core hydrogen burning stars in the CMD
ZAHB: zero-age horizontal branch: the locus of core helium burning stars in the CMD, at the beginning of equilibrium helium burning

1 The Milky Way System: A Global Perspective

It is a capital mistake to theorize before one has data.
Sherlock Holmes

We will see in the later sections that our ideas about the general characteristics of globular cluster systems are going to be severely limited, and even rather badly biased, if we stay only within the Milky Way. But the GCS of our own Galaxy is quite correctly the starting point in our journey. It is not the largest such system; it is not the most metal-poor or metal-rich; it is probably not the oldest; and it is certainly far from unique. It is simply the one we know best, and it has historically colored all our ideas and mental images of what we mean by "globular clusters", and (even more importantly) the way that galaxies probably formed.

1.1 A First Look at the Spatial Distribution

Currently, we know of 147 objects within the Milky Way that are called globular clusters (Harris 1996a). They are found everywhere from deep within the Galactic bulge out to twice the distance of the Magellanic Clouds. Figure 2 shows the spatial distribution of all known clusters within ~ 20 kiloparsecs of the Galactic center, and (in an expanded scale) the outermost known clusters. To plot up these graphs, I have already assumed a "distance scale"; that is, a specific prescription for converting apparent magnitudes of globular cluster stars into true luminosities. As discussed in Sect. 2, this prescription is

$$M_V(\mathrm{HB}) = 0.15\,[\mathrm{Fe/H}] + 0.80$$

where [Fe/H] represents the cluster heavy-element abundance (metallicity) and $M_V(\mathrm{HB})$ is the absolute V magnitude of the horizontal branch in the color–magnitude diagram (abbreviated CMD; see the Appendix for a sample cluster in which the principal CMD sequences are defined). For metal-poor clusters in which RR Lyrae stars are present, by convention $M_V(\mathrm{HB})$ is identical to the mean luminosity of these RR Lyraes. For metal-richer clusters in which there are only red HB stars and no RR Lyraes, $M_V(\mathrm{HB})$ is equal to the mean luminosity of the red side of the horizontal branch (RHB). More will be said about the calibration of this scale in Sect. 2; for now, we will simply use it to gain a broad picture of the entire system.

Throughout this section, the numbers (X, Y, Z) denote the usual distance coordinates of any cluster relative to the Sun: X points from the Sun in

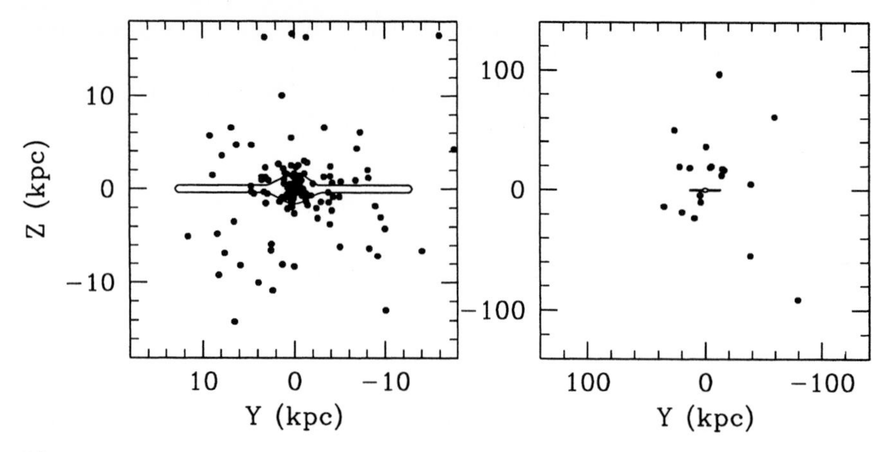

Fig. 2. *Left panel:* Spatial distribution of the inner globular cluster system of the Milky Way, projected onto the YZ plane. Here the Sun and Galactic center are at $(0,0)$ and we are looking in along the X–axis toward the center. *Right panel:* Spatial distribution in the YZ plane of the outer clusters

toward the Galactic center, Y points in the direction of Galactic rotation, and Z points perpendicular to the Galactic plane northward. The coordinates (X,Y,Z) are defined as:

$$X = R \cos b \cos \ell\,, \quad Y = R \cos b \sin \ell\,, \quad Z = R \sin b\,, \tag{1}$$

where (ℓ, b) are the Galactic longitude and latitude and R is the distance of the cluster from the Sun. In this coordinate system the Sun is at $(0,0,0)$ kpc and the Galactic center at $(8,0,0)$ kpc (see below).

In *very* rough terms, the GCS displays spherical symmetry – at least, as closely as any part of the Galaxy does. Just as Harlow Shapley did in the early part of this century, we still use it today to outline the size and shape of the Galactic *halo* (even though the halo field stars outnumber those in globular clusters by at least 100 to 1, the clusters are certainly the easiest halo objects to find). But we can see as well from Fig. 2 that the whole system is a *centrally concentrated* one, with the spatial density ϕ (number of clusters per unit volume in space) varying as $\phi \sim R_{gc}^{-3.5}$ over most of the halo (Fig. 3). Unfortunately, one immediate problem this leaves us is that more than half of our globular clusters can be studied only by peering in toward the Galactic center through the heavy obscuration of dust clouds in the foreground of the Galactic disk.[1] Until recent years our knowledge of these

[1] For this reason, the YZ plane was used for the previous figure to display the large-scale space distribution. Our line of sight to most of the clusters is roughly parallel to the X–axis and thus any distance measurement error on our part will skew the estimated value of X much more than Y or Z. Of all possible projections, the YZ plane is therefore the most nearly "error-free" one.

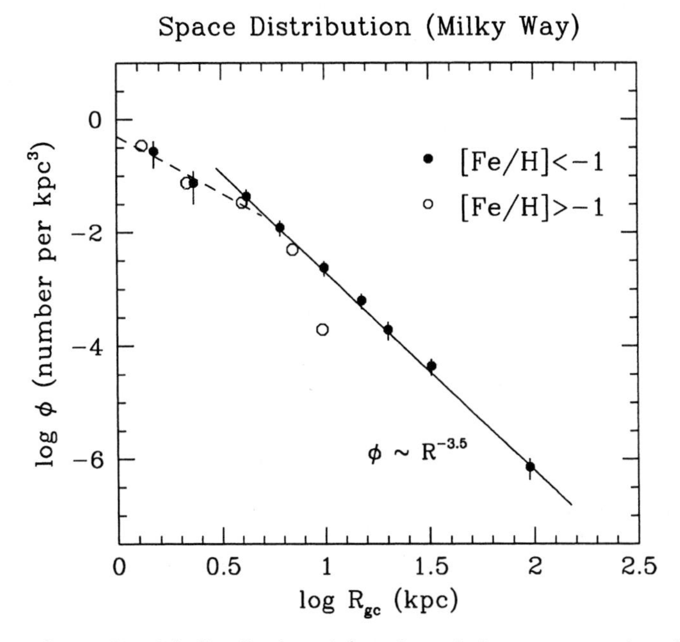

Fig. 3. Spatial distribution ϕ (number of clusters per unit volume) as a function of Galactocentric distance R_{gc}. The metal-poor subpopulation ([Fe/H] < -1) is shown in *solid dots*, the metal-richer subpopulation ([Fe/H] > -1) as *open symbols*. For $R_{gc} \gtrsim 4$ kpc (*solid line*), a simple power-law dependence $\phi \sim R_{gc}^{-3.5}$ matches the spatial structure well, while for the inner bulge region, ϕ flattens off to something closer to an R^{-2} dependence. Notice that the metal-richer distribution falls off steeply for $R_{gc} \gtrsim 10$ kpc. This plot implicitly (and wrongly!) assumes a spherically symmetric space distribution, which smooths over any more detailed structure; see the discussion below

heavily reddened clusters remained surprisingly poor, and even today there are still a few clusters with exceptionally high reddenings embedded deep in the Galactic bulge about which we know almost nothing (see the listings in Harris 1996a).

However, progress over the years has been steady and substantial: compare the two graphs in Fig. 4. One (from the data of Shapley 1918) is the very first 'outside view' of the Milky Way GCS ever achieved, and the one used by Shapley to estimate the centroid of the system and thus – again for the first time – to determine the distance from the Sun to the Galactic center. The second graph shows us exactly the same plot with the most modern measurements. The data have improved dramatically over the intervening 80 years in three major ways: (1) The sample size of known clusters is now almost twice as large as Shapley's list. (2) Shapley's data took no account of reddening, since the presence and effect of interstellar dust was unknown

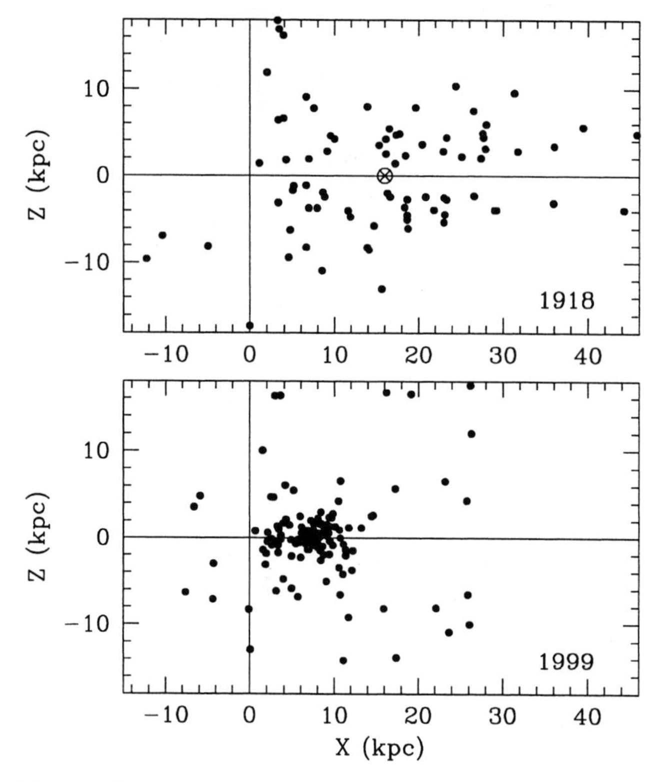

Fig. 4. *Upper panel:* The spatial distribution of the Milky Way clusters as measured by Shapley (1918). The Sun is at (0,0) in this graph, and Shapley's estimated location of the Galactic center is marked at (16,0). *Lower panel:* The spatial distribution in the same plane, according to the best data available today. The tight grouping of clusters near the Galactic center (now at (8,0)), and the underlying symmetry of the system, are now much more obvious

then; the result was to overestimate the distances for most clusters and thus to elongate their whole distribution along the line of sight (roughly, the X-axis). (3) The fundamental distance scale used by Shapley – essentially, the luminosity of the RR Lyraes or the tip of the red-giant branch – was about one magnitude brighter than the value adopted today; again, the result was to overestimate distances for almost all clusters. Nevertheless, this simple diagram represented a breakthrough in the study of Galactic structure; armed with it, Shapley boldly argued both that the Sun was far from the center of the Milky Way, and that our Galaxy was much larger than had been previously thought.

The foreground reddening of any given cluster comes almost totally from dust clouds in the Galactic disk rather than the bulge or halo, and so reddening correlates strongly with Galactic latitude (Fig. 5). The basic cosecant–law

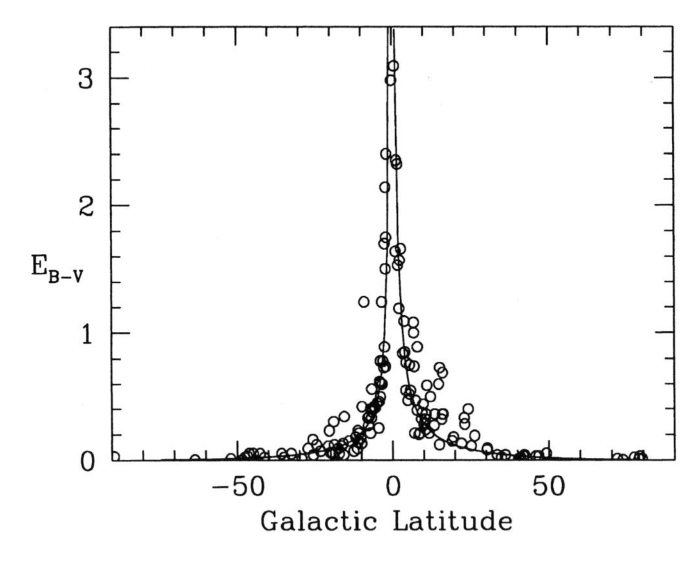

Fig. 5. The foreground reddening of globular clusters E_{B-V} plotted against Galactic latitude b (in degrees). The equations for the cosecant lines are given in the text

dependence of E_{B-V} shows how very much more difficult it is to study objects at low latitudes. Even worse, such objects are also often afflicted with severe contamination by field stars and by differential (patchy) reddening. The equations for the reddening lines in Fig. 5 are:

Northern Galactic hemisphere ($b > 0$): $E_{B-V} = 0.060\ (\csc |b| - 1)$,
Southern Galactic hemisphere ($b < 0$): $E_{B-V} = 0.045\ (\csc |b| - 1)$.

Individual globular clusters have been known for at least two centuries. Is our census of them complete, or are we still missing some? This question has been asked many times, and attempted answers have differed quite a bit (e.g. Racine & Harris 1989; Arp 1965; Sharov 1976; Oort 1977; Barbuy et al. 1998). They are luminous objects, and easily found anywhere in the Galaxy as long as they are not either (a) *extremely* obscured by dust, or (b) too small and distant to have been picked up from existing all-sky surveys. Discoveries of faint, distant clusters at high latitude continue to happen occasionally as lucky accidents, but are now rare (just five new ones have been added over the last 20 years: AM-1 [Lauberts 1976; Madore & Arp 1979], Eridanus [Cesarsky et al. 1977], E3 [Lauberts 1976], Pyxis [Irwin et al. 1995], and IC 1257 [Harris et al. 1997a]). Recognizing the strong latitude effect that we see in Fig. 5, we might make a sensible estimate of missing heavily reddened clusters by using Fig. 6. The number of known clusters per unit latitude angle b rises exponentially to lower latitude, quite accurately as $n \sim exp^{-|b|/14^{0}}$ for $2^{\circ} \lesssim b \lesssim 40^{\circ}$. It is only the first bin ($|b| < 2^{\circ}$) where incompleteness appears to be important;

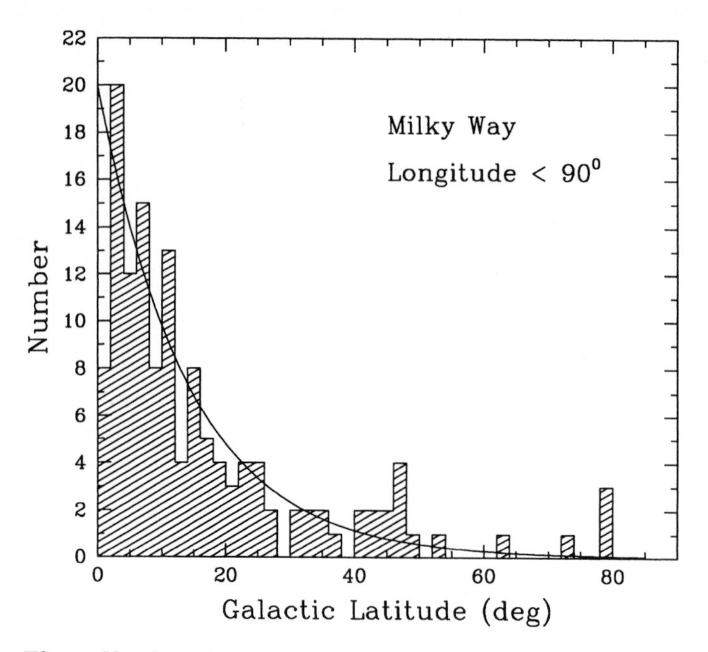

Fig. 6. Number of globular clusters as a function of Galactic latitude $|b|$, plotted in 2° bins; only the clusters within ±90° longitude of the Galactic center are included. An exponential rise toward lower latitude, with an e–folding height of 14°, is shown as the *solid line*

$\sim$ 10 additional clusters would be needed there to bring the known sample back up to the curve.

Combining these arguments, I estimate that the total population of globular clusters in the Milky Way is $N = 160 \pm 10$, and that the existing sample is now likely to be more than 90% complete.

1.2 The Metallicity Distribution

The huge range in *heavy-element abundance* or *metallicity* among globular clusters became evident to spectroscopists half a century ago, when it was found that the spectral lines of the stars in most clusters were remarkably weak, resembling those of field subdwarf stars in the Solar neighborhood (e.g., Mayall 1946; Baum 1952; Roman 1952). Morgan (1956) and Baade (1958) at the landmark Vatican conference suggested that their compositions might be connected with Galactocentric location R_{gc} or Z. These ideas culminated in the classic work of Kinman (1959a,b), who systematically investigated the correlations among composition, location, and kinematics of subsamples within the GCS. By the beginning of the 1960's, these pioneering studies had been used to develop a prevailing view in which (a) the GCS possessed a *metallicity gradient*, with the higher-metallicity clusters residing only in the inner bulge

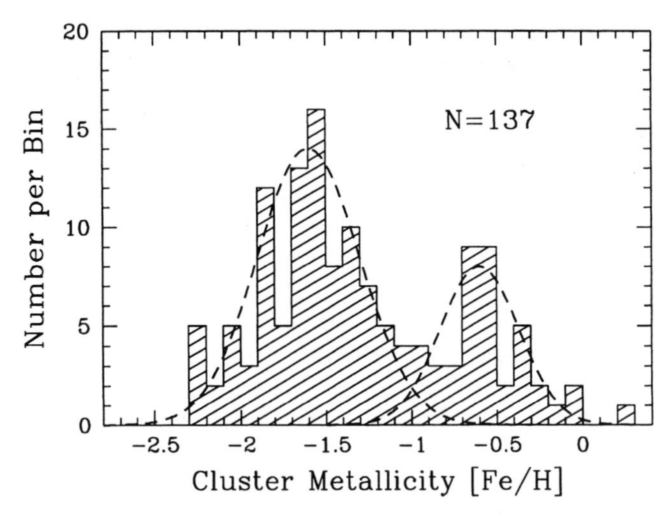

Fig. 7. Metallicity distribution for 137 Milky Way globular clusters with measured [Fe/H] values. The metallicities are on the Zinn–West (1984) scale, as listed in the current compilation of Harris (1996a). The bimodal nature of the histogram is shown by the two Gaussian curves whose parameters are described in the text

regions and the average metallicity of the system declining steadily outwards; (b) the metal-poor clusters were a dynamically 'hot' system, with large random space motions and little systemic rotation; (c) the metal-richer clusters formed a 'cooler' subsystem with significant overall rotation and lower random motion.

All of this evidence was thought to fit rather well into a picture for the formation of the Galaxy that was laid out by Eggen, Lynden-Bell, & Sandage (1962 [ELS]). In their model, the first stars to form in the protogalactic cloud were metal-poor and on chaotic, plunging orbits; as star formation continued, the remaining gas was gradually enriched, and as it collapsed inward and spun up, subsystems could form which were more and more disk-like. The timescale for all of this to take place could have been no shorter than the freefall time of the protogalactic cloud (a few 10^8 y), but might have been significantly longer depending on the degree of pressure support during the collapse. If pressure support was important, then a clear metallicity gradient should have been left behind, with cluster age correlated nicely with its chemical composition. The rough age calibrations of the globular clusters that were possible at the time (e.g., Sandage 1970) could not distinguish clearly between these alternatives, but were consistent with the view that the initial collapse was rapid.

This appealing model did not last – at least, not in its original simplicity. With steady improvements in the database, new features of the GCS emerged. One of the most important of these is the *bimodality of the cluster metallicity distribution*, shown in Fig. 7. Two rather distinct metallicity groups clearly exist, and it is immediately clear that the simple monolithic–collapse model for the formation of the GCS will not be adequate. To avoid prejudicing our view of these two subgroups as belonging to the Galactic halo, the disk, the bulge, or something else, I will simply refer to them as the *metal-poor component* (MPC) and the *metal-rich component* (MRC). In Fig. 7, the MPC has a fitted centroid at [Fe/H] $= -1.6$ and a dispersion $\sigma = 0.30$ dex, while the MRC has a centroid at [Fe/H] $= -0.6$ and dispersion $\sigma = 0.2$. The dividing line between them I will adopt, somewhat arbitrarily, at [Fe/H] $= -0.95$ (see the next section below).[2]

The distribution of [Fe/H] with location is shown in Fig. 8. Clearly, the dominant feature of this diagram is the *scatter* in metallicity at any radius R_{gc}. Smooth, pressure–supported collapse models are unlikely to produce a result like this. But can we see any traces at all of a metallicity gradient in which progressive enrichment occurred? For the moment, we will ignore the half-dozen remote objects with $R_{gc} > 50$ kpc (these "outermost-halo" clusters probably need to be treated separately, for additional reasons that we will see below). For the inner halo, a *small* net metallicity gradient is rather definitely present amidst the dominant scatter. Specifically, within both the MPC and MRC systems, we find $\Delta[\mathrm{Fe/H}]/\,\Delta \log R_{gc} = -0.30$ for the restricted region $R_{gc} \lesssim 10$ kpc; that is, the heavy-element abundance scales as $(Z/Z_{\odot}) \sim R^{-0.3}$. At larger R_{gc}, no detectable gradient appears.

[2] Note that the [Fe/H] values used throughout my lectures are ones on the "Zinn–West" (ZW) metallicity scale, the most frequently employed system through the 1980's and 1990's. The large catalog of cluster abundances by Zinn & West (1984) and Zinn (1985) was assembled from a variety of abundance indicators including stellar spectroscopy, color–magnitude diagrams, and integrated colors and spectra. These were calibrated through high-dispersion stellar spectroscopy of a small number of clusters obtained in the pre-CCD era, mainly from the photographic spectra of Cohen (see Frogel et al. 1983 for a compilation). Since then, a much larger body of spectroscopic data has been built up and averaged into the ZW list, leading to a somewhat heterogeneous database (for example, see the [Fe/H] sources listed in the Harris 1996a catalog, which are on the ZW scale). More recently, comprehensive evidence has been assembled by Carretta & Gratton (1997) that the ZW [Fe/H] scale is nonlinear relative to contemporary high-resolution spectroscopy, even though the older abundance indicators may still provide the correct *ranking* of relative metallicity for clusters. The problem is also discussed at length by Rutledge et al. (1997). Over the range containing most of the Milky Way clusters ([Fe/H] $\lesssim -0.8$) the scale discrepancies are not large (typically ± 0.2 dex at worst). But at the high-[Fe/H] end the disagreement becomes progressively worse, with the ZW scale overestimating the true [Fe/H] by ~ 0.5 dex at near-solar true metallicity. At the time of writing these chapters, a completely homogeneous metallicity list based on the Carretta–Gratton scale has not yet been constructed. Lastly, it is worth emphasizing that the quoted metallicities for globular clusters are almost always based on spectral features of the highly evolved red giant stars. Eventually, we would like to base [Fe/H] on the (much fainter) unevolved stars.

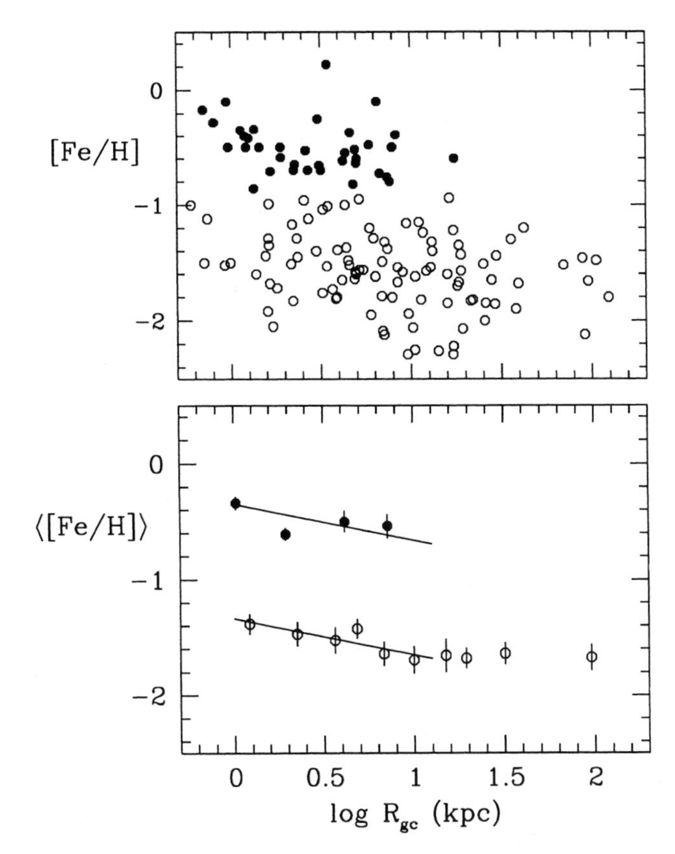

Fig. 8. [Fe/H] plotted against Galactocentric distance R_{gc}. *Upper panel*: Individual clusters are plotted, with MRC objects as *solid symbols* and MPC as *open symbols*. *Lower panel*: Mean [Fe/H] values for radial bins. Both MRC and MPC subsystems exhibit a slight gradient Δ[Fe/H]/ $\Delta\log R_{gc} = -0.30$ for $R_{gc} \lesssim 10$ kpc, as shown by the *solid lines*. For the more distant parts of the halo, no detectable mean gradient exists

These features – the large scatter and modest inner-halo mean gradient – have been taken to indicate that the inner halo retains a trace of the classic monolithic rapid collapse, while the outer halo is dominated by chaotic formation and later accretion. They also helped stimulate a very different paradigm for the early evolution of the Milky Way, laid out in the papers of Searle (1977) and Searle & Zinn (1978 [SZ]). Unlike ELS, they proposed that the protoGalaxy was in a clumpy, chaotic, and non-equilibrium state in which the halo-star (and globular cluster) formation period could have lasted over many Gigayears. An additional key piece of evidence for their view was to be found in the connections among the *horizontal-branch morphologies* of the globular clusters, their locations in the halo, and their metallicities (Fig. 9).

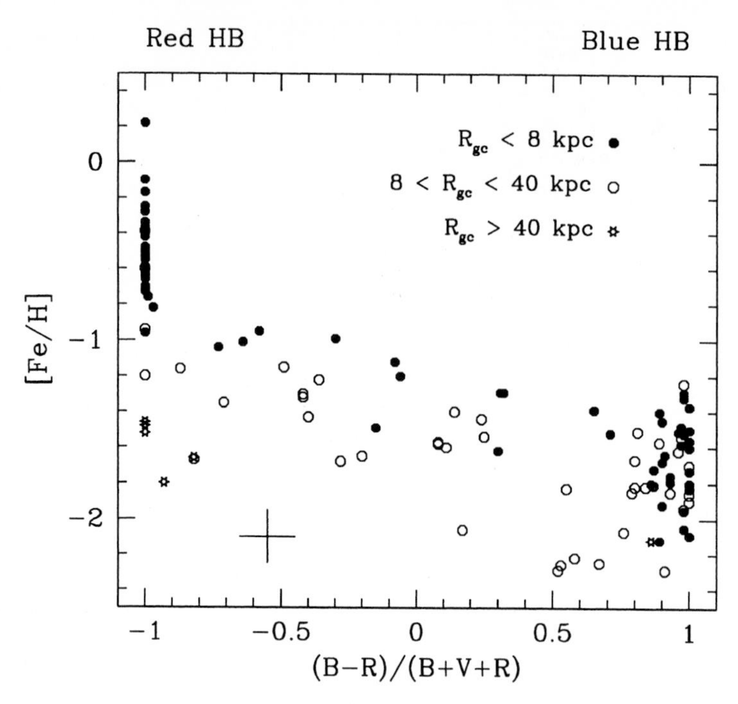

Fig. 9. Metallicity versus horizontal branch type for globular clusters. The HB ratio $(B-R)/(B+V+R)$ (Lee et al. 1994) is equal to −1 for clusters with purely red HBs, increasing to +1 for purely blue HBs. A typical measurement uncertainty for each point is shown at lower left. Data are taken from the catalog of Harris (1996a)

They noted that for the inner-halo clusters ($R_{gc} \lesssim R_0$), there was generally a close correlation between HB type and metallicity, as if all these clusters were the same age and HB morphology was determined only by metallicity. (More precisely, the same type of correlation would be generated if there were a one-to-one relation between cluster metallicity and age; i.e. if metallicity determined both age and HB type together. However, Lee et al. (1994) argue from isochrone models that the inner-halo correlation is nearly what we would expect for a single-age sequence differing only in metallicity.) In general, we can state that the morphology of the CMD is determined by several "parameters" which label the physical characteristics of the stars in the cluster. The *first parameter* which most strongly controls the distribution of stars in the CMD is commonly regarded to be metallicity, i.e. the overall heavy-element abundance. But quantities such as the HB morphology or the color and steepness of the giant branch do not correlate uniquely with only the metallicity, so more parameters must come into play. Which of these is most important is not known. At various times, plausible cases have been made that the dominant *second parameter* might be cluster age, helium abundance, CNO–group elements, or other factors such as mass loss or internal stellar rotation.

By contrast, for the intermediate- and outer-halo clusters the correlation between [Fe/H] and HB type becomes increasingly scattered, indicating that other parameters are affecting HB morphology just as strongly as metallicity. The interpretation offered by SZ was that the principal "second parameter" is age, in the sense that the *range in ages* is much larger for the outer-halo clusters. In addition, the progressive shift toward redder HBs at larger Galactocentric distance (toward the left in Fig. 9) would indicate a trend toward lower mean age in this interpretive picture.

From the three main pieces of evidence (a) the large scatter in [Fe/H] at any location in the halo, (b) the small net gradient in mean [Fe/H], and (c) the weaker correlation between HB type and metallicity at increasing R_{gc}, SZ concluded that the entire halo could not have formed in a pressure-supported monolithic collapse. Though the inner halo could have formed with some degree of the ELS–style formation, the outer halo was dominated by chaotic formation and even accretion of fragments from outside. They suggested that the likely formation sites of globular clusters were within large individual gas clouds (to be thought of as protogalactic 'fragments'), within which the compositions of the clusters were determined by very local enrichment processes rather than global ones spanning the whole protogalactic potential well. Although a large age range is not *necessary* in this scheme (particularly if other factors than age turn out to drive HB morphology strongly), a significant age range would be much easier to understand in the SZ scenario, and it opened up a wide new range of possibilities for the way halos are constructed. We will return to further developments of this picture in later sections. For the moment, we will note only that, over the next two decades, much of the work on increasingly accurate age determination and composition analysis for globular clusters all over the Galactic halo was driven by the desire to explore this roughed-out model of 'piecemeal' galaxy formation.

1.3 The Metal-Rich Population: Disk or Bulge?

Early suggestions of distinct components in the metallicity distribution were made by, e.g., Marsakov & Suchkov (1976) and Harris & Canterna (1979), but it was the landmark paper of Zinn (1985) which firmly identified two distinct subpopulations *and* showed that these two groups of clusters also had distinct kinematics and spatial distributions. In effect, *it was no longer possible to talk about the GCS as a single stellar population.* Our next task is, again, something of a historically based one: using the most recent data, we will step through a classic series of questions about the nature of the MRC and MPC.

The spatial distributions of the MPC and MRC are shown in Fig. 10. Obviously, the MRC clusters form a subsystem with a much smaller scale size.

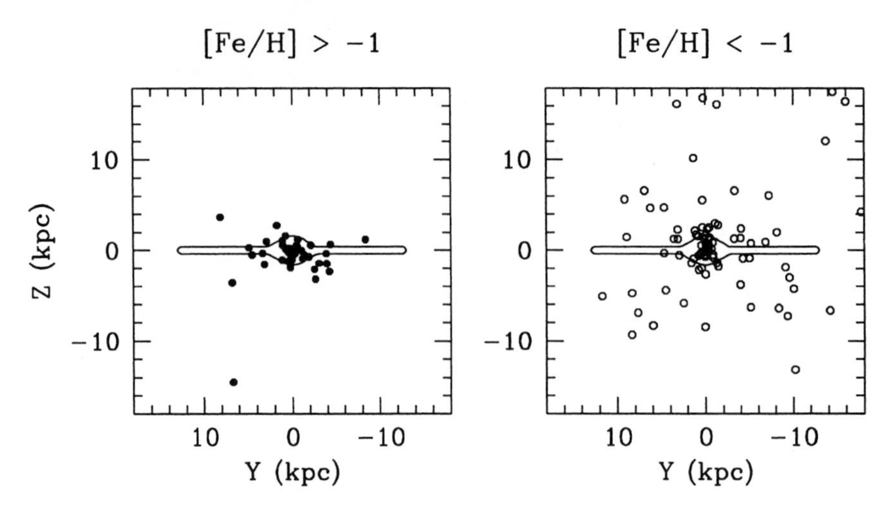

Fig. 10. Spatial distribution projected on the YZ plane for the metal-rich clusters (*left*) with [Fe/H]> -0.95, and the metal-poor clusters (*right*) with [Fe/H]< -0.95. In the left panel, the most extreme outlying point is Palomar 12, a "transition" object between halo and disk

Since the work of Zinn (1985) and Armandroff (1989), the MRC has conventionally been referred to as a "disk cluster" system, with the suggestion that these clusters belonged spatially and kinematically to the thick disk. This question has been re-investigated by Minniti (1995) and Côté (1999), who make the case that they are better associated with the Galactic bulge. A key observation is the fact that the relative number of the two types of clusters, $N_{\rm MRC}/N_{\rm MPC}$, *rises steadily inward to the Galactic center*, in much the same way as the bulge-to-halo-star ratio changes inward, whereas in a true "thick-disk" population this ratio should die out to near-zero for $R_{\rm gc} \lesssim 2$ kpc.

The MRC space distribution is also not just a more compact version of the MPC; rather, it appears to be genuinely flattened toward the plane. A useful diagnostic of the subsystem shape is to employ the angles (ω, θ) defining the cluster location on the sky relative to the Galactic center (see Zinn 1985 and Fig. 11). Consider a vector from the Galactic center to the cluster as seen projected on the sky: the angular length of the vector is ω, while the orientation angle between ω and the Galactic plane is θ:

$$\cos\omega = \cos b \cos\ell \,, \quad \tan\theta = \tan b \csc\ell \,. \tag{2}$$

In Fig. 12, the (ω, θ) point distributions are shown separately for the MRC and MPC subsystems. Both graphs have more points at smaller ω, as is expected for any population which is concentrated toward the Galactic center. However, any spherically symmetric population will be uniformly distributed in the azimuthal angle θ, whereas a flattened (disk or bulge) population will be biased toward small values of θ. The comparison test must also recognize

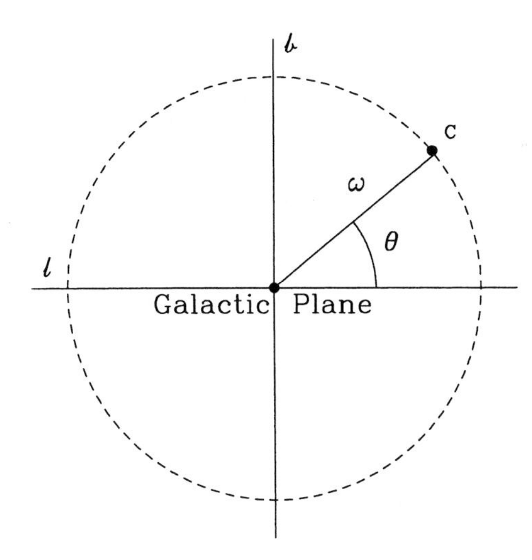

Fig. 11. Definition of the angles ω, θ given in the text: the page represents the plane of the sky, centered on the Galactic center. The Galactic longitude and latitude axes (ℓ, b) are drawn in. The distance from the Galactic center to the cluster C subtends angle ω, while θ is its orientation angle to the Galactic plane

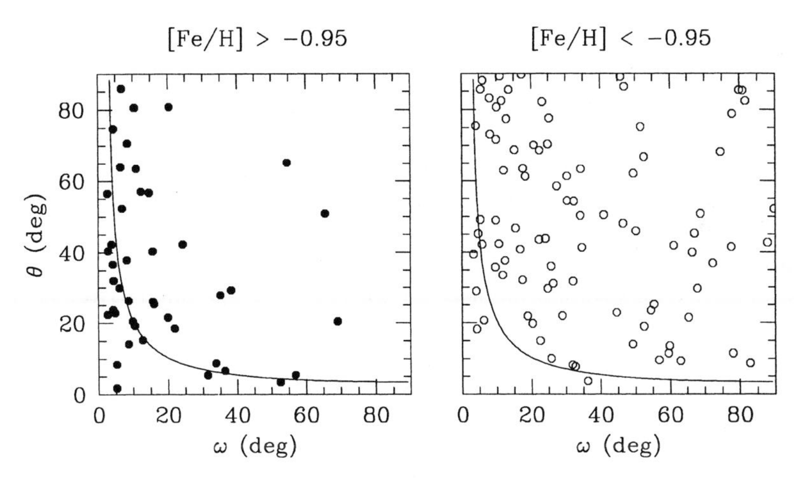

Fig. 12. Spatial distribution diagnostics for the MRC (*left*) and MPC (*right*) clusters. Here ω is the angle between the Galactic center and the cluster as seen on the sky, and θ is the angle between the Galactic plane and the line joining the Galactic center and the cluster. A line of constant Galactic latitude ($b = 3.5$ degrees) is shown as the *curved line* in each figure. Below this line, the foreground reddening becomes large and incompleteness in both samples is expected

Table 1. Spatial flattening parameters for bulge clusters

	$\langle\theta\rangle$ (deg)	$\langle\|Z\|\rangle/\langle\|Y\|\rangle$	$\langle\|Z\|\rangle/\langle\sqrt{X^2+Y^2}\rangle$
MRC ($\omega < 20^o$)	40.0 ± 4.8	0.81 ± 0.20	0.37 ± 0.08
MPC ($\omega < 20^o$)	56.8 ± 4.5	1.63 ± 0.39	0.68 ± 0.13

the probable incompleteness of each sample at low latitude: for $|b| \lesssim 3°$, the foreground absorption becomes extremely large, and fewer objects appear below that line in either diagram.

A marked difference between the two samples emerges (Table 1) if we simply compare the mean $\langle\theta\rangle$ for clusters within 20° of the center (for which the effects of reddening should be closely similar on each population). A population of objects which has a spherical spatial distribution *and is unaffected by latitude incompleteness* would have $\langle\theta\rangle = 45°$, whereas sample incompleteness at low b would bias the mean $\langle\theta\rangle$ to higher values. Indeed, the MPC value $\langle\theta\rangle = 57°$ is consistent with that hypothesis – that is, that low-latitude clusters are missing from the sample because of their extremely high reddenings. However, the MRC value $\langle\theta\rangle = 40°$ – which must be affected by the *same* low-latitude incompleteness – can then result only if it belongs to an intrinsically flattened distribution. A Kolmogorov–Smirnov test on the θ-distribution confirms that these samples are different at the $\sim 93\%$ confidence level, in the sense that the MRC is more flattened.

Another way to define the same result is to compare the linear coordinates Z, Y, and $\sqrt{X^2+Y^2}$ (Table 1). The relevant ratios Z/Y and $Z/\sqrt{X^2+Y^2}$ are half as large for the MRC as for the MPC, again indicating a greater flattening to the plane.

Our tentative conclusion from these arguments is that the *inner* MRC – the clusters within $\omega \sim 20°$ or 3 kpc of the Galactic center – outline something best resembling a *flattened bulge population.* Kinematical evidence will be added in Sect. 4.

1.4 The Distance to the Galactic Center

As noted above, Shapley (1918) laid out the definitive demonstration that the Sun is far from the center of the Milky Way. His first estimate of the distance to the Galactic center was $R_0 = 16$ kpc, only a factor of two different from today's best estimates (compare the history of the Hubble constant over the same interval!). In the absence of sample selection effects and measurement biases, Shapley's hypothesis can be written simply as $R_0 = \langle X\rangle$ where the mean X–coordinate is taken over the entire globular cluster population (indeed, the same relation can be stated for any population of objects centered at the same place, such as RR Lyraes, Miras, or other standard candles).

But of course the sample mean $\langle X \rangle$ is biased especially by incompleteness and nonuniformity at low latitude, as well as distortions in converting distance modulus $(m-M)_V$ to linear distance X: systematic errors will result if the reddening is estimated incorrectly or if the distance–scale calibration for $M_V(\mathrm{HB})$ is wrong. Even the random errors of measurement in distance modulus convert to asymmetric error bars in X and thus a systematic bias in $\langle X \rangle$. One could minimize these errors by simply ignoring the "difficult" clusters at low latitude and using only low-reddening clusters at high latitude. However, there are not that many high-halo clusters ($N \sim 50$), and they are widely spread through the halo, leaving an uncomfortably large and irreducible uncertainty of $\sim \pm 1.5$ kpc in the centroid position $\langle X \rangle$ (see, e.g., Harris 1976 for a thorough discussion).

A better method, outlined by Racine & Harris (1989), is to use the *inner* clusters and to turn their large and different reddenings into a partial advantage. The basic idea is that, to first order, the great majority of the clusters we see near the direction of the Galactic center are *at the same true distance* R_0 – that is, they are *in* the Galactic bulge, give or take a kiloparsec or so – despite the fact that they may have wildly different *apparent* distance moduli.[3] This conclusion is guaranteed by the strong central concentration of the GCS (Fig. 2) and can be quickly verified by simulations (see Racine & Harris). For the inner clusters, we can then write $d \simeq R_0$ for essentially all of them, and thus

$$(m-M)_V \equiv (m-M)_0 + A_V \simeq \mathrm{const} + R \cdot E_{B-V} \tag{3}$$

where $R \simeq 3.1$ is the adopted ratio of total to selective absorption. Now since the horizontal-branch magnitude V_{HB} is a good indicator of the cluster apparent distance modulus, varying only weakly with metallicity, a simple graph of V_{HB} against reddening for the inner globular clusters should reveal a straight-line relation with a (known) slope equal to R:

$$V_{\mathrm{HB}} \simeq M_V(\mathrm{HB}) + 5\log(R_0/10\mathrm{pc}) + R \cdot E_{B-V} \ . \tag{4}$$

The observed correlation is shown in Fig. 13. Here, the "component" of V_{HB} projected onto the X–axis, namely $V_{\mathrm{HB}} + 5\log(\cos\omega)$, is plotted against reddening. As we expected, it resembles a distribution of objects which are all at the same *true* distance d (with some scatter, of course) but with different amounts of foreground reddening. There are only 4 or 5 obvious outliers which

[3] It is important to realize that the clusters nearest the Galactic center, because of their low Galactic latitude, are reddened *both* by local dust clouds in the Galactic disk near the Sun *and* by dust in the Galactic bulge itself. In most cases the contribution from the nearby dust clouds is the dominant one. Thus, the true distances of the clusters are almost uncorrelated with foreground reddening (see also Barbuy et al. 1998 for an explicit demonstration). Clusters on the far side of the Galactic center are readily visible in the normal optical bandpasses unless their latitudes are $\lesssim 1^\circ$ or 2°.

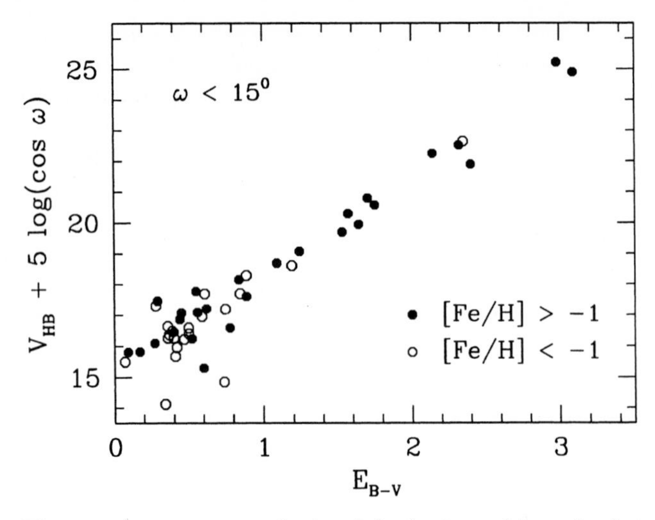

Fig. 13. Apparent magnitude of the horizontal branch plotted against reddening, for all globular clusters within $\omega = 15°$ of the Galactic center. MRC and MPC clusters are in *solid* and *open symbols*

are clearly well in front of or behind the Galactic bulge. The quantity we are interested in is the *intercept* of the relation, i.e. the value at zero reddening. This intercept represents the distance modulus of an *unreddened* cluster which is *directly at the Galactic center.*

We can refine things a bit more by taking out the known second-order dependence of A_V on E_{B-V}, as well as the dependence of $V_{\rm HB}$ on metallicity. Following Racine & Harris, we define a linearized HB level as

$$V_{\rm HB}^C = V_{\rm HB} + 5\,\log(\cos\omega) - 0.05\,E_{B-V}^2 - 0.15\,([{\rm Fe/H}] + 2.0)\,. \qquad (5)$$

The correlation of $V_{\rm HB}^C$ with E_{B-V} is shown in Fig. 14. Ignoring the 5 most deviant points at low reddening, we derive a best-fit line

$$V_{\rm HB}^C = (15.103 \pm 0.123) + (2.946 \pm 0.127)\,E_{B-V} \qquad (6)$$

with a remaining r.m.s. scatter of ± 0.40 in distance modulus about the mean line. The slope of the line $\Delta V/\Delta E_{B-V} \sim 3$ is just what it should be if it is determined principally by reddening differences *that are uncorrelated with true distance.* The intercept is converted into the distance modulus of the Galactic center by subtracting our distance scale calibration $M_V({\rm HB}) = 0.50$ at $[{\rm Fe/H}] = -2.0$. We must also remove a small geometric bias of 0.05 ± 0.03 (Racine & Harris) to take account of the fact that our line-of-sight cone defined by $\omega < 15°$ has larger volume (and thus proportionally more clusters) beyond the Galactic center than in front of it. The error budget will also include $\Delta(m - M) \sim \pm 0.1$ (internal) due to uncertainty in the reddening law, and (pessimistically, perhaps) a ± 0.2 mag external uncertainty in the distance

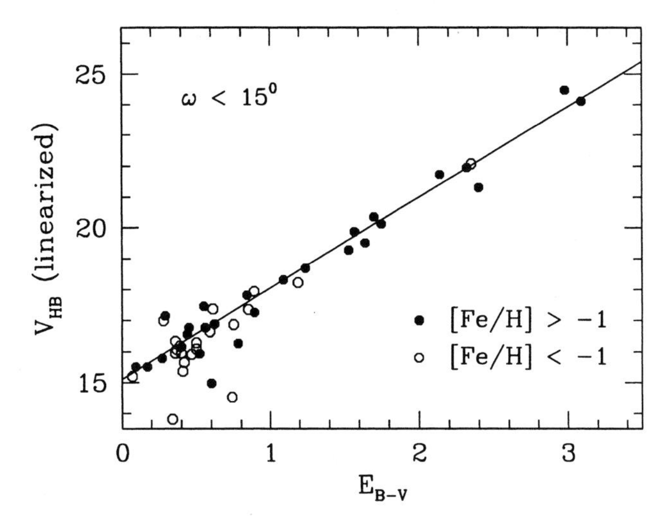

Fig. 14. Apparent magnitude of the horizontal branch plotted against reddening, after projection onto the X–axis and correction for second-order reddening and metallicity terms. The equation for the best-fit line shown is given in the text; it has a slope $R \sim 3$ determined by foreground reddening. The intercept marks the distance modulus to the Galactic center

scale zeropoint. In total, our derived distance modulus is $(m - M)_0(\mathrm{GC}) = 14.55 \pm 0.16$ (int) ± 0.2 (ext), or

$$R_0 = 8.14 \text{ kpc } \pm 0.61 \text{ kpc (int) } \pm 0.77 \text{ kpc (ext)} . \tag{7}$$

It is interesting that the dominant source of uncertainty is in the luminosity of our fundamental standard candle, the RR Lyrae stars. By comparison, the intrinsic cluster-to-cluster scatter of distances in the bulge creates only a ± 0.35 kpc uncertainty in R_0.

This completes our review of the spatial distribution of the GCS, and the definition of its two major subpopulations. However, before we go on to discuss the kinematics and dynamics of the system, we need to take a more careful look at justifying our fundamental distance scale. That will be the task for the next section.

2 The Distance Scale

The researches of many commentators have already thrown much darkness on this subject, and it is probable that, if they continue, we shall soon know nothing at all about it.

Mark Twain

About 40 years ago, there was a highly popular quiz show on American television called "I've Got a Secret". On each show, three contestants would come in and all pretend to be the same person, invariably someone with an unusual or little-known occupation or accomplishment. Only one of the three was the real person. The four regular panellists on the show would have to ask them clever questions, and by judging how realistic the answers sounded, decide which ones were the imposters. The entertainment, of course, was in how inventive the contestants could be to fool the panellists for as long as possible. At the end of the show, the moderator would stop the process and ask the real contestant to stand up, after which everything was revealed.

The metaphor for this section is, therefore, "Will the real distance scale please stand up?" In our case, however, the game has now gone on for a century, and there is no moderator. For globular clusters and Population II stars, there are several routes to calibrating distances. These routes do not agree with one another; and the implications for such things as the cluster ages and the cosmological distance scale are serious. It is a surprisingly hard problem to solve, and at least some of the methods we are using must be wrong. But which ones, and how?

The time-honored approach to calibrating globular cluster distances is to measure some identifiable sequence of stars in the cluster CMD, and then to establish the luminosities of these same types of stars in the Solar neighborhood by trigonometric parallax. The three most obvious such sequences (see the Appendix) are:

- *The horizontal branch*, or RR Lyrae stars: In the V band, these produce a sharp, nearly level and thus almost ideal sequence in the CMD. The problem is in the comparison objects: field RR Lyrae variables are rare and uncomfortably distant, and thus present difficult targets for parallax programs. There is also the nagging worry that the field halo stars may be astrophysically different (in age or detailed chemical composition) from those in clusters, and the HB luminosity depends on many factors since it represents a rather advanced evolutionary stage. The HB absolute magnitude almost certainly depends weakly on metallicity. It is usual to parametrize this effect simply as $M_V(\mathrm{HB}) = \alpha\,[\mathrm{Fe/H}] + \beta$, where (α, β) are to be determined from observations – and, we hope, with some guidance from theory.[4]

[4] As noted in the previous section, I define V_{HB} as the mean magnitude of the horizontal-branch stars without adjustment. Some other authors correct V_{HB} to the slightly fainter level of the "zero-age" unevolved ZAHB.

- *The unevolved main sequence (ZAMS)*: modern photometric tools can now establish highly precise main sequences for any cluster in the Galaxy not affected by differential reddening or severe crowding. As above, the problem is with the comparison objects, which are the unevolved halo stars or "subdwarfs" in the Solar neighborhood. Not many are near enough to have genuinely reliable parallaxes even with the new *Hipparcos* measurements. This is particularly true for the lowest-metallicity ones which are the most relevant to the halo globular clusters; and most of them do not have accurate and detailed chemical compositions determined from high-dispersion spectroscopy.
- *The white dwarf sequence*: this faintest of all stellar sequences has now come within reach from *Hubble Space Telescope (HST)* photometry for a few clusters. Since its position in the CMD is driven by different stellar physics than is the main sequence or HB, it can provide a uniquely different check on the distance scale. Although such stars are common, they are so intrinsically faint that they must be *very* close to the Sun to be identified and measured, and thus only a few comparison field-halo white dwarfs have well established distances.

These classic approaches each have distinct advantages and problems, and other ways have been developed to complement them. In the sections below, I provide a list of the current methods which seem to me to be competitive ones, along with their results. Before we plunge into the details, I stress that this whole subject area comprises a vast literature, and we can pretend to do nothing more here than to select recent highlights.

2.1 Statistical Parallax of Field Halo RR Lyraes

Both the globular clusters and the field RR Lyrae stars in the Galactic halo are too thinly scattered in space for almost any of them to lie within the distance range of direct trigonometric parallax. However, the radial velocities and proper motions of the field RR Lyraes can be used to solve for their luminosity through statistical parallax. In principle, the trend of luminosity with metallicity can also be obtained if we divide the sample up into metallicity groups.

An exhaustive analysis of the technique, employing ground-based velocities and Lick Observatory proper motions, is presented by Layden et al. (1996). They use data for a total of 162 "halo" (metal-poor) RR Lyraes and 51 "thick disk" (more metal-rich) stars in two separate solutions, with results as shown in Table 2. Recent solutions are also published by Fernley et al. (1998a), who use proper motions from the *Hipparcos* satellite program; and by Gould & Popowski (1998), who use a combination of Lick ground-based and *Hipparcos* proper motions. These studies are in excellent agreement with one another, and indicate as well that the metallicity dependence of M_V(RR) is small. The statistical–parallax calibration traditionally gives lower-luminosity

Table 2. Statistical parallax calibrations of field RR Lyrae stars

Region	M_V(RR)	[Fe/H]	Source
Halo	0.71 ± 0.12	-1.61	Layden et al. 1996
Halo	0.77 ± 0.17	-1.66	Fernley et al. 1998a
Halo	0.77 ± 0.12	-1.60	Gould & Popowski 1998
Disk	0.79 ± 0.30	-0.76	Layden et al. 1996
Disk	0.69 ± 0.21	-0.85	Fernley et al. 1998a

results than most other methods, but if there are problems in its assumptions that would systematically affect the results by more than its internal uncertainties, it is not yet clear what they might be. The discussion of Layden et al. should be referred to for a thorough analysis of the possibilities.

2.2 Baade–Wesselink Method

This technique, which employs simultaneous radial velocity and photometric measurements during the RR Lyrae pulsation cycle, is discussed in more detail in this volume by Carney; here, I list only some of the most recent results. A synthesis of the data for 18 field RR Lyrae variables over a wide range of metallicity (Carney, Storm, & Jones 1992) gives

$$M_V(\mathrm{RR}) = (0.16 \pm 0.03)\,[\mathrm{Fe/H}] + (1.02 \pm 0.03)\,. \tag{8}$$

As Carney argues, the uncertainty in the *zeropoint* of this relation quoted above is only the internal uncertainty given the assumptions in the geometry of the method; the external uncertainty is potentially much larger. However, the *slope* is much more well determined and is one of the strongest aspects of the method if one has a sample of stars covering a wide metallicity range (see also Carney's lectures in this volume, and Fernley et al. 1998b for additional discussion of the slope α).

The Baade–Wesselink method can also be applied to RR Lyraes that are directly in globular clusters; although these are much fainter than the nearest field stars and thus more difficult to observe, at least this approach alleviates concerns about possible differences between field RR Lyraes and those in clusters. Recent published results for four clusters are listed in Table 3 (from Liu & Janes 1990; Cohen 1992; and Storm et al. 1994a,b). The third column of the table gives the measured M_V(RR), while for comparison the fourth column gives the expected M_V from the field-star equation above. Within the uncertainties of either method, it is clear that the statistical parallax and Baade–Wesselink measurements are in reasonable agreement.

Table 3. Baade–Wesselink calibrations of RR Lyrae stars in four clusters

Cluster	[Fe/H]	M_V(BW)	M_V(eqn)
M92	-2.3	0.44, 0.64	0.65
M5	-1.3	0.60	0.81
M4	-1.2	0.80	0.83
47 Tuc	-0.76	0.71	0.90

2.3 Trigonometric Parallaxes of HB Stars

The *Hipparcos* catalog of trigonometric parallaxes provides several useful measurements of field HB stars for the first time (see Fernley et al. 1998a; Gratton 1998). One of these is RR Lyrae itself, for which $\pi = (4.38 \pm 0.59)$ mas, yielding $M_V(\mathrm{RR}) = 0.78 \pm 0.29$ at $[\mathrm{Fe/H}] = -1.39$. The red HB star HD 17072 (presumably a more metal-rich one than RR Lyrae) has a slightly better determined luminosity at $M_V(\mathrm{HB}) = 0.97 \pm 0.15$. Finally, Gratton (1998) derives a parallax-weighted mean luminosity for $\sim$ 20 HB stars of $M_V(\mathrm{HB}) = 0.69 \pm 0.10$ at a mean metallicity $\langle[\mathrm{Fe/H}]\rangle = -1.41$, though of course the parallaxes for any individual HB star in this list are highly uncertain. At a given metallicity, these HB luminosities tend to sit $\sim$ 0.1 to 0.2 mag higher than the ones from statistical parallax and Baade–Wesselink.

2.4 Astrometric Parallax

We turn next to distance calibration methods of other types, which can be used secondarily to establish M_V(HB).

An ingenious method applying directly to clusters without the intermediate step of field stars, and without requiring any knowledge of their astrophysical properties, is that of "astrometric parallax": the internal motions of the stars within a cluster can be measured either through their radial velocity dispersion $\sigma(v_r)$, or through their dispersion in the projected radial and tangential proper motions $\sigma(\mu_r, \mu_\theta)$ relative to the cluster center. These three internal velocity components can be set equal through a simple scale factor involving the distance d,

$$\sigma(v_r) = \mathrm{const} \cdot d \cdot \sigma(\mu) \tag{9}$$

and thus inverted to yield d, independent of other factors such as cluster metallicity and reddening. The two μ-dispersions can also be used to model any radial anisotropy of the internal motions, and thus to adjust the scaling to $\sigma(v_r)$.

This method is in principle an attractive and powerful one, though the available measurements do not yet reach a level of precision for individual clusters that is sufficient to confirm or rule out other approaches definitively. A preliminary summary of the current results by Rees (1996) gives distances for five intermediate-metallicity clusters (M2, M4, M5, M13, M22, with a mean $\langle[\mathrm{Fe/H}]\rangle = -1.46$) equivalent to $M_V(\mathrm{HB}) = 0.63 \pm 0.11$. For one low-metallicity cluster (M92, at $[\mathrm{Fe/H}] = -2.3$), he finds $M_V(\mathrm{HB}) = 0.31 \pm 0.32$. On average, these levels are $\sim 0.1 - 0.2$ mag brighter than the results from statistical parallax or Baade–Wesselink.

2.5 White Dwarf Sequences

Recently Renzini et al. (1996) have used deep *HST* photometry to establish the location of the WD sequence in the low-metallicity cluster NGC 6752 and to match it to five DA white dwarfs in the nearby field. The quality of the fit is remarkably tight even given the relatively small number of stars. The derived distance modulus corresponds to $M_V(\mathrm{HB}) = 0.52 \pm 0.08$ at a cluster metallicity $[\mathrm{Fe/H}] = -1.55$. The critical underlying assumption here is that the *mass* of the white dwarfs in the cluster – which is the most important determinant of the WD sequence luminosity – has the same canonical value $\simeq 0.6\ \mathrm{M}_\odot$ as the field DA's.

In a comparably deep photometric study of the nearby cluster M4, Richer et al. (1995) take the argument in the opposite direction: by using the heavily populated and well defined WD sequence along with a distance derived from main sequence fitting, they derive the WD mass, which turns out to be $\simeq 0.50 - 0.55\ \mathrm{M}_\odot$. A third deep white dwarf sequence has been measured for NGC 6397 by Cool et al. (1996), again with similar results, and *HST*-based results for other clusters are forthcoming. Thus at the present time, it appears that the fundamental distance scale from WDs is consistent with the range of numbers from the other approaches and deserves to be given significant weight. We can look forward, in a few years time, to a much more complete understanding of the relative WD vs. ZAMS distance scales and to a stronger contribution to the zeropoint calibration. Still deeper observations will, eventually, be able to find the faint-end termination of the WD sequence and place completely new observational limits on the cluster ages.

2.6 Field Subdwarf Parallaxes and Main Sequence Fitting

The technique which has generated the most vivid recent discussion (and controversy) centers on the matching of nearby halo main-sequence stars (subdwarfs) to cluster main sequences. It was widely expected that the *Hipparcos* project would, for the first time, supply a large number of high-quality trigonometric parallaxes for low-metallicity stars in the Solar neighborhood and would essentially solve the distance scale problem at a level which could claim to being definitive. Unfortunately, this hope has not been borne out.

The whole problem in the fitting procedure is essentially that any given collection of subdwarfs does not automatically give us a "sequence" which can

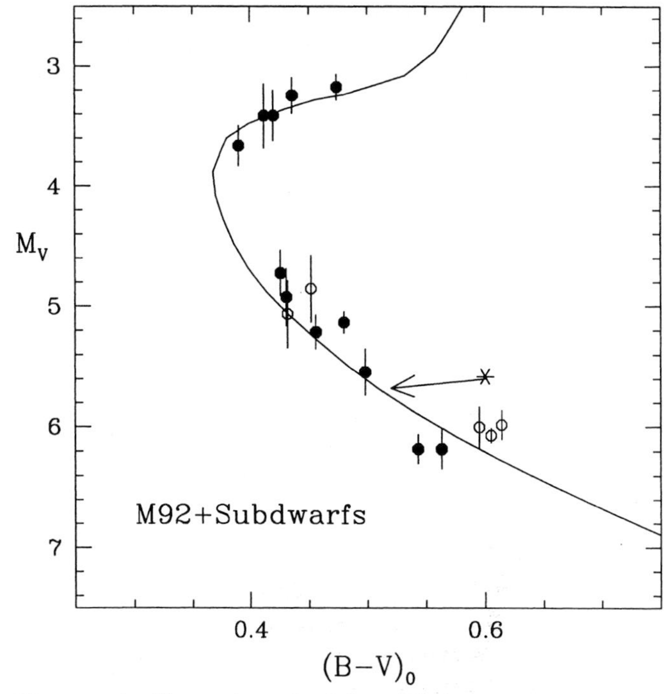

Fig. 15. An illustration of subdwarf fitting to a cluster main sequence. Nearby metal-poor subdwarfs (Pont et al. 1998), shown as the *dots*, are superimposed on the fiducial sequence for the metal-poor cluster M92 (Stetson & Harris 1988), for an assumed reddening $E(B-V) = 0.02$ and a distance modulus $(m-M)_V = 14.72$. The location of each star on this diagram must be adjusted to the color and luminosity it would have at the metallicity of M92 ([Fe/H] $= -2.2$). For a typical subdwarf at [Fe/H] ~ -1.6 (*starred symbol*), the size of the color and luminosity corrections is indicated by the *arrow*. The luminosity and color corrections follow the bias prescriptions in Pont et al. Known or suspected *binary* stars are plotted as *open circles*

then be matched immediately to a globular cluster. The individual subdwarfs all have different distances (and thus parallax uncertainties) and metallicities. All of them have to be relocated in the CMD back to the positions they would have *at the metallicity of the cluster*, and various biases may exist in the measured luminosities (see below). The more distant or low-latitude ones may even have small amounts of reddening, and the sample may also include undetected binaries. Thus before any fit to a given cluster can be done, a fiducial main sequence must be constructed out of a collection of subdwarfs which *by definition* is heterogeneous.

Figure 15 illustrates the procedure. The luminosity M_V of a given subdwarf, calculated directly from its raw trigonometric parallax and apparent magnitude (starred symbol in the figure), is adjusted by an amount ΔM_V

for various sample bias corrections as described below. Next, the raw color index $(B-V)$ is adjusted by an amount $\Delta(B-V)$ to compensate for the difference in metallicity between subdwarf and cluster, and also for any reddening difference between the two. Usually $\Delta(B-V)$ is negative since most of the known subdwarfs are more metal-rich than most of the halo globular clusters, and the main sequence position becomes bluer at lower metallicity. The change of color with metallicity is normally calculated from theoretical isochrones; although this is the only point in the argument which is model dependent, it is generally regarded as reliable to ± 0.01 for the most commonly used indices such as $(B-V)$ or $(V-I)$ (the differential color shifts with metallicity are quite consistent in isochrones from different workers, even if the absolute positions may differ slightly).

The greatest concerns surround (a) the believed absolute accuracy of the published parallaxes, and (b) the degree to which bias corrections should be applied to the measured luminosities. These biases include, but are not limited to, the following effects:

- The Lutz–Kelker (1973) effect, which arises in parallax measurement of any sample of physically identical stars which are scattered at different distances. Since the volume of space sampled increases with distance, there will be more stars at a given π that were scattered inward by random measurement error from larger distances than outward from smaller distances. The deduced luminosity M_V of the stars therefore tends statistically to be too faint, by an amount which increases with the relative measurement uncertainty σ_π/π (see Hanson 1979 and Carretta et al. 1999 for a comprehensive discussion and prescriptions for the correction).
- The binary nature of some of the subdwarfs, which (if it lurks undetected) will bias the luminosities upward.
- The strong increase of σ_π with V magnitude (fainter stars are more difficult to measure with the same precision). This effect tends to remove intrinsically fainter stars from the sample, and also favors the accidental inclusion of binaries (which are brighter than single stars at the same parallax).
- The metallicity distribution of the known subdwarfs, which is asymmetric and biased toward the more common higher-metallicity (redder) stars. In any selected sample, accidental inclusion of a higher-metallicity star is thus more likely than a lower-metallicity one, which is equivalent to a mean sample luminosity that is too high at a given color.

It is evident that the various possible luminosity biases can act in opposite directions, and that a great deal of information about the subdwarf sample must be in hand to deal with them correctly. Four recent studies are representative of the current situation. Reid (1997) uses a sample of 18 subdwarfs with $\sigma_\pi/\pi < 0.12$ along with the Lutz–Kelker corrections and metallicity adjustments to derive new distances to five nearby clusters of low reddening. Gratton et al. (1997) use a different sample of 13 subdwarfs, again with

$\sigma_\pi/\pi < 0.12$, and exert considerable effort to correct for the presence of binaries. They use Monte Carlo simulations to make further (small) corrections for parallax biases, and derive distances to nine nearby clusters. When plotted against metallicity, these define a mean sequence

$$M_V(\mathrm{HB}) = (0.125 \pm 0.055)\,[\mathrm{Fe/H}] + (0.542 \pm 0.090) \tag{10}$$

which may be compared (for example) with the much fainter Baade–Wesselink sequence listed earlier. Pont et al. (1998) employ still another sample of 18 subdwarfs and subgiants with $\sigma_\pi/\pi \lesssim 0.15$ and do more Monte Carlo modelling to take into account several known bias effects simultaneously. They find that the net bias correction ΔM_V is small – nearly negligible for [Fe/H] ~ -1 and only $+0.06$ for [Fe/H] ~ -2. They derive a distance only to M92, the most metal-poor of the standard halo clusters, with a result only slightly lower than either Reid or Gratton et al. found. Lastly, a larger set of 56 subdwarfs drawn from the entire *Hipparcos* database is analyzed by Carretta et al. (1999), along with a comprehensive discussion of the bias corrections. Their results fall within the same range as the previous three papers.

Regardless of the details of the fitting procedure, the basic effect to be recognized is that the *Hipparcos* parallax measurements for the nearby subdwarfs tend to be a surprising ~ 3 milliarcseconds smaller than previous ground-based measurements gave. This difference then translates into brighter luminosities by typically $\Delta M_V \sim 0.2 - 0.3$ mag (see Gratton et al.). At the low-metallicity end of the globular cluster scale ([Fe/H] $\simeq -2.2$, appropriate to M92 or M15), the *Hipparcos*-based analyses yield $M_V(\mathrm{HB}) \simeq 0.3 \pm 0.1$, a level which is $\gtrsim 0.3$ mag brighter than (e.g.) from statistical parallax or Baade–Wesselink.

This level of discrepancy among very different methods, each of which seems well defined and persuasive on its own terms, is the crux of the current distance scale problem. Do the *Hipparcos* parallaxes in fact contain small and ill-understood errors of their own? Is it valid to apply Lutz–Kelker corrections – or more generally, other types of bias corrections – to single stars, or small numbers of them whose selection criteria are poorly determined and inhomogeneous? And how many of the subdwarfs are actually binaries?

The one subdwarf for which no luminosity bias correction is needed (or in dispute) is still Groombridge 1830 (HD 103095), by far the nearest one known. As an instructive numerical exercise, let us match this one star *alone* to the cluster M3 (NGC 5272), which has essentially the same metallicity and is also unreddened. Its *Hipparcos*-measured parallax is $\pi = (109.2 \pm 0.8)$ mas, while the best ground-based compilation (from the Yale catalog; see van Altena et al. 1995) gives $\pi = (112.2 \pm 1.6)$ mas. The photometric indices for Gmb 1830, from several literature sources, are $V = 6.436 \pm 0.007$, $(B-V) = 0.75 \pm 0.005$, $(V-I) = 0.87 \pm 0.01$, giving $M_V = 6.633 \pm 0.016$ with no significant bias corrections. Its metallicity is [Fe/H] $= -1.36 \pm 0.04$ (from a compilation of several earlier studies) or -1.24 ± 0.07 from the data of Gratton et al. (1997). This is nearly identical with [Fe/H] $= -1.34 \pm 0.02$ for M3 (Carretta & Gratton

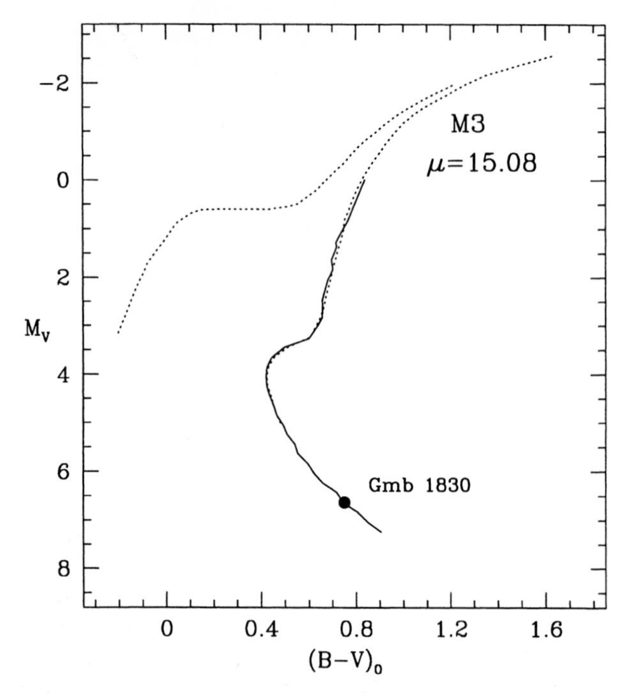

Fig. 16. Main sequence fit of the nearest subdwarf, Groombridge 1830, to the globular cluster M3. The cluster and the subdwarf have nearly identical metallicities and are unreddened. The *solid line* gives the deep main sequence and subgiant data for M3 from Stetson (1998), while the *dotted line* defining the brighter sections of the CMD is from Ferraro et al. (1997). The resulting distance modulus for M3 is $(m - M)_0 = 15.08 \pm 0.05$

1997). Gmb 1830 can safely be assumed to be unreddened, and the foreground reddening for M3 is usually taken as $E(B - V) = 0.00$ (Harris 1996a) and is in any case unlikely to be larger than 0.01. Thus the color adjustments to Gmb 1830 are essentially negligible as well. No other degrees of freedom are left, and we can match the star directly to the M3 main sequence at the same color to fix the cluster distance modulus. The result of this simple exercise is shown in Fig. 16. It yields $M_V(\mathrm{HB}) = 0.59 \pm 0.05$, which is ~ 0.2 mag fainter than the level obtained by Reid (1997) or Gratton et al. (1997) from the entire sample of subdwarfs.

Clearly, it is undesirable to pin the entire globular cluster distance scale (and hence the age of the universe) on just one star, no matter how well determined. Nevertheless, this example illustrates the fundamental uncertainties in the procedure.

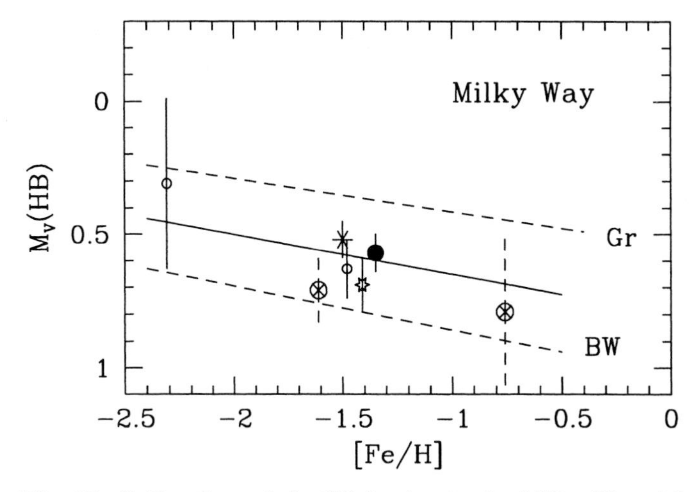

Fig. 17. Calibrations of the HB luminosity for Milky Way globular clusters. The *upper dashed line* (Gr) is the *Hipparcos* subdwarf calibration from Gratton et al. (1997), and the *lower dashed line* (BW) is the Baade–Wesselink calibration for field RR Lyraes from Carney et al. (1992), as listed in the text. Other symbols are as follows: *Solid dot:* Main sequence fit of Groombridge 1830 to M3. *Large asterisk:* Fit of white dwarf sequence in NGC 6752 to nearby field white dwarfs. *Small open circles:* Astrometric parallaxes, from Rees (1996) in two metallicity groups. *Large circled crosses:* Statistical parallax of field RR Lyrae stars, from Layden et al. (1996). *Open star:* Mean trigonometric parallax of field HB stars. Finally, the *solid line* is the adopted calibration, $M_V(\mathrm{HB}) = 0.15\,[\mathrm{Fe/H}] + 0.80$

2.7 A Synthesis of the Results for the Milky Way

The upper and lower extremes for the globular cluster distance scale as we now have them are well represented by the Baade–Wesselink field RR Lyrae calibration given by (8) and the Gratton et al. *Hipparcos*-based subdwarf fits of (10). These are combined in Fig. 17 along with the results from the other selected methods listed above. Also notable is the fact that the slope of the relation is consistently near $\alpha \simeq 0.15$ (see also Carney in this volume). To set the zeropoint, I adopt a line passing through the obvious grouping of points near [Fe/H] ~ -1.4, and about halfway between the two extreme lines. This relation (solid line in Fig. 17) is

$$M_V(\mathrm{HB}) = 0.15\,[\mathrm{Fe/H}] + 0.80\,. \tag{11}$$

Realistically, what uncertainty should we adopt when we apply this calibration to measure the distance to any particular object? Clearly the error is dominated not by the internal uncertainty of any one method, which is typically in the range ± 0.05 to 0.10 mag. Instead, it is dominated by the external level of disagreement between the methods. How much weight one should put

Table 4. Field RR Lyrae stars in the LMC

Location	$\langle V\rangle_{RR}$	Source
NGC 1783 field	19.25 ± 0.05	Graham 1977
NGC 2257 field	19.20 ± 0.05	Walker 1989
NGC 1466 field	19.34 :	Kinman et al. 1991
NGC 2210 field	19.22 ± 0.11	Reid & Freedman 1994
MACHO RRd's	19.18 ± 0.02	Alcock et al. 1997

on any one method has often been a matter of personal judgement. As a compromise – perhaps a pessimistic one – I will use $\sigma(M_V) = \pm 0.15$ mag as an estimate of the true external uncertainty of the calibration at any metallicity.

A comprehensive evaluation of the distance scale, concentrating on the subdwarf parallax method but also including a long list of other methods, is given by Carretta et al. (1999). Their recommended HB calibration – tied in part to the distance to the LMC measured by both Population I and II standard candles – corresponds to $M_V(\mathrm{HB}) = 0.13$ [Fe/H] $+0.76$, scarcely different from (11) above. (NB: note again that $M_V(\mathrm{HB})$ is subtly different from both $M_V(\mathrm{ZAHB})$ and $M_V(\mathrm{RR})$: the ZAHB is roughly 0.1 mag fainter than the mean HB because of evolutionary corrections, and the mean level of the RR Lyraes is about 0.05 mag brighter than the ZAHB for the same reason. As noted previously, I use the mean HB level without adjustments.)

2.8 Comparisons in the LMC and M31

Extremely important external checks on the globular cluster distance scale can be made through the Cepheids and other Population I standard candles, once we go to Local Group galaxies where both types of indicators are found at common distances. By far the most important two "testing grounds" are the Large Magellanic Cloud and M31, where several methods can be strongly tested against one another.

For the LMC, RR Lyrae variables are found in substantial numbers both in its general halo field and in several old globular clusters. The field-halo variables have mean V magnitudes as listed in Table 4, from five studies in which statistically significant numbers of variables have been measured. Using a foreground absorption for the LMC of $E(B-V) = 0.08 \pm 0.01$ and $A_V = 0.25 \pm 0.03$, I calculate a weighted mean dereddened magnitude $\langle V_0\rangle = 18.95 \pm 0.05$ (the mean is driven strongly by the huge MACHO sample, though the other studies agree closely with it). The mean metallicity of the field variables appears to be near [Fe/H] $\simeq -1.7$ (see van den Bergh 1995, and the references in the table). Thus our adopted Milky Way calibration

Table 5. A summary of distance calibrations for the LMC

Method	$(m-M)_0$
Cepheids (Clusters, BW)	18.49 ± 0.09
Mira PL relation	18.54 ± 0.18
SN1987A ring (4 recent analyses)	18.51 ± 0.07
RR Lyraes (clusters, field)	18.44 ± 0.15

would give $M_V(\mathrm{RR}) = 0.55 \pm 0.15$ (estimated external error) and hence a true distance modulus $(m-M)_0(\mathrm{LMC}) = 18.40 \pm 0.15$.

Well determined mean magnitudes are also available for RR Lyrae stars in seven LMC globular clusters (Walker 1989; van den Bergh 1995). Using the same foreground reddening, we find an average dereddened RR Lyrae magnitude for these clusters of $\langle V_0 \rangle = 18.95 \pm 0.05$. Their mean metallicity in this case is $[\mathrm{Fe/H}] \simeq -1.9$, thus from our Milky Way calibration we would predict $M_V(\mathrm{RR}) = 0.52 \pm 0.15$ and hence $(m-M)_0(\mathrm{LMC}) = 18.44 \pm 0.15$. The cluster and field RR Lyrae samples are in substantial agreement. Gratton (1998) and Carretta et al. (1999) suggest, however, that the *central bar* of the LMC could be at a different distance – perhaps as much as 0.1 mag further – than the average of the widely spread halo fields. Unfortunately, it is the LMC bar distance that we really want to have, so this contention introduces a further level of uncertainty into the discussion.

How do these RR Lyrae-based distance estimates compare with other independent standard candles, such as the LMC Cepheids or the SN1987A ring expansion? These methods themselves are not without controversy (for more extensive reviews, see, e.g., van den Bergh 1995; Fernley et al. 1998a; Gieren et al. 1998; or Feast 1998). Fundamental parallax distances to the Hyades and Pleiades can be used to establish main sequence fitting distances to Milky Way open clusters containing Cepheids, which then set the zeropoint of the Cepheid period–luminosity relation and hence the distance to the LMC. The Baade–Wesselink method can also be adapted to set distances to Cepheids in the nearby field. The SN1987A ring expansion parallax is an important new independent method, but here too there are disagreements in detail about modelling the ring geometry (cf. the references cited above). A brief summary of the most accurate methods, drawn from Fernley et al. (1998a) and Gieren et al. (1998), is given in Table 5. Although the individual moduli for these methods (as well as others not listed here) range from ~ 18.2 up to 18.7, it seems to me that an adopted mean $(m-M)_0(\mathrm{LMC}) = 18.5 \pm 0.1$ is not unreasonable. For comparison, the comprehensive review of Carretta et al. (1999) recommends $(m-M)_0 = 18.54 \pm 0.04$.

The step outward from the LMC to M31 can be taken either by comparing the mean magnitudes of the halo RR Lyrae variables in each galaxy, by the Cepheids in each, or by the RGB tip stars:

- *RR Lyraes*: In the M31 halo, the sample of RR Lyraes found by Pritchet & van den Bergh (1987) has $\langle V_0 \rangle(\text{M31}) = 25.04 \pm 0.10$ and thus there is $\Delta(m-M)_0(\text{M31-LMC}) = 6.09 \pm 0.11$, or $(m-M)_0 = 24.59 \pm 0.15$.
- *Cepheids*: Two studies employing optical photometry give $\Delta(m-M)_0 = 5.92 \pm 0.10$ (Freedman & Madore 1990) or 6.07 ± 0.05 (Gould 1994) from different prescriptions for matching the P-L diagrams in the two galaxies. The recent study of Webb (1998), from *JHK* near-infrared photometry which is less affected by reddening and metallicity differences, gives $\Delta(m-M)_0 = 5.92 \pm 0.02$ and thus $(m-M)_0 = 24.42 \pm 0.11$.
- *Red Giant Branch Tip*: A precise method which is more or less independent of both the Cepheids and RR Lyraes is the luminosity of the old red giant branch tip (TRGB) of the halo stars (essentially, the luminosity of the core helium flash point), which for metal-poor populations has a nearly constant luminosity $M_I = -4.1 \pm 0.1$ (Lee et al. 1993a; Harris et al. 1998b). For a wide sample of the M31 halo field giants, Couture et al. (1995) find an intrinsic distance modulus $(m-M)_0 = 24.5 \pm 0.2$.
- *Other methods*: Other useful techniques include the luminosities of disk carbon stars (Brewer et al. 1995), surface brightness fluctuations of globular clusters (Ajhar et al. 1996), and luminosities of the old red clump stars (Stanek & Garnavich 1998). These give results in exactly the same range $(m-M)_0 \sim 24.4 - 24.6$. Holland (1998) has used theoretical isochrone fits to the red giant branches of 14 halo globular clusters in M31 to obtain $(m-M)_0 = 24.47 \pm 0.07$.

Combining all of these estimates, I will adopt $(m-M)_0(\text{M31}) = 24.50 \pm 0.14$. Putting back in the M31 foreground reddening $E(B-V) = 0.09 \pm 0.03$ (van den Bergh 1995) then gives an apparent distance modulus $(m-M)_V(\text{M31}) = 24.80 \pm 0.15$.

To bring a last bit of closure to our discussion, we can finally test our Milky Way globular cluster distance scale against the mean HB levels that are directly observed in the globular clusters of M31. Fusi Pecci et al. (1996) have carried out homogeneous reductions for *HST* images of eight M31 clusters, with the results as shown in Fig. 18. The unweighted least-squares line defined by these eight points[5] is

$$V_{\text{HB}}(\text{observed}) = (0.096 \pm 0.078)\,[\text{Fe/H}] + (25.56 \pm 0.09)\,. \tag{12}$$

[5] This is *not* the same line derived by Fusi Pecci et al.; both their slope and zeropoint are slightly different. The reason for the difference is that they adjust the raw V_{HB} values to the somewhat fainter unevolved ZAHB position. Here, I use the directly observed V_{HB} without adjustment.

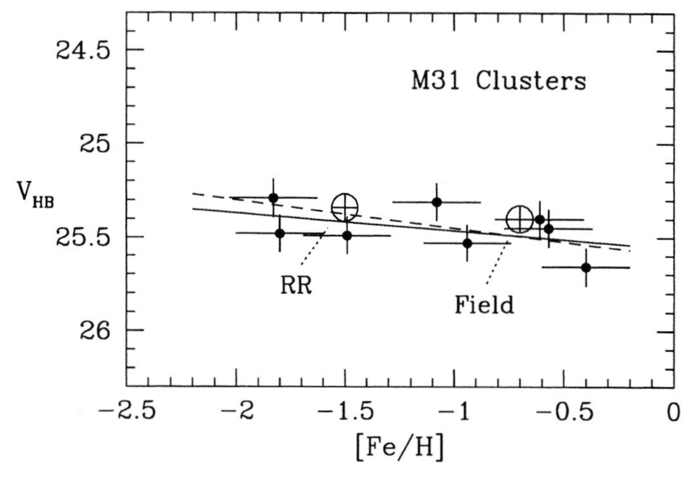

Fig. 18. Horizontal branch levels for globular clusters in M31, plotted against cluster metallicity. *Solid dots:* $V_{\rm HB}$ values for eight clusters from Fusi Pecci et al. (1996). *Large circled crosses:* Mean magnitudes for the halo field RR Lyraes (van den Bergh 1995) and the red HB stars near cluster G1 (Rich et al. 1996). The *solid line* shows the mean relation for the clusters as defined in the text, while the *dashed line* is our fiducial Milky Way relation added to the distance modulus $(m-M)_V(\mathrm{M31}) = 24.80$

For comparison, if we take our fiducial Milky Way relation and transport it outward by our best-estimate distance modulus $(m-M)_V = 24.8$, we obtain

$$V_{\rm HB}(\text{predicted}) = M_V(\text{HB}) + 24.80 = 0.15\,[\text{Fe/H}] + (25.60 \pm 0.15)\,. \quad (13)$$

These two relations are remarkably consistent with one another, and give additional confidence that our fundamental distance scale is not likely to be wrong by more than the tolerance that we have claimed.

3 The Milky Way System: Kinematics

A hypothesis or theory is clear, decisive, and positive, but it is believed by no one but the man who created it. Experimental findings, on the other hand, are messy, inexact things which are believed by everyone except the man who did that work.

Harlow Shapley

Much information about the origin and history of the Milky Way GCS is contained in the cluster space motions or *kinematics*. Armed with this kind of information along with the spatial distributions and cluster metallicities (Sect. 1), we can make considerably more progress in isolating recognizable subsystems within the GCS, and in comparing the clusters with other types of halo stellar populations.

The seminal work in kinematics of the GCS is to be found in the pioneering study of Mayall (1946), a paper which is just as important in the history of the subject as the work by Shapley (1918) on the cluster space distribution.[6] Other landmarks that progressively shaped our prevailing view of the GCS kinematics are to be found in the subsequent work of Kinman (1959a,b), Frenk & White (1980), and Zinn (1985).

3.1 Coordinate Systems and Transformations

The basic question in GCS kinematics is to determine the relative amounts of *ordered* motion of the clusters (net systemic rotation around the Galactic center) and *random* internal motion. The ratio of these quantities must depend on their time and place of origin, and thus on such measurables as cluster age, spatial location, or metallicity.

The first attempts at kinematical solutions (Mayall, Kinman, and others) used the simplest traditional formalism in which the Solar motion $\boldsymbol{U}$ was calculated relative to the clusters or various subsets of them. Formally, if v_r equals the radial velocity of the cluster relative to the Solar Local Standard of Rest, then

$$v_r = U \cos \lambda \tag{14}$$

where λ is the line-of-sight angle to the cluster (defined in Fig. 19). A graph of v_r against $(\cos \lambda)$ should yield a straight-line solution through the origin with slope U. For true halo objects with little or no systemic rotation, U must therefore be approximately equal to V_0(LSR), the rotation speed of the Solar Local Standard of Rest around the Galactic center; more strictly, U relative to the GCS must represent a lower limit to V_0 except in extreme scenarios where the halo, or part of it, is in retrograde rotation. Mayall, in the very first attempt to do this, derived $U \simeq 200$ km s^{-1}, a value scarcely different from the modern solutions of $\simeq 180$ km s^{-1}. The simple Solar motion approach, however, gives little information about the net motions of the many clusters for which the line of sight from Sun to cluster is roughly at right angles to the $\boldsymbol{V_0}$ vector. Instead, we will move directly on to the modern formalism, as laid out (e.g.) in Frenk & White (1980) and in many studies since.

Referring to Fig. 19, let us consider a cluster which has a Galactocentric distance R and a rotation speed $V(R)$ around the Galactic center. Its radial velocity relative to the Solar LSR is then

$$v_r = V \cos \psi - V_0 \cos \lambda \tag{15}$$

[6] Mayall's paper is essential reading for any serious student of the subject. Now half a century old, it stands today as a remarkable testament to the author's accomplishment of a major single piece of work in the face of several persistent obstacles. It also typifies a brutally honest writing style that is now rather out of fashion.

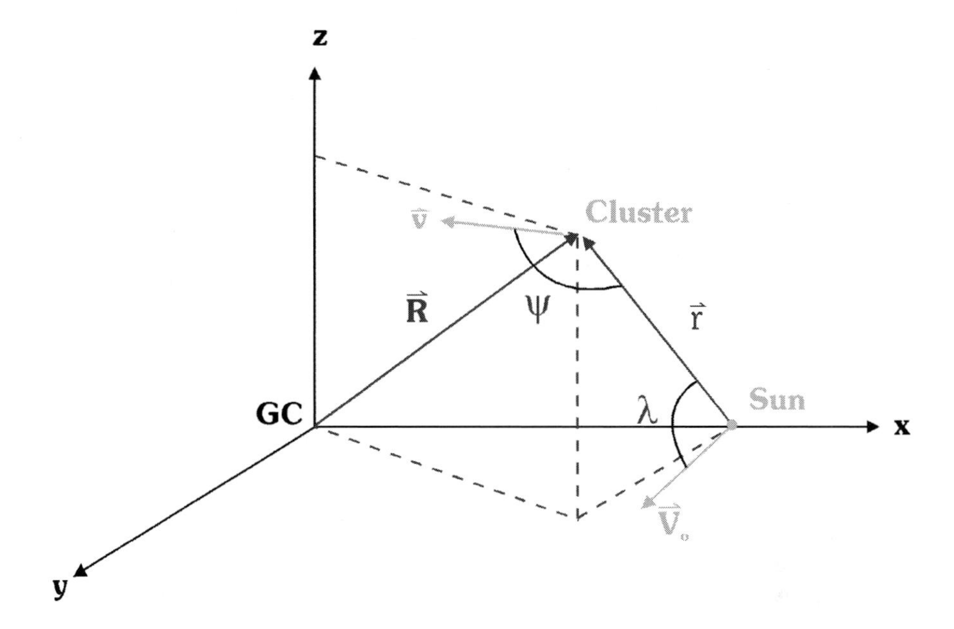

Fig. 19. Geometry for the rotational motions of the Sun and a cluster around the Galactic center *GC*

where λ is the angle between the Solar motion vector $\boldsymbol{V_0}$ and the line of sight $\boldsymbol{r}$ to the cluster; and ψ is the angle between $\boldsymbol{r}$ and the rotation vector $\boldsymbol{V}$ of the cluster. We have

$$\boldsymbol{V_0} \cdot \boldsymbol{r} = V_0 \, r \, \cos\lambda \tag{16}$$

which gives, after evaluating the dot product,

$$\cos\lambda = \cos b \cdot \sin\ell \,. \tag{17}$$

Similarly, $\cos\psi$ can be evaluated from the dot product

$$\boldsymbol{V} \cdot \boldsymbol{r} = V \, r \, \cos\psi \tag{18}$$

which eventually gives

$$\cos\psi = \frac{R_0 \cos b \, \sin\ell}{((r \cos b \, \sin\ell)^2 + (R_0 - r \cos b \, \cos\ell)^2)^{1/2}} \,. \tag{19}$$

The equation of condition for V is then

$$V \cos\psi = v_r + V_0 \cos\lambda \tag{20}$$

where v_r is the directly measured radial velocity of the cluster (relative to the LSR!); and (λ, ψ) are known from the distance and direction of the cluster.

We explicitly *assume* $V_0 = 220$ km s^{-1} for the Solar rotation. The quantity on the right-hand side of the equation is the radial velocity of the cluster relative to a stationary point at the Sun (i.e., in the rest frame of the Galactic center). Thus when it is plotted against $\cos\psi$, we obtain a straight-line relation with slope V (the net rotation speed of the group of clusters) and intercept zero.

Frenk & White (1980) demonstrate that an unbiased solution for V is obtained by adding the weighting factor $\cos\psi$,

$$V = \frac{\langle \cos\psi(v_r + V_0 \cos\lambda)\rangle}{\langle \cos^2\psi\rangle} \pm \frac{\sigma_{\rm los}}{(\Sigma \cos^2\psi)^{1/2}} \tag{21}$$

where $\sigma_{\rm los}$, the "line of sight" velocity dispersion, is the r.m.s. dispersion of the data points about the mean line.

Which of the parameters in the above equation can potentially generate significant errors in the solution? As we will see below, typically $\sigma_{\rm los} \sim 100$ km s^{-1}, whereas the measurement uncertainties in the radial velocities of the clusters are $\varepsilon(v_r) \lesssim 5$ km s^{-1} (cf. Harris 1996a). The radial velocity measurements themselves thus do not contribute anything important to uncertainties in V or $\sigma_{\rm los}$. In addition, λ depends only on the angular location of the cluster on the sky and is therefore virtually error-free. The last input parameter is the angle ψ: *Uncertainties in the estimated distances* r can affect ψ severely – and asymmetrically – as is evident from inspection of Fig. 19, and these can then be translated into biases in V and $\sigma_{\rm los}$. This point is also stressed by Armandroff (1989). In turn, the uncertainty $\varepsilon(m-M)_V$ in distance modulus is most strongly correlated with cluster reddening (larger reddening increases both the absolute uncertainty in the absorption correction A_V, and the amount of *differential reddening*, which makes the identification of the CMD sequences less precise). A rough empirical relation

$$\varepsilon(m-M)_V \simeq 0.1 + 0.4\,E(B-V) \tag{22}$$

represents the overall effect reasonably well.

Clearly, the clusters near the Galactic center and Galactic plane – which preferentially include the most metal-rich clusters – will be the most severely damaged by this effect. Since the dependence of ψ on r is highly nonlinear, the error bars on $\cos\psi$ can be large and asymmetric for such clusters. An example is shown in the kinematics diagram of Fig. 20. By contrast, the same diagram for any subset of the high-halo clusters is, point by point, far more reliable and thus considerably more confidence can be placed in the $(V, \sigma_{\rm los})$ solution for such subgroups. Fortunately, however, the high numerical weights given to the clusters at large $(\cos\psi)$, which are exactly the objects that have the lowest reddenings and the most reliable distance estimates, make the solution for rotation speed V more robust than might at first be expected.

3.2 The Metal-Rich Clusters: Bulge-Like and Disk-Like Features

Somewhat contrary to historical tradition, let us first investigate the kinematics of the MRC clusters.

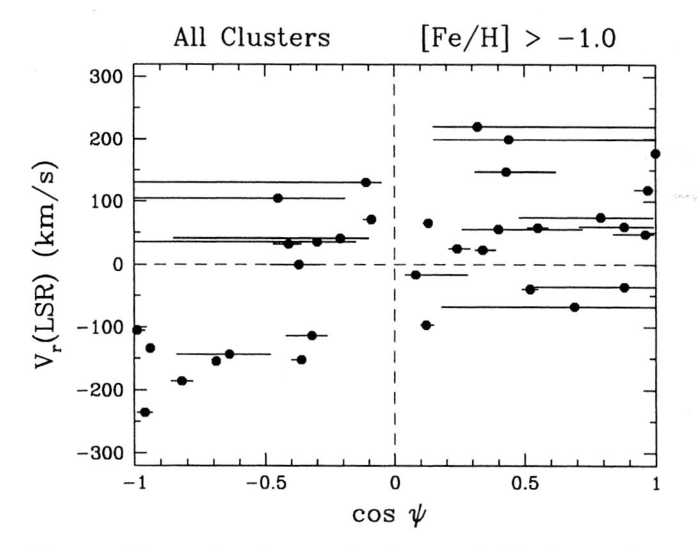

Fig. 20. Kinematics diagram for the metal-rich clusters in the Milky Way. Here $v_r(\mathrm{LSR}) = v_r + V_0 \cos \lambda$ is the radial velocity of the cluster relative to a stationary point at the Solar LSR. The horizontal error bars on each point show how each cluster can shift in the diagram due solely to uncertainties in its estimated distance, as given by (22)

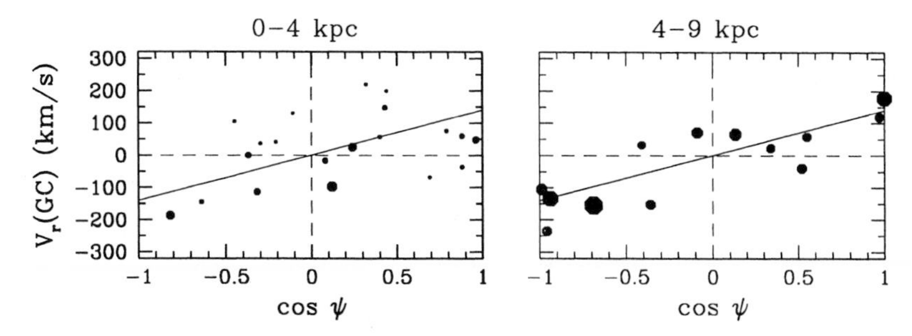

Fig. 21. Kinematics diagrams for the metal-rich Milky Way clusters. *Left panel:* Normalized radial velocity against position angle for the innermost MRC clusters ($R_{\mathrm{gc}} < 4$ kpc). The relative statistical weights for each point are indicated by symbol size (see text). *Right panel:* Outer MRC clusters (4 kpc $< R_{\mathrm{gc}} <$ 9 kpc). Note the larger weights on these points because of their lower reddenings and more accurate distance measurements. The same *solid line* (V = 140 km s^{-1} rotation speed) is plotted in both panels, although the formal solution for the innermost clusters is $V \sim 86$ km s^{-1}; see the text

In Fig. 21, we see the kinematics diagrams for the inner ($R_{\mathrm{gc}} < 4$ kpc) and outer (4 kpc $< R_{\mathrm{gc}} <$ 9 kpc) MRC clusters. The best-fit numerical solutions, listed in Table 6, show healthy rotation signals and moderately low rms dispersions for both, although V(rot) is clearly higher (and σ_{los} lower)

Table 6. Mean rotation velocities of subsets of clusters

Sample	Subgroup	n	V (km/s)	$\sigma_{\rm los}$ (km/s)
MRC	All [Fe/H] > -1	33	118±26	89±11
MRC	$R_{\rm gc} = 0 - 4$ kpc	20	86±40	99±15
MRC	$R_{\rm gc} = 4 - 9$ kpc	13	147±27	66±12
MPC	All [Fe/H] < -1	89	30±25	121± 9
MPC	$R_{\rm gc} = 0 - 4$ kpc	28	56±37	122±16
MPC	$R_{\rm gc} = 4 - 8$ kpc	19	12±31	79±12
MPC	$R_{\rm gc} = 8 - 12$ kpc	12	26±63	148±29
MPC	$R_{\rm gc} = 12 - 20$ kpc	14	−97±110	132±24
MPC	$-2.30 <$ [Fe/H] < -1.85	17	139±57	114±19
MPC	$-1.85 <$ [Fe/H] < -1.65	19	41±55	142±22
MPC	$-1.65 <$ [Fe/H] < -1.50	21	−35±59	134±20
MPC	$-1.50 <$ [Fe/H] < -1.32	17	−12±56	106±17
MPC	$-1.32 <$ [Fe/H] < -1.00	17	31±32	80±13
MPC	All [Fe/H] < -1.70	30	80±43	130±16
MPC	BHB, $R_{\rm gc} > 8$ kpc	20	55±58	115±17
MPC	RHB, $R_{\rm gc} > 8$ kpc	18	−39±83	158±26
MPC	RHB excl. N3201	17	32±88	149±24

for the outer sample. The inner subgroup is what we discussed in Sect. 1 as Minniti's (1995) bulge-like population. The velocities for these clusters are replotted against Galactic longitude in Fig. 22, following Minniti (1995) and Zinn (1996), in which it can be seen that they match well with the net rotation speed of the RGB stars in the bulge. When we add this evidence (not conclusive by itself!) to the space distribution discussed in Sect. 1, it seems likely that the inner MRC clusters are plausibly interpreted as a flattened bulge population with a rotation speed near $V \sim 90$ km s^{-1}.

The outer subgroup (4 to 9 kpc; second panel of Fig. 21) more nearly resembles what Zinn (1985) and Armandroff (1989) first suggested to be a "thick disk" population. The issue is discussed at length in other recent papers by Armandroff (1993), Norris (1993), and Zinn (1996). If this identification is correct, it would be highly suggestive that there is a genuine disk subsystem within the Milky Way GCS which formed along with the thick-disk stars; if so, it should then be possible to set the formation epoch of the thick disk quite accurately by the chronology of these clusters.

Although this interpretation of the data is well known, it is not quite ironclad. The well determined rotation speed of the outer MRC clusters, $V =$

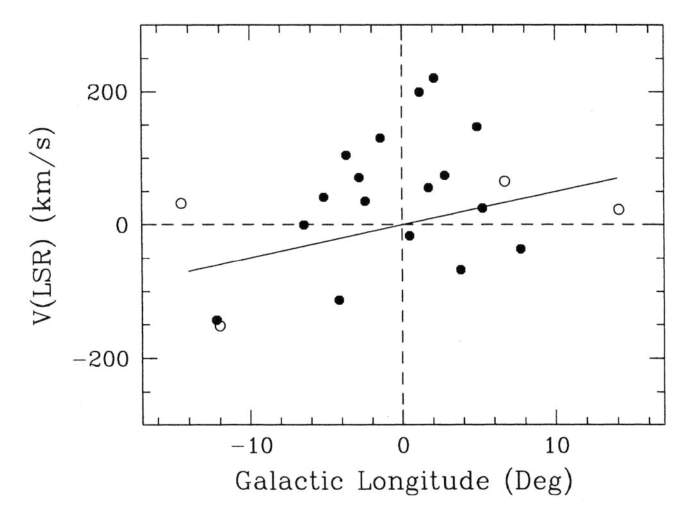

Fig. 22. Normalized radial velocity against Galactic longitude, for the MRC clusters within $\sim 15^{\circ}$ of the Galactic center. *Filled circles* are clusters within 4 kpc of the center; *open circles* are ones with $R_{gc} > 4$ kpc. The *solid line* represents the rotation curve of the Galactic bulge from red giant stars (Minniti 1995; Zinn 1996)

147 ± 27 km s^{-1}, is noticeably less than normally quoted values for the thick-disk stars, which are near $V \simeq 180$ km s^{-1} (cf. Armandroff 1989; Norris 1993; Majewski 1993). It is tempting to imagine that the outer MRC clusters may be the remnants of a "pre-thick-disk" epoch of star formation, during which their parent Searle–Zinn gaseous fragments had not fully settled into a disklike configuration, still preserving significant random motions. The leftover gas from this period would have continued to collapse further into the thick disk and (later) the old thin disk, with progressively larger rotation speeds.

At the same time, it is plainly true that some individual clusters have disk-like orbital motions (e.g., Cudworth 1985; Rees & Cudworth 1991; Dinescu et al. 1999). Burkert & Smith (1997) have gone further to suggest that the outer MRC clusters form a disklike subsystem, while the inner low-luminosity ones form an elongated bar-like structure (see also Côté 1999). However, a serious concern is that the distance estimates for the inner low-luminosity bulge clusters are likely to be more badly affected by extreme reddening and field contamination than for the luminous clusters in the same region of the bulge. The apparent elongation of the set of low-luminosity clusters along the X axis may therefore be an artifact. Better photometry and cleaner CMDs for these objects, possibly from near-infrared observations, are needed to clear up the problem.

Why should we try so hard to relate the globular clusters to the halo field stars? There is a nearly irresistible temptation to force some given subset of the globular clusters to correspond *exactly* with some population of field stars that look similar according to their kinematics, metallicity distribution, and space distribution. But the more we find out about these subsystems in detail, the harder it is to make such precise correpondences. As will be discussed later (Sect. 8), the formation of massive star clusters must be a relatively rare and inefficient process within their progenitor gas clouds. If the globular clusters formed first out of the densest gas clumps, the remaining gas (which in fact would be the majority of the protostellar material) would have had plenty of opportunity to collide with other gaseous fragments, dissipate energy, and take up new configurations before forming stars. By then, it would have lost its "memory" of the earlier epoch when the globular clusters formed, and would essentially behave as a different stellar population (Harris 1998). In short, there seems to be no compelling reason to believe that subgroups of the GCS should be cleanly identified with any particular field-star population. This concern will surface again in the next section.

3.3 The Metal-Poor Clusters

We will now turn to the MPC clusters, which form the majority of the Milky Way globular cluster system. One minor correction we need to make before proceeding is to note that the four clusters believed to belong to the Sagittarius dwarf (NGC 6715, Arp 2, Ter 7, Ter 8; see Da Costa & Armandroff 1995) all have similar space motions and locations: we will keep only NGC 6715 as the "elected representative" for Sagittarius and discard the other three. Other correlated moving groups involving similarly small numbers of clusters have been proposed to exist (Lynden–Bell & Lynden–Bell 1995; Fusi Pecci et al. 1995), but these are much less certain than Sagittarius, and for the present we will treat the remaining clusters as if they are all uncorrelated.

Plotting all the MPC clusters at once in the kinematics diagram (first panel of Fig. 23), we see that as a whole it is totally dominated by random motion with no significant mean rotation. As before, the symbol size denotes the relative uncertainty in $\cos\psi$; since most of the objects here have low reddenings and well determined distances, they have much more accurately fixed locations in the diagram.

However, throwing them all into the same bin is guaranteed to obscure the existence of any distinct subsystems. The other panels of Fig. 23 show the sample broken into five bins of Galactocentric distance, with rather arbitrarily chosen boundaries. Again, the individual solution parameters are listed in Table 6. A small positive rotation signal is present in the innermost 4 kpc zone, whereas no significant rotation appears in any of the other bins. As is evident from the figure, for clusters more distant than $R_{gc} \gtrsim 15$ kpc the range of $\cos\psi$ becomes so small that no valid solution for V can be performed. In the $4-8$ kpc bin, the dispersion is distinctly lower than in any of the other bins; the meaning of this anomaly is unclear (particularly since it is accompanied by zero rotation), and the possibility that it is simply a statistical fluctuation cannot be ruled out given the small numbers of points.

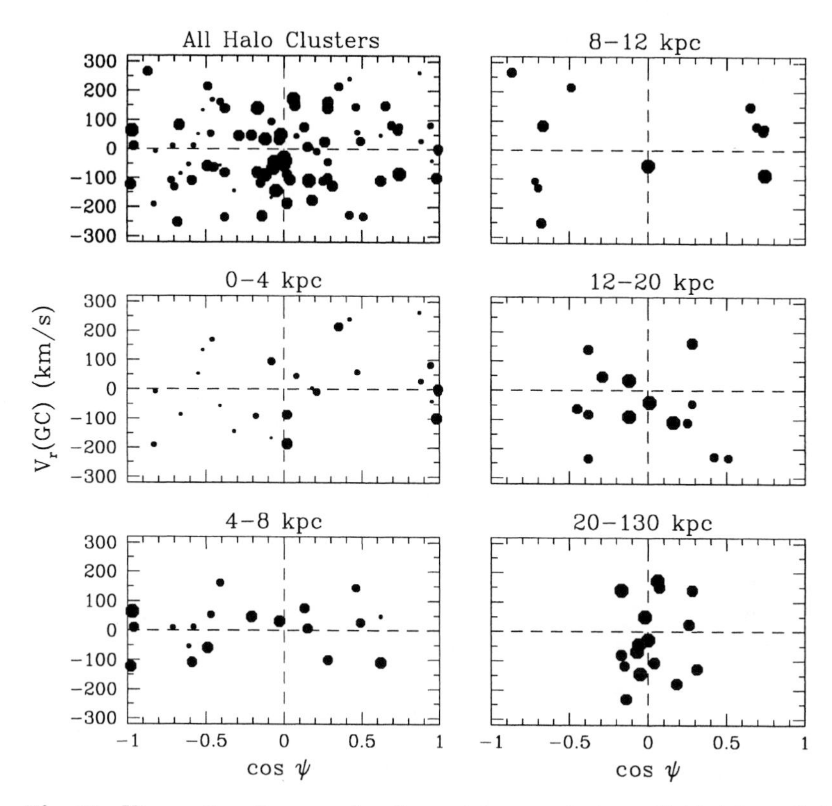

Fig. 23. Kinematics diagrams for the metal-poor clusters, divided into Galactocentric distance intervals. Larger points indicate clusters with lower reddenings and thus better determined distances and ψ values

Grouping the clusters by metallicity is also instructive. In most Galaxy formation models, we might anticipate that this version would be closer to a chronological sequence where the higher-metallicity objects formed a bit later in the enrichment history of their parent gas clouds. A sample of this breakdown is shown in Fig. 24, where now we exclude the six most remote clusters ($R_{gc} > 50$ kpc) that have no effect on the solution. *No significant net rotation is found* for any metallicity subgroup except for the lowest-metallicity bin. For the intermediate-metallicity groups, notice (Table 6) the slight (but not statistically significant!) dip into net retrograde rotation. The middle bin in particular is influenced strongly by the single object NGC 3201 (point at uppermost left). This particular cluster has a uniquely strong influence on the kinematical solution because of its location near the Solar antapex and large positive radial velocity; it carries the highest statistical weight of any cluster in the entire sample. More will be said about this interesting and somewhat deceptive subgroup below.

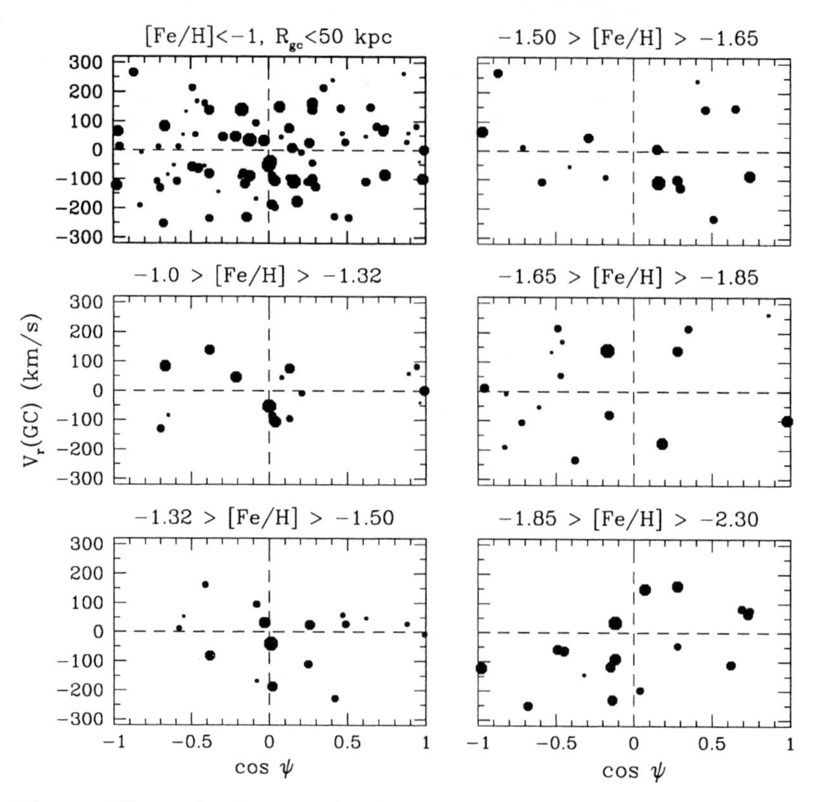

Fig. 24. Kinematics diagrams for the metal-poor clusters, divided by [Fe/H] intervals. Larger points indicate clusters with lower reddenings and thus better determined distances and ψ values

A potentially more important trend, which does not depend on a single object, shows up in the very lowest metallicity bin. The metal-poorest clusters exhibit a strong and significant net rotation. To find out where this signal is coming from, in Fig. 25 we combine all clusters more metal-poor than [Fe/H] $= -1.7$ and relabel them by distance as well. As noted above, we find that the objects with $R_{\rm gc} \gtrsim 15$ kpc contribute little to the solution for V (but do affect the dispersion); it is the mid- to inner-halo objects which drive the rotation solution, even though some have low weight because of uncertain distances. The result from Fig. 25 is $V = (80 \pm 43)$ km s^{-1} – a surprisingly large positive value, considering that we would expect this set of clusters to be the oldest ones in the Galaxy, and that no other subgroup of low-metallicity clusters displays any significant net rotation. The analysis of three-dimensional space motions by Dinescu et al. (1999) for a selection of these clusters yields a similar result. They obtain $V = (114 \pm 24)$ km s^{-1} for the metal-poor clusters with $R < 8$ kpc, formally in agreement with the solution from the radial velocities alone.

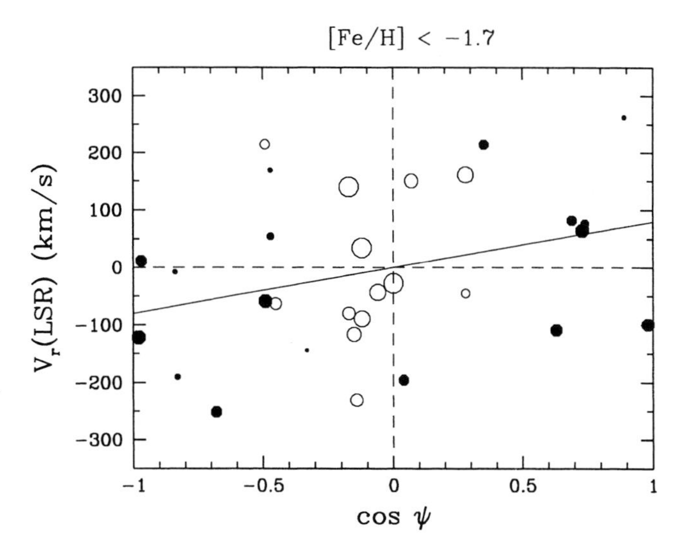

Fig. 25. Kinematics diagram for the extreme low-metallicity clusters ([Fe/H] < −1.7). The *solid line* indicates the formal solution $V = 80$ km s^{-1} for this group of objects. *Filled symbols* are ones within 15 kpc of the Galactic center, *open symbols* are ones outside 15 kpc

It seems necessary to conclude that the inner halo (0–4 kpc) has a rotation speed of $\sim 80 - 100$ km s^{-1} *regardless of metallicity:* the MPC and MRC clusters move alike. Are we seeing here the traces of an ELS-style formation epoch of collapse and spin-up in the inner halo, which (as SZ first claimed) would have been less important further out in the halo?

Still another way to plot this trend is shown in Fig. 26. Here, we start with the list of all 94 clusters with [Fe/H] < -0.95 and known velocities, sort them in order of metallicity, and solve for rotation V using the first 20 clusters in the list. We then shift the bin downward by one object (dropping the first one in the list and adding the 21st) and redo the solution. We shift the bin down again, repeating the process until we reach the end of the list. Clearly any one point is not at all independent of the next one, but this moving-bin approach is an effective way to display any global trends with changing metallicity. What we see plainly is the clear net rotation of the lowest-metallicity subpopulation, which smoothly dies away to near-zero rotation for [Fe/H] $\gtrsim -1.7$. (NB: The apparently sudden jump into retrograde rotation at [Fe/H] $\simeq -1.6$ is, once again, due to NGC 3201, which enters the bin in that range. If this cluster is excluded, the net rotation stays strictly near zero across the entire range.)

A minor additional point (shown in the lower panel of Fig. 26) is that the mean galactocentric radius decreases slightly as the metallicity of the bin shifts from the most metal-poor objects to the less metal-poor end. This trend is simply the result of the fact that there is a small metallicity gradient

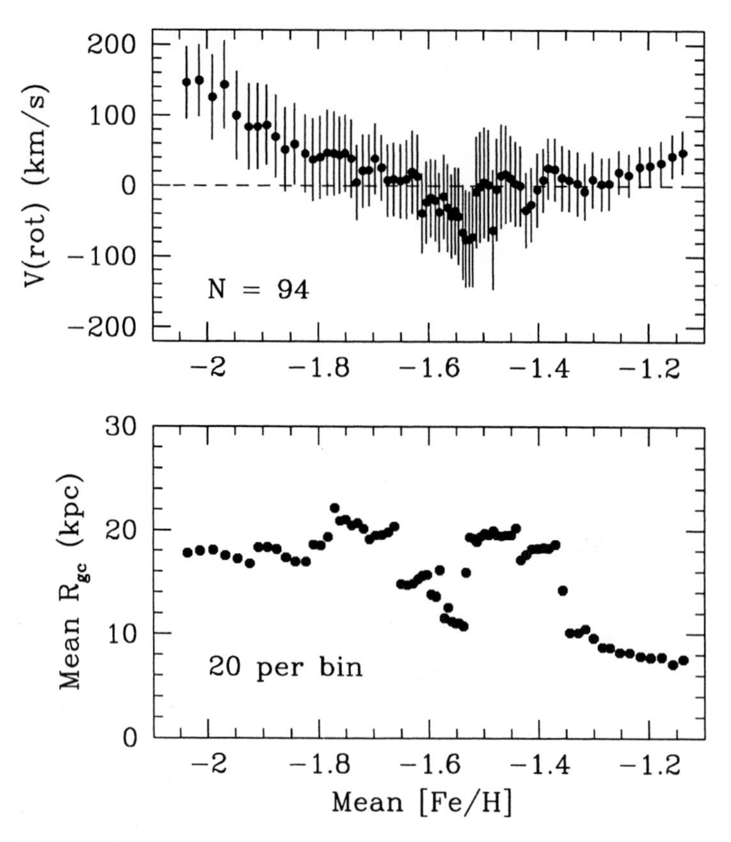

Fig. 26. Rotation speed plotted against mean [Fe/H], for the "moving bin" calculation described in the text. Over the interval $-2.3 \lesssim$ [Fe/H] $\lesssim -1.7$, note the smooth decline in V. Each bin contains 20 clusters

in the MPC system (Sect. 1), with the most metal-poor objects located more frequently at larger radii.

3.4 Retrograde Motion: Fragments and Sidetracks

Though the ELS-style formation picture may still hold some validity for the Galaxy's inner halo ($R_{\rm gc} \lesssim R_0$), very different ideas began to emerge for the outer halo especially in the literature of the past decade. Numerous pieces of evidence, as well as theoretical ideas, arose to suggest that much of the halo might have been *accreted* in the form of already-formed satellite fragments, each one of which would now be stretched out around the halo in a thin tidal streamer (see, for example, Majewski et al. 1996; Johnston 1998; Grillmair 1998, for review discussions with extensive references).

The particular relevance of these ideas to the globular clusters began with a comment by Rodgers & Paltoglou (1984) that the clusters in the metallicity range $-1.4 > [Fe/H] > -1.7$ not only had a small anomalous retrograde rotation, but that most of them also had similar horizontal-branch morphologies. These objects included clusters like M3, NGC 3201, NGC 7006, and several others with HBs that are well populated across the RR Lyrae instability strip. By contrast, intermediate-metallicity clusters like M13 (with extreme blue HBs) did not show this collective retrograde rotation. Rodgers & Paltoglou speculated that the "anomalous" group might have had a common origin in a small satellite galaxy that was absorbed by the Milky Way on a retrograde orbit.[7] They suggested that by contrast, the M13-type clusters with prograde or near-zero rotation were the "normal" ones belonging to the Milky Way halo from the beginning. The broader idea extending beyond this particular subset of clusters was that many individual ancestral satellites of the Galaxy might exist, and might still be identified today: to quote Rodgers & Paltoglou, "To be identified now as a component of the galactic outer halo, a parental galaxy must have produced a significant number of clusters in which a small range of metallicity is dominant and must have sufficiently distinct kinematics".

This idea was pursued later in an influential paper by Zinn (1993a) and again by Da Costa & Armandroff (1995). To understand it, we need to refer back to the HB morphology classification diagram of Fig. 9. In this diagram, the "normal" relation between HB type and metallicity is defined by the objects within $R_{gc} \lesssim 8$ kpc. The M3-type clusters further out in the halo and with generally redder HBs fall to the left of this normal line. The Rodgers & Paltoglou sample is drawn from the metallicity range -1.4 to -1.7, and indeed it can be seen that most of the clusters in that narrow horizontal cut across Fig. 9 belong to the red-HB group. Zinn, using the *assumption* that HB morphology is driven primarily by age for a given [Fe/H], called the normal ($[Fe/H] < -0.8$, blue-HB) clusters the Old Halo and the redder-HB ones the Younger Halo (though the latter group is not intended to be thought of as "young" in an absolute sense). If interpreted this way through typical HB models (e.g. Lee et al. 1994) – that is, if age is the dominant second parameter in Fig. 9 – then the Younger Halo clusters would need to be anywhere from ~ 2 to 5 Gyr younger than the Old Halo.

Zinn compared the kinematics of these two groups, finding $V = -64 \pm 74$ km s^{-1} for the Younger Halo and $V = 70 \pm 22$ km s^{-1} for the Old Halo, as well as a noticeably lower dispersion σ_{los} for the Old Halo. Developing the SZ formation picture further from these results, Zinn concluded "It seems likely ... that some of the outlying [protogalactic, gaseous] fragments escaped

[7] It should be noted that *individual* clusters with retrograde orbits are certainly not unusual: since the halo velocity dispersion is high and it is basically a pressure-supported system (high random motions and low overall rotation speed), there will be a large mix of both prograde and retrograde orbits to be found. The issue here is that it is hard to see how a *collective* retrograde motion of an entire identifiable group of clusters could have arisen in any other way than accretion after the main *in situ* star formation phase of the halo.

destruction, remained in orbit about the collapsed Galaxy, and evolved into satellite dwarf galaxies ... it is proposed that such satellite systems were the sites of the formation of the Younger Halo clusters".

Almost simultaneously, van den Bergh (1993a,b) used a different type of graphical analysis of kinematics to isolate rather similar subgroups, one of which (corresponding roughly to the Zinn Younger Halo) he postulates to have retrograde-type orbits. Van den Bergh went even further along the same line to envisage a single large ancestral fragment for these: "... the hypothetical ancestral galaxy that formed ... clusters with M3-like color-magnitude diagrams merged with the main body of the protoGalaxy on a plunging retrograde orbit".

It appears to me that these interpretations are quite risky, and that we need to take a fresh look at the actual data upon which they are built. There are at least two serious problems:

- The interpretation of HB morphology as a fair indication of cluster age (Lee et al. 1994; Chaboyer et al. 1996) is not proven; in fact, more recent evidence based directly on deep main-sequence photometry of clusters in each of the two groups suggests just the opposite in at least some cases. The clusters M3 and M13, with similar chemical compositions and different HB morphologies, form a classic "second parameter" pair. Precise differential main sequence fitting (Catelan & de Freitas Pacheco 1995; Johnson & Bolte 1998; Grundahl et al. 1998) indicates that these clusters have the same age to within the ~ 1 Gyr level which is the current precision of the technique. The still more extreme second-parameter trio NGC 288/362/1851 (e.g., Stetson et al. 1996; Sarajedini et al. 1997) may exhibit an age range of ~ 2 Gyr. Other recent studies of red-HB clusters in the outermost Milky Way halo (Stetson et al. 1999) and in the LMC and Fornax dwarf (e.g., Johnson et al. 1999; Olsen et al. 1998; Buonanno et al. 1998) with moderately low metallicities show age differences relative to M3 and M92 that are modest, at most 2 Gyrs and often indistinguishable from zero. In short, the hypothesis that cluster age is the dominant second parameter may indeed work for some clusters but does not seem to be consistent with many others. Additional combinations of factors involving different ratios of the heavy elements, mixing, rotation, helium abundance, or mass loss will still need to be pursued much more carefully (cf. the references cited above).
  We should not continue to use the terms "old" and "young" for these groups of clusters; in what follows, I will refer to them instead as the blue-HB and red-HB groups.
- The "retrograde rotation" of the red-HB group (Table 6) is not statistically significant. We also need to ask how it arises in the first place. The formally negative rotation of the red-HB group is driven *very strongly* by the single object NGC 3201, which (as we saw above) has a uniquely powerful influence on the kinematical solutions for any group it is put

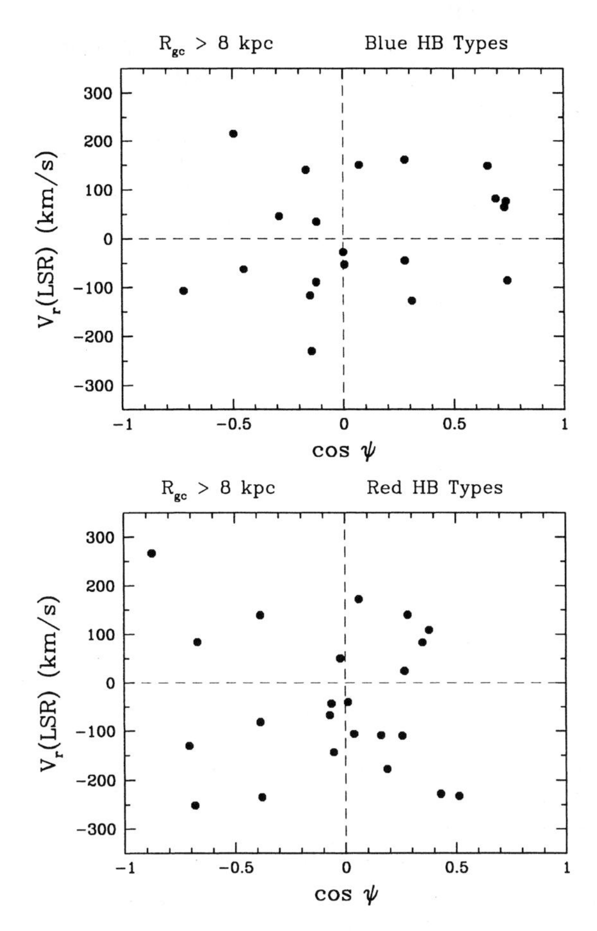

Fig. 27. *Upper panel:* Rotation solution for the outer-halo (8 – 40 kpc) clusters with blue HB types. *Lower panel:* Rotation solution for outer-halo clusters with red HB types; note NGC 3201 at upper left

into. Taking NGC 3201 out of the sample (see Table 6) turns out to change $V(\text{rot})$ by a full $+70$ km s^{-1}, changing the retrograde signature to a prograde one.[8] Neither the prograde or retrograde value is, however, significantly different from zero.

Some of these points are demonstrated further in Fig. 27. Here, we show the kinematics diagrams for the two groups of clusters. To make the groups

[8] We can, of course, treat the BHB group similarly by removing the single most extreme point (in this case, NGC 6101) and redoing the solution. $V(\text{rot})$ changes from (55 ± 58) to (88 ± 54) km s^{-1}, a statistically insignificant difference. This test verifies again that NGC 3201 has a uniquely strong influence on whatever set of objects it is included with.

as strictly comparable as possible, we draw each one *strictly from the same zone of the halo*, 8 kpc $< R_{gc} <$ 40 kpc, and ignore the blue-HB clusters in the inner halo. We see from the figure that neither sample has a significant rotation, either prograde or retrograde, whether or not we choose to remove NGC 3201 (though it is evident from the graphs just how influential that one cluster is).

True retrograde orbits are notoriously hard to deduce from radial velocities alone (NGC 3201 is one of the rare exceptions). This type of analysis would benefit greatly from reliable knowledge of *absolute proper motions* of these clusters, from which we can deduce their full three-dimensional space motions. Proper motions (μ_δ, μ_α) now exist in first-order form for almost 40 clusters, from several recent studies notably including Cudworth & Hansen (1993); Odenkirchen et al. (1997); and Dinescu et al. (1999); see Dinescu et al. for a synthesis of all the current results with extensive references. We can employ these to make useful classifications of orbital types (clearly prograde, clearly retrograde, or plunging) and the general range of orbital eccentricities. In the new orbital data summarized by Dinescu et al. (1999), we find 10 BHB clusters and 6 RHB clusters with $R_{gc} > 8$ kpc. For the BHB subset, the mean orbital eccentricity is $\langle e \rangle = 0.66 \pm 0.06$ and energy is $\langle E \rangle = -(5.2 \pm 0.9) \times 10^4$ km^2 s^{-2}. For the RHB subset, these numbers are $\langle e \rangle = 0.65 \pm 0.07$ and $\langle E \rangle = -(5.2 \pm 0.9) \times 10^4$ km^2 s^{-2}. In the BHB group, we find 5 prograde orbits, 3 retrograde, and 2 "plunging" types; in the RHB group, there are 3 prograde, 1 retrograde, and 2 plunging.

All these comparisons suggest to me that the two groups have no large collective differences in orbital properties; the rather modest differences in mean rotation speed are driven strongly by small-sample statistics. In addition, the normal assumption of an approximately isotropic orbital distribution for the halo clusters still seems to be quantitatively valid.

Where does this analysis leave the search for remnants of accreted satellites in the halo? My impression – perhaps a pessimistic one – is that distinct moving groups have proven almost impossible to find (if they exist in the first place) from the analyses of globular cluster motions. Once we start subdividing our meager total list of halo clusters by all the various parameters such as metallicity, spatial zones, or CMD morphology, the selected samples quickly become too small for statistically significant differences to emerge. The one outstanding exception is of course the four Sagittarius clusters, which are a physically close group that has not yet been tidally stretched out all around the halo. But even here, one suspects that they would not yet have been unambiguously realized to be part of a single system if their parent dwarf galaxy had not called attention to itself. If Sagittarius is a typical case of an accreted satellite, then we could reasonably expect that any others in the past would have brought in similarly small numbers of globular clusters – one, two, or a handful at a time – and thus extremely hard to connect long after the fact. My conclusion is that, if accreted remnant satellites in the halo are to be identified from this type of analysis, they will have to emerge from the study of halo field stars (e.g., Majewski et al. 1996), for which vastly larger and more statistically significant samples of points can be accumulated.

Three-dimensional space velocities would also, of course, be immensely valuable in the search for physically connected moving groups of clusters. Majewski (1994) notes that "the key test of common origin must come with orbital data derived from complete space velocities for these distant objects". Direct photometric searches for "star streams" trailing ahead of or behind disrupted satellites, have been proposed (see the reviews of Grillmair 1998; Johnston 1998). However, numerical experiments to simulate tidal stripping indicate that these disrupted streams would be thinly spread across the sky, enough so that they would be visible only for the largest satellites such as Sagittarius and the LMC.

3.5 Orbits in the Outermost Halo

Finding traces of disrupted satellites that are still connected along an orbital stream should be easiest in the outermost halo where the satellites originally resided and where the orbital timescales are the longest. It has been proposed several times, for example, that the Magellanic Clouds are connected along a great circle with other objects including the dwarf spheroidal satellites Draco, Ursa Minor, and Carina, and the small clusters Palomar 12 and Ruprecht 106 (Kunkel & Demers 1975, 1977; Lynden-Bell 1976). A similar stream comprising Fornax, Leo I and II, Sculptor, Palomar 3, Palomar 4, and AM-1 has been proposed (Majewski 1994; Lynden-Bell & Lynden-Bell 1995; Fusi Pecci et al. 1995). Correlation analyses of the radial velocities and locations of all the globular clusters and dwarf satellites have been carried out (Lynden-Bell & Lynden-Bell 1995; Fusi Pecci et al. 1995), with the result that several possible orbital "groups" have been proposed, usually containing just three or four clusters each. At this stage it is unclear how real any of these groups might be. (It is noteworthy, however, that this analysis successfully connected the Sagittarius clusters before the dwarf galaxy itself was found.)

From the viewpoint of space distribution alone, it is unquestionably true that the objects beyond $R_{gc} \sim 60$ kpc are not isotropically distributed around the Galaxy. Most of these objects do lie moderately close to a single plane (the Fornax-Leo-Sculptor stream) which is nearly perpendicular to the Galactic plane. (It should be kept in mind that their distribution is "planar" only in a relative sense; the thickness of the plane is about 50 kpc and the diameter about an order of magnitude larger.) This stream is shown in Fig. 28 (adapted from Majewski 1994), where we are looking at it in an orientation which minimizes the side-to-side spread in locations: we select a new axis $X' = X \cos\theta + Y \sin\theta$ where θ is the coordinate rotation angle between the X, X' axes and now X, Y are measured relative to the Galactic center. For the Fornax-Leo-Sculptor stream, Majewski (1994) finds $\theta \simeq 50°$. An alternate, and intriguing, interpretation of the strongly prolate distribution of these

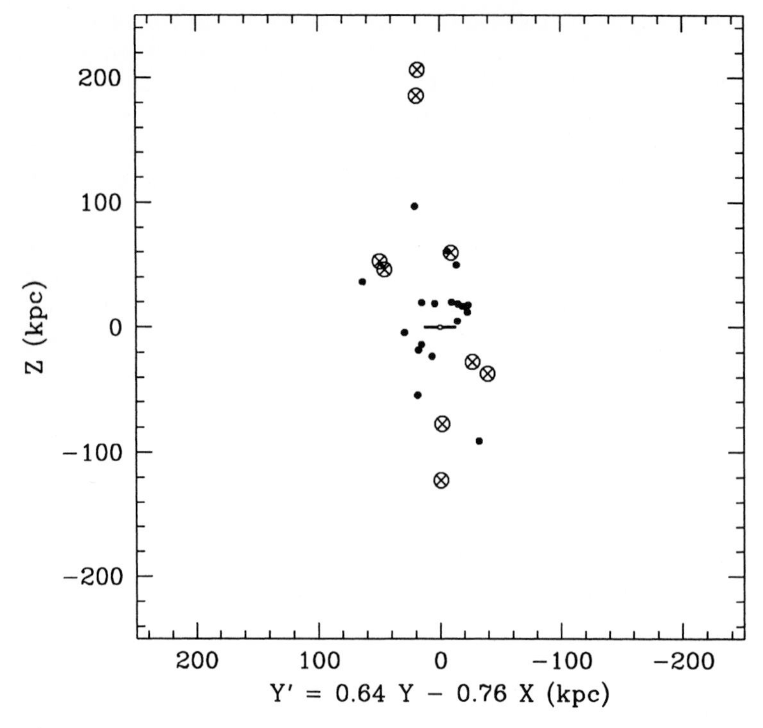

Fig. 28. Locations of remote satellites around the Milky Way: *solid dots* are halo clusters, *crosses* are dwarf galaxies. The X' axis is rotated 50° counterclockwise from the normal X-axis, so that we are looking nearly parallel to the proposed Fornax-Leo-Sculptor plane. Note that X, Y here are measured relative to the Galactic center

satellites is that they delineate the shape of the outermost dark matter halo of the Milky Way (see Hartwick 1996 for a kinematical analysis and complete discussion of this possibility).

3.6 Some Conclusions

A summary of the essential points that we have discussed in this section may be helpful.

- Mean rotation speeds and orbital velocity dispersions can be usefully estimated from cluster radial velocities and properly designed kinematics diagrams. The main sources of uncertainty in these plots are (a) uncertainties in the measured cluster distances, which are important for highly reddened clusters in the inner halo; and (b) the small numbers of points in any one subsample. Unfortunately, we can do nothing to improve our sample size of *clusters*, but the kinematics of the halo can also be studied through much larger samples of *field stars*.

- The metal-rich (MRC) clusters have a strong systemic rotation with two weakly distinguishable subcomponents: an inner ($0-4$ kpc) bulge-like system with $V(\mathrm{rot}) \sim 90$ km s^{-1}, and an outer ($4-8$ kpc) system with $V(\mathrm{rot}) \sim 150$ km s^{-1} somewhat more like the thick disk.
- The most metal-poor (MPC) clusters, those with [Fe/H] $\lesssim -1.7$, have a systemic prograde rotation of $V(\mathrm{rot}) \sim 80-100$ km s^{-1}, somewhat like the MRC bulge population.
- *Individual* clusters with retrograde orbits certainly exist, but:
- There are no unambiguous subgroups of clusters that have *systemic* retrograde rotation that are identifiable on the basis of Galactocentric distance, metallicity, or HB morphology. Previous suggestions of such retrograde groups seem to have arisen because of the unfortunate and uniquely strong influence of the single cluster NGC 3201 – again, a consequence of the small samples of objects and the effects of rare outliers on statistical distributions.
- To first order, the mid-to-outer halo MPC clusters can reasonably be described as forming a system with small net rotation and roughly isotropic orbit distribution. With the exception of the Sagittarius clusters, no "accreted satellite" remnant groups have yet been reliably identified *from the clusters alone*. The main hope for identifying such groups lies with the analysis of field-star populations.
- Several of the outermost clusters ($R_{\mathrm{gc}} > 50$ kpc) may be part of an extremely large-scale orbital stream (the Fornax-Leo-Sculptor stream).
- We urgently need more and better three-dimensional space motions for the halo clusters, measured through accurate absolute proper motions. Such information will allow us to determine the systemic rotation $V(\mathrm{rot})$ of the *outer* halo; the degree of anisotropy of the orbital distribution; the true fraction of retrograde orbits; and the identification of true orbital families and tidal streams.

4 The Milky Way System: Dynamics and Halo Mass Profile

There is no illusion more dangerous than the belief that the progress of science is predictable.

Freeman Dyson

4.1 The Orbit Distribution

It is obvious from the large line-of-sight velocity dispersions quoted in the previous section that the globular clusters do not have circular or true disklike orbits as a group (though a few individual ones may). Neither do they have plunging, purely radial orbits as a group, since the large tangential motions of

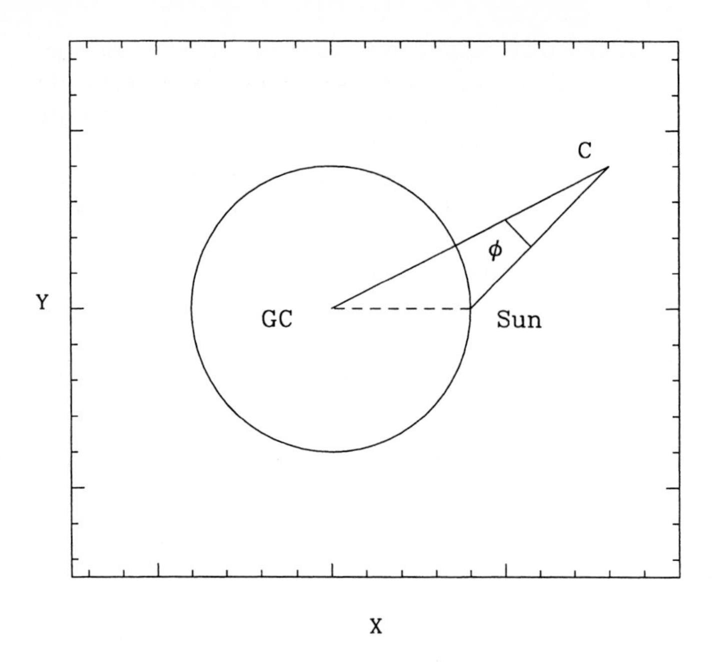

Fig. 29. Geometric definition of the angle φ between Sun and Galactic center (GC), as seen from the globular cluster C. The circle on the Galactic XY plane has radius R_0

many of them are obvious. From the near-uniformity of the observed $\sigma_{\rm los}$ in any direction through the halo, we normally assume the orbital distribution to be *isotropic* (that is, $\sigma_U \simeq \sigma_V \simeq \sigma_W$ along any three coordinate axes). The best current evidence (Dinescu et al. 1999, from measurement of the three-dimensional space motions of the clusters) continues to support the isotropic assumption.

Another diagnostic of the cluster orbits which was used early in the subject (e.g., von Hoerner 1955; Kinman 1959b) and recently revived by van den Bergh (1993a,b) is the velocity ratio

$$u_0 = \frac{v_r({\rm LSR})}{V(r)} \tag{23}$$

where $V(r)$ is the rotation speed for circular orbits at distance r from the Galactic center; and v_r is the radial velocity of the cluster relative to a stationary point at the Sun, as used in the previous section. Now also let φ be the angle between Sun and Galactic center (GC), as seen from the cluster,

$$\cos\varphi = \frac{r - R_0 \cos b \cos \ell}{(r^2 + R_0^2 - 2rR_0 \cos b \cos \ell)^{1/2}} \tag{24}$$

(see Fig. 29). If the cluster is on a circular orbit, then clearly $v_r({\rm LSR}) =$

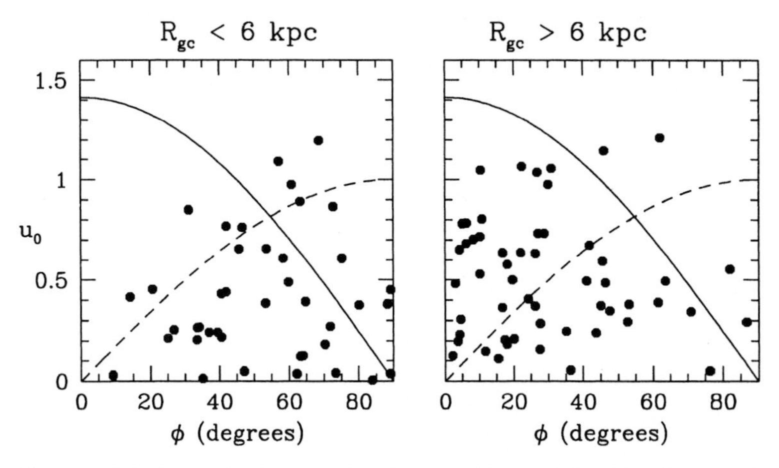

Fig. 30. Velocity-ratio diagram for cluster orbits as defined in the text. Any cluster on a purely circular orbit around the GC would lie on the *dashed line* $u_0 = \sin\varphi$. The *solid line*, $u_0 = 2^{1/2}\cos\varphi$, would be the locus of clusters on purely radial escape orbits if there were no additional halo mass outside their current location

$V\sin\varphi$. However, if it is on a purely radial orbit with respect to the GC, then $v_r(\mathrm{LSR}) = V\cos\varphi$, and if $u_0 \gtrsim \sqrt{2}$ for such cases then nominally the cluster would be on an escape orbit *if* there were no additional halo mass outside its current location. Figure 30 shows the distribution of ratios u_0 against φ for the clusters in two different areas of the halo, where we have assumed $V_0 \equiv 220$ km s^{-1} for all R_{gc}. Objects near the left-hand side of the diagram must be on strongly radial orbits, while ones on the right-hand side must be on more nearly circular orbits. The wide range of locations across the diagram confirms that neither purely radial nor circular orbits are dominant, and the fact that almost all the points stay comfortably below the upper envelope $u_0 = \sqrt{2}$ indicates – as it should – that they are well within the limits of bound orbits.

4.2 The Mass of the Halo: Formalism

Early in the history of this subject, it was realized that the radial velocities of the globular clusters provided an extremely effective way to estimate the mass profile of the Galaxy out to large distances. Roughly speaking, the velocity dispersion σ_v of a group of clusters at distance r from the Galactic center reflects the total mass $M(r)$ enclosed within r, $GM(r)/r \sim \sigma_v^2$. Knowing σ_v, we may then invert this statement and infer $M(r)$. The first comprehensive attempt to do this was by Kinman (1959b), with later and progressively more sophisticated analyses by (among others) Hartwick & Sargent (1978), Lynden-Bell et al. (1983), Little & Tremaine (1987), and Zaritsky et al. (1989). The limits to our knowledge of the halo mass profile are now set by our lack of knowledge of the true three-dimensional space motions of the clusters.

Interested readers should see Fig. 6 of Kinman (1959b), where the cluster velocity dispersions are used to plot the $M(r)$ profile in much the same way as is done here. Kinman's graph provides the earliest clear-cut evidence that I am aware of that $M(r)$ grows linearly out to at least $r \sim 20$ kpc in our Galaxy, thus showing the existence of its "dark-matter" halo – although it was not interpreted as such at the time. It was only 15 to 20 years later that astronomers routinely accepted the dominance of unseen matter in galaxies, although the key observations were in front of them, in basically correct form, far earlier.

Turning the first-order virial-theorem argument above into a quantitative formalism can be done in a variety of ways. Always, a major source of uncertainty is that we observe only one component of the true space velocity. Here, I will set up the formal analysis of the particle velocity distribution with an approach adapted from Hartwick & Sargent (1978); it has the advantage of physical clarity, and of displaying the key geometric parameters of the mass distribution in explicit form. Readers should see Little & Tremaine (1987) and Zaritsky et al. (1989) for alternate modern formulations based on Bayesian statistics.

Consider a system of particles orbiting in the halo with number density in phase space $f(x_i)$ with position and velocity coordinates x_i $(i = 1, \ldots, 6)$. The particles are imagined to be dominated by internal random motions (that is, they have a high radial velocity dispersion relative to the Galactic center, which provides the main support of the system against gravity), and the distribution is explicitly assumed time-independent, $\partial f/\partial t = 0$. At this point we also assume a roughly spherical mass distribution so that the potential energy is $U = U(r)$. We impose the appropriate physical boundary condition that $f \rightarrow 0$ as $x_i \rightarrow \infty$ for any coordinates or velocities. Finally, we adopt spherical polar coordinates (r, θ, φ) and velocity components (R, Θ, Φ).

The collisionless Boltzmann equation for such a system is then

$$\frac{df}{dt} = 0 = \dot{r}\frac{\partial f}{\partial r} + \dot{R}\frac{\partial f}{\partial R} + \dot{\Theta}\frac{\partial f}{\partial \Theta} + \dot{\Phi}\frac{\partial f}{\partial \Phi} \tag{25}$$

(see, for example, Chp. 4 of Binney & Tremaine 1987), and where we have already dropped any derivatives $\partial/\partial\theta, \partial/\partial\varphi$ from spherical symmetry. We can write the various force components as derivatives of the potential energy,

$$F_r = \frac{-\partial U}{\partial r} = \dot{R} - \frac{\Theta^2}{r} - \frac{\Phi^2}{r} , \tag{26}$$

$$F_\theta = 0 = -\frac{1}{r}\frac{\partial U}{\partial \theta} = \dot{\Theta} + \frac{R\Theta}{r} - \frac{\Phi^2}{r}\cot\theta , \tag{27}$$

$$F_\varphi = 0 = -\frac{1}{r\sin\theta}\frac{\partial U}{\partial \varphi} = \dot{\Phi} + \frac{R\Phi}{r} + \frac{\Theta\Phi}{r}\cot\theta . \tag{28}$$

When we use these to substitute expressions in (25) for $\dot{R}, \dot{\Theta}, \dot{\Phi}$, the Boltzmann equation becomes

$$0 = R\frac{\partial f}{\partial r} + \Big(\frac{\Theta^2 + \Phi^2}{r} - \frac{dU}{dr}\Big)\frac{\partial f}{\partial R} + \frac{1}{r}\Big(\Phi^2 \cot\theta \; - \; R\,\Theta\Big)\frac{\partial f}{\partial \Theta}$$
$$- \frac{1}{r}(R\Phi + \Theta\,\Phi\;\cot\theta)\frac{\partial f}{\partial \Phi} \equiv g(r, R, \Theta, \Phi)\,. \tag{29}$$

Next take the first R-moment of this equation and integrate over all velocities:

$$0 = \int_R \int_\Theta \int_\Phi g \;\cdot\; R\, dR\, d\Theta\, d\Phi \tag{30}$$

and also define the number density $\nu(r)$ (number of particles per unit volume of space) as

$$\nu \equiv \int_R \int_\Theta \int_\Phi f \cdot dR\, d\Theta\, d\Phi\;, \tag{31}$$

and in general define the mean of any quantity Q as $\langle Q\rangle \equiv \frac{1}{\nu}\int Q\, f\, dR\, d\Theta\, d\Phi$. When we carry out the integration in (30), we find that several of the terms conveniently vanish either from symmetry or from our adopted boundary conditions (we will leave this as a valuable exercise for the reader!), and the remainder reduces to

$$0 = \frac{d}{dr}(\nu\langle R^2\rangle) - \frac{\nu}{r}\langle\Theta^2 + \Phi^2\rangle + \nu\frac{dU}{dr} + \frac{2\,\nu}{r}\langle R^2\rangle\,. \tag{32}$$

Now we can put in the radial force component

$$\frac{dU}{dr} = \frac{G\,M(r)}{r^2} \tag{33}$$

where $M(r)$ is the mass contained within radius r, and we reduce (32) further to

$$G\,M(r) = r\,\langle R^2\rangle\left(\frac{\langle\Theta^2 + \Phi^2\rangle}{\langle R^2\rangle} - \frac{r\,d\nu}{\nu\,dr} - \frac{r}{\langle R^2\rangle}\frac{d\langle R^2\rangle}{dr} - 2\right). \tag{34}$$

The second and third terms in the brackets of (34) can be simplified for notation purposes if we denote them as α and β, such that the density of the GCS is assumed to vary with radius as $\nu \sim r^\alpha$, and the radial velocity dispersion as $\langle R^2\rangle \sim r^\beta$. We can also recognize that the halo has some net rotation speed Θ_0 such that the tangential velocity can be broken into rotational and peculiar (random) parts,

$$\Theta = \Theta_0 + \Theta_p \tag{35}$$
$$\langle\Theta^2\rangle = \Theta_0^2 + \langle\Theta_p^2\rangle \tag{36}$$

and finally for notation purposes we define the *anisotropy parameter*

$$\lambda \equiv \frac{\langle \Theta_p^2 + \Phi^2 \rangle}{\langle R^2 \rangle} . \tag{37}$$

For purely radial orbits, clearly $\lambda \to 0$, and for isotropically distributed orbits where all three random velocity components are similar, $\lambda = 2$. With these various abbreviations, the simplified moment of the Boltzmann equation takes the final form that we need,

$$G\, M(r) = \langle r \rangle \langle R^2 \rangle \left(\lambda - \alpha - \beta + \frac{\Theta_0^2}{\langle R^2 \rangle} - 2 \right) . \tag{38}$$

The quantities on the right-hand side represent the characteristics of our tracer population of particles (the globular clusters), while the single quantity on the left ($M(r)$) represents the gravitational potential to which they are reacting. This form of the equation displays rather transparently how the deduced mass $M(r)$ depends on the radial velocity dispersion $\langle R^2 \rangle$ of a set of points with mean Galactocentric radius $\langle r \rangle$; on the orbit anisotropy λ and the rotation speed Θ_0; and on the density distribution ν and radial dependence of the velocities β. It is easier to achieve a given radial velocity dispersion if the cluster orbits are more purely radial (smaller λ, thus smaller enclosed mass M). Similarly, a steeper halo profile (more negative α or β) or a larger systemic rotation speed Θ_0 will lead to a larger mass for a given velocity dispersion.

Can we simplify this relation any further? On empirical grounds we expect from our accumulated evidence about the Galaxy to find roughly isotropic orbits $\lambda \sim 2$, a nearly isothermal halo and thus constant velocity dispersion $\beta \sim 0$, and a small halo rotation $\Theta_0^2 \ll \langle R^2 \rangle$. Thus as a first-order guess at the mass distribution we might expect very crudely

$$G\, M(r) \simeq -\alpha \, \langle r \rangle \, \langle R^2 \rangle . \tag{39}$$

At this point, we could now select groups of clusters at nearly the same radial range $\langle r \rangle$, calculate their velocity dispersion $\langle R^2 \rangle$, and immediately deduce the enclosed mass $M(r)$. We will see below, however, that it will be possible to refine this guess and to place somewhat better constraints on the various parameters.

4.3 The Mass of the Halo: Results

To calculate the mass profile $M(r)$ for the Milky Way, we will now use the radial velocity data for all the MPC clusters ([Fe/H] < -0.95), which form a slowly rotating system dominated by internal random motions. We will, of course, find that we rapidly run out of clusters at large r, just where we are most interested in the mass distribution. To help gain statistical weight, we will therefore add in the data for nine satellites of the Milky Way, recognizing

Table 7. Dwarf satellites of the Milky Way

Satellite	ℓ	b	V_{HB}	E_{B-V}	μ_0	[Fe/H]	v_r	Sources
LMC	280.47	−32.89	19.20	0.06	18.54		245.0	1,2
Ursa Minor	104.95	44.80	19.90	0.01	19.37	−2.2	−247.4	3,4
Sculptor	287.54	−83.16	20.10	0.03	19.46	−1.8	109.9	5,6
Draco	86.36	34.71	20.10	0.02	19.54	−2.1	−293.3	4,7
Sextans	243.50	42.27	20.36	0.03	19.75	−2.05	227.9	8,9
Carina	260.11	−22.22	20.65	0.03	19.96	−1.52	223.1	10,11
Fornax	237.10	−65.65	21.25	0.02	20.64	−1.4	53.0	12,13
Leo II	220.17	67.23	22.18	0.02	21.52	−1.9	76.0	14,15
Leo I	225.98	49.11	22.75:	0.02	22.18	−2.0	285.0	16,17

Sources: (1) This paper (Sect. 2) (2) NASA Extragalactic Database (NED) (3) Nemec et al. 1988 (4) Armandroff et al. 1995 (5) Da Costa 1984 (6) Queloz et al. 1995 (7) Carney & Seitzer 1986 (8) Suntzeff et al. 1993 (9) Mateo et al. 1995 (10) Smecker-Hane et al. 1994 (11) Mateo et al. 1993 (12) Smith et al. 1996 (13) Mateo et al. 1991 (14) Mighell & Rich 1996 (15) Vogt et al. 1995 (16) Lee et al. 1993b (17) Zaritsky et al. 1989

fully (see the previous section) that the orbital motions of some of these may well be correlated. These nine include the Magellanic pair (LMC+SMC) plus the dwarf spheroidals more remote than $\sim$ 50 kpc from the Galactic center (Draco, Ursa Minor, Carina, Sextans, Sculptor, Fornax, Leo I and II). Furthermore, we will also put in data for 11 remote RR Lyrae and horizontal-branch stars in the halo in the distance range 40 kpc $< r <$ 65 kpc, as listed by Norris & Hawkins (1991). To keep track of the possibility that these latter halo *field stars* might have (say) a different anisotropy parameter λ than the clusters, we will keep them in a separate bin from the clusters.

The relevant data for the dwarf satellites are summarized in Table 7. Successive columns list the Galactic latitude and longitude in degrees; the measured horizontal-branch magnitude V_{HB} of the old-halo or RR Lyrae stars; the foreground reddening; the intrinsic distance modulus and mean metallicity of the stars; and the heliocentric radial velocity (km s^{-1}). For all but the LMC and Leo I, the distance is calculated from V_{HB} and our adopted prescription $M_V(\mathrm{HB}) = 0.15[\mathrm{Fe/H}] + 0.80$. For Leo I, the distance estimate relies on the I-band magnitude of the RGB tip.

The velocity dispersion data and calculated mass profile are listed in Table 8. Here, the MPC clusters have been divided into radial bins with roughly a dozen objects per bin; successive columns list the mean Galactocentric radius, number of clusters in the bin, mean radial velocity dispersion, and enclosed mass determined as described below.

Table 8. Radial velocity dispersion profile

$\langle R_{gc}\rangle$ (kpc)	n	$\langle R^2\rangle^{1/2}$	$M(r)$ (10^{11} $M_\odot$)	Comment
1.33 ± 0.12	14	133 ± 24	0.17 ± 0.03	
2.38 ± 0.10	8	136 ± 32	0.31 ± 0.08	
3.83 ± 0.14	12	99 ± 19	0.30 ± 0.06	
5.55 ± 0.20	9	80 ± 18	0.31 ± 0.07	
7.55 ± 0.26	10	127 ± 27	0.88 ± 0.19	
10.4 ± 0.28	9	145 ± 33	1.40 ± 0.32	
16.3 ± 0.71	15	135 ± 24	1.88 ± 0.34	
29.2 ± 2.1	10	141 ± 30	3.93 ± 0.88	
53.8 ± 2.5	11	112 ± 22	4.94 ± 1.03	RR Lyraes
95.1 ± 8.1	6	109 ± 30	8.97 ± 2.61	Outer GCs only
71.0 ± 6.2	5	113 ± 34	6.95 ± 2.22	GCs + dSph
131.0 ± 20.7	10	87 ± 19	8.28 ± 2.19	GCs + dSph

In Fig. 31, we first show the r.m.s. velocity dispersion $\langle R^2\rangle^{1/2}$ plotted against Galactocentric distance. The velocities R are the radial velocities of the clusters after removal of the LSR motion V_0, i.e. they are the same as $\sigma_{\rm los}$ used in the discussion of kinematics in Sect. 3. From the graph, we see that σ stays nearly uniform at ~ 130 km s^{-1} for $r \lesssim 50$ kpc and then declines gradually at larger distances. An extremely crude first-order model for this behavior would be a simple isothermal halo for which the velocity dispersion $\sigma \sim$ const and $M(r)$ increases in direct proportion to r, out to some truncation limit r_h beyond which the density is arbitrarily zero. At larger distances the velocity dispersion would then decline as $\sigma \sim r^{-1/2}$. For a choice of halo 'limiting radius' $r_h \simeq 60$ kpc, this oversimplified model actually represents the data points quite tolerably. Whether or not we exclude the field-star point (labelled RR) makes little difference, suggesting that there are probably no gross intrinsic differences between the clusters and field stars.

For our final calculation of the mass profile $M(r)$, we employ (38) and add in what we know about the space distribution and kinematics of our test particles, the globular clusters:

- From Fig. 3, we see that the space density exponent α steepens smoothly from an inner-halo value of $\simeq -2$ up to -3.5 in the outer halo.
- For the anisotropy parameter, we have $\lambda = 1.2 \pm 0.3$, determined directly from the three-dimensional space motions summarized by Dinescu et al. (1999). This value represents a mild anisotropy biased in the radial direction, and appears not to differ significantly with location in the halo, at least for the metal-poor clusters that we are using.

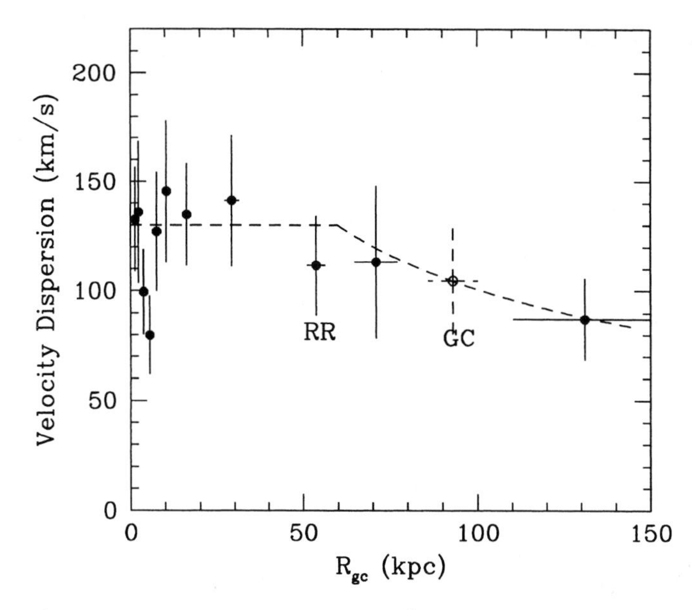

Fig. 31. Radial velocity dispersion (line-of-sight relative to a fixed point at the Sun) plotted against Galactocentric distance. The point labelled 'RR' is for the 11 field RR Lyrae and HB stars described in the text. The point labelled 'GC' is for the six most remote clusters without including any of the dwarf satellite galaxies. The *dashed line* shows the expected trend for the velocity dispersion if the halo were ideally isothermal out to a radius $r_h \sim 60$ kpc, and then truncated abruptly there

- The rotation velocity Θ_0 is $\simeq 80$ km s^{-1} for $R \lesssim 8$ kpc and $\simeq 20$ km s^{-1} for $R \gtrsim 8$ kpc (see Sect. 3 above).
- Lastly, we adopt $\beta \simeq 0$ since no strong variation of velocity dispersion with radius is evident.

The resulting mass profile is shown in Fig. 32. As expected, $M(r)$ grows linearly for $r \lesssim 40$ kpc but then increases less steeply, reaching $\sim (8 \pm 2) \times 10^{11}$ $M_\odot$ at $R \sim 100$ kpc.

An extra useful comparison – at least for the inner part of the halo – is our knowledge of the rotation speed of the Galactic *disk*, $V_0 = 220$ km s^{-1} roughly independent of radius. Within the region of disk/halo overlap, we can then write

$$G\,M = r\,V_0^2 \tag{40}$$

which becomes $M(r) = 1.13 \times 10^{10}$ $M_\odot$ r(kpc). This relation matches the GCS dispersion data in Fig. 32 extremely well for $r \lesssim 50$ kpc.

With the foregoing arguments, we have been able to extend the mass profile for the Milky Way outward to about three times further than in Kinman's original attempt 40 years ago. However, our suggestion that the halo density

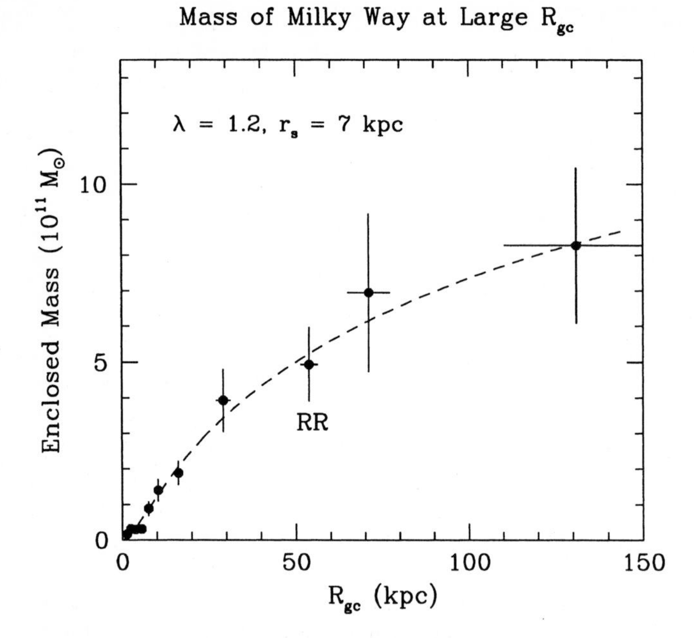

Fig. 32. Mass profile for the Milky Way halo, with the "best fit" set of parameters discussed in the text. The *dashed line* is an NFW model halo for a characteristic radius $r_s = 7$ kpc

begins to "die" somewhere around $r_h \sim 60$ kpc is risky, since we do not know if our model assumptions apply to the outermost objects. The motions of several of these clusters and dwarf satellites may be *correlated* (see Sect. 3), which would invalidate our adopted values for λ, Θ_0, and $\langle R^2 \rangle$. Perhaps equally important is that the most remote single object in our list, Leo I, may not be bound to the Galaxy (see Lee et al. 1993b; Zaritsky et al. 1989). Of the dozen most remote objects, Leo I has both the largest distance ($R_{\rm gc} = 277$ kpc) and largest velocity ($v_r = 175$ km s^{-1}) and thus carries significant weight in the solution by itself. There is no way to rule out the likely possibility that it is simply a dwarf moving freely within the Local Group, like others at similar distances from M31. To *marginally* bind Leo I to the Milky Way would require $M(\text{total}) \gtrsim 10^{12}$ M$_\odot$ within 100 kpc, which in turn would be marginally inconsistent with the mass given by all the other remote satellites *if* their motions are independent and roughly isotropic.

A widely used analytic model for dark-matter halos with some basis in cosmological N-body simulations is that of Navarro et al. (1996; denoted NFW), giving a density profile

$$\varrho(r) = \frac{\varrho_0}{\frac{r}{r_s}\left(1 + \frac{r}{r_s}\right)^2} \tag{41}$$

where the free parameter r_s is a scale radius chosen to fit the galaxy concerned. Integration of this spherically symmetric profile yields for the enclosed mass

$$M(r) = \text{const} \left(\ln\left(1 + \frac{r}{r_s}\right) - \frac{(r/r_s)}{\left(1 + \frac{r}{r_s}\right)} \right) . \tag{42}$$

An illustrative mass profile for this model is shown in Fig. 32 as the dashed line; for the Milky Way halo, a suitable scale radius is evidently near $r_s \sim 7$ kpc. At large radius, the NFW model profile falls off as $\varrho \sim r^{-3}$ and the enclosed mass $M(r)$ keeps growing logarithmically with r. Whether or not the real Milky Way halo genuinely agrees with this trend, or whether there is a steeper cutoff in the density profile past $R_{\rm gc} \sim 60$ kpc, is impossible to say at present. Some guidance may be obtained from measurements of the halo surface density of M31 along its minor axis by Pritchet & van den Bergh (1994), who find that the halo light drops more steeply (more like $\varrho \sim r^{-5}$ at large projected radius) than the NFW model. If the same is true for the Milky Way, then we might anticipate that the halo mass 'converges' somewhere near $8 \times 10^{11}\ M_\odot$.

A promising avenue to determining the total mass of the Galaxy more accurately would be to obtain true three-dimensional space motions for the outermost satellites. Absolute proper motions have been obtained for the Magellanic Clouds (e.g., Kroupa & Bastian 1997a,b; Jones et al. 1994), leading to a total mass estimate for the Galaxy (Lin et al. 1995) of $\sim 5.5 \times 10^{11}\ M_\odot$ within 100 kpc, consistent with what we have derived here. However, similar data for several of the much more distant ones will be needed for a reliable answer to emerge.

This rather frustrating state of uncertainty is where we will have to leave the subject.

5 The Milky Way System: The Luminosity Function and Mass Distribution

Generally, researchers don't shoot directly for a grand goal ... [but] our piecemeal efforts are worthwhile only insofar as they are steps toward some fundamental question.

Martin Rees

The *integrated luminosity* of a globular cluster is the cumulative light of all its stars, and as such it is a direct indicator of the total cluster mass as long as we know the total mass-to-light ratio. We will see later (Sect. 8) that the distribution of cluster masses is a major clue to understanding their process of formation. In this section, we will investigate what this distribution looks like, and see how it encouraged the first outward steps to comparisons of globular cluster systems in other galaxies.

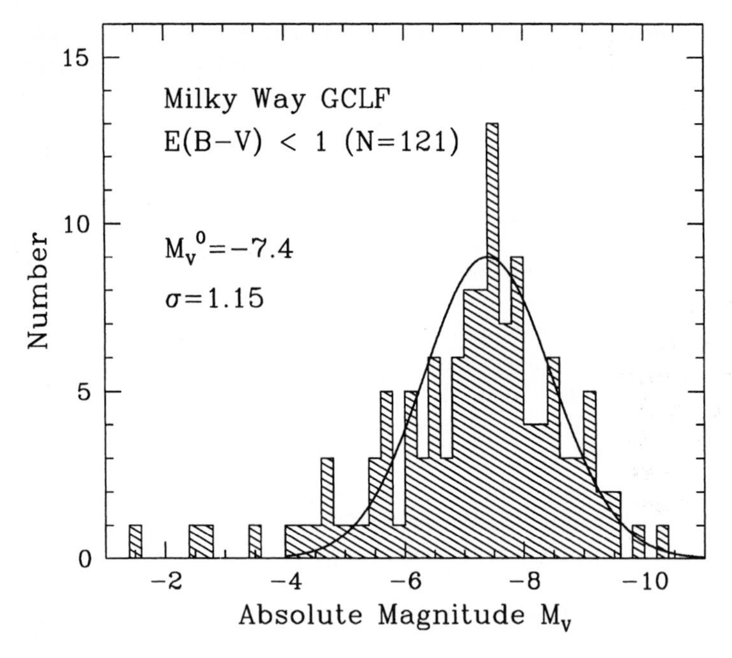

Fig. 33. Number of globular clusters per 0.2-magnitude bin, for the Milky Way. A Gaussian curve with mean $M_V = -7.4$ and standard deviation $\sigma = 1.15$ mag is superimposed to indicate the degree of symmetry of the distribution

5.1 Defining the GCLF

For a whole ensemble of globular clusters, the relative number *per unit magnitude* is called the globular cluster luminosity function or GCLF.[9] This distribution is plotted in Fig. 33 for 121 Milky Way clusters with moderately reliable data (Harris 1996a; of 140 clusters with measured total magnitudes and distances, we reject 19 with extremely high reddenings). Formally, the total cluster magnitude M_V^T is independent of foreground absorption, as it is simply the difference between the apparent total magnitude and apparent distance modulus, $M_V^T = V^T - (m-M)_V$. However, the quality of both V^T and the measured distance obviously degrades with increased reddening, for the reasons discussed in Sect. 1. We note in passing that the zeropoint of the luminosity scale varies directly with the adopted zeropoint of the RR Lyrae distance scale ($M_V(\mathrm{HB}) = 0.50$ at [Fe/H] $= -2.0$; see Sect. 2), but the overall shape of the distribution is nearly independent of the [Fe/H] coefficient of the distance scale since clusters of all metallicities lie at all luminosities. For example, arbitrarily adopting $\alpha = 0$ for the slope of the RR Lyrae luminosity

[9] Do not confuse the GCLF with the number of stars per unit magnitude within one cluster, which is the *stellar* luminosity function.

calibration would change the peak point of the GCLF by less than 0.05 mag and have negligible effect on the standard deviation.

The GCLF is strikingly simple: unimodal, nearly symmetric, and rather close to a classic Gaussian shape. Very luminous clusters are rare, but (perhaps counterintuitively) so are faintest ones. The GCLF that we see today *must* be a combined result of the initial mass spectrum of cluster formation, and the subsequent $\gtrsim 12$ Gigayears of dynamical evolution of all the clusters within their parent galaxy. Do the features of the distribution depend in any obvious way on other observable quantities, or on which galaxy they are in? Is it a "universal" function? There are two immediate reasons for raising these questions: the first, which we will take up again in Sect. 7, is to use the GCLF as a standard candle for extragalactic distance determination. The second and more astrophysically important reason, which we will explore further in Sects. 8 and 9, is to investigate how globular clusters form.

5.2 Correlations Within the Milky Way

An obvious starting point for the Milky Way clusters is to compare the GCLFs for the two major subpopulations (MRC and MPC). These are shown in Fig. 34. At first glance, the MRC distribution is the broader of the two, favoring fainter clusters a bit more once we take into account the different total numbers. (An earlier comparison with less complete data is made by Armandroff 1989.) But an obvious problem with such a statement is that we are comparing groups of clusters with rather different *spatial distributions*: all the MRC clusters are close to the Galactic center, but roughly half the MPC clusters are in the mid-to-outer halo, where they have had the luxury to be relatively free of dynamical erosion due to bulge shocking, dynamical friction, or even disk shocking. If we compare clusters of both types in the same radial zone, we obtain something like what is shown in Fig. 35. Excluding the thinly populated low-luminosity tail of the distribution ($M_V^T > -6$) and using only the clusters within $R_{gc} < 8$ kpc, we find that the GCLFs are virtually identical. In short, we have little reason to believe that cluster luminosity depends strongly on metallicity.

We have a much better *a priori* reason to expect the GCLF to depend on Galactocentric distance, because the known processes of dynamical erosion depend fairly sensitively on the strength of the Galactic tidal field. Thus in an idealized situation where all clusters were born alike, they would now have masses (luminosities) that were clear functions of location after enduring a Hubble time's worth of dynamical shaking. Other factors will, of course, enter to confuse the issue (the degree of orbital anisotropy might also change with mean R_{gc} and thus modify the shocking rates; and in the absence of any quantitative theory of cluster formation, we might also imagine that the mass spectrum at formation could depend on location within the protoGalaxy).

The distribution of luminosity M_V^T with location is shown in Fig. 36. Two features of the distribution are apparent to eye inspection: (a) the average

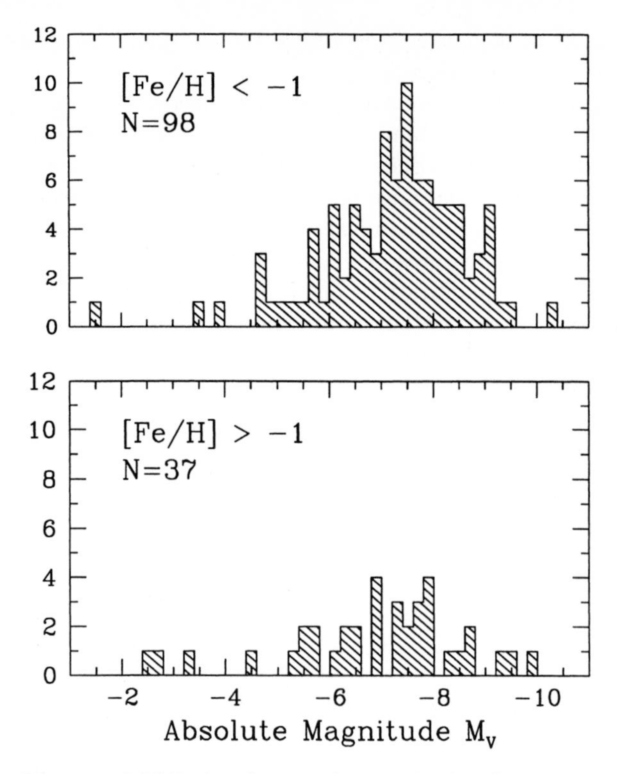

Fig. 34. GCLFs for the metal-poor clusters (*upper panel*) and metal-rich clusters (*lower panel*) in the Milky Way. There are five additional clusters not shown here which have measured absolute magnitudes, but unknown metallicities

luminosity does change with location, rising gradually to a peak somewhere around $\log R_{gc} \sim 0.9$ and then declining again further outward; and (b) there is a progressive outward "spreading" of the distribution over a larger M_V^T range. These trends are roughly quantified in Table 9, where the mean and standard deviation of the GCLF are listed for six rather arbitrarily selected radial bins. The last line gives the resulting parameters for the entire combined sample excluding only the faintest few clusters ($M_V^T > -4.5$; cf. the Gaussian curve with these parameters in Fig. 33). Other recent discussions of the observations are given by Kavelaars & Hanes (1997), Gnedin (1997), and Ostriker & Gnedin (1997), who also find radial trends in the GCLF peak at the $\pm 0.2 - 0.3$ magnitude level and assert that these are likely to be due to differences in their rates of dynamical evolution (assuming, of course, a similar initial mass spectrum).

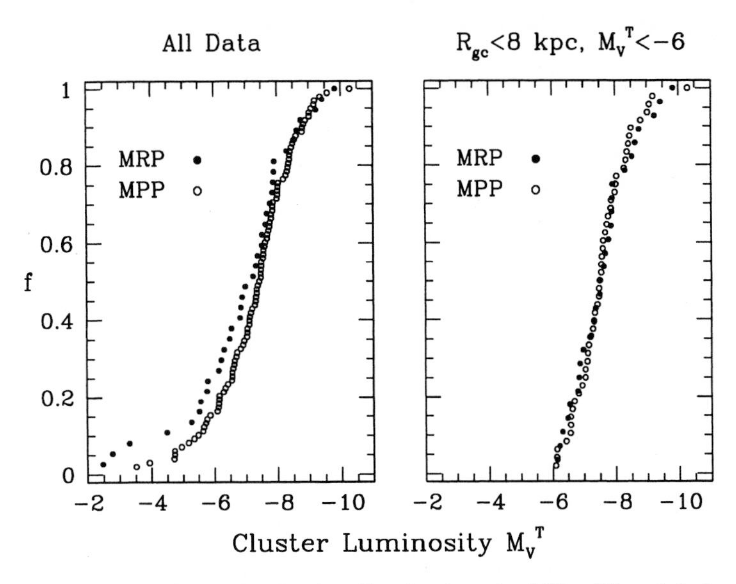

Fig. 35. Cumulative luminosity distributions for Milky Way globular clusters, separated by metallicity. The metal-rich population (here labelled MRP) is shown as the *solid dots*, and the metal-poor population (MPP) as *open circles*. The fraction of the total in each group is plotted against M_V^T. *Left panel:* all clusters with known luminosities are included. The difference between the two distributions is mainly in the low-luminosity tail. *Right panel:* Clusters within 8 kpc of the Galactic center and more luminous than $M_V = -6$. No difference is apparent

Table 9. GCLF parameters versus Galactocentric distance

R_{gc} Range	N	$\langle M_V^T \rangle$	$\sigma(M_V)$
0 – 2 kpc	26	-7.28 ± 0.18	0.93 ± 0.12
2 – 5 kpc	40	-7.41 ± 0.17	1.08 ± 0.12
5 – 8 kpc	20	-7.46 ± 0.27	1.21 ± 0.18
8 – 15 kpc	17	-7.97 ± 0.22	0.91 ± 0.15
15 – 42 kpc	22	-7.06 ± 0.24	1.11 ± 0.16
> 60 kpc	6	-5.91 ± 0.76	1.86 ± 0.52
All R_{gc}	131	-7.40 ± 0.11	1.15 ± 0.08

5.3 Dynamical Effects

Any star cluster has a limited lifetime. Even if it is left in isolation, the slow relaxation process of star-star encounters will eject individual stars, while the more massive stars (and binaries) will sink inward in the potential well, driving a net energy flow outward and a steady increase in central concentration.

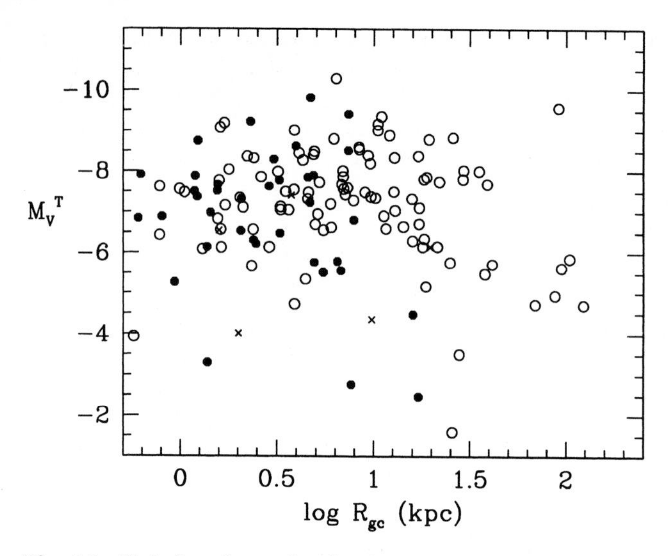

Fig. 36. Globular cluster luminosity versus location in the Galaxy, plotted as absolute integrated magnitude M_V^T versus log R_{gc}. MRC clusters are *filled symbols*; MPC clusters are *open symbols*; and clusters with unknown metallicities are *crosses*

The existence of a tidal cutoff imposed by the Galaxy will serve to enhance the process of stellar evaporation, and move the cluster more rapidly toward core collapse and eventual dissolution. An early phase of cluster evolution that is not yet well understood is its first $\sim 10^7 - 10^8$ y, during which its initial population of high-mass stars evolves. In this stage, the residual gas left after cluster formation as well as most of the material in the massive stars will be ejected through stellar winds and supernova, thus expanding the cluster, enhancing tidal losses, and (possibly) leading it down the road to rapid disruption.

Closer in to the Galactic center, other mechanisms will become much more important. If a cluster is on a high-eccentricity orbit, the impulsive shock received near perigalactic passage will pump energy into its stellar orbits, hastening the overall expansion of the outer envelope of the cluster and thus its loss of stars to the field. In a disk galaxy, clusters on high-inclination orbits will feel similar impulsive shocks, with similar destructive results; while clusters that stay completely within the disk may encounter shocks from giant molecular clouds (GMCs) that have masses similar to theirs. Lastly, if a cluster already relatively close to the galactic center finds itself unluckily in a rotating bar-like potential, it can be forced into a chaotic orbit which can at some point take it right through the nucleus of the galaxy and thus dissolve it at a single stroke (Long et al. 1992). Clusters that are less massive, or structurally more diffuse, are more subject to these strong kinds of tidal disruptions.

Very close in to the galactic center, the classic effect of dynamical friction also takes hold (Tremaine et al. 1975). As the cluster moves through the galactic bulge, the field stars are drawn towards it as it passes through. A slight density enhancement of stars is created just behind the cluster, and this acts as a slow gravitational brake on its motion, causing the cluster orbit to spiral in to eventual destruction in the nucleus. This effect works faster on more massive clusters, and it is likely that any clusters (or dwarf galaxies) with $M \gtrsim 10^8$ $M_\odot$ would not have survived if they passed anywhere within the disk or bulge of the Galaxy (cf. the references cited below).

The long list of erosive mechanisms makes the galaxy sound like a dangerous place indeed for a star cluster to live. Clearly all of these processes are operating simultaneously, but the strength and rapidity of each of them depends strongly on the cluster's location within the halo or disk, its orbital eccentricity, and its mass and initial structure. The overall concept that the clusters we see today are the hardiest "survivors" of some larger original population was introduced by Fall & Rees (1977) and has dominated the thinking in this subject since then. Ideally, we would like to perform a comprehensive numerical simulation of an entire GCS in which *all* the various mechanisms are treated realistically over 10^{10} y, and thus find out what types of initial cluster distributions lead to GCSs like the one we see today. This is still a formidable task, not just because of sheer demands on computing power, but also because the initial conditions are poorly known. (What was the original distribution of cluster masses and structures? How important are possible differences in the stellar mass function (IMF) at formation? What was the initial space distribution of clusters like, and what about their starting distribution of orbits?) The potential range of parameter space is huge, and must eventually be linked to a complete theory of cluster formation.

Nevertheless, steady progress has been made, to the point where several excellent attempts at evolutionary syntheses have now been made. Recent examples include Capriotti & Hawley (1996); Gnedin & Ostriker (1997); Murali & Weinberg (1997a,b,c); Vesperini (1997, 1998); Vesperini & Heggie (1997); Baumgardt (1998); and Okazaki and Tosa (1995), among others. These papers are extremely valuable for illuminating the relative importance of the various mechanisms at different places in the host galaxy.

- At short distances ($R_{gc} \lesssim$ 1 to 2 kpc), dynamical friction will remove the most massive clusters. At larger distances, dynamical friction becomes negligible for the typical globular clusters we see today.
- In the bulge and inner halo regions ($R_{gc} \lesssim 6$ kpc), bulge shocking and disk shocking are the dominant effects, acting particularly to destroy lower-mass clusters ($M \lesssim 10^5$ $M_\odot$) and to partially erode higher-mass ones.
- At still larger distances out in the halo, all the processes generally weaken to timescales longer than a Hubble time, and the slow mechanism of tidal evaporation becomes, somewhat by default, the dominant one.

How much of the current GCLF shape has been *produced* by sheer dynamical evolution is still unclear. It is reasonable to suspect that many more clusters at low masses must have existed initially, but that the higher-mass ones have been more immune to removal. Interestingly, an initial power-law mass distribution function such as $dN/dM \sim M^{-2}$, with many low-mass objects, evolves quickly into something resembling the traditional Gaussian in $(dN/d\log M)$, and once the GCLF acquires this symmetric Gaussian-like form, it seems able to maintain that form for long periods (cf. Vesperini 1998; Murali & Weinberg 1997c; Baumgardt 1998). Most of the lowest-mass objects ($M \lesssim 10^4$ $M_\odot$) are disrupted even if they were initially present in large numbers. At higher masses, individual clusters evolve to lower mass and thus slide downward through the GCLF, rather slowly for high-mass clusters and more quickly for the smaller objects at the low-mass tail. But the shape of the overall distribution and the critical parameters (turnover point and dispersion) change rather slowly. Much more work remains to be done with different initial conditions to see how robust these first predictions are.

5.4 Analytic Forms of the GCLF

Since the GCLF was first defined observationally, a variety of simple functions have been applied to it for interpolation purposes. By far the most well known of these is the Gaussian in number versus magnitude,

$$\phi(m) = \frac{1}{(2\pi)^{1/2}\sigma} e^{-(m-m_0)^2/2\sigma^2} \tag{43}$$

where $\phi(m)$ reaches a peak at the *turnover point* m_0 and has a *dispersion* σ. (*NB:* Strictly defined, ϕ is the *relative* number of clusters – the number per unit magnitude, as a fraction of the total cluster population over all magnitudes – so that $\int \phi(m)dm = 1$. Often, however, it is plotted just as number of clusters per magnitude bin, in which case the proportionality constant in front of the Gaussian exponential must be renormalized.)

It must be stressed that this "Gaussian paradigm" has no physical basis other than the rough evolutionary scenarios sketched out above. It was adopted – rather informally at first – as a fitting function for the Milky Way and M31 globular cluster systems during the 1970's by Racine, Hanes, Harris, de Vaucouleurs and colleagues; it was finally established as the preferred analytical fitting function in two papers by Hanes (1977) and de Vaucouleurs (1977). Later, it became clear that the globular clusters in other galaxies consistently showed the same basic GCLF features that we have already discussed for the Milky Way, and the Gaussian description continued to be reinforced. There is, however, no astrophysical model behind it, and it remains strictly a descriptive function that has only the advantages of simplicity and familiarity.

Another analytic curve that is just as simple as the Gaussian is the t_5 function (Secker 1992),

$$\phi(m) = \frac{8}{3\sqrt{5}\pi\sigma}\left(1 + \frac{(m-m_0)^2}{5\sigma^2}\right)^{-3} . \tag{44}$$

Like the Gaussian, t_5 is a symmetric function with two free fitting parameters (m_0, σ) but differs slightly from the Gaussian primarily in the wings of the distribution. Secker's objective tests with the Milky Way and M31 GCLFs showed that t_5 is slightly superior to the Gaussian in providing a close match to the data. Applications to GCLFs in giant elliptical galaxies have also shown the same thing (Secker & Harris 1993; Forbes 1996a,b; Kissler et al. 1994). The Gaussian model, however, has the strong advantage of familiarity to most readers.

More careful inspection of the full GCLF reveals, of course, that it is *not* symmetric about the turnover point – the long tail on the faint end of the distribution has no equivalent on the bright end, and cannot be matched by a single Gaussian or t_5 curve. To take account of these features, other authors have tried more complex polynomial expansions. Abraham & van den Bergh (1995) introduced

$$h_j = const \sum_{i=1}^{N} e^{-x_i^2/2} H_j(x_i) \tag{45}$$

where H_j is the jth-order Hermite polynomial, $x_i = (m_i - m_0)/\sigma$, and σ is the Gaussian dispersion. Here (h_3, h_4) are significantly nonzero for (e.g.) the Milky Way GCLF and represent the skewness and non-Gaussian shape terms. Still another approach is used by Baum et al. (1995, 1997), as an attempt to reproduce the combined Milky Way and M31 GCLFs. They adopt an asymmetric hyperbolic function

$$\log \frac{n}{n_0} = -(a/b)\,\left(b^2 + (m-A)^2\right)^{1/2} + g(m-A) \tag{46}$$

where A is nearly equal to the turnover point m_0 for mild asymmetry, and a, b, g are parameters to be determined by the (different) bright-end and faint-end shapes of the GCLF.

It is trivially true that analytic curves with more free parameters will produce more accurate fits to the data. But none of these functions leads to any immediate insight about the astrophysical processes governing the cluster luminosities and masses, and we will not pursue them further.

5.5 The LDF: Power-Law Forms

In all of the fitting functions mentioned above, the GCLF is plotted as the number of clusters per unit *magnitude*. This way of graphing the function (dating back at least to Hubble's time) turned out to be a historically unfortunate decision. A more physically oriented choice would have been to plot the number per unit *luminosity* (or equivalently, the number per unit mass once we multiply by the mass-to-light ratio). We will call this latter form the *luminosity distribution function* or LDF. If the masses of globular clusters are governed by a fairly simple process of formation, then we could reasonably expect on physical grounds that the LDF might look like a simple *power law* in number per unit mass. Power-law distributions of mass or size, of the differential form $dN/dM \sim M^{-\alpha}$, are produced by many phenomena such as turbulence spectra, accretional growth, impact cratering, and so forth; and we will see strong evidence later in our discussion that the same result is true for star clusters that have recently formed in active galaxies.

Early suggestions that the LDF of globular clusters did in fact obey a power-law form were made by Surdin (1979) and Racine (1980) but these were unfortunately ignored for many years. The point was rediscovered by Richtler (1992). The power-law formulation of the LDF was pursued further by Harris & Pudritz (1994) and connected for the first time to a specific physical model of cluster formation in giant gas clouds (resembling Searle-Zinn fragments; we will return to a discussion of formation modelling in Sects. 8 and 9).

How does the LDF relate to our more familiar GCLF? Considerable confusion between them persists in the literature, even though the difference between them is mathematically trivial. The GCLF histogram uses bins in magnitude ($\log L$), whereas the LDF uses bins in L itself. Obviously, equal intervals in $\log L$ are not equal intervals in L, so the number distribution per bin has a different shape in each case. The power-law LDF is essentially a plot of $N(L)dL$, whereas the Gaussian-like GCLF $\phi(M_V)$ is a plot of $N(\log L)d(\log L)$. Thus the two functions scale as

$$\phi(M_V) \sim \frac{dN}{d\log L} \sim L\frac{dN}{dL} \sim L^{1-\alpha} .$$

The relation between the two forms is discussed in detail by McLaughlin (1994).

A power-law fit to the Milky Way LDF is shown in Fig. 37, adapted from Harris & Pudritz (1994). An exponent $\alpha_1 = 1.8 \pm 0.2$ provides an entirely acceptable fit for almost the entire observed range of cluster masses. It is only over the lowest $\sim$ 10% (i.e. $M \lesssim 10^5$ $M_\odot$) that the curve predicts far more clusters than are actually present. For this lower mass range, the data follow a much shallower power law near $\alpha_0 \sim 0.2 \pm 0.2$, at least in the Milky Way and M31 (Harris & Pudritz 1994; McLaughlin 1994). It is precisely at this rather abrupt slope change between α_0 and α_1 at $\sim 10^5$ $M_\odot$ that the classic turnover point of the GCLF lies. About half the numbers of clusters are fainter than that point, and about half are brighter. But it is empirically true that the GCLF $\phi(M_V)$ is roughly *symmetric* about the turnover point. In that case, we then require the exponents α_0, α_1 to be related by $(1 - \alpha_0) \simeq (\alpha_1 - 1) \simeq 0.8 \pm 0.2$.

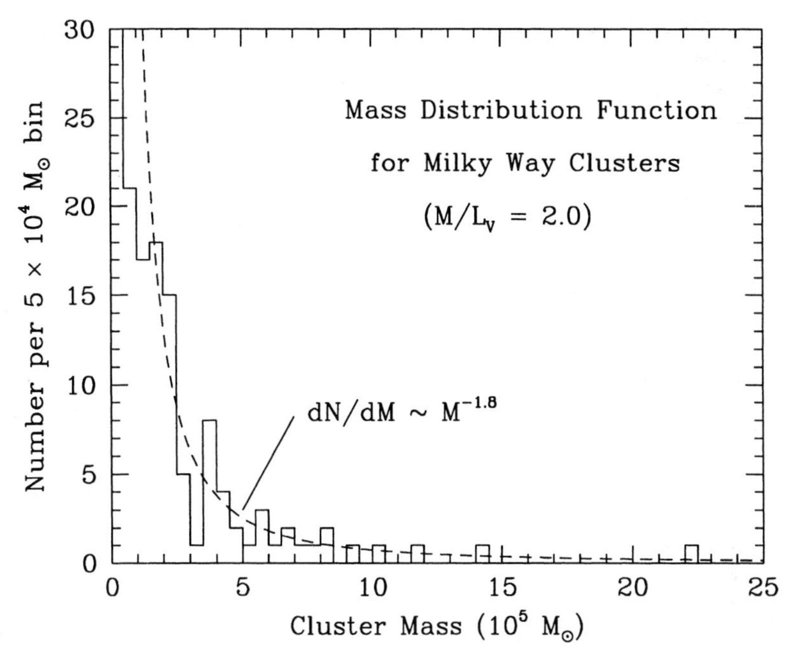

Fig. 37. Number of globular clusters per unit mass, plotted in linear form. The directly observed cluster luminosity L_V in solar units, given by $L_V = 10^{0.4(M_{V,\odot}-M_V^T)}$ has been converted to mass $M/\mathrm{M}_\odot$ with constant mass-to-light ratio $M/L_V = 2.0$. A power law $dN/dM \sim M^{-1.8\pm0.2}$ matches the data for $M \gtrsim 2 \times 10^5\ \mathrm{M}_\odot$ but continues upward too steeply for lower masses

As noted earlier, we should expect that the LDF (or GCLF) as we see it today is a relic of *both* the mass spectrum of cluster formation and the subsequent dynamical evolution of the system. If we look at the distinctive and simple shape of the LDF – a double power law with a fairly sharp transition at $\sim 10^5\ \mathrm{M}_\odot$ – it is natural and extremely tempting to speculate that the *upper* mass range (α_1) represents the mass distribution laid down at formation, while the *lower* range (α_0) is created by the long-term effects of dynamical evolution, which should gradually carve away the more vulnerable lower-mass clusters. However, there are hints from observations of young globular clusters in interacting and merging galaxies that the initial mass distribution *already* has a shallow slope at the low-mass end right from the start; see Sect. 9 below. We are still some crucial steps away from understanding the correct full set of initial conditions for the LDF.

If this rough scenario of formation combined with evolution is indeed on the right track, then we should expect that large numbers of low-mass clusters have been destroyed over the past Hubble time. But does that mean that a large fraction of the Galactic halo was built from dissolved globular clusters? No! The important point here is that these small clusters contain a *small fraction of the total mass* in the whole cluster system, even if they are numerous. The big clusters simply outweigh them by large factors. (For example, ω Centauri, the most luminous cluster in the Milky Way, contains about 8% of the total GCS mass all by itself. The clusters more massive than 10^5 $M_\odot$ contain almost 95% of the total.) Harris & Pudritz (1994) demonstrate that if the original mass distribution followed the α_1 slope all the way down to low mass, then only about 30% of the total mass of the GCS would be lost if *every* cluster less massive than the present-day turnover point were destroyed.

Accounting for the fact that the clusters at all masses will be at least partially damaged by dynamical erosion (cf. the references cited above), it is reasonable to guess that roughly half the total original mass in the GCS has been escaped from the clusters to join the field halo population (see also Gnedin & Ostriker 1997; Murali & Weinberg 1997c; Vesperini 1998, for similar mass ratio estimates from dynamical simulations). The field halo stars outnumber the cluster stars by roughly 100:1, so most of the field stars must have originated in star-forming regions that never took the form of bound star clusters.

5.6 Comparisons with M31

A crucial step in constraining formation models for globular clusters in the early protogalaxies – as well as dealing with the more practical matter of using them as standard candles – is to start comparing the Milky Way GCS with those in other galaxies. The first and most obvious test is with M31, as the nearest large galaxy reasonably similar to ours.

M31 contains a globular cluster population at least twice as large as the Milky Way's, and thus more than all the other Local Group galaxies combined. The first $\sim$ 150 were found by Hubble (1932) in a photographic survey. Major additions to the list of globular cluster candidates were made in subsequent photographic surveys by Baade (published in Seyfert & Nassau 1945), Vetešnik (1962), Sargent et al. (1977), Crampton et al. (1985), and by the Bologna consortium in the early 1980's, leading to their complete catalog (Battistini et al. 1987) which lists $\sim$ 700 candidates of various quality rankings. About 250 - 300 of these are almost certainly globular clusters, while another $\sim$ 300 are almost certainly various contaminants; the remainder (mostly projected on the M31 disk) have not yet been adequately classified. The true GCS population total in M31 is thus probably around 300, but is simply not yet known to better than $\sim$ 30%.

The published surveys used a wide range of strategies including object image morphology, brightness and color, and low-resolution spectral characteristics. The object lists from the different surveys overlap considerably in their discoveries and rediscoveries of clusters, but also contain numerous contaminants. Projected on the disk of M31, there are also open clusters, compact HII regions, and random clumps of bright stars which can masquerade as globular clusters. In the vast halo beyond the disk area, the main interlopers are faint, distant background galaxies which may have the round and compact morphology of globular clusters. The published surveys cover the M31 region rather thoroughly out to projected distances of only $r \sim 20$ kpc; a few more remote clusters have been found (Mayall & Eggen 1953; Kron & Mayall 1960), but the outer halo – a huge projected area on the sky – has yet to be systematically searched. If the Milky Way situation is any indication, not many new remote ones can be expected. A new CCD imaging survey primarily covering the disk of M31 by Geisler et al. is underway, and promises to push the detection limits to the faint end of the GCLF.

The true globulars in M31 must be found one by one. The techniques that have proven most useful include:

- *Radial velocity* (Huchra et al. 1991): Since $v_r(\mathrm{M31}) = -350$ km s^{-1}, any objects with strongly positive v_r are certainly background galaxies or (occasionally) Milky Way halo stars.
- *Image structure* (Racine 1991; Racine & Harris 1992): high resolution imaging with a large telescope can produce a definitive classification: if a candidate object is resolved into stars, it is a cluster. Under $\sim 0\farcs5$ resolution with the CFHT, color-magnitude diagrams for several halo clusters were successfully achieved which showed directly for the first time that the M31 clusters were indeed "normal" globulars like the familiar ones in the Milky Way (Heasley et al. 1988; Christian & Heasley 1991; Couture et al. 1995). With the greater resolving power of the *HST*, deeper and more precise CMDs for the M31 clusters have now become possible (Rich et al. 1996; Fusi Pecci et al. 1996; Holland et al. 1997).
- *Color indices*: The integrated colors of globular clusters occupy a fairly narrow region in a two-color plane such as $(U-B, B-V)$, $(B-V, V-R)$, etc. Other types of objects – galaxies particularly – usually have very different colors and can be eliminated. Reed et al. (1992, 1994) show that two-thirds of the background galaxies can be cleanly rejected this way.

Using a combination of all three of these techniques, Reed et al. (1994) constructed an almost completely clean sample of halo clusters (not projected on the M31 disk, thus free of internal reddening differences) which provides the best database we have for comparison with the Milky Way GCS.

In most other respects – metallicity distribution and kinematics – the M31 globular cluster system presents much the same story as does the Milky Way system, with small differences. Huchra et al. (1991) define MRC and MPC populations separated at [Fe/H] $= -0.8$ which strongly resemble the analogous ones in the Milky Way. The MRC is the more centrally concentrated of the two, and has a healthy overall rotation reaching ~ 150 km s^{-1} at a projected radius in the disk $r \simeq 5$ kpc. The MPC is more spatially extended and has much smaller net rotation ($\lesssim 50$ km s^{-1}). It is, however, also true that the overall scale size of the M31 system is bigger than that of the Milky Way, and moderately metal-rich clusters can be found at surprisingly large distances (the luminous cluster G1 = Mayall II is the prime example, with

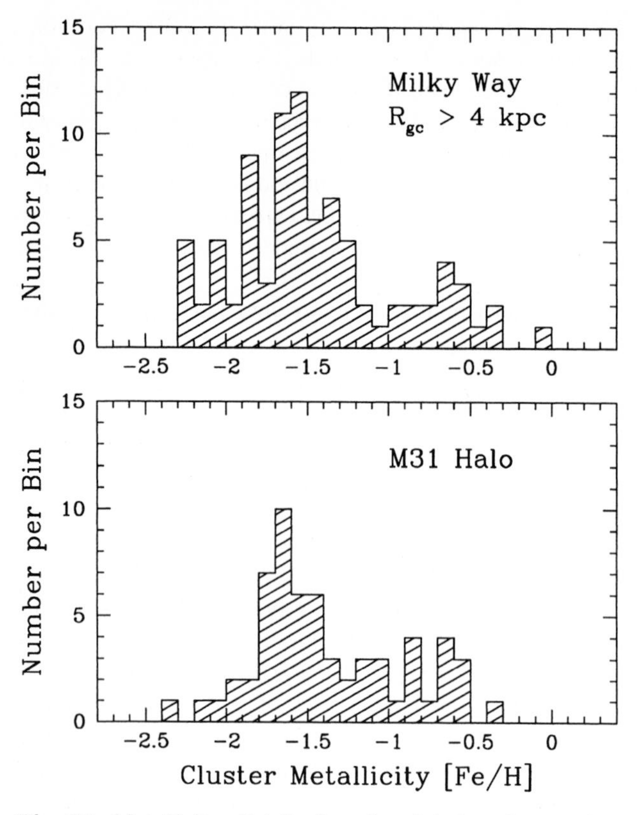

Fig. 38. Metallicity distribution for globular clusters in two galaxies. (**a**) In the *upper panel*, the distribution is shown for the Milky Way clusters with $R_{gc} > 4$ kpc. (**b**) In the *lower panel*, the distribution for the M31 halo is shown, with data drawn from Reed et al. (1994). The M31 sample is only slightly more weighted to MRC objects

[Fe/H] ~ -1 at a projected distance of 40 kpc). The *relative* numbers of clusters at different metallicities, at least in the halo, are not strongly different from the [Fe/H] distribution in the Milky Way (see Fig. 38). Huchra et al. (1991) find relatively more metal-rich objects overall, including those in the disk, but it is easy to recognize the same bimodal form of the metallicity distribution, broadened by observational scatter.

There is still much work to be done to understand the characteristics of the M31 GCS at the same level of detail as we have for the Milky Way, and readers should refer to the papers cited above for further discussion.

The first GCLF comparison between M31 and the Milky Way was done by Hubble himself, in his 1932 discovery paper. At that time, the standard distance modulus in use for M31 was $(m-M) = 22$, almost three magnitudes smaller than today's best estimates. To make matters worse, the adopted luminosity for the RR Lyrae stars then was $M_{pg} \simeq M_B = 0.0$, almost a magnitude brighter than today's calibrations. In other words, the M31 clusters were being measured as much too faint in absolute magnitude, the Milky Way clusters too bright, and the combination left almost no overlap between the two GCLFs. Hubble relied on the similar *form* of the distributions as much as on their luminosity levels, and successfully concluded that "among known types of celestial bodies, the objects in M31 find their closest analogy in globular clusters".

The next serious GCLF comparison was in the landmark paper of Kron & Mayall (1960), which presented a comprehensive new set of integrated magnitudes and colors for globular clusters in several Local Group galaxies. By then, the basic distance scale issues in the Local Group had been settled (at least, the M31 discrepancy had been reduced to ~ 0.5 mag, rather than the 2.5-mag difference used during Hubble's time), and the true similarity between the globular cluster systems in M31 and the Milky Way clearly emerged. It was, by that time, also evident that the peak point M_V^0 of the GCLF was not just an artifact of incomplete observations, but was a real feature of the GCSs. Kron & Mayall's paper represents the *first explicit use of the GCLF turnover point as a standard candle for distance determination.* In all respects it is the same approach as we use today (Jacoby et al. 1992).

What does the M31/Milky Way comparison look like in modern terms? Unfortunately, we have to restrict our match to the halos of each, since the available list of objects projected on the disk of M31 is still too ill-defined (it is too contaminated with non-globulars, too incomplete at faint magnitudes, and photometrically too afflicted with random errors and poorly determined differential absorption). However, the halo sample (Reed et al. 1994) gives us an excellent basis for comparison: it is clean, complete down to a magnitude level well past the turnover point, unaffected by differential reddening, and well measured by modern CCD photometry.

We first need to worry a bit more, though, about which part of the Milky Way system we should use for comparison. We have already seen (Fig. 36 and Table 9) that its GCLF parameters depend on location. Another way to display it, using the running mean approach defined in our kinematics discussion, is shown in Fig. 39. The average luminosity (and also the turnover point, which is the GCLF median) rises smoothly outward from the Galactic center to a maximum at $R_{\rm gc} \simeq 9$ kpc, then declines again. How can we make a valid comparison in the face of this amount of internal variation, which seems to vitiate the whole use of the GCLF as a standard candle?

But wait! The Milky Way is unique in the sense that it is the only galaxy for which we have the full three-dimensional information on the GCS space distribution. As an indicator of dynamical effects on the system or even small differences in the typical cluster mass at formation, Fig. 39 is of great interest on its own merits. But it should not be applied as it stands to any other galaxy. Instead, we should look at the Milky Way as if we were far outside

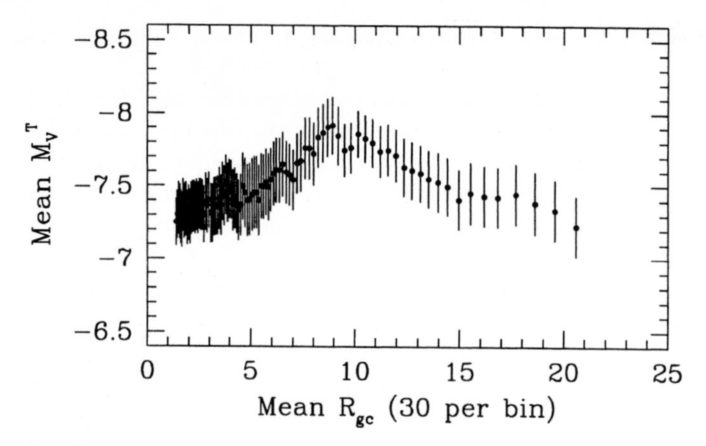

Fig. 39. Mean absolute magnitude for Milky Way globular clusters as a function of Galactocentric distance. Each point represents the mean $\langle M_V^T \rangle$ for the 30 clusters centered at the given distance; the next point outward is the same mean where the innermost cluster in the previous bin has been dropped and the next cluster outward has been added

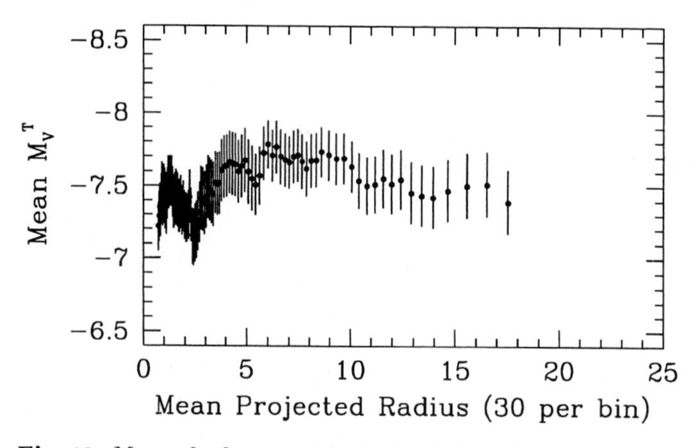

Fig. 40. Mean absolute magnitude for Milky Way globular clusters as a function of two-dimensional projected distance from the Galactic center, $r_p = (Y^2 + Z^2)^{1/2}$. Each point represents the mean $\langle M_V^T \rangle$ for the 30 clusters centered at the given distance; the next point outward is the same mean where the innermost cluster in the previous bin has been dropped and the next cluster outward has been added

it and could see only the distances of the clusters projected on the sky. The best projection to use is $r_p = \sqrt{Y^2 + Z^2}$, dropping the X-axis which is most affected by internal distance errors (see Sect. 1 above). When we do this, and again take running means of cluster luminosity, we get the result shown in Fig. 40. Rather surprisingly, we see that the large-scale global variation has largely been smoothed out, and the biggest part of it (the first outward rise) has been compressed down to the innermost ~ 2 kpc projected onto the Galactic bulge.

Somewhat arbitrarily, I will take the region $r_p > 3$ kpc (containing 75 clusters) as the fiducial Milky Way sample. If we were to view the Milky Way at the same inclination angle to the disk as we see M31, this cutoff in projected distance would correspond roughly to the inner distance limits in the M31 halo sample. Over this range, there is little variation in the Milky Way mean cluster luminosity, and we can be more encouraged to try it out as a standard candle. We have just seen that this uniformity is something of an illusion! Larger internal differences are being masked, or washed out, by the projection effect from three to two dimensions. But the Galaxy can hardly be unique in this respect. We must therefore suspect that *the same smoothing may well be happening for the GCLF in any large galaxy that we look at*, where much of the information on the true amount of internal variation with position has simply been lost. To my knowledge, this point has not been realized, or used, in any previous application of the Milky Way GCLF as a distance indicator.

Having set up the fairest comparison sample that we can manufacture from the Milky Way, we can finally match it up with M31. The result is shown in Fig. 41. These two GCLFs are remarkably similar. The M31 sample differs only in the lack of faint clusters ($M_V^T \gtrsim -5.5$), for which the existing surveys are incomplete. Fitting Gaussian or t_5 functions to both galaxies yields turnover levels of $M_V^0 = -7.68 \pm 0.14$ (for the Milky Way projected-halo sample) and $M_V^0 = -7.80 \pm 0.12$ (for the M31 halo sample). These numbers are not significantly different. Let us turn the argument around: if we had used the M31 GCLF to *derive* a distance modulus, we would have obtained $(m-M)_V = V_0(\text{M31}) - M_V^0(\text{Milky Way}) = (17.00 \pm 0.12) - (-7.68 \pm 0.14) = (24.68 \pm 0.18)$. Thus at the $\sim 0.10 - 0.15$ magnitude level of precision, the GCLF turnovers are similar and the method seems to work much better than we had any right to expect.

In Sect. 7, we will start testing this procedure for much more remote galaxies, and eventually become bold enough to derive the Hubble constant with it. Next, though, we need to take our first steps beyond the Local Group and find out what globular cluster systems look like in galaxies of very different types.

6 An Overview of Other Galaxies: Basic Parameters

First gather the facts; then you can distort them at your leisure.
Mark Twain

Globular cluster systems have now been discovered and studied to some degree in more than a hundred galaxies. These cover the entire range of Hubble types from irregulars to ellipticals, the luminosity range from tiny dwarf ellipt-

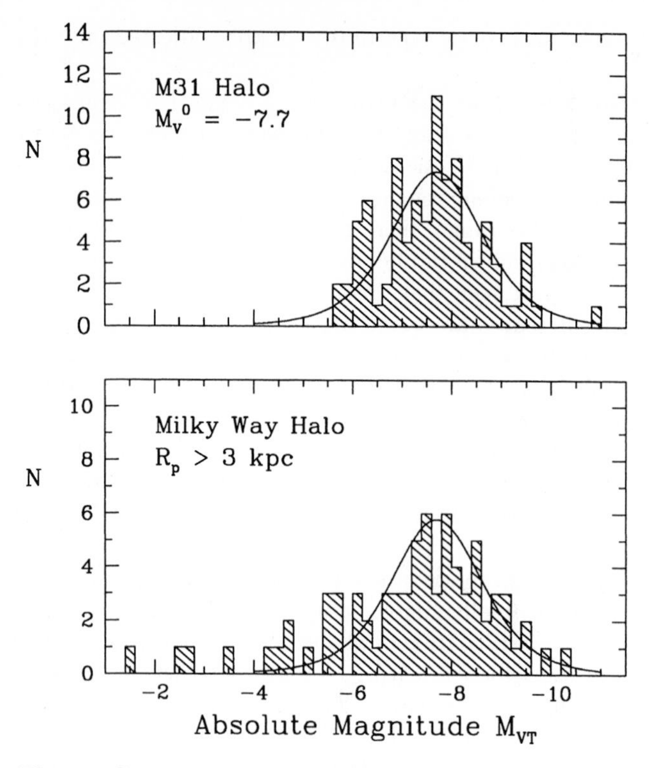

Fig. 41. Comparison of the GCLFs in M31 (*upper panel*) and the Milky Way (*lower panel*). The Milky Way sample is defined from clusters with projected (2-D) distances larger than 3 kpc as discussed in the text. The M31 halo sample, from Reed et al. (1994), has been shifted to absolute magnitude assuming our previously derived distance modulus $(m-M)_V = 24.80$ (Sect. 2)

icals up to supergiant cD's, and environments from isolated "field" galaxies to the richest Abell clusters.

With the best imaging tools we have at present (the *HST* cameras), individual globulars can be resolved into their component stars rather easily for galaxies within the Local Group, i.e. at distances $\lesssim 1$ Mpc. With increasing difficulty, resolution of clusters into stars can also be done for galaxies up to several Megaparsecs distant (a color-magnitude diagram has been obtained for a halo cluster in the giant elliptical NGC 5128 at $d \simeq 4$ Mpc; see G.Harris et al. 1998). But for still more remote galaxies, we see the presence of the globular cluster population only as an excess of faint, small objects concentrated around the galaxy center (Fig. 42).

Studying the GCSs in most galaxies then becomes more of a statistical business, with the genuine clusters seen against a background of "contaminating" field objects (usually a combination of foreground stars and faint, compact background galaxies). Identifying *individual* globulars one by one

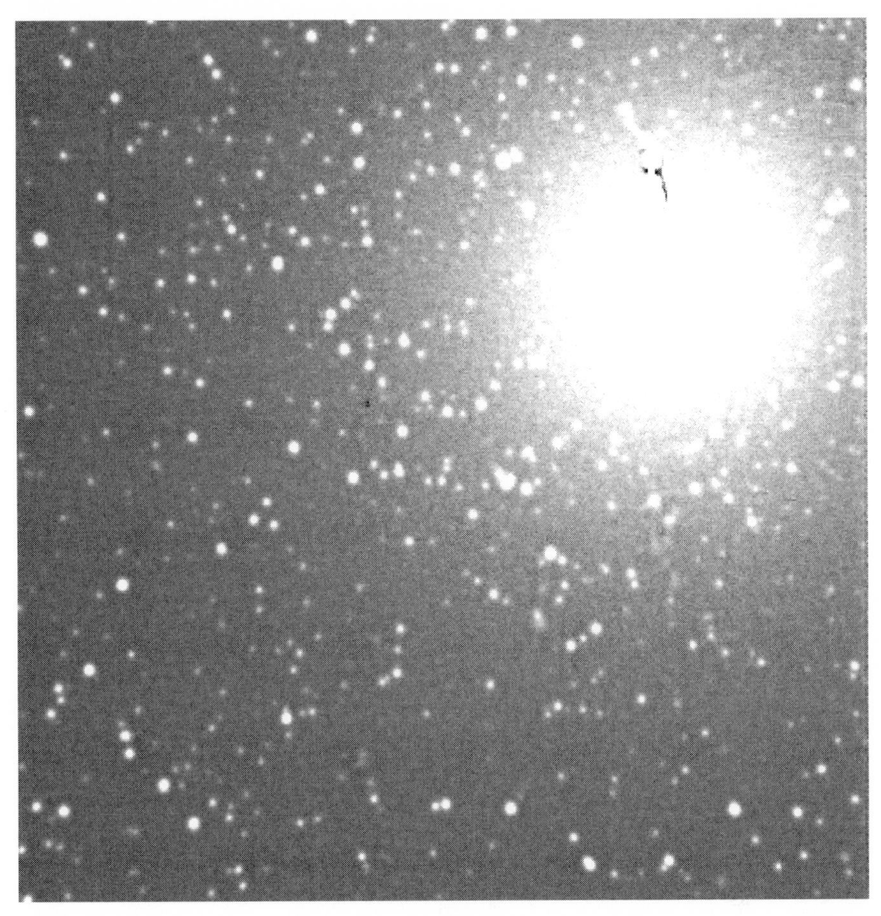

Fig. 42. A deep R-band image of the Virgo giant elliptical M87, with its swarm of globular clusters shown as the hundreds of faint starlike images around it. Traces of the nuclear jet can be seen extending upward from the galaxy center. The region shown is about 10 kpc on a side. This image was acquired with the High Resolution Camera at the Canada-France-Hawaii Telescope (see Harris et al. 1998a)

can be done – usually radial velocity measurement is a definitive separator when combined with the integrated magnitudes and colors – but only with observational efforts that are greatly more time consuming.

What characteristics of a GCS can we measure? The list of quantities, given below, is almost the same as for the Milky Way. Our only serious restriction is that the depth and quality of the information we can gather inevitably becomes more limited at larger distances.

Total populations: Simplest of all measurable quantities is the number of globular clusters present in the galaxy. Naively, we might expect that $N_{\rm cl}$ should rise in direct proportion to the galaxy luminosity (or mass). That seemed to be the case in the early days of the subject (Hanes 1977; Harris & Racine 1979), with the single exception of M87, which was recognized from the beginning to be anomalous, However, it is now realized that there is a great deal of real scatter from one galaxy to another around this basic proportionality relation. Understanding the total population size or global cluster formation efficiency is one of the most challenging questions in the entire subject, leading us far into issues of galaxy formation and evolution.

Metallicities: The normal assumption or "null hypothesis" of GCS work is that the old-halo clusters we see in other galaxies basically resemble the familiar ones in the Milky Way. This assumption has been fully borne out in the various Local Group members where detailed comparisons of stellar content have been possible. We can then use the integrated cluster colors in some reasonably sensitive index like $(V-I)$, $(C-T_1)$, etc., to estimate the *metallicity distribution function* (MDF) of the system, since metallicity determines the integrated color of old stellar systems much more strongly than other factors such as age. With quite a lot more effort, we can use the absorption line indices in their integrated spectra to do the same thing. Where it has been possible to do both, the color and spectral approaches for estimating metallicity agree well both at low dispersion (e.g., Racine et al. 1978; Huchra et al. 1991; Brodie & Huchra 1991) and in the finer detail that has been achieved more recently (e.g., Cohen et al. 1998; Jablonka et al. 1992, 1996).

Luminosities: The *luminosity distribution function* (LDF; see Sect. 5) of the GCS is the visible signature of the cluster mass distribution. As discussed earlier, the LDF we see today should be the *combined* result of the mass spectrum at formation, and the subsequent effects of dynamical evolution in the galactic tidal field.

Spatial distribution: The GCS in any galaxy is a centrally concentrated subsystem, generally following the structure of the visible halo light. However, particularly in giant ellipticals the GCS often traces a somewhat shallower radial falloff than the halo, and in extreme cases (cD galaxies) it may be closer to representing the more extended dark-matter potential well. Many recent studies have attempted to correlate the MDF and LDF with the spatial distribution, thus extracting more clues to the system formation and evolution.

Radial velocity distribution: In principle, much the same types of kinematic and dynamical studies of the Milky Way GCS can be carried out in any galaxy for which we can acquire a large enough set of cluster radial velocities. However, the internal precisions of the velocity measurements need to be $\sim \pm 50$ km s^{-1} for the internal dynamics of the halo to be adequately studied, and acquiring absorption-line velocities for large samples of objects as faint as those in Virgo and beyond has been difficult. With the advent of the new generation of 8-m and 10-m optical telescopes, this type of work has now been started in earnest by several groups.

6.1 Defining and Measuring Specific Frequency

The total population of clusters in a galaxy is usually represented by the *specific frequency* S_N, the number of clusters per unit galaxy luminosity (Harris & van den Bergh 1981; Harris 1991):

$$S_N = N_{\rm cl} \cdot 10^{0.4(M_V^T + 15)} \tag{47}$$

where M_V^T is the integrated absolute magnitude of the host galaxy and $N_{\rm cl}$ is the total number of clusters. This definition can be rewritten in terms of the visual luminosity of the galaxy L_V in solar units,

$$S_N = 8.55 \times 10^7 \, \frac{N_{\rm cl}}{(L_V/L_\odot)} \, . \tag{48}$$

Estimating S_N for a given galaxy is therefore a simple process in principle, but requires two kinds of completeness corrections. If the imaging coverage of the galaxy is spatially incomplete, then radial extrapolations have to be made to estimate $N_{\rm cl}$. Similarly, if the photometric limit reaches only part way down the GCLF (as is almost invariably the case), then an extrapolation in magnitude is also needed, starting with an assumed distance to the galaxy (Fig. 43). *By convention*, the GCLF shape is assumed to be Gaussian (see the previous section) for purposes of estimating the total population. In most galaxies, it is often the case that the faint limits of the observations turn out to be somewhere near the GCLF turnover.

There are two obvious ways in which the predicted value of $N_{\rm cl}$ can go wrong. If the fainter half of the GCLF has a very different shape from the brighter half that is directly observed, then we would end up miscalculating the total $N_{\rm cl}$. And if the limit of observations falls well short of even the turnover point, then the extrapolation from N(obs) to $N_{\rm cl}$ can be uncomfortably large even if the assumption of symmetry is valid. Thus it seems that the specific frequency is a rather uncertain number.

Or is it? The procedure is actually not as risky as it first looks, for two reasons:

- We calculate $N_{\rm cl}$ essentially by using the Gaussian-like shape of the GCLF to determine the number of clusters *on the bright half*, and then doubling it. In most galaxies we never see the faint half, and never use it. In other words, the specific frequency is really a ratio which compares the number of *bright* clusters in different galaxies. Thus the first rule of specific frequency is:

S_N *measures the number of clusters brighter than the GCLF turnover* V_0.

- Despite this reassurance, S_N would still be an invalid quantity if the *absolute* magnitude of the turnover differed wildly from one galaxy to the next – or, indeed, if there were no turnover at all. But by all available evidence (introduced in Sect. 5, and discussed further in Sect. 7 below),

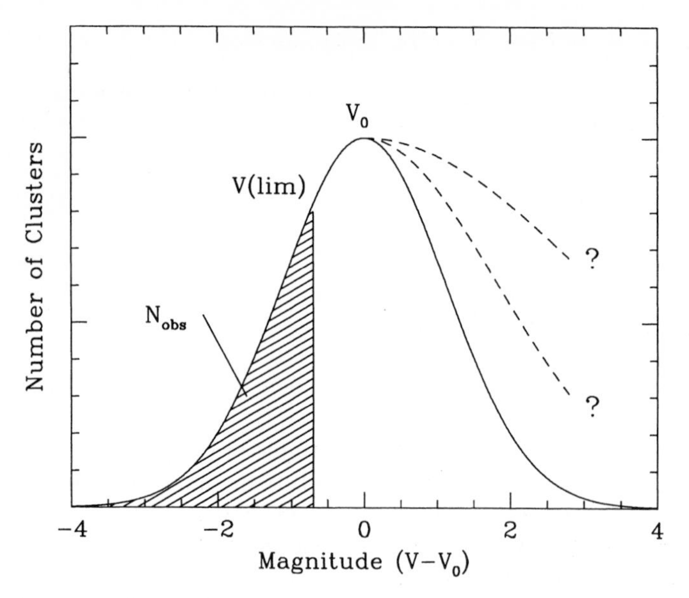

Fig. 43. Calculation of total cluster population. The GCLF is assumed to have a Gaussian-like form shown by the *solid curve*, with turnover point at apparent magnitude V_0. The limiting magnitude of the photometry is $V_{\rm lim}$, so that the total observed population of clusters $N_{\rm obs}$ is given by the *shaded area*. The total population $N_{\rm cl}$ over the entire GCLF is then $N_{\rm cl} = N({\rm obs})/F$ where the completeness fraction F is the shaded area divided by the total area under the Gaussian. If the unobserved faint half of the GCLF had a different shape (*dashed lines*), the total population estimate would be affected significantly, but the number of *bright* clusters (more luminous than the turnover) would not

the GCLF has amazingly similar parameters from place to place. Perhaps against all *a priori* expectations, the GCLF shape is the closest thing to a universal phenomenon that we find in globular cluster systems. Thus we have our second rule,

S_N provides a valid basis for comparison among galaxies because of the universality of the GCLF.

In summary, we can go ahead and use S_N knowing that it has reasonable grounding in reality.

The estimated S_N is fairly insensitive to the assumed galaxy distance d, because any change in d will affect both the calculated galaxy luminosity and total cluster population in the same sense (see Harris & van den Bergh 1981). However, it is sensitive to mistakes in the assumed limiting magnitude $V_{\rm lim}$, or in the background contamination level. Suppose that for a distant galaxy you count N faint starlike objects around the galaxy, and N_b background objects in an adjacent field of equal area down to the same limiting magnitude. By hypothesis, the excess $N_o = (N - N_b)$ is the globular cluster population, and the uncertainty is

$$N_o \pm \Delta_N = (N - N_b) \pm \sqrt{N + N_b}\ . \tag{49}$$

The total over all magnitudes is $N_{\mathrm{cl}} = N_o/F$. Now suppose that the uncertainty ΔV in the limit V_{lim} translates into an uncertainty ΔF in the completeness factor F: we can then show

$$\frac{\Delta S_N}{S_N} = \left(\frac{(N + N_b)}{(N - N_b)^2} + \left(\frac{\Delta F}{F} \right)^2 \right)^{1/2} . \tag{50}$$

Numerical trials with this relation show that to produce S_N estimates that are no more uncertain than (say) 20%, we need to have observations reaching $V_{\mathrm{lim}} \gtrsim (V_0 - 1)$, i.e. to within a magnitude of the turnover or fainter. If the raw counts are *very* dominated by background contamination, the situation may be worse. And if the observations fall short of the turnover by 2 magnitudes or more, the relative uncertainty $\Delta S_N / S_N$ starts increasing dramatically and the estimates become quite rough.

6.2 Specific Frequency: Trends and Anomalies

Let us now turn to some of the results for specific frequencies. Elliptical galaxies are the simplest to work with, and make up by far the biggest share of the database for globular cluster systems. Figure 44 shows the current results for E galaxies over all luminosities, from dwarfs to supergiant cD's (data are taken from the compilations of Blakeslee et al. 1997; Harris et al. 1998a; Miller et al. 1998; and a few recent individual studies).

From this simple graph we can already draw several conclusions. First, over a range of almost 10^4 in galaxy luminosity L, the mean specific frequency is nearly constant; that is, to first order the total number of clusters rises in nearly direct proportion to parent galaxy luminosity, $N_t \sim L$.

Second, there is *significant scatter* at all L. For giant ellipticals, S_N in individual galaxies ranges from a high near ~ 15 to a low near ~ 1 or perhaps even lower. For dwarf ellipticals, the range is even larger, with S_N(max) near 30. This scatter extends far beyond the internal uncertainties in estimating S_N, and must certainly be real. It was suspected to exist from the earliest samples of E galaxies (Hanes 1977; Harris & van den Bergh 1981), and later studies from more comprehensive samples and deeper photometry have only confirmed and extended the first estimates of the range in S_N that real galaxies exhibit. The specific frequency is a parameter which differs by as much as a factor of *twenty* between galaxies which have otherwise similar structures, luminosities, and metallicities. This is one of the most remarkable results to emerge from the study of globular cluster systems. Although some plausible ideas are beginning to emerge (Sect. 8), it still lacks a compelling theoretical explanation.

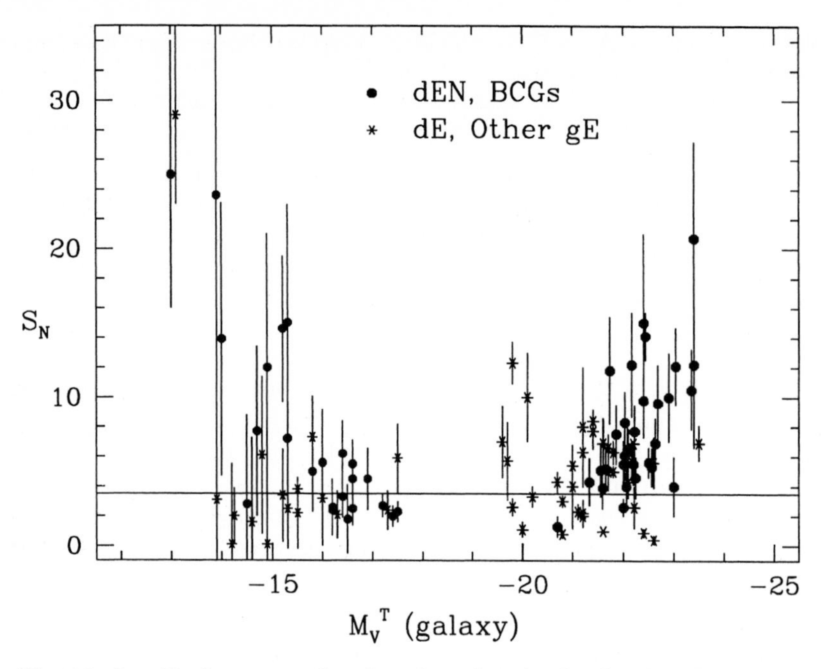

Fig. 44. Specific frequency S_N plotted against luminosity for elliptical galaxies. *Solid symbols* are for cD-type giants (brightest cluster ellipticals) and nucleated dwarf ellipticals, while *starred symbols* are for normal gE's and non-nucleated dwarfs. The baseline "normal" level is at $S_N = 3.5$; see text. The gap in the range $M_V^T \sim -18$ to -20 is a selection effect; no globular cluster systems have been studied for galaxies in that range

Third, significant correlations of S_N with other galaxy properties do exist. The strongest and most obvious connection is with *environment*. At the high-luminosity end, it appears that there is something special about the giant ellipticals that sit at the centers of large clusters of galaxies – the "brightest cluster galaxies" (BCGs) which often have cD-type structures (high luminosities, along with extended envelopes of stellar material that appear to follow the potential well of the cluster as a whole). These particular galaxies have the highest known specific frequencies among gE galaxies. The prototype of this class is M87, the Virgo cluster cD and the center of the biggest concentration of galaxies in the Virgo region, which has an extremely well determined $S_N = 14.1 \pm 1.6$ (Harris et al. 1998a) almost three times larger than the mean for other Virgo ellipticals. Since cD's are few and far between, it took many years for a significant sample of globular cluster system observations to be built up for them, and for a long time M87 was regarded as being virtually unique (see, e.g., Hanes 1977; Harris & Smith 1976; Harris & van den Bergh 1981; Harris 1988a for the initial historical development). Since then, many more BCGs have been studied, and a clear correlation of S_N with luminosity has emerged: the more luminous BCGs have higher specific frequencies (Blakeslee 1997; Harris et al. 1998a). Since the brighter BCGs tend to be found in more populous clusters of galaxies, the hint is that denser, richer environments lead to higher specific frequencies.

Even without the BCGs, a similar conclusion would emerge from the rest of the ellipticals. It was first suggested by Harris & van den Bergh (1981) that the ellipticals in small, sparse groups of galaxies or in the "field" had systematically lower S_N than those in richer systems like Virgo or Fornax. Larger samples confirmed this. For E's in small groups, S_N is typically $\sim 1-3$, while in larger groups (Fornax, Virgo, and above) we find $S_N \simeq 5$ (Harris 1991).

Until the past few years, not much was known about globular cluster systems in *dwarf* ellipticals, the small galaxies at the opposite end of the luminosity scale. But they, too, exhibit a large S_N range and some intriguing correlations which have been revealed by new surveys (Durrell et al. 1996a,b; Miller et al. 1998). There appears to be a dichotomy between *nucleated* dE's (those with distinct central compact nuclei) and non-nucleated dE's. The dE's present a fairly simple story, with a mean $\langle S_N \rangle \simeq 2$ independent of luminosity and with not much scatter. In striking contrast, the dE,N systems show a clear correlation of S_N with luminosity, in the opposite sense to the BCGs: less luminous dE,N's have higher specific frequencies. The most luminous dwarfs of both types have similarly low specific frequencies, but at progressively lower L, the specific frequency in dE,N's steadily increases, reaching the highest values at the low-L end.

We would like to understand why the specific frequency displays such a large range. How can otherwise-similar galaxies make (or keep) vastly different numbers of old-halo star clusters? Speculations began as soon as the phenomenon was discovered, concentrating first on the environmental connection and on the "anomaly" of the BCGs (e.g., Harris & Smith 1976; van den Bergh 1977; Harris & Racine 1979). Many other ideas entered the game later on. We will discuss these in the last two sections; but for the moment, we will say only that no single explanation or mechanism seems able to produce the full range of specific frequencies seen amongst all the ellipticals. It is a remarkably simple phenomenon, but remains a hard one to explain.

By contrast with the ellipticals, disk and spiral galaxies so far present a much more homogeneous picture. The Sb/Sc/Irr systems, to within factors of two, have specific frequencies similar to that of the Milky Way, in the range $S_N \simeq 0.3 - 1.0$ (Harris 1991; Kissler-Patig et al. 1999). (In fact, given the difficulty in measuring S_N in disk-type galaxies, where the total numbers of clusters are much lower than in gE galaxies to start with, and where the disk light and dust add further confusion, it is possible that the nominal differences in specific frequency among disk galaxies are entirely due to observational scatter.) Apparently, the spirals have not experienced the same range of formation processes or evolutionary histories that the ellipticals have.

If we take these numbers at face value, it would seem the spirals and irregulars are much less efficient at forming globular clusters than are most ellipticals. But an obvious difficulty in making the comparison is that these late-type galaxies have much higher proportions of "young" stellar populations which make them more luminous than ellipticals of the same mass. To correct for this effect, it has become customary to adjust the total luminosity of the galaxy to the "age-faded" value that it would have if all its stars evolved passively to $\gtrsim 10$ Gyr, like those of ellipticals (see Sect. 9 below). This correction must be done on an individual basis for each spiral or irregular, and adds a further uncertainty to the comparison. On average, this "renormalized" specific frequency falls in the range $S_N \sim 2 \pm 1$, which is similar to the typical values for dE (non-nucleated) galaxies, or large E galaxies in sparse groups, or S0 galaxies (which are disk systems free of dust or young stars). They still fall well short of the $S_N \sim 5$ level associated with gE members of rich groups, which in turn are lower than most BCGs.

These comparisons make it tempting to suggest that there is a "natural" level for S_N which is somewhere in the range $\sim 2-4$, applying to spirals, S0's, dwarf ellipticals, and many large ellipticals in a wide range of environments (refer again to Fig. 44, where a baseline $S_N = 3.5$ is shown). It would not be reasonable to expect this level to be *exactly* the same in all these galaxies, because the numbers of clusters are determined by formation efficiencies and dynamical evolution which are, at some level, stochastic processes. Some scatter is also introduced in the measurement process (see above), which in the worst cases can leave S_N uncertain by 50% or so. The major mystery has always been the extreme situations which go far beyond these normal cases: the BCGs, and the nucleated dwarfs.

6.3 Metallicity Distributions

An important trace of the early history of any galaxy is left behind in the metallicity distribution of its halo stars. Unfortunately, almost all galaxies are too remote for individual stars to be resolved, so what we know about their chemical composition is indirect, relying only on various averages over the MDF. This is where the GCS gives us a distinct advantage: the globular clusters are old-halo objects that can be found *one by one* in galaxies far too distant for any individual stars to be studied, including many unusual galaxy types. In these galaxies, we can derive a full *distribution function* of metallicity for the GCS, and not just the mean metallicity (Harris 1995). Insofar as the GCS represents the halo field-star population, we can use this MDF to deduce the early chemical enrichment history of the system.

In Fig. 45, MDFs are shown for a representative sample of galaxies which cover the presently known range of mean metallicities. The metallicity range correlates strongly with galaxy size. In the dE's, almost all the clusters are low-metallicity objects, like the MPC clusters in the halo of our Milky Way with an average near [Fe/H] $\simeq -1.6$ (see Fig. 7). This observation fits in

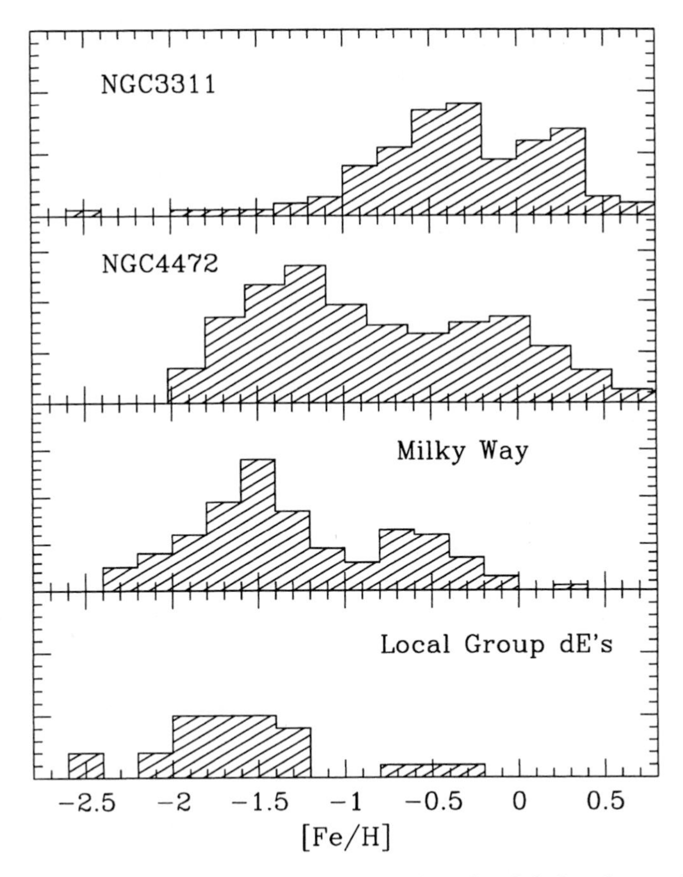

Fig. 45. Metallicity distribution functions for globular clusters in selected galaxies. The *top panel* shows the MDF for the cD galaxy in the Hydra I cluster (Secker et al. 1995); the *second panel* shows the Virgo giant elliptical NGC 4472 (Geisler et al. 1996); and the ıbottom panel is a composite of the clusters in all the Local Group dwarf ellipticals (Harris 1991; Da Costa & Armandroff 1995)

well with standard views of the early evolution of dwarf ellipticals, in which a small, isolated protogalactic gas cloud undergoes a single major burst of star formation, but ejects a large fraction of its gas in the process (e.g., Dekel & Silk 1986; Babul & Rees 1992). Since its tiny potential well cannot hold the gas ejected by the first round of stellar winds and supernovae, the heavy-element enrichment cannot proceed to completion and the "effective yield" of the enrichment is much lower than normal, leaving only metal-poor stars behind (Hartwick 1976). What is therefore more surprising is that these small ellipticals have any metal-rich clusters at all: two with [Fe/H] $\gtrsim -1$ are probably members of the Local Group dE's, and there are clear hints that others can be found, albeit in small numbers, with similar metallicities (see, e.g., Durrell et al. 1996a for the Virgo dwarfs). How did these few relatively metal-rich clusters arise in circumstances that are strongly biased against normal metal enrichment? The answers are not yet clear. They may simply represent rare instances where unusually dense pockets of the proto-dE got an early start and held its gas long enough for the local enrichment to proceed up to higher levels than normal. Alternately, they may represent a somewhat later and more minor epoch of star formation driven by late infall of gas or by the triggering of whatever residual gas was left in the system.

In somewhat larger and more complex galaxies – the Milky Way, M31, and normal ellipticals – a metal-poor component is usually present at roughly the same metallicity level as we find in the dwarfs, but a much more significant higher-metallicity population also appears and the MDF as a whole begins to look very broad. At the upper end of the scale, in some high-luminosity ellipticals such as NGC 3311 in the Hydra I cluster and IC 4051 in Coma, the MPC almost disappears and we are left with only a metal-rich GCS (e.g., Secker et al. 1995; Woodworth & Harris 2000). The relative proportions of MPC and MRC can differ quite noticeably from one host galaxy to another, even among otherwise similar galaxies, and in some ellipticals the MDF is narrow and not easily described as a mixture of metal-poor and metal-rich components (e.g., Ajhar et al. 1994; Kissler-Patig et al. 1997a).

The large differences in MDFs, coupled with the amazingly similar *luminosity* distribution functions of globular clusters in all galaxies, already put important constraints on formation models for globular clusters. Clearly, the GCS formation process must be a robust one which gives the same cluster mass spectrum *independent of the metallicity of the progenitor gas clouds.* In addition, the MDFs already challenge our traditional, Milky-Way-bred notions that a "globular cluster" is prototypically a massive, old, *metal-poor* star cluster. It is not. By sheer weight of numbers and high specific frequency, a large fraction of all the globular clusters in the universe reside in giant ellipticals, and many of these are metal-rich, extending up to (and beyond) solar metallicity.

When the entire range of galaxies is plotted, we find a correlation of mean GCS metallicity with galaxy luminosity (Brodie & Huchra 1991; Harris 1991; Ashman & Zepf 1998; Forbes et al. 1996b). The equation

$$\langle[\mathrm{Fe/H}]\rangle = -0.17\, M_V^T - 4.3 \qquad (51)$$

matches the overall trend accurately for the ellipticals. However, the correlation is much closer for the dE's than for the giant ellipticals, which exhibit a large galaxy-to-galaxy scatter in mean [Fe/H] and almost no trend with M_V^T. The reason for this large scatter appears (Forbes et al. 1997) to be that this mean correlation ignores the large variety of mixtures between the MPC and MRC parts of the MDF that are found from one galaxy to another. It seems too much of an oversimplification to think of the entire MDF as a unit.

Another general result valid for most E galaxies is that the GCS mean metallicity is *lower* than the galaxy halo itself by typically 0.5 dex. That is, the same scaling rule of metallicity versus total size applies to the galaxy itself and to the GCS, but with the GCS offset to lower metallicity (Brodie & Huchra 1991; Harris 1991). The initial interpretation of this offset (cf. the references cited) was that the GCS formed slightly earlier in sequence than most of the halo stars, and thus was not as chemically enriched. This view dates from a time when it was thought that there was a single, fairly sharply defined formation epoch for the clusters. As we will see below, however, the story cannot be quite that simple for most large galaxies.

6.4 Substructure: More Ideas About Galaxy Formation

We have already discussed the *bimodal* structure of the MDF for the Milky Way clusters: they fall into two rather distinct subgroups (MPC, MRC), and the MDF itself can be well matched analytically by a simple combination of two Gaussians. For giant E galaxies, it is easily possible to obtain MDFs built out of hundreds and even thousands of clusters, and the same sorts of statistical analyses can readily be applied. However, it was only during the past decade that MDFs for these galaxies became internally precise enough that bimodal, and even multimodal, substructure began to emerge from the obviously broad color distributions. Observationally, the most important breakthrough in this field was the employment of highly sensitive photometric indices, especially the Washington $(C-T_1)$ index (Geisler & Forte 1990). With it, the intrinsic metallicity-driven color differences between clusters stood out clearly above the measurement scatter for the first time, and CCD photometry of large samples of clusters could be obtained. (Other well known color indices such as $(B-V)$ or $(V-I)$ are only half as sensitive to metallicity as $(C-T_1)$ or $(B-I)$. Although it is still possible to obtain precise MDFs from them, the demands for high precision photometry are more stringent, and were generally beyond reach until the present decade; see Ashman & Zepf 1998).

On the analytical side, better statistical tools were brought to bear on the MDFs (Zepf & Ashman 1993; Ashman et al. 1994; Zepf et al. 1995). These studies revealed that the color distributions of the clusters in giant E galaxies, which were initially described simply as "broad", could be matched better as bimodal combinations of Gaussians strongly resembling the ones for the Milky Way. Improvements in the quality of the data have tended to confirm these conclusions, with the multimodal character of the color distribution standing out more clearly (e.g., Whitmore et al. 1995; Geisler et al. 1996; Forbes et al. 1998; Puzia et al. 1999). Since the integrated colors of globular clusters vary linearly with [Fe/H] for [Fe/H] $\lesssim -0.5$ (Couture et al. 1990; Geisler & Forte 1990), a bimodal color distribution translates directly into a bimodal MDF.

But is a bimodal MDF a clear signature of two major, distinct formation epochs in these gE galaxies, analogous to the ones postulated for the Milky Way? Zepf and Ashman have repeatedly interpreted the bimodality in terms of their merger model for elliptical galaxies (Ashman & Zepf 1992), in which the MPC clusters are assumed to be the ones formed in the first star formation burst, and the MRC ones are due to later bursts driven by mergers and accretions which bring in new supplies of gas. This scenario will be discussed further in Sect. 9 below. Meanwhile, other authors have noted that bimodality is, although common, not a universal phenomenon in E galaxies. In many other cases, the GCS color distribution is closer to a unimodal one and remarkably narrow, with typical width σ[Fe/H] $\simeq 0.3$ (e.g., Ajhar et al. 1994; Kissler-Patig et al. 1997a; Elson et al. 1998). Furthermore, in cases where the MDF is approximately unimodal, the *mean* metallicity of the clusters is not always the same between galaxies: some are rather metal-poor (like those in the Milky Way halo), while others (notably the Coma giant IC 4051 or the Hydra cD NGC 3311; see Secker et al. 1995; Woodworth & Harris 2000) are strikingly metal-rich, with a peak at [Fe/H] $\simeq -0.2$ and clusters extending well above solar abundance. In such galaxies, it is puzzling that there would be little or no trace of any first-generation metal-poor stellar population.

Forbes et al. (1997) provide an analysis of all the available MDFs for giant ellipticals which suggest an interesting pattern in the mean metallicities of the two modes. The peak of the MRC falls consistently at [Fe/H](MRC) $= -0.2$ for the most luminous gE's, with little scatter. By contrast, the peak of the MPC (on average ~ 1 dex lower) shows considerable galaxy-to-galaxy scatter. The cluster sample sizes in some of these galaxies are small, and the identifications of the mode locations are debatable in some cases; but their basic conclusion seems sound, and may turn into a strong constraint on more advanced formation models. They argue that a two-phase *in situ* burst is the best interpretation to generate the basic features of these MDFs.

The presence or absence of bi- or multi-modality seems to correlate with little else. Generally valid statements seem to be that the cD-type (BCG) galaxies have the broadest MDFs (bimodal or multimodal); normal ellipticals can have broad or narrow MDFs according to no pattern that has yet emerged. The sample of well determined MDFs is, however, not yet a large one, and could be considerably expanded with studies of more ellipticals in more environments and over a wider range of sizes.

A superb illustration of what can be obtained from such studies is found in the work by Geisler et al. (1996) and Lee et al. (1998) on the Virgo giant NGC 4472. With CCD imaging and Washington filters, they obtained accurate $(C - T_1)$ indices for a deep and wide-field sample of globular clusters around this gE galaxy. Two diagrams taken from their study are shown in Figs. 46 and 47. A plot of cluster color or metallicity against galactocentric radius (Fig. 46) reveals distinct MRC and MPC subpopulations which also follow different spatial distributions. The redder MRC objects follow a radial distribution that is similar to that of the halo light of the galaxy, and their mean color is strikingly similar to that of the halo (Fig. 47). By contrast, the bluer MPC clusters – equally numerous – follow a much more extended spatial

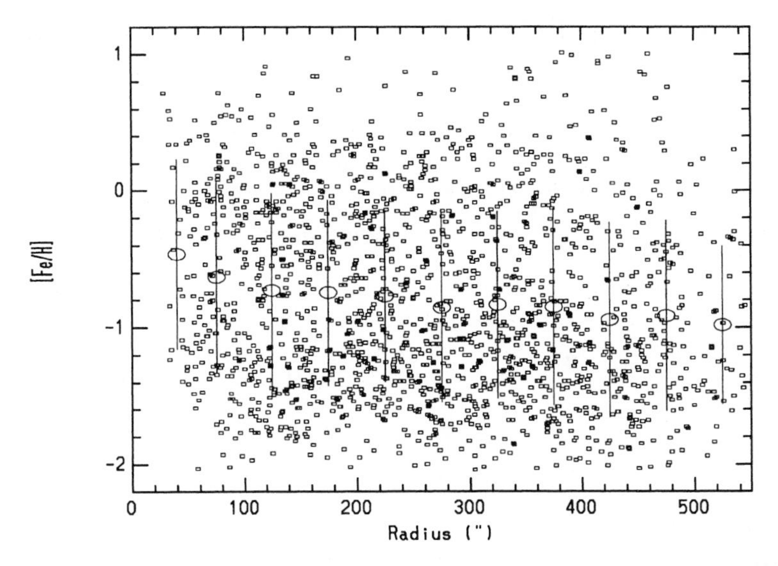

Fig. 46. Metallicity vs. radius for globular clusters in NGC 4472, from Geisler et al. (1996). Note the bimodal distribution in metallicity, with the redder (more metal-rich) population more centrally concentrated. Figure courtesy Dr. D. Geisler

structure and are more metal-poor than the MRC clusters or the galaxy halo by fully 1 dex.

The GCS *as a whole* displays a radial metallicity gradient, with the mean color decreasing steadily outward (Fig. 47). Yet neither the MPC or MRC subgroups exhibit significant changes in mean color with radius by themselves. The gradient in the GCS as a whole is, therefore, in some sense an artifact! It is a simple consequence of the different radial distributions of the two subpopulations: the MRC clusters dominate the total GC numbers at small radii, while the MPC clusters dominate at large radii, so that the mean color of all clusters combined experiences a net outward decrease.

This same observational material also allows us to place interesting limits on the *specific frequency for each of the two subgroups*. Taking the bimodal MDF at face value, let us suppose that NGC 4472 formed in two major starbursts. By hypothesis, the earlier one produced the MPC clusters along with some halo light (i.e., field stars) at the same metallicity. The later and stronger burst formed the MRC clusters, with more field stars and with associated clusters at the same (higher) metallicity.

Geisler et al. (1996) estimate that the total number of MPC clusters is $N_{\rm MPC} = 3660$. Their mean color is $(C - T_1) = 1.35 \pm 0.05$ (see Fig. 47). Similarly, for the MRC clusters they estimate $N_{\rm MRC} = 2440$, with a mean color $(C - T_1) = 1.85 \pm 0.05$. But now, the mean color of the *halo light* is $(C - T_1) = 1.85 \pm 0.05$, exactly the same as that of the MRC clusters. Thus under our assumptions, *the vast majority of the halo stars must belong to the*

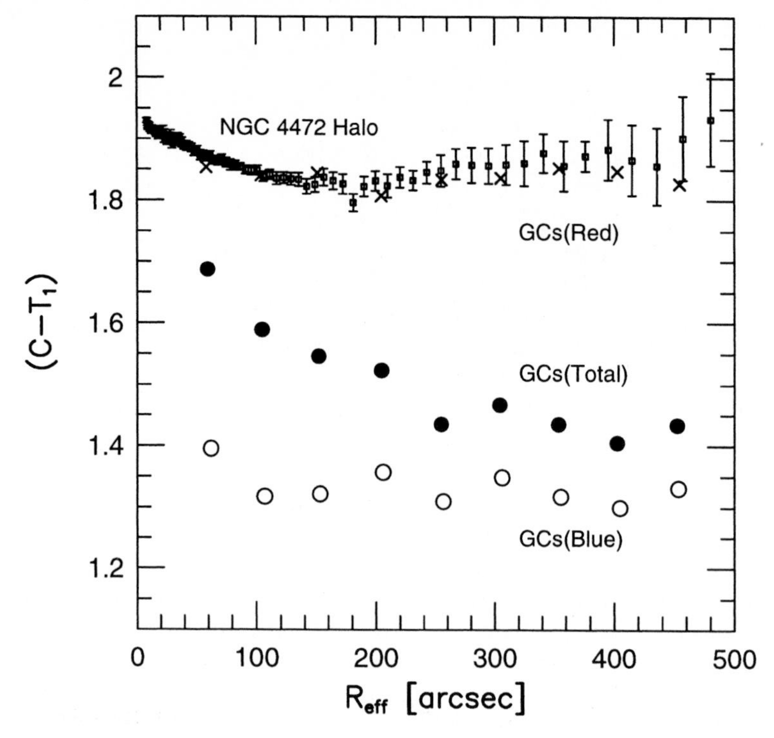

Fig. 47. Mean color $(C - T_1)$ vs. radius for globular clusters in NGC 4472, from Lee et al. (1998). Plotted separately are the red (metal-rich) globular clusters as *crosses*; blue (metal-poor) clusters as *open circles*; and the integrated color of the NGC 4472 halo light (*small boxes* with error bars). The mean color of all clusters combined (red + blue) is shown as the *solid dots*. Figure courtesy Dr. D. Geisler

second, more metal-rich formation epoch; otherwise, their integrated color would lie distinctly between the two groups of clusters. A straightforward calculation shows that if the MPC halo light (which by hypothesis has a color $(C - T_1) \simeq 1.35$) makes up more than about 6% of the total galaxy light, then the integrated color of the whole halo will be bluer than $(C - T_1) = 1.80$, which would bring it outside the error bars of the observations.

Turning this calculation around, we conclude that the MRC starburst made up $\gtrsim 94\%$ of the stellar population of the galaxy.

Finally, we can convert these numbers into specific frequencies. The integrated magnitude of the whole galaxy is $V^T(\text{N4472}) = 8.41$. Splitting it in the proportions estimated above, we then have $V^T(\text{MRC}) \simeq 8.48$, and $V^T(\text{MPC}) \gtrsim 11.30$. Thus the metal-richer component has

$$S_N(\text{MRC}) = 2.4 \pm 0.3$$

while a *lower limit* for the metal-poor burst is

$$S_N(\mathrm{MPC}) \gtrsim 50 \ !$$

The specific frequency of the first, metal-poor starburst must have been extremely high – higher, in fact, than in any galaxy as a whole that we know of today. Either the conversion rate of gas into bound globular clusters was outstandingly efficient in the initial burst, or a great deal of the initial gas present was ejected or unused for star formation during the burst. As we will see later, the latter explanation currently seems to be the more likely one (see also Forbes et al. 1997 for a similar argument). One possibility (Harris et al. 1998a) is that the initial metal-poor gas formed the MPC clusters that we now see, but was then prevented from forming its normal proportion of stars by the first major burst of supernovae and the development of a galactic wind. A large part of this gas – now enriched by the first starburst – later underwent dissipational collapse, most of it then being used up in the second burst. Contrarily, the specific frequency of the second starburst – which produced most of the galaxy's stars – was quite modest, falling well within the "normal" range mentioned previously for many kinds of galaxies.

This discussion operates within the context of a generic "in situ" model of formation, i.e., one in which the galaxy formed out of gas from within the protogalaxy. However, the relative specific frequencies in the MPC and MRC subgroups would be the same in any other scheme; they depend only on the assumption that the MPC clusters and metal-poor halo light go together, and that the MRC clusters and metal-rich halo light go together.

Several other galaxies appear to present a story with strong similarities to that in NGC 4472, such as NGC 1399 (Ostrov et al. 1998; Forbes et al. 1998) and M87 itself (Whitmore et al. 1995; Kundu et al. 1999). Two major subgroups dominate the MDF, each of which displays little or no metallicity gradient in itself. The MRC is more centrally concentrated, giving rise to a net [Fe/H] gradient in the whole GCS. Conversely, in galaxies with clearly unimodal MDFs, none so far show any clear evidence for metallicity gradients. In summary, the presence or absence of gradients in halo metallicity appears to connect strongly with the form of the MDF. Each separate stage of cluster formation generated clusters at similar metallicities all across the potential well of the galaxy, and it is only the different radial concentrations of these components that gives rise to an overall gradient in the total GCS.

Another elliptical galaxy of special interest is NGC 5128, the dominant galaxy in the small, nearby Centaurus group ($d = 3.9$ Mpc). The importance of this galaxy is that it is the only giant elliptical in which we have been able to directly compare the MDF of the *halo stars* with the *clusters*. G. Harris et al. (1999) have used deep *HST*/WFPC2 photometry in V and I to obtain direct color-magnitude photometry of the red-giant stars in the outer halo of NGC 5128, from which they generate an MDF by interpolation within standard RGB evolutionary tracks. The comparison between the two MDFs (clusters and halo stars) is shown in Fig. 48.

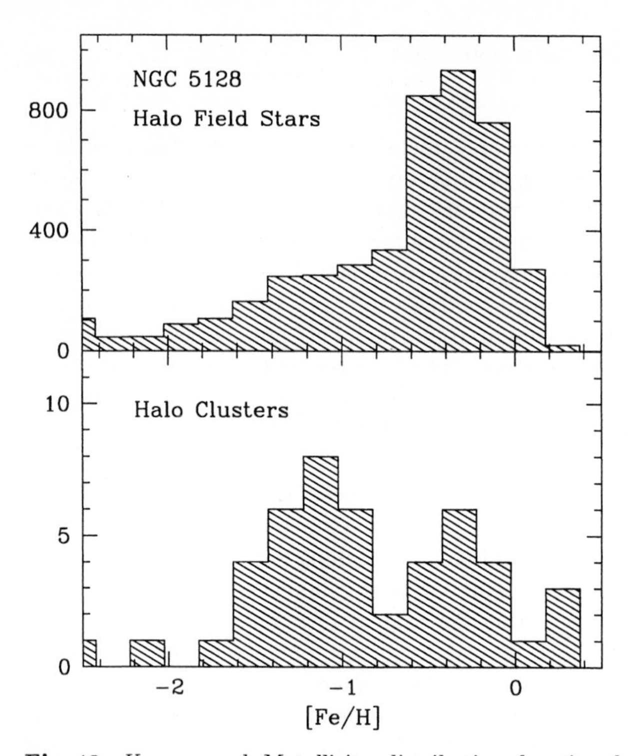

Fig. 48. *Upper panel:* Metallicity distribution function for red-giant stars in the outer halo of NGC 5128, at a projected location 20 kpc from the galaxy center. *Lower panel:* MDF for the globular clusters in the halo of NGC 5128 more distant than 4′ (4.5 kpc) from the galaxy center. Data are from G. Harris et al. (1992, 1999)

The NGC 5128 halo stars display an MDF with at least two major components; roughly two-thirds of the stars are in the narrow metal-rich component located at [Fe/H](peak)$= -0.3$ and with dispersion σ[Fe/H]$\simeq 0.25$. Remarkably, the metal-rich part of the bimodal cluster MDF has the same location and the same dispersion. Its specific frequency (that is, the ratio of MRC clusters to MRC stars) is S_N(MRC) $\simeq 1.5$. By contrast, the metal-poor component makes up about a third of the halo stars but about two-thirds of the clusters, so that its specific frequency is S_N(MPC) $\simeq 4.3$. This is, however, only a local estimate for one spot in the halo. The global value of S_N(MPC) across the entire galaxy would be larger if the inner parts of the halo contain relatively more MRC stars; that is, if the halo has a mean metallicity gradient.

G. Harris et al. (1999) argue that the most likely interpretation of the early history of this galaxy is an *in situ* formation model much like the one outlined above: two rather distinct stages of star formation, in which the first (metal-poor) one left most of the gas unconverted, but slightly enriched from the first, low-efficiency round of star formation. The later (metal-richer) burst then converted most of the gas and produced the main visible bulk of the galaxy. Though later accretions of small satellites must have played some role in building up NGC 5128 (one gas-rich accretion has clearly occurred recently to fuel the starburst activity within the inner ~ 5 kpc), these do not seem to have affected the outer-halo regions.

NGC 5128 may of course not be typical of all ellipticals. But interactions of the type it is now undergoing are now realized to be fairly commonplace for large galaxies, so there is every reason for optimism that we can use it to learn about the early evolution of many giant ellipticals.

The analysis of GCS metallicity distributions has been one of the most productive routes to understanding cluster formation and the early histories of galaxies. I urge interested readers to see the extensive discussion of Ashman & Zepf (1998) for more of the history of MDFs and their analysis.

6.5 Radial Velocities and Dynamics

If we want to study the dynamics of the halo in a distant galaxy, then globular clusters give us the same advantage over halo field stars as they did for the metallicity distributions: because we can identify them one by one, we can build up the actual velocity *distribution function* rather than just a luminosity-weighted mean. Potentially, we can use cluster velocities in remote galaxies to determine (a) the kinematic differences between MPC and MRC clusters, where they are present; (b) the mass distribution $M(r)$ and the amount of dark matter; (c) the orbital distribution and the degree of anisotropy; and (d) the presence (or absence) of "intergalactic" globular clusters, i.e. clusters moving freely in the potential well of the galaxy cluster as a whole.

Obtaining the necessary velocity measurements is a demanding job, requiring the biggest available optical telescopes and *large* samples of clusters. Early velocity measurements were accomplished for a few dozens of clusters in three giant ellipticals, M87 (Mould et al. 1987, 1990; Huchra & Brodie 1987), NGC 4472 (Mould et al. 1990), and NGC 5128 (H.Harris et al. 1988). These studies were consistent with the expected results that the velocity distributions were roughly isotropic and that the velocity dispersion was nearly uniform with radius, thus $M(r) \sim r$. However, more recent studies – with higher-quality data and significantly larger samples – have begun to reveal more interesting features. For M87, Cohen & Ryzhov (1997) have used a sample of ~ 200 clusters extending out to $r \sim 30$ kpc to suggest that the velocity dispersion *rises* with radius, indicating $M(r) \sim r^{1.8}$. The cluster velocities thus suggest the presence of an extensive amount of halo mass which bridges the mass profile of the central cD galaxy to the larger-scale mass distribution as determined from the hot X-ray gas on 100-kpc scales. Still larger samples of cluster velocities for M87 are in progress, and may be able to provide first hints on the velocity anisotropy parameters.

Sharples et al. (1998) have published the first stages of a study of similar scale on NGC 4472, the other Virgo supergiant. Other notable studies for disk galaxies include the recent work on the Sombrero Sa galaxy (NGC 4594) by Bridges et al. (1997) and on NGC 3115 by Kavelaars (1998). For NGC 1399, the central cD galaxy in Fornax, several dozen cluster velocities have now been obtained (Grillmair et al. 1994; Minniti et al. 1998; Kissler-Patig 1998; Kissler-Patig et al. 1999). They find that the GCS velocity dispersion at $r \gtrsim 20$ kpc is noticeably higher than that of the inner halo stars or clusters (as deduced from the integrated light and planetary nebulae) but similar instead to the population of *galaxies* around NGC 1399, suggesting that many of the globular clusters in the cD envelope may belong to the Fornax potential as a whole rather than the central elliptical. Some contamination from neighboring ellipticals is also a possibility, and considerably more datapoints will be needed to sort out the alternatives (see Kissler-Patig et al. 1999).

Here we end our overview of globular cluster systems in different galaxies. After a brief detour into the Hubble constant (next section), we will return in the last two sections to a discussion of current ideas about globular cluster formation and the early history of galaxies.

7 The GCLF and the Hubble Constant

The only goal of science is the diminution of the distance between present knowledge and truth.

Steven Goldberg

Globular cluster systems are astrophysically most important for what they can tell us about galaxy formation. Confronted with the rich variety of observational information we now have for GCSs in many galaxies, and the range of implications it all has for galaxy formation, it is somewhat surprising to recall that they were historically first regarded as attractive for their potential as *extragalactic distance indicators* – that is, standard candles. In this section, we will take a brief look at the history of attempts to use globular clusters as standard candles; discuss the basic technique in its contemporary form; work through the empirical calibration issues; and finally, see how it is applied to remote galaxies and derive a new estimate of H_0.

7.1 Origins

The brightest globular clusters are luminous ($M_V \lesssim -11$) and thus detectable at distances far beyond the Local Group – particularly in giant ellipticals with populous GCSs that fill up the bright end of the cluster luminosity distribution. M87, the central cD in the Virgo cluster, was the first such galaxy to attract attention. Attempts to use the brightest clusters began with the discovery paper by Baum (1955), who first noted the presence of globular clusters around M87 visible on deep photographic plates. In several later papers (Sandage 1968; Racine 1968; van den Bergh 1969; de Vaucouleurs 1970; Hodge 1974), the mean magnitudes of these few brightest clusters were used to estimate the distance to M87, under the assumption that their intrinsic luminosities were the same as those of Mayall II (the brightest cluster in M31), or ω Centauri (the brightest in the Milky Way), or some average of the most luminous clusters in the Local Group galaxies. All of these attempts were eventually abandoned after it became clear that the brightest clusters drawn from a huge statistical sample – like the M87 GCS – would be more luminous than those drawn from the much smaller Milky Way and M31 samples, even if their GCLFs were basically similar (which was itself an unproven assumption).

The modern approach to employing the GCLF begins with the work of Hanes (1977), who carried out a large photographic survey of the globular cluster systems in several Virgo ellipticals. The photometric limits of this material still fell well short of the GCLF turnover, but the basic principle was established that *the entire GCLF* had considerably more information than just its bright tip, and could be matched in its entirety with the calibrating GCSs in the Milky Way or M31. The Gaussian interpolation model for the GCLF was also employed in essentially the same way we use it today. With the benefit of hindsight (see the discussion of Harris 1988b), we can see from Hanes' analysis that he would have correctly predicted the Virgo GCLF turnover magnitude if he had known the right value of the GCLF dispersion σ for these ellipticals. Somewhat deeper photographic photometry for additional Virgo ellipticals was obtained by Strom et al. (1981) and Forte et al. (1981), with similar results.

The subject – like most other areas of observational astronomy – was revolutionized by the deployment of the enormously more sensitive CCD cameras, beginning in the mid-1980's. At last, the anticipated GCLF turnover was believed to be within reach of the new CCD cameras on large telescopes. Once again, M87 was the first target: long exposures with a first-generation CCD camera by van den Bergh et al. (1985) attained a photometric limit of $B \simeq 25.4$. They did indeed reach the turnover point, though they could not definitively prove it, since the photometric limit lay *just* past the putative turnover. Still deeper B-band data were obtained by Harris et al. (1991) for three other Virgo ellipticals, which finally revealed that the turnover had been reached and passed, with the data exhibiting a clearly visible downturn extending 1.5 mag past the peak. For the first time, it was possible to argue on strictly observational grounds that the GCLF had the same fundamental shape in E galaxies as in the Milky Way and M31. With the advent of the Hubble Space Telescope era in the 1990's, considerably more distant targets have come within reach, extending to distances where galactic motions are presumed to be dominated by the cosmological Hubble flow and peculiar motions are negligible.

7.2 The Method: Operating Principles

In its modern form, the GCLF is the simplest of standard candles that apply to remote galaxies. For the purposes of this section, we will use the classic Gaussian-like form of the luminosity distribution (number of clusters per unit magnitude). The observational goal is nothing more than *to find the apparent magnitude V^0 of the turnover point.* Once an absolute magnitude M_V^0 is assumed, the distance modulus follows immediately. The precepts of the technique are laid out in Secker & Harris (1993) and in the reviews of Jacoby et al. (1992) and Whitmore (1997). Briefly, the basic attractions of the GCLF method are as follows:

- M_V^0 is more luminous than any other stellar standard candle except for supernovae. With the *HST* cameras ($V_{\rm lim} \gtrsim 28$), its range extends to $d \sim$ 120 Mpc and potentially further.
- Globular clusters are old-halo objects, so in other galaxies they are as free as possible from problems associated with dust and reddening inside the target galaxy.
- They are nonvariable objects, thus straightforward to measure (no repeat observations are necessary).
- They are most numerous in giant E galaxies which reside at the centers of rich galaxy clusters. These same objects are the ones which are the main landmarks in the Hubble flow, thus concerns about peculiar motions or interloping galaxies are minimized.

Clearly, it shares at least some of these advantages with other techniques based on old stellar populations that work at somewhat shorter range: the planetary nebula luminosity function, surface brightness fluctuations, and the RGB tip luminosity (Jacoby et al. 1992; Lee et al. 1993a).

Having listed its attractions, we must also be careful to state the concerns and potential pitfalls. There are two obvious worries arising from the astrophysical side. *First*, globular clusters are small *stellar systems* rather than individual stars. We cannot predict their luminosities starting from a secure basis in stellar physics, as we can do for (e.g.) Cepheids, planetary nebulae, or RGB tip stars. Indeed, to predict the luminosity distribution of globular clusters, we would first have to know a great deal about how they form. But understanding their formation process almost certainly involves complex, messy gas dynamics (see Sect. 8 below), and at the moment, we have no such complete theory on hand. In any case, we might well expect *a priori* that clusters would form with different typical masses or mass distributions in different environments, such as at different locations within one galaxy, or between galaxies of widely different types.

Second, globular clusters are $\sim 10 - 15$-Gyr-old objects, and as such they have been subjected to a Hubble time's worth of dynamical erosion within the tidal fields of their parent galaxies. Since the efficiencies of these erosive processes also depend on environment (Sect. 5), shouldn't we expect the GCLFs to have evolved into different shapes or mean luminosities in different galaxies, even if they started out the same?

In the absence of direct observations, these theoretical expectations seem formidable. But we should not mistake the relative roles of theory and experiment: that is, arguments based on whatever is the current state of theory should not prevent us from going out and discovering what the real objects are like. For the distance scale, the fundamental issue (Jacoby et al. 1992) can be simply stated: *Any standard candle must be calibrated strictly on observational grounds; the role of theory is to explain what we actually see.* Theory may give us an initial motivation or overall physical understanding of a particular standard candle, but the only way that our carefully constructed distance scale can be independent of changes in the astrophysical models is to build it purely on measurement.

At the same time, we must recognize the challenges as honestly as we can. If we are to use the GCLF as a standard candle, we must have clear evidence that the turnover magnitude M_V^0 is in fact the same from galaxy to galaxy.

More precisely, we must be confident that the behavior of M_V^0 is *repeatable* from galaxy to galaxy. This is, of course, not a black-and-white statement but rather a matter of degree: like any other empirical standard candle, M_V^0 cannot be a perfect, ideally uniform number. But is it a "constant" at the level of, say, ± 0.1 magnitude? ± 0.2 mag? or worse? This is the practical question which determines how interesting the GCLF actually is as a distance indicator, and which must be settled empirically.

7.3 Calibration

We calibrate the turnover luminosity M_V^0 by measuring it in several other nearby galaxies whose distances are well established from precise stellar standard candles. But just using the Milky Way and M31 (Sect. 5) will not do. We will be particularly interested in using the GCLF in remote giant ellipticals, and these are galaxies of quite a different type than our nearby spirals.

The closest large collections of E galaxies are in the Virgo and Fornax clusters. Fortunately, these are near enough that their distances can be measured through a variety of stellar standard candles, and so these two clusters must be our main proving grounds for the GCLF calibration. Here, I will use galaxy distances established from four different methods which have sound physical bases and plausible claims to precisions approaching ± 0.1 magnitude in distance modulus: (a) the period-luminosity relation for Cepheids; (b) the luminosity function for planetary nebulae (PNLF); (c) surface brightness fluctuations for old-halo stellar populations (SBF); and (d) the red-giant branch tip luminosity (TRGB). For extensive discussions of these (and other) methods, see Jacoby et al. (1992) and Lee et al. (1993a). In Table 10, recent results from these four methods are listed for several galaxy groups and individual galaxies with globular cluster systems. In most cases, the mutual agreements

Table 10. Distance moduli for nearby galaxy groups

Galaxy Group	$(m - M)_0$	Method	Sources	Mean
Virgo Cluster	30.99 ± 0.08	Cepheids	1,2,3,4	30.97 ± 0.04
	30.98 ± 0.18	TRGB	5	
	30.84 ± 0.08	PNLF	6,7	
	31.02 ± 0.05	SBF	8,9,10,11	
Fornax Cluster	31.35 ± 0.07	Cepheids	12	31.27 ± 0.04
	31.14 ± 0.14	PNLF	13	
	31.23 ± 0.06	SBF	8	
Leo I Group	30.01 ± 0.19	Cepheids	14	30.17 ± 0.05
	30.30 ± 0.28	TRGB	15	
	30.10 ± 0.08	PNLF	16,17	
	30.20 ± 0.05	SBF	8,11,18	
Coma I Group	30.08 ± 0.08	PNLF	19	30.08 ± 0.07
	30.08 ± 0.07	SBF	20	
Coma II Group	30.54 ± 0.05	PNLF	19	30.81 ± 0.14
	30.95 ± 0.07	SBF	8,20	
NGC 4365	31.73 ± 0.10	SBF	8	31.73 ± 0.10
NGC 3115	30.29 ± 0.20	TRGB	21	30.16 ± 0.10
	30.17 ± 0.13	PNLF	22	
	29.9 ± 0.25	SBF	21,22	
NGC 4594	29.74 ± 0.14	PNLF	23	29.70 ± 0.10
	29.66 ± 0.08	SBF	22	

Sources: (1) Ferrarese et al. 1996 (2) Pierce et al. 1994 (3) Saha et al. 1996a (4) Saha et al. 1996b (5) Harris et al. 1998b (6) Jacoby et al. 1990 (7) Ciardullo et al. 1998 (8) Tonry et al. 1997 (9) Neilsen et al. 1997 (10) Pahre & Mould 1994 (11) Morris & Shanks 1998 (12) Madore et al. 1998 (13) McMillan et al. 1993 (14) Graham et al. 1997 (15) Sakai et al. 1997 (16) Ciardullo et al. 1989 (17) Feldmeier et al. 1997 (18) Sodemann & Thomsen 1996 (19) Jacoby et al. 1996 (20) Simard & Pritchet 1994 (21) Kundu & Whitmore 1998 (22) Ciardullo et al. 1993 (23) Ford et al. 1996

among these methods are good, and bear out their claimed accuracies in the references listed.

The final column of the table gives the adopted mean distance modulus for each group, along with the *internal* r.m.s. uncertainty of the mean. As a gauge of the true (external) uncertainty, we can note that to within ± 0.1 in distance modulus, the absolute zeropoints of each technique are consistent with the Local Group (LMC and M31) distance scale discussed in Sect. 2 above.

Next, we need to have well established *apparent magnitudes* V^0 for the GCLF turnover levels in as many galaxies as possible. The most straightforward numerical technique is to start with the observed GCLF (corrected for background contamination and photometric incompleteness; see the Appendix) and fit any of the adopted interpolation functions to it – usually the Gaussian, but others such as the t_5 function have been used too. The best-fit function gives the nominal apparent magnitude V^0 of the turnover point. Secker & Harris (1993) define a more advanced maximum-likelihood procedure for fitting the raw data (that is, the list of detected objects in the field, sorted by magnitude) to the adopted function, convolved with the photometric error and completeness functions and added to the observed background LF. Both approaches have proved to generate valid results, though the latter method provides a more rigorous understanding of the internal uncertainties.

To determine V^0, we need to have GCLF photometry extending clearly past the turnover: the deeper the limit, the more precisely we can identify it independent of assumptions about the shape or dispersion of the GCLF as a whole. (It is important to note here that we do *not* necessarily want to use a fitting function which will match the entire GCLF, which may or may not be asymmetric at magnitudes far out in the wings. The entire goal of the numerical exercise is *to estimate the magnitude of the turnover point as accurately as possible*; thus, we want a fitting function which will describe the peak area of the GCLF accurately and simply. In other words, it is to our advantage to use a simple, robust function which will not be overly sensitive to the behavior of the GCLF in the far wings. The Gaussian and t_5 functions, with just two free parameters, meet these requirements well.)

The results for E galaxies with well determined GCLF turnovers are listed in Tables 11 and 12, while Table 13 gives the same results for several disk galaxies. (Note that the turnover luminosities for the Milky Way and the Local Group dE's are already converted to absolute magnitude.) The fourth column in each table gives the magnitude limit of the photometry relative to the turnover level; obviously, the larger this quantity is, the more well determined the turnover point will be. The absolute magnitude of the turnover in each galaxy is obtained by subtraction of the intrinsic distance moduli in Table 10, and subtraction of the foreground absorption A_V. Fortunately, A_V is small in most cases, since almost all the galaxies listed here are at high latitude.

The values of M_V^0 in the individual galaxies are shown in Figs. 49 and 50, and the mean values are listed in Table 14. The results for the giant ellipticals are particularly important, since these act as our calibrators for the more remote targets. From the first entry in Table 14, we see that the gE galaxy-to-galaxy scatter in M_V^0 is at the level of ± 0.15 mag *without any further corrections* due to environment, metallicity, luminosity, or other possible

Table 11. GCLF turnover magnitudes for giant E galaxies

Galaxy Group	Galaxy	V^0(turnover)	$V(\lim) - V^0$	Sources
Virgo Cluster	N4472	23.87 ± 0.07	$\simeq 1.5$	1,2,3
	N4478	23.82 ± 0.38	2.7	4
	N4486	23.71 ± 0.04	$\simeq 2.1$	5,6,7,8,9
	N4552	23.70 ± 0.30	0.7	2
	N4649	23.66 ± 0.10	$\simeq 1.8$	1
	N4697	23.50 ± 0.20	1.2	10
Fornax Cluster	N1344	23.80 ± 0.25	1.0	11
	N1374	23.52 ± 0.14	0.5	12
	N1379	23.92 ± 0.20	1.0	12,13
	N1399	23.86 ± 0.06	1.0	11,12,14,15
	N1404	23.94 ± 0.08	1.0	11,15,16
	N1427	23.78 ± 0.21	-0.2	12
Leo I Group	N3377	22.95 ± 0.54	1.3	17
	N3379	22.41 ± 0.42	1.3	17
Coma I Group	N4278	23.23 ± 0.11	1.6	18
Coma II Group	N4494	23.34 ± 0.18	1.7	18,19
NGC 4365	N4365	24.42 ± 0.18	0.8	1,2,20

Sources: (1) Secker & Harris 1993 (2) Ajhar et al. 1994 (3) Lee et al. 1998 (4) Neilsen et al. 1997 (5) Harris et al. 1991 (6) McLaughlin et al. 1994 (7) Whitmore et al. 1995 (8) Harris et al. 1998a (9) Kundu et al. 1999 (10) Kavelaars & Gladman 1998 (11) Blakeslee & Tonry 1996 (12) Kohle et al. 1996 (13) Elson et al. 1998 (14) Bridges et al. 1991 (15) Grillmair et al. 1999 (16) Richtler et al. 1992 (17) Harris 1990b (18) Forbes 1996b (19) Fleming et al. 1995 (20) Forbes 1996a

parameters. Much the same scatter emerges if we use only the gE galaxies within one cluster (Virgo or Fornax) where they are all at a common distance (cf. Harris et al. 1991; Jacoby et al. 1992; Whitmore 1997 for similar discussions). This all-important quantity determines the intrinsic accuracy that we can expect from the technique. Clearly, part of the dispersion in M_V^0 must be due simply to the statistical uncertainty in determining the apparent magnitude of the turnover from the observed GCLF (which is typically ± 0.1 mag at best; see below), and part must be due to uncertainties in the adopted distances to the calibrating galaxies (which again are likely to be ± 0.1 mag at best). When these factors are taken into account, the raw observed scatter in the turnover magnitudes is encouragingly small.

In summary, I suggest that the directly observed dispersion in the turnover luminosity gives a reasonable estimate of the precision we can expect from

Table 12. GCLF turnover magnitudes for dwarf E galaxies

Galaxy Group	Galaxy	V^0(turnover)	$V(\lim) - V^0$	Sources
Virgo Cluster	8 dE's	24.1 ± 0.3	0.7	1
NGC 3115	DW1	23.1 ± 0.3	1.4	2
Local Group	NGC 147	-5.99 ± 0.92	2:	3
	NGC 185	-6.49 ± 0.71	2:	3
	NGC 205	-7.27 ± 0.27	2:	3
	Fornax	-7.06 ± 0.95	3:	3
	Sagittarius	-6.28 ± 1.21	4:	4

Sources: (1) Durrell et al. 1996a (2) Durrell et al. 1996b (3) Harris 1991 (4) This paper

Table 13. GCLF turnover magnitudes for disk galaxies

Group	Galaxy	Type	V^0(turnover)	$V(\lim) - V^0$	Sources
Fornax	N1380	S0	23.92 ± 0.20	1.1	1,2
NGC 3115	N3115	S0	22.37 ± 0.05	0.7	3
Virgo SE	N4594	Sa	23.3 ± 0.3	1.0	4
Coma I:	N4565	Sb	22.63 ± 0.21	0.8	5
M81	M81	Sb	20.30 ± 0.3	2	6
Local Group	M31	Sb	17.00 ± 0.12	2	7,8
Local Group	Milky Way	Sbc	-7.68 ± 0.14	5	9
Local Group	M33	Sc	17.74 ± 0.17	2	10
Local Group	LMC	Im	11.13 ± 0.32	3	10

Sources: (1) Blakeslee & Tonry 1996 (2) Kissler-Patig et al. 1997b (3) Kundu & Whitmore 1998 (4) Bridges & Hanes 1992 (5) Fleming et al. 1995 (6) Perelmuter & Racine 1995 (7) Reed et al. 1994 (8) Secker 1992 (9) This paper (Sect. 5) (10) Harris 1991

the technique: for giant E galaxies with well populated GCLFs, the expected uncertainty in the resulting distance modulus is near ± 0.15 mag.

One remaining anomaly within the set of gE galaxies is a slight systematic discrepancy between the Fornax and Virgo subsamples. For the six Virgo ellipticals by themselves, we have $\langle M_V^0 \rangle \simeq -7.26 \pm 0.07$, while for the six Fornax ellipticals, $\langle M_V^0 \rangle \simeq -7.47 \pm 0.07$. These differ formally by (0.21 ± 0.10), significant at the two-standard-deviation level. Why? If the GCLF turnover is, indeed, fundamentally similar in these rich-cluster ellipticals and subject only to *random* differences from one galaxy to another, then this discrepancy

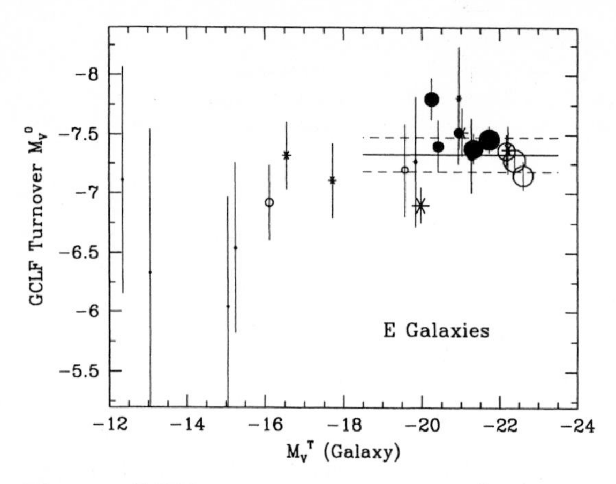

Fig. 49. GCLF turnover luminosity M_V^0 for elliptical galaxies, plotted against galaxy luminosity M_V^T. *Solid dots* are ellipticals in the Fornax cluster, *open circles* are Virgo ellipticals, and *asterisks* are ellipticals in smaller groups. Symbol size goes in inverse proportion to the internal uncertainty in the turnover (smaller symbols have larger random errors). The horizontal *solid line* indicates the mean $\langle M_V^0 \rangle$ for the giant ellipticals, with the ± 0.15 galaxy-to-galaxy range indicated by the *dashed lines*

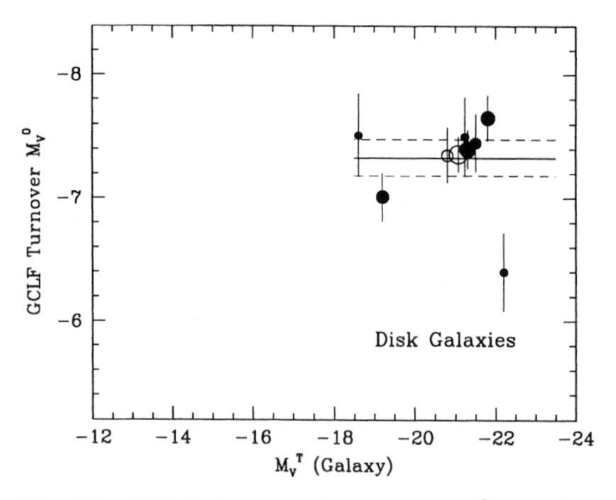

Fig. 50. GCLF turnover luminosity M_V^0 for disk galaxies. *Solid dots* are spirals (Sa to Im types) and *open circles* are S0's. The horizontal lines are taken from the previous figure, and indicate the mean and standard deviation for giant ellipticals. Most of the large disk galaxie sit slightly above the mean line for the ellipticals. The anomalously low point is the Sa galaxy NGC 4594

would suggest that we have either overestimated the distance to Fornax, or underestimated the distance to Virgo, or some combination of both. But the stellar standard candles listed above agree quite well with one another in each cluster. This puzzling discrepancy is not large; but it suggests, perhaps, that

Table 14. Final GCLF turnover luminosities

Galaxy Type	N	Mean M_V^0	rms scatter
Giant Ellipticals	16	-7.33 ± 0.04	0.15
Dwarf Ellipticals	14	-6.90 ± 0.17	0.6:
All Disk Galaxies	9	-7.46 ± 0.08	0.22
S0 and Sb	6	-7.57 ± 0.08	0.20

NB: The mean for the giant ellipticals excludes NGC 4278, at $M_V^0 = -6.9$; its distance modulus is probably suspect.

the external uncertainty in the GCLF turnover method may be closer to ± 0.2 mag.

What of the other types of galaxies? From the evidence so far, dwarf ellipticals have turnover luminosities that are *fainter* by $\sim 0.3 - 0.4$ mag in M_V than in the giants. Quite obviously, though, measuring the turnover in any one dwarf is a risky business because of the small sample size (perhaps only one or two dozen globular clusters per galaxy even in the best cases; see Durrell et al. 1996a,b). Many must be averaged together to beat down the individual statistical uncertainties.

In the disk galaxies, the turnover may be slightly *brighter* (by ~ 0.2 mag) than in gE's, though the nominal difference is not strongly significant. This latter result, if real, may be tangible evidence that dynamical evolution of globular clusters in disk galaxies has been somewhat stronger due to disk shocking, which would remove a higher proportion of the fainter clusters. The one strikingly anomalous case is NGC 4594, with a much fainter turnover level than average. Although its distance seems relatively well determined (PNLF, SBF), the GCLF turnover magnitude relies on only one small-field CCD study and may be suspect. This galaxy is the nearest giant edge-on Sa and should be studied in much more detail.

7.4 Functional Fitting and the Role of the Dispersion

We see that the absolute magnitude of the turnover point is reasonably similar in widely different galaxies. Now, what can we say about the *dispersion* of the GCLF? Specifically, in our Gaussian interpolation model, is the standard deviation σ_G reasonably similar from one galaxy to another?

An important side note here is that in practice, σ_G really represents the shape of the *bright half* of the GCLF, since in most galaxies beyond the Local Group we do not have data that extend much beyond the turnover point itself. Thus if the relative numbers of faint clusters were to differ wildly from one type of galaxy to another, we would not yet have any way to see it (nor would it matter for the standard-candle calibration). However, to test the uniformity

Table 15. GCLF dispersion measurements

Ellipticals		**Disks**	
Galaxy	σ_G	Galaxy	σ_G
N1344	1.35 ± 0.18	Milky Way	1.15 ± 0.10
N1379	1.55 ± 0.21	M31	1.06 ± 0.10
N1399	1.38 ± 0.09	M33	1.2 :
N1404	1.32 ± 0.14	N1380	1.30 ± 0.17
N4278	1.21 ± 0.09	N3115	1.29 ± 0.06
N4365	1.49 ± 0.20	N4565	1.35 ± 0.22
N4472	1.47 ± 0.08		
N4478	1.16 ± 0.21		
N4486	1.40 ± 0.06		
N4494	1.09 ± 0.11		
N4636	1.35 ± 0.06		
N4649	1.26 ± 0.08		
N5846*	1.34 ± 0.06		

*Source for NGC 5846: Forbes et al. 1996a

of σ_G, we want to use only the calibrating galaxies for which the limit of the photometry is clearly fainter than the turnover. If the data fall short of V^0, or just barely reach it, then it is generally not possible to fit a Gaussian curve to the data and solve *simultaneously* for both V^0 and σ_G; the two parameters are correlated, and their error bars are asymmetric.

This latter numerical problem was already realized in attempts to fit the first deep CCD data in M87 (van den Bergh et al. 1985; Hanes & Whittaker 1987), and is also discussed at length in Harris (1988b) and Secker & Harris (1993). The reason for the asymmetry can be seen immediately if we refer again to Fig. 43: if the observations do not extend past the turnover, then there are no faint-end data points to constrain the upper limits on either σ_G or V^0, and a statistically good fit can be obtained by choices of these parameters that may be much larger than the true values. By contrast, values that are much too *small* are ruled out by the well determined bright-end observations. The net result is unfortunately that *both the dispersion and the turnover tend to be overestimated* if both are allowed to float in the fitted solution.

Best-fit values of σ_G are listed in Table 15 for most of the same galaxies listed above. Ellipticals are listed on the left, and disk galaxies on the right. For the six disk galaxies, the weighted mean is $\langle\sigma_G\rangle = 1.21 \pm 0.05$.

For 12 ellipticals (excluding NGC 4478, which is a peculiar tidally truncated companion of NGC 4472), we obtain $\langle \sigma_{\rm G} \rangle = 1.36 \pm 0.03$.

An interesting comparison of this mean value can be obtained from the results of GCLFs in 14 BCG galaxies from the surface brightness fluctuation study of Blakeslee et al. (1997). They find $\langle \sigma_{\rm G} \rangle = 1.43 \pm 0.06$. (In their SBF analysis, only the few brightest globular clusters are actually resolved on the raw images, but the fluctuation contribution due to the fainter unresolved ones must be numerically removed before the fluctuation signal from the halo light can be determined. They assume that the GCLF follows a Gaussian shape with an assumed M_V^0 equal to that of M87, and then solve for the dispersion.)

In summary, a mean value $\sigma_{\rm G} = 1.4 \pm 0.05$ appears to match most giant ellipticals rather well, and $\sigma_{\rm G} = 1.2 \pm 0.05$ will match most spirals.

Some common-sense prescriptions can now be written down for the actual business of fitting an interpolation function to an observed GCLF. Starting with the observations of cluster numbers vs. magnitude, your goal is simply to *estimate the turnover point as accurately as possible.* Choose a simple, robust interpolation function which will match the center of the distribution and don't worry about the extreme wings. But should you try to solve for both V^0 and $\sigma_{\rm G}$, which are the two free parameters in the function? This depends completely on how deep your photometry reaches. Experience shows that if you have fully corrected your raw data for photometric incompleteness and subtracted off the contaminating background LF, and you clearly see that your data reach *a magnitude or more past the turnover point*, then you can safely fit one of the recommended functions (Gaussian or t_5) to it and solve for both parameters. However, if your photometric limit falls short of the turnover, or does not go *clearly* past it, then your best course of action is to *assume* a value for the dispersion and solve only for the turnover magnitude. This approach will introduce some additional random uncertainty in V^0, but will considerably reduce its systematic uncertainty.

The actual function fitting process can be developed into one in which the assumed model (Gaussian or t_5) is convolved with the photometric completeness and measurement uncertainty functions (see the Appendix), added to the background LF, and then matched to the raw, uncorrected LF. A maximum-likelihood implementation of this approach is described in Secker & Harris (1993).

Putting these results together, we now have some confidence *on strictly empirical grounds* that the turnover luminosity in gE galaxies has an observational scatter near ± 0.15 mag, and a Gaussian dispersion $\sigma_{\rm G} \simeq 1.4 \pm 0.05$. These statements apply to the central cD-type galaxies in Virgo and Fornax, as well as to other gE's in many groups and clusters. The GCLFs in dwarf ellipticals and in disk galaxies are noticeably, but not radically, different in mean luminosity and dispersion.

This is all the evidence we need to begin using the GCLF as a standard candle for more remote ellipticals. The near-uniformity of the GCLF luminosity and shape, in an enormous range of galaxies, is a surprising phenomenon on astrophysical grounds, and is one of the most remarkable and fundamentally important characteristics of globular cluster systems.

7.5 The Hubble Constant

To measure H_0, we need GCLF measurements in some target galaxies that are much more distant than our main group of calibrators in Virgo and Fornax. Such observations are still a bit scarce, but the numbers are steadily growing. Our preferred route will be the classic one through the "Hubble diagram". We start with Hubble's law for redshift v_r and distance d:

$$v_r = H_0\, d \tag{52}$$

or in magnitude form where d is measured in Mpc and v_r in km s^{-1},

$$5 \log v_r \;=\; 5 \log H_0 + (m - M)_0 - 25 \,. \tag{53}$$

Now substitute the apparent magnitude of the GCLF turnover, $V^0 = M_V^0 + (m - M)_0$, and we obtain

$$\log v_r \;=\; 0.2V^0 + \log H_0 - 0.2M_V^0 - 5 \,. \tag{54}$$

Thus a plot of $\log v_r$ against apparent magnitude V^0 for a sample of giant elliptical galaxies should define a straight line of slope 0.2. The zeropoint (intercept) is given by $\langle \log v_r - 0.2V^0 \rangle = \log H_0 - 0.2M_V^0 - 5$, where the mean in brackets is taken over the set of observed data points. Once we insert our adopted value of M_V^0, the value of the Hubble constant H_0 follows immediately.

Relevant data for a total of 10 galaxies or groups ranging from the Virgo cluster out to the Coma cluster (the most remote system in which the GCLF turnover has been detected) are listed in Table 16 and plotted in Fig. 51. This figure is the first published "Hubble diagram" based on globular cluster luminosities, and it has been made possible above all by the recent *HST* photometry of a few remote ellipticals.

In the Table, the entries for Virgo and Fornax are the mean $\langle V^0 \rangle$ values taken from Table 11 above. The cosmological recession velocities $v_r = cz$ for each target assume a Local Group infall to Virgo of 250 ± 100 km s^{-1} (e.g., Ford et al. 1996; Hamuy et al. 1996; Jerjen & Tammann 1993, among many others). For the mean radial velocities of the clusters, especially Virgo and Fornax, see the discussions of Colless & Dunn (1996), Girardi et al. (1993), Huchra (1988), Binggeli et al. (1993), Mould et al. (1995), and Hamuy et al. (1996). The Coma cluster ellipticals (IC 4051 and NGC 4874, and the lower limit for NGC 4881) provide especially strong leverage on the result for H_0, since they are easily the most distant ones in the list, and the correction of the cluster velocity to the cosmological rest frame is only a few percent.

Table 16. GCLF Turnover Levels in Remote Galaxies

Cluster	Galaxy	v_r(CMB) (km s^{-1})	V^0	Sources
Virgo	6 gE's	1300	23.73 ± 0.03	1
Fornax	6 gE's	1400	23.85 ± 0.04	1
NGC 5846	NGC 5846	2300	25.08 ± 0.10	2
Coma	IC 4051	7100	27.75 ± 0.20	3,4
Coma	NGC 4874	7100	27.82 ± 0.12	5
Coma	NGC 4881	7100	> 27.6	6
A 262	NGC 705	4650	26.95 ± 0.3	7
A 3560	NGC 5193	4020	26.12 ± 0.3	7
A 3565	IC 4296	4110	26.82 ± 0.3	7
A 3742	NGC 7014	4680	26.87 ± 0.3	7

Sources: (1) This section (2) Forbes et al. 1996a (3) Baum et al. 1997 (4) Woodworth & Harris 2000 (5) Kavelaars et al. 2000 (6) Baum et al. 1995 (7) Lauer et al. 1998

Encouragingly, however, the points for all the objects fall on the best-fit line to within the combined uncertainties in V^0 and v_r.

The last four entries in Table 16, from Lauer et al. (1998), are derived from directly resolved globular cluster populations measured from I-band *HST* imaging. These provide important verification that the cluster populations around BCG's are well within reach out to the Coma distance.

The straight average of the datapoints for the first five entries in the table (the ones with resolved GCLF turnovers) gives $\langle \log v_r - 0.2V^0 \rangle = -1.664 \pm 0.018$. Putting in $M_V^0 = -7.33 \pm 0.04$ from Table 14, we obtain $H_0 = (74 \pm 4)$ km s^{-1} Mpc^{-1}. The quoted error of course represents only the internal uncertainty of the best-fit line. The true uncertainty is dominated by the absolute uncertainty in the fundamental distance scale (Sect. 2), which we can estimate (perhaps pessimistically) as ± 0.2 mag once we add all the factors in the chain from parallaxes through the Milky Way to the Virgo/Fornax calibrating region. (For comparison, the scatter of the points about the mean line in Fig. 51 is ± 0.25 mag.) A ± 0.2-mag error in M_V^0 translates into $\Delta H_0 = \pm 7$. Thus our end result for H_0 is

$$H_0 = (74 \pm 4 \,[\mathrm{int}], \pm 7 \,[\mathrm{ext}])\,\mathrm{km\,s^{-1}Mpc^{-1}}\,. \tag{55}$$

Taking the mean of all 9 points in the table, including the turnovers deduced from the SBF analysis, would have yielded $H_0 = 72$.

What are the ultimate limits of this distance scale technique? With the *HST* cameras, many other gE galaxies (BCGs in a variety of Abell clusters) can be added to the graph in Fig. 51 out to a limit which probably approaches

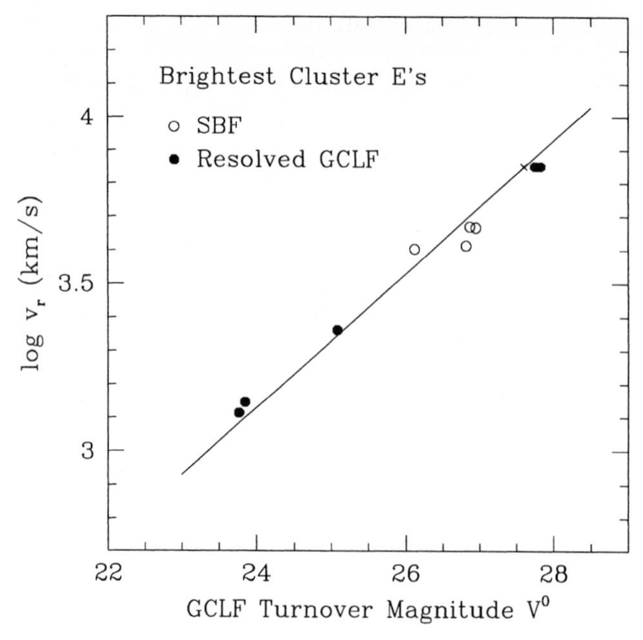

Fig. 51. Hubble diagram for globular cluster luminosity functions. The cosmological recession velocity v_r is plotted against the apparent magnitude of the GCLF turnover, for 10 brightest cluster galaxies or groups of galaxies. *Solid dots* are ones in which the GCLF has been directly resolved down to the turnover point. *Open dots* are ones in which the turnover level has been deduced by a fit to the surface brightness fluctuation function; see text. The *cross* is the lower limit for the Coma elliptical NGC 4881. The best-fit straight line (with equal weights to all the solid dots) yields a Hubble constant $H_0 = 74 \pm 8$

$cz \sim 10,000$ km s^{-1}. With two or three times as many points, the random uncertainty of the fitted line zeropoint can then be reduced to ± 2 km s^{-1} Mpc^{-1}. Similarly, if the true distance uncertainty to the Fornax and Virgo calibrators can be reduced eventually to ± 0.1 mag, then the total error (internal + external) in H_0 will be reduced to about 7%, making it competitive with any of the other methods in the literature. A more detailed discussion of the uncertainties is given by Whitmore (1997).

This approach to measuring H_0 with the GCLF is the most defensible one on astrophysical grounds. We are deliberately comparing galaxies of strictly similar types (giant ellipticals) over a range of distances, so that we can plausibly argue that the intrinsic differences in their GCLFs, due to any differences in the globular cluster system formation or evolution, will be minimized. Nevertheless, if we wish to be a bit more audacious, we can take a bigger leap of faith by pinning our assumed turnover luminosity M_V^0 *to the Milky Way alone*, arguing that there is no compelling evidence as yet that M_V^0(gE) is systematically different from M_V^0(spiral) (Table 14). This assumption would allow us to go directly from our own Galaxy to the Hubble constant in a single leap, bypassing any of the other steps through the Local Group, Virgo, or Fornax. If we do this with the two Coma ellipticals, using the relevant numbers listed above we obtain $H_0 \simeq 56-65$ depending on whether we adopt $M_V^0 = -7.4$ from the entire Milky Way sample or -7.68 from the $r_p > 3$ kpc projected halo sample (Sect. 5). It is not clear which we should do. In addition, the internal errors are significantly larger than before since only a single galaxy with a rather small GCS population is being used to calibrate the luminosity. However, this Milky Way route should be considered only as an interesting numerical exercise: there is no believable "principle of universality" for GCLFs that we can invoke here, and the true systematic differences between ellipticals and spirals are quantities which must be worked out on observational grounds.

The way that the Hubble constant affects various cosmological parameters is well known and will not be reviewed here (see the textbook of Peebles 1993 or the review of Carroll & Press 1992). For $H_0 \simeq 70$, the Hubble expansion time is $H_0^{-1} = 14.0$ Gyr. If the total mass density has its closure value of $\Omega = 1$, then the true age of the universe is $\tau = (2/3)H_0^{-1} = 9.3$ Gyr, which falls short of the currently calibrated maximum ages of the oldest stars by 3 to 5 Gyr. However, there are strong experimental indications that the overall mass density (dark or otherwise) is only $\Omega_M \simeq 0.1 - 0.3$, such as from the virial masses of rich clusters of galaxies at large radius (Carlberg et al. 1996), the abundances of the light elements (e.g., Mathews et al. 1996), the number density evolution of rich clusters of galaxies (e.g., Bahcall et al. 1997), or the power spectrum of the cosmic microwave background (e.g., Lineweaver et al. 1998).

If $\Lambda = 0$ (no vacuum energy density term) and there are no other terms to add to Ω(global) (Carroll & Press 1992), then the true age of the universe for $\Omega \sim 0.2$ would be $\tau \simeq 13$ Gyr. A value in that range is in reasonable agreement with contemporary estimates of the ages of the oldest stars in the galaxy, measured either by globular cluster ages from isochrone fitting (e.g., VandenBerg et al. 1996; Chaboyer et al. 1998; Carretta et al. 1999), or by thorium radioactive-decay age dating of metal-poor halo stars (Cowan et al. 1997). However, early results from the Hubble diagram analysis of distant supernovae favor a nonzero Ω_Λ and a combined sum $(\Omega_M + \Omega_\Lambda) \sim 1$ (Perlmutter et al. 1997; Riess et al. 1998), though on strictly observational grounds the case is still open. Should we take the somewhat cynical view that those who are hunting for large Ω_Λ are (to quote Erasmus) "looking in utter darkness for that which has no existence"? That would be premature. Many possibilities still exist for additional contributions to Ω, modified inflation models, and so on. The debate is being pursued on many fronts and is certain to continue energetically.

8 Globular Cluster Formation: In Situ Models

If we knew *what we were doing, it wouldn't be research.*
Anonymous

Understanding how globular clusters form – apparently in similar ways in an amazingly large variety of parent galaxies – is a challenging and long-standing problem. Though it still does not have a fully fleshed-out solution, remarkable progress has been made in the last decade toward understanding the times and places of cluster formation.

The scope of this problem lies in the middle ground between galaxy formation and star formation, and it is becoming increasingly clear that we will need elements of both these upper and lower scales for the complete story to emerge. At the protogalactic ($\sim$ 100 kpc) scale, the key question appears to be: How is the protogalactic gas organized? Assuming it is clumpy, what is the characteristic mass scale and mass spectrum of the clumps? Then, at the next level down ($\sim$ 1 kpc), we need to ask how protoclusters form within a single one of these gas clouds. Finally, at the smallest scales ($\lesssim$ 0.1 pc), we ask how the gas within protoclusters turns itself into stars. At *each* level it is certain that the answers will involve complex gas dynamics, and full numerical simulations covering the entire $\gtrsim 10^{10}$ dynamic range in mass and length with equal and simultaneous precision are still formidable tasks. We can, however, hope to explore some partial answers. In these next two sections, we will discuss some of the current ideas for massive star cluster formation and ask how successful they are at matching the observations we have now accumulated.

8.1 Summarizing the Essential Data

The first theoretical ideas directly relevant to globular cluster formation (in the literature before about 1992) were usually based on the concept that GC formation was in some way a "special" Jeans-mass type of process that belonged to the pre-galactic era (e.g., most notably Peebles & Dicke 1968; Fall & Rees 1985). Such approaches were very strongly driven by the characteristics of the globular clusters in the Milky Way alone, which as we have seen are massive, old, (mostly) metal-poor, and scattered through the halo. Ashman & Zepf (1998) provide an excellent overview of these early models.

All these early models run into severe difficulties when confronted with the rich range of GCS properties in other galaxies, along with the visible evidence of newly formed globular-like clusters in starburst galaxies (Sect. 9 below). For example, traditional models which assumed that globular clusters formed out of *low-metallicity* gas must now be put aside; the plain observational fact is that many or most of the globular clusters in giant E galaxies – and many in large spirals – have healthy metallicities extending up to solar abundance and perhaps even higher. Similarly, no theory can insist that globular clusters are all "primordial" objects in the sense that they formed *only* in the early universe; a wealth of new observations of colliding and starburst galaxies give compelling evidence that $\sim 10^5 - 10^6$ $M_\odot$ clusters can form in today's universe under the right conditions.

This remarkable new body of evidence has dramatically changed our thinking about cluster formation. It is hard to avoid the view that globular cluster formation is not particularly special, and is in fact linked to the more general process of star cluster formation at any mass or metallicity (e.g. Harris 1996b). Let us summarize the key observational constraints:

- *The Luminosity Distribution Function (LDF):* The number of clusters per unit mass (or luminosity) is rather well approximated by a simple empirical power law $dN/dL \sim L^{-1.8\pm0.2}$ for $L \gtrsim 10^5 L_\odot$. As far as we can tell, this LDF shape is remarkably independent of cluster metallicity, galactocentric distance, parent galaxy type, or other factors such as environment or specific frequency. For smaller masses ($M \lesssim 10^5$ $M_\odot$), dN/dL becomes more nearly constant with L, with a fairly sharp changeover at $10^5 L_\odot$ (the turnover point of the GCLF). As we saw in the previous section, the GCLF turnover is similar enough from place to place that it turns out to be an entirely respectable standard candle for estimating H_0.
- *The Metallicity Distribution Function:* The number of clusters at a given metallicity differs significantly from one galaxy to another. In the smallest dwarf galaxies, a simple, single-burst model leaving a low-metallicity population gives a useful first approximation. In large spiral galaxies and in many large ellipticals, clearly bimodal MDFs are present, signalling at least a two-stage (or perhaps multi-stage) formation history. And in some giant ellipticals such as NGC 3311 (Secker et al. 1995) or IC 4051 (Woodworth & Harris 2000), the MDF is strongly weighted to the high-[Fe/H] end, with the metal-poor component almost completely lacking. What sequence of star formation histories has generated this variety?
- *Specific Frequencies:* The classic "S_N problem" is simply stated: Why does the relative number of clusters differ by more than an order of magnitude among otherwise-similar galaxies (particularly elliptical galaxies)? Or is there a hidden parameter which, when included, would make the true cluster formation efficiency a more nearly universal ratio?
- *Continuity of cluster parameters:* Aside from the points mentioned above, one obvious observational statement we can now make (Harris 1996b) is that in the 3-parameter space of cluster mass, age, and metallicity (M, τ, Z), we can find star clusters within *some* galaxy with almost every possible combination of those parameters. The Milky Way is only one of the diverse cluster-forming environments we can choose to look at. In the physical properties of the clusters themselves, there are no sudden transitions and no rigid boundaries in this parameter space.

8.2 The Host Environments for Protoclusters

What framework can we assemble to take in all of these constraints? Having been forced to abandon the view that globular cluster formation is a special, early process, let us make a fresh start by taking the opposite extreme as a guiding precept:

All types of star clusters are fundamentally similar in origin, and we will not invoke different formation processes on the basis of mass, age, or metallicity.

The immediate implication of this viewpoint is that we should be able to learn about the formation of globular clusters by looking at the way star clusters are forming today, both in the Milky Way and elsewhere. This same point was argued on an empirical basis in a series of papers by Larson (e.g., 1988, 1990a,b, 1993, 1996) and has now turned into the beginnings of a more quantitative model by Harris & Pudritz (1994) and McLaughlin & Pudritz (1996); see also Elmegreen & Falgarone (1996) and Elmegreen & Efremov (1997) for an approach which differs in detail but starts with the same basic viewpoint. In these papers, we can find the salient features of cluster formation which are relevant to this new basis for formation modelling:

- Star clusters are seen to form out of the very densest clumps of gas within giant molecular clouds (GMCs).
- In general, the mass contained within any one protocluster is a small fraction (typically 10^{-3}) of the total mass of its host GMC. The formation of bound star clusters, in other words, is an unusual mode which seems to require a *large surrounding reservoir of gas.*
- Many or most *field stars* within the GMC are also expected to form within small groups and associations, as recent high-resolution imaging studies of nearby star forming regions suggest (e.g., Zinnecker et al. 1993; Elmegreen et al. 1999). Most of these clumps are likely to become quickly unbound (within a few Myr) after the stars form, presumably because much less than 50% of the gas within the clump was converted to stars before the stellar winds, ultraviolet radiation, and supernova shells generated by the young stars drive the remaining gas away. Observationally, we see that typically *within one GMC* only a handful ($\sim 1 - 10$) of protoclusters will form within which the star formation efficiency is high enough to permit the cluster to remain gravitationally bound over the long term. This empirical argument leads us to conclude that on average, perhaps $\lesssim 1\%$ of the host GMC mass ends up converted into bound star clusters; a much higher fraction goes into what we can call "distributed" or field-star formation.[10]

[10] As we will see later, McLaughlin (1999) arrives at a fundamentally similar conversion ratio of ~ 0.0025 by comparing the total mass in globular clusters to total galaxy mass (stars plus gas) for giant E galaxies.

- The larger the GMC, the more massive the typical star cluster we find in it. In the Orion GMC, star clusters containing $10^2 - 10^3$ $M_\odot$ have recently formed, within a GMC of $\sim 10^5$ $M_\odot$. But, for example, in the much more massive 30 Doradus region of the LMC, a $\gtrsim 2 \times 10^4$ $M_\odot$ cluster (R136) has formed; this young object can justifiably be called a young and more or less average-sized globular cluster. Still further up the mass scale, in merging gas-rich galaxies such as the Antennae (see Sect. 9 below), $\gtrsim 10^5$ $M_\odot$ young star clusters have formed within the $10^7 - 10^8$ $M_\odot$ gas clouds that were accumulated by collisional shocks during the merger.
- A GMC, as a whole, has a clumpy and filamentary structure with many embedded knots of gas and denser gas cores. Its internal pressure is dominated by the energy density from turbulence and weak magnetic field; direct thermal pressure is only a minor contributor. (That is, the internal motions of the gas within the GMC are typically an order of magnitude higher than would be expected from the temperature of the gas alone; other sources of energy are much more important.) Thus, the GMC lifetime as a gaseous entity is at least an order of magnitude longer than would be expected from radiative cooling alone; the GMC cannot cool and collapse until the internal magnetic field leaks away (e.g., Carlberg & Pudritz 1990; McKee et al. 1993), unless external influences cause it to dissipate or disrupt sooner. The gas within the GMC therefore has plenty of time to circulate, and the dense cores have relatively large amounts of time to grow and eventually form stars.
- The dense gas cores within GMCs are particularly interesting for our purposes, because they are the candidates for proto star clusters. Their mass spectrum should therefore at least roughly resemble the characteristic power-law mass distribution function that we see for the star clusters themselves. And indeed, they do – perhaps better than we could have expected: the *directly observed* mass distribution functions of the gaseous clumps and cores within GMCs follow $dN/dM \sim M^{-\alpha}$, with mass spectral index α in the range $1.5 - 2.0$. The same form of the mass distribution function, and with exponent in the same range, is seen for young star clusters in the LMC, the Milky Way, and the interacting galaxies within which massive clusters are now being built (see below). As we have already seen, the luminosity distribution function for globular clusters more massive than $\sim 10^5$ $M_\odot$ follows the same law, with minor variations from one galaxy to another. On physical grounds, the extremely high star formation efficiency ($\sim 50\%$ or even higher) necessary for the formation of a bound star cluster is the connecting link that guarantees the similarity of the mass distributions – the input mass spectrum of the protoclusters, and the emergent mass spectrum of the young star clusters (see Harris & Pudritz 1994).

The clues listed above provide powerful pointers toward the view that globular clusters formed within GMCs by much the same processes that we see operating today within gas-rich galaxies. The single leap we need to make from present-day GMCs to the formation sites of globular clusters is simply one of mass scale. Protoglobular clusters are necessarily in the range $\sim 10^4 - 10^6$ $M_\odot$. Then by the scaling ratios mentioned above, they must have formed within very large GMCs – ones containing $\sim 10^7 - 10^9$ $M_\odot$ of gas and having linear sizes up to ~ 1 kpc (Harris & Pudritz 1994). These postulated "supergiant" molecular clouds or SGMCs are larger than even the most massive GMCs found in the Local Group galaxies today by about one order of magnitude. But in the pregalactic era, they must have existed in substantial numbers within the potential wells of the large protogalaxies, as well as being scattered in sparser numbers between galaxies.

The SGMCs, in size and mass, obviously resemble the pregalactic 'fragments' invoked two decades ago by Searle (1977) and Searle & Zinn (1978). Their reason for doing so was driven by the need for appropriate environments in which place-to-place differences in local chemical enrichment could arise, thus producing a globular cluster system with a large internal scatter in metallicity and little or no radial gradient. These same dwarf-galaxy-sized gas clouds also turn out to be just what we need to produce star clusters with the right mass scale and mass spectrum. Whether we call them SGMCs, protogalactic subsystems, or pregalactic fragments, is a matter only of terminology (Harris 1996b).

We might wish to claim that globular cluster formation does preferentially belong to a "special" epoch – the early universe of protogalaxies. The grounds for this claim are simply that this was the epoch when by far the most gas was available for star formation, and SGMCs could be assembled in the largest numbers. Many Gigayears later, in today's relatively star-rich and gas-poor universe, most of the gas is in the form of (a) rather small GMCs (within spiral and irregular galaxies), which can produce only small star clusters, (b) the much lower-density ISM within the same galaxies, and (c) hot X-ray halo gas in giant ellipticals and rich clusters of galaxies, within which star formation cannot take place. Globular-sized cluster formation can still happen, but only in the rare situations where sufficiently large amounts of relatively cool gas can be assembled.

In short, the populations of globular clusters in galactic halos can be viewed as byproducts of the star formation that went on in their highly clumpy protogalaxies. Direct observations of high-redshift galaxies confirm the basic view that large galaxies form from hierarchical merging of smaller units (e.g., Pascarelle et al. 1996; Madau et al. 1996; Steidel et al. 1996; van den Bergh et al. 1996; Glazebrook et al. 1998; and references cited there). Even the smaller systems at high redshift appear to be undergoing star formation both before and during their agglomerations into larger systems. One can scarcely improve on Toomre's (1977) prescient remark that there was almost certainly "a great deal of merging of sizeable bits and pieces (including many lesser galaxies) early in the career of every major galaxy".

It is also apparent that, if any one of these SGMCs were to avoid amalgamation into a larger system and were left free to evolve on its own, it would end up as a normal dwarf galaxy – either spheroidal or irregular, depending on what happens later to its gas supply. The identification of the dwarf ellipticals that we see today as leftover "unused" pieces, some of them with globular clusters of their own, is also a natural step (Zinn 1980, 1993b; Mateo 1996). However, it seems considerably riskier to assume further that the halo of the Milky Way, or other large galaxies, was simply built by the accretion of dwarfs that had *already* formed most of their stars (e.g. Mateo 1996). Much of the merging and amalgamation of these building blocks must have happened early enough that they were still mostly gaseous – as indeed, some still are today (see below).

8.3 A Growth Model for Protoclusters

If we have convinced ourselves that globular clusters need large, local reservoirs of gas within which to form, how does the host SGMC actually convert a small portion of its gas into protoclusters? A full understanding of the process must surely plunge us deeply into the complex business of gas magnetohydrodynamics. Yet at its basis, the driving mechanism can reasonably be expected to be a simple one. The most successful, and most quantitative, approach we have at present is the model of McLaughlin & Pudritz (1996) based on the precepts in Harris & Pudritz (1994). The basis of this model is that the dense clumps of gas circulating within the GMC will build up into protoclusters by successive collision and agglomeration. Collisional growth is a well understood process which arises in many comparable situations (such as planetesimal growth in the protosolar nebula, or the buildup of a cD galaxy from its smaller neighbors). It is, in addition, important to note that the clumpy, filamentary structure of the typical GMC and its high internal motions will guarantee that collisional agglomeration will be taking place regardless of whatever else is happening within the cloud. Furthermore, the long lifetime of the GMC against cooling and collapse (see above) suggests that the growth process will have a sensibly long time to work.

We now briefly outline the essential steps in the collisional growth process, as developed especially by Field & Saslaw (1965), Kwan (1979), and McLaughlin & Pudritz (1996) for protoclusters. It is schematically outlined in Fig. 52. The GMC is idealized as containing a large supply of small gas particles of mass m_0 (small dots) which circulate within the cloud. As a very rough estimate of their mass range, we might perhaps think of the m_0's as physically resembling the $\sim$ 100 $M_\odot$ dense cores found in the Milky Way GMCs. Whenever two clouds collide, they stick together and build up larger clumps. Eventually, when a clump gets large enough, it terminates its growth by going into star formation and turning into a star cluster; the unused gas from the protocluster will be ejected back into the surrounding GMC, thus partly repopulating the supply of m_0's. At all times, the total mass in the

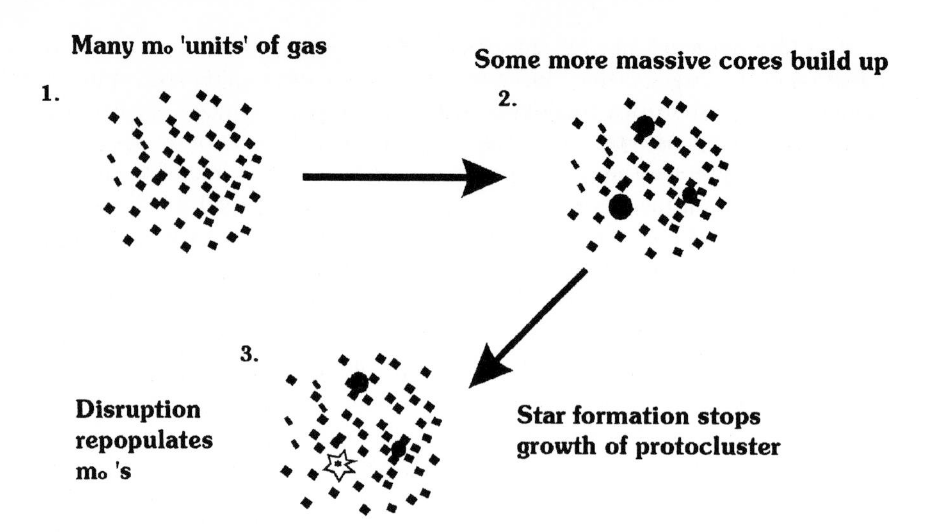

Fig. 52. Schematic illustration of the growth of protoclusters within a GMC by collisional agglomeration

protoclusters is assumed to be much less ($\lesssim 1\%$) than the total GMC mass, so the supply of m_0's is always large.

The growth of the protoclusters can then be followed as the sum of gains and losses, through a rate equation which generates a clump mass spectrum dn/dm. The number at a given mass m *decreases* whenever (a) a cloud at m combines with another at m' to form a bigger one; or (b) a cloud at m turns into stars, at a rate determined by the cooling timescale (denoted τ_m). Conversely, the number at m *increases* whenever (c) two smaller clouds m', m'' combine to form one cloud at m, or (d) larger clouds disrupt to form smaller ones, according to a "replenishment" spectrum $r(m)$. The sum of all four processes operating together as time goes on creates the output mass spectrum. Small clouds always vastly outnumber the larger ones, so that in a statistical sense, the larger ones almost always grow by absorbing much smaller clouds. By contrast, mutual collisions between two already-massive clouds are relatively rare.

The basic theory of Field & Saslaw (1965), which assumes an initial population of identical m_0's and velocities, and simple geometric collisions with no disruption or cooling, yields a characteristic distribution $dn/dm \sim m^{-1.5}$. Kwan's (1979) development of the model shows that a range of mass exponents (mostly in the range $\sim 1.5 - 2.0$) can result depending on the way the internal cloud structure and cloud velocity distribution vary with m. McLaughlin & Pudritz (1996) further show the results of including the star formation timescale and disruption processes. The detailed shape of the emergent mass spectrum is controlled by two key input parameters:

(a) The first parameter is the dependence of cloud lifetime on mass, which is modelled in this simple theory as $\tau_m \sim m^c$ for some constant exponent $c < 0$. In this picture, more massive clouds – or at least the dense protocluster regions within them – should have equilibrium structures with shorter dynamical timescales and shorter expected lifetimes before beginning star formation, thus $c < 0$. Cloud growth in this scenario can be thought of as a stochastic race against time: large clouds continue to grow by absorbing smaller ones, but as they do, it becomes more and more improbable that they can continue to survive before turning into stars. The consequence is that at the high-mass end, the slope of the mass spectrum dn/dm gradually steepens.

(b) The second parameter is the cloud lifetime τ against star formation divided by the typical cloud-cloud collision time, which is $\tau_0 \simeq m_0/(\varrho\sigma_0 v_0)$. Here ϱ is the average mass density of the clouds, σ_0 is the collision cross section of two clouds at m_0, and v_0 is the typical relative velocity between clouds. Denote $\tau_\star$ as a fiducial cloud lifetime, and then define the timescale ratio $\beta = \tau_\star/\tau_0$. For the collisional growth model to give the result that we need, we must have $\beta \gg 1$. That is, the internal timescale of the cloud governing how soon it can cool, dissipate, and go into star formation, must be significantly longer than the cloud-cloud collision time. If it is not (i.e. suppose $\beta \sim 1$), it would mean that the protocluster clouds would turn into stars roughly as fast as they could grow by collision, and thus the emergent star clusters would all have masses not much larger than m_0 itself. The larger the value of β, the shallower the slope of the spectrum dn/dm will be, and the further up to higher mass it will extend (though at high mass, it will get truncated by the decrease of τ_m, as noted above).

In summary, the timescale ratio β influences the basic slope of the power-law mass spectrum, while the exponent c determines the upper-end falloff of the spectrum slope and thus the upper mass limit of the distribution. Changes in other features of the model, such as the initial mass distribution or the details of the replenishment spectrum $r(m)$, turn out to have much less important effects. (For example, rather than assuming all the clouds to have the same mass m_0, one could assume some initial range in masses. However, the main part of the emergent mass spectrum that we are interested in is where m is orders of magnitude larger than m_0, where the memory of the initial state has been thoroughly erased by the large number of collisions. See McLaughlin & Pudritz for additional and much more detailed discussion.)

Fits of this model to the observed LDFs of the well observed globular cluster systems in the Milky Way, M31, and M87 show remarkably good agreement for the clusters *above the turnover point* (see Figs. 53 and 54). The exponent c is $\simeq -0.5$ for all of them, while the ratio β is $\simeq 115$ for M87 but $\simeq 35$ for the steeper LDFs in the two spiral galaxies.

Notably, this simple theory reproduces the upper $\sim 90\%$ of the cluster mass distribution extremely well, but it does not reproduce the abrupt flattening of the LDF at the low-mass end. A natural suspicion is, of course, that the

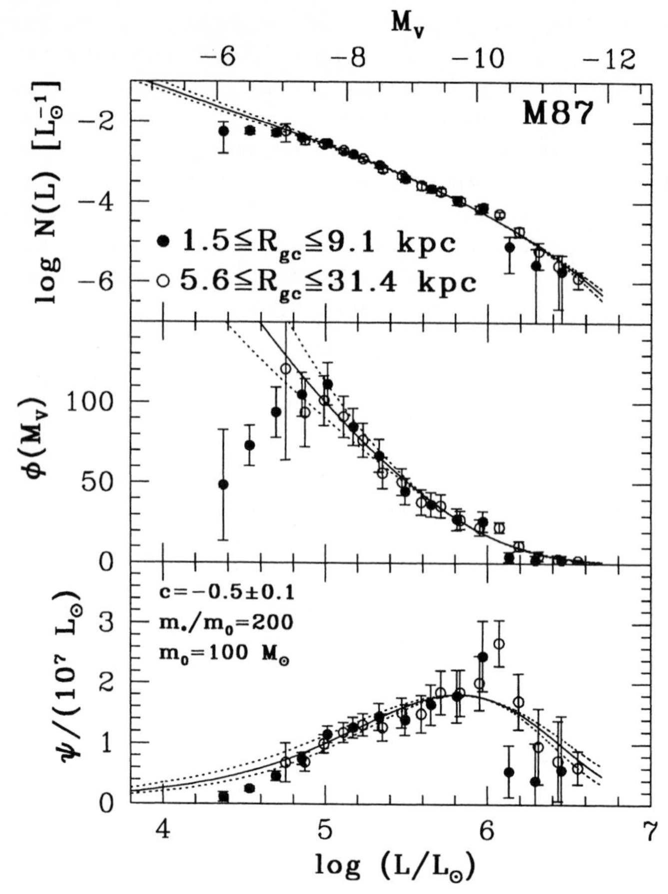

Fig. 53. Fit of the collisional-growth model of the mass distribution function to the LDF for M87 (McLaughlin 1999, private communication). The *top panel* shows the LDF itself, for a collisional growth model with "particle size" $m_0 = 100\ M_\odot$ and timescale exponent $c = -1/2$ (see text). *Solid dots* show the observed LDF for globular clusters in the inner halo of M87, open circles for the outer halo. The *middle panel* shows the luminosity distribution in its more traditional form as the number per unit magnitude (GCLF); and the *bottom panel* shows the luminosity-weighted GCLF (essentially, the amount of integrated light contained by all the clusters in each bin). Note that the model line in each case fits the data well for $\log(L/L_\odot) \gtrsim 4.7$ (the upper 90% of the mass range) but predicts too many clusters at fainter levels

observed distribution contains the combined effects of more than 10^{10} years of dynamical erosion on these clusters in the tidal field of the parent galaxy, which would preferentially remove the lower-mass ones. In other words, the formation model *must* predict "too many" low-mass clusters compared with the numbers we see today. A valuable test of this idea would therefore be to

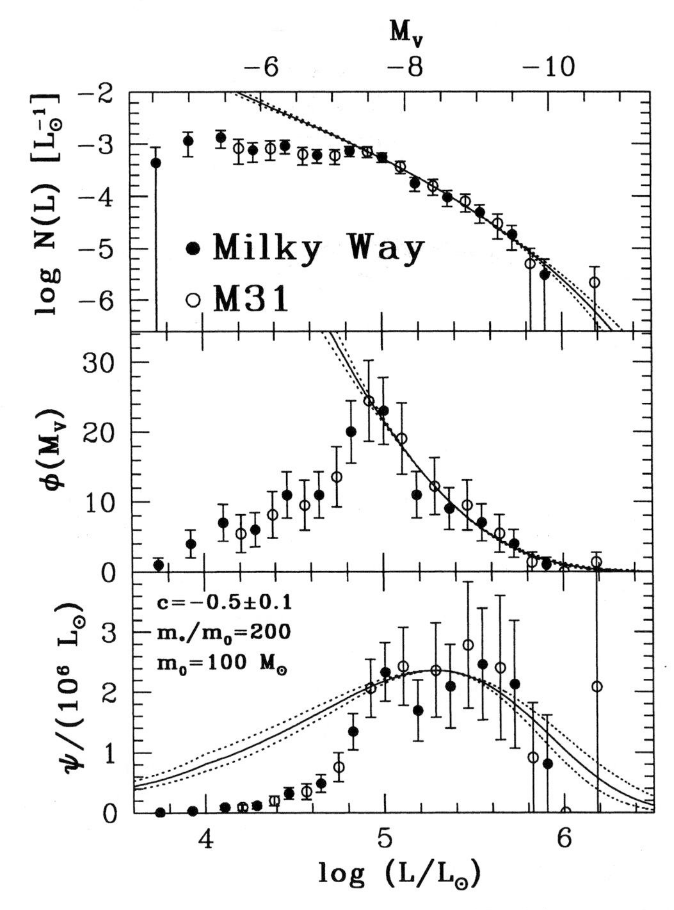

Fig. 54. Fit of the collisional-growth model of the mass distribution function to the LDFs for M31 and the Milky Way (McLaughlin 1999, private communication). Panels are the same as in the previous figure

compare the model with an LDF for a much younger set of clusters which can plausibly be assumed to have a small dispersion in age and which have had much less time to be damaged by dynamical evolution.

The best available such data at present are for the newly formed star clusters in recent mergers such as NGC 4038/4039 (Whitmore & Schweizer 1995) and NGC 7252 (Miller et al. 1997). An illustrative fit of the collisional-growth theory to these LDFs is shown in Fig. 55. Here, the low-mass end of the cluster mass distribution is more obviously present in significant numbers, and the overall distribution provides a closer global match to the theory. Encouragingly, the gradual steepening of the LDF toward the high-mass end is present here as well, just as we expect from the model.

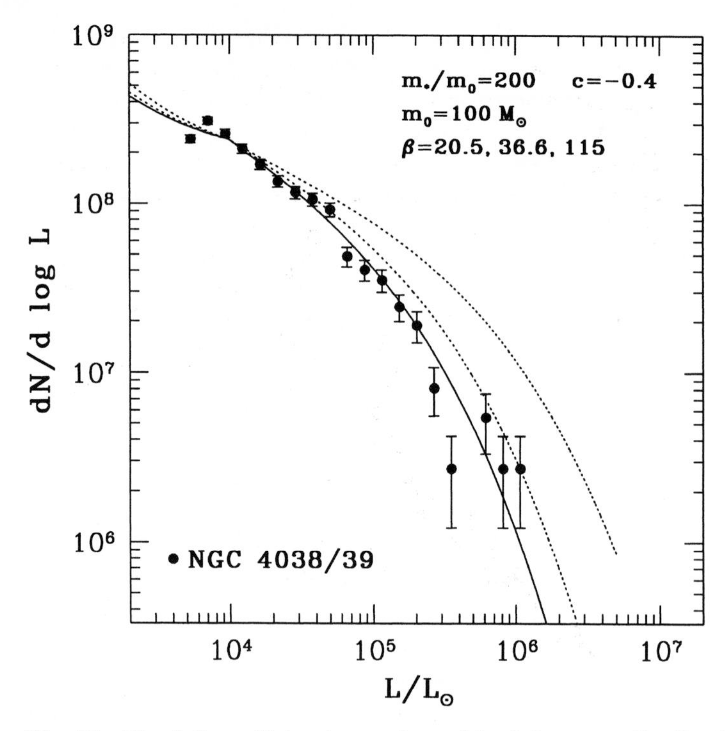

Fig. 55. Fit of the collisional-growth model of the mass distribution function to the young star clusters in the merging galaxies NGC 4038/39 (McLaughlin 1999, private communication). The *curves* running from left to right correspond to the three β-values listed

The fit shown in Fig. 55 is deceptively good, however, because it assumes that the clusters shown have a *single age*, i.e. it assumes all of them were formed in one short-duration burst. In this view, any differences in the measured colors of the clusters are ascribed as due to internal random differences in reddening rather than age. For a relatively young merger like NGC 4038/39, the clusters must surely have differences in both reddening and age, but clearly separating out these two factors is difficult. If the clusters formed in one burst, the few-Myr age differences would generate unimportant scatter in the LDF. At the opposite extreme, the ages could range over the entire $\sim$ 200 Myr duration of the merger event. From the $(U-V, V-I)$ color distribution and the presence of HII regions around many clusters, Whitmore & Schweizer (1995) argue that their age range is probably 3 to 30 Myr if they have solar metallicity, and that reddening differences probably make the estimated age range look artificially broad. Meurer (1995) provides a brief but perceptive discussion of the effect of any age range on the LDF, showing that its shape can be systematically distorted by the different rates at which younger and older clusters will fade over the same length of time.

A more extensive analysis is presented by Fritze-von Alvensleben (1998, 1999), who uses a different set of cluster models than Whitmore & Schweizer, and a different assumed metallicity of $0.5Z_{\odot}$. She deduces from the mean $(V-I)$ colour of the sample a mean cluster age of 200 Myr, similar to the estimated time of the last pericenter passage of the two galaxies (Barnes 1988). Notably, she also finds that by keeping only the most compact objects (effective radii $R_{\rm eff} < 10$ pc), the resulting LDF is then proportionately less populated at the *faint* end, curving over more strongly than would be expected from the model of Fig. 55. This effect is enhanced even further if the ages of the clusters are *individually* estimated from their $(V-I)$ colors (assuming, perhaps wrongly, that all of them have the same reddening). If all the clusters are individually age-faded to 12 Gyr, the resulting GCLF strongly resembles the classic Gaussian in number per unit magnitude, with the expected turnover at $M_V \sim -7$.

Fritze-von Alvensleben's analysis provides interesting evidence that the GCLF of globular clusters may take on its "standard" Gaussian-like form in number per unit magnitude at a very early stage. We can speculate that most of the faint clusters that should theoretically form in large numbers will appear in diffuse clumps that dissolve quickly away into the field during the first few $\sim 10^8$ y, leaving the low-mass end of the distribution depleted as the observations in all old galaxies demand.

Much remains to be investigated in more detail. For example, in the formation model outlined above, the key quantities (c, β) are free parameters to be determined by the data; more satisfactorily, we would like to understand their numerical ranges from first principles more accurately than in the present rough terms. In addition, it is not clear what determines the $\gtrsim 10^3$ ratio of host GMC mass to typical embedded cluster mass, or what this ratio might depend on. Nevertheless, this basic line of investigation appears to be extremely promising.

8.4 The Specific Frequency Problem: Cluster Formation Efficiency

Another of the outstanding and longest-standing puzzles in GCS research, as described above, is the S_N problem: in brief, why does this simple parameter differ so strongly from place to place in otherwise similar galaxies?

Let us first gain an idea of what a "normal" specific frequency means in terms of the mass fraction of the galaxy residing in its clusters. Define an efficiency parameter e as number of clusters per unit mass,

$$e = \frac{N_{\rm cl}}{M_g} \tag{56}$$

where $N_{\rm cl}$ is the number of clusters and M_g is the total gas mass in the protogalaxy that ended up converted into stars (that is, into the visible light). Then we have S_N = const $\times$ e or

$$e = \frac{S_N}{8.55 \times 10^7 (M/L)_V} \tag{57}$$

where $(M/L)_V \sim 8$ is the visual mass-to-light ratio for the typical old-halo stellar population. Most of the reliable S_N measurements are for E galaxies, and for a baseline average $S_N^0 \simeq 3.5$ (Sect. 6), we immediately obtain a fiducial efficiency ratio $e_0 \simeq 5.1 \times 10^{-9} M_\odot^{-1}$, or $e_0^{-1} \sim 2 \times 10^8$ $M_\odot$ per cluster. For an average cluster mass $\langle M_{\rm cl} \rangle = 3 \times 10^5$ $M_\odot$ (from the Milky Way sample), the typical mass fraction in globular clusters is $e_0 \langle M_{\rm cl} \rangle = 0.0015$, or 0.15%. Although this ratio is encouragingly close to what we argued empirically for GMCs in the previous section, we would have to allow for galaxy-to-galaxy differences of factors of 5 or so both above and below this mean value, in order to accommodate all the E galaxies we know about. Invoking simple differences in the efficiency with which gas was converted into star clusters remains a possibility, but is an uncomfortably arbitrary route.

The alternative possibility is to assume that the initial cluster formation efficiency was more or less *the same in all environments*, but that the higher-S_N galaxies like M87 did not use up all their initial gas supply (in a sense, we should view such galaxies not as "cluster-rich" but instead as 'field-star poor"). This view has been raised by Blakeslee (1997), Blakeslee et al. (1997), Harris et al. (1998a), and Kavelaars (1999), and is developed in an extensive analysis by McLaughlin (1999). Simply stated, this view requires that the globular clusters formed in numbers that were in direct proportion to the *total available gas supply* within the whole protogalaxy, and *not* in proportion to the amount of gas that actually ended up in stars of all types.

McLaughlin (1999) defines a new efficiency parameter as a ratio of masses as follows:

$$\varepsilon = \frac{M_{\rm cl}}{M_\star + M_{\rm gas}} \tag{58}$$

where $M_\star$ is the mass now in visible stars, while $M_{\rm gas}$ is the remaining mass in or around the galactic halo. This residual gas was, by hypothesis, originally part of the protogalaxy. In most large galaxies, $M_{\rm gas} \ll M_\star$. However, for the giant E galaxies with large amounts of hot X-ray halo gas, the additional $M_{\rm gas}$ factor can be quite significant, especially for cD galaxies and other BCGs (brightest-cluster galaxies). Under this hypothesis, the observed specific frequency represents the *proportion of unused or lost initial gas mass* (Harris et al. 1998a):

$$f_M(\text{lost}) = \frac{M_{\rm gas}}{M_\star + M_{\rm gas}} = \left(1 - \frac{S_N^0}{S_N}\right) . \tag{59}$$

If, for example, we adopt a baseline $S_N^0 = 3.5$, then a high-S_N BCG like M87 would have $f_M \sim 0.7$, implying that a startlingly high amount – almost three-quarters – of its initial protogalactic mass went unconverted during star formation. Blakeslee (1997) hypothesizes that much of the gas in the original distribution of pregalactic clouds may have been stripped away to join the general potential well of the galaxy cluster during the violent virialization stage. Harris et al. (1998a) suggest instead that a large amount of gas within the proto-BCG may have been expelled outward in a galactic wind during the first major, violent burst of star formation. Both of these mechanisms may have been important, particularly in rich clusters of galaxies. In either case, this ejected or stripped gas would now occupy the halo and intracluster medium in the form of the well known hot gas detectable in X-rays.

By a detailed analysis of three giant E galaxies with high-quality surface photometry and data for X-ray gas and globular cluster populations (NGC 4472 and M87 in Virgo, and NGC 1399 in Fornax), McLaughlin (1999) finds that the mass ratio ε is much the same in all three, at $\langle\varepsilon\rangle = 0.0025 \pm 0.007$. This result also turns out to hold at the *local* as well as the *global* level, at any one radius of the halo as well as averaged over the whole galaxy. M87, with the highest specific frequency, also has the most halo gas; its proportions of stellar and gas mass are to first order similar. For most galaxies (non-BCGs), the halo gas makes only a minor contribution and S_N is a more nearly correct representation of the mass ratio ε.

M87 is only one of many BCGs, and these galaxies as a class are the ones with systematically high S_N and high X-ray luminosity. The direct observations of their cluster populations (see Harris et al. 1998a for a summary of the data) show that the total cluster population scales with visual galaxy luminosity as $N_{\rm cl} \sim L_V^{1.8}$; while the X-ray luminosity scales as $L_X \sim L_V^{2.5\pm0.5}$. With the *very* crude (and, in fact, incorrect) assumption that the X-ray gas mass scales directly as L_X, we would then expect that the ratio of gas mass to stellar mass increases with galaxy size roughly as $(M_{\rm gas}/M_\star) \sim L_V^{1.5}$, which turns out to match the way in which S_N systematically increases with luminosity for BCGs.

McLaughlin (1999) analyzes these scaling relations in considerably more detail, putting in the fundamental-plane relations for gE galaxies (scale size, internal velocity dispersion, and mass-to-light ratio as functions of L_V) as well as the way that the gas mass scales with X-ray luminosity, temperature, and halo scale size. When these are factored in, he obtains

$$S_N \sim \frac{N_{\rm cl}}{L_{\rm gal}} \sim \varepsilon\,(1 + \frac{M_{\rm gas}}{M_\star}) L_{\rm gal}^{0.3} \,. \tag{60}$$

This correlation is shown as the model line in Fig. 56. The last term $L_{\rm gal}^{0.3}$ accounts for the systematic increase in mass-to-light ratio with galaxy luminosity: bigger ellipticals have more mass per unit light and thus generated more *clusters* per unit light under the assumption that ε (the cluster mass fraction) was constant. The term in parentheses $(1 + \frac{M_{\rm gas}}{M_\star})$ accounts for the presence of

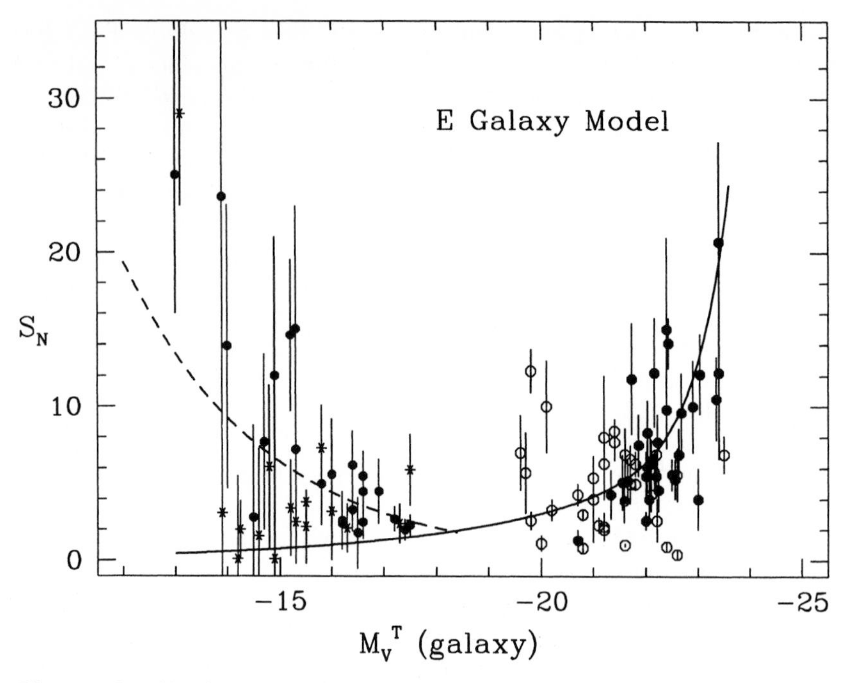

Fig. 56. Specific frequency against luminosity for elliptical galaxies. The model line (*solid curve*, from McLaughlin 1999), assumes $\varepsilon = 0.0025 =$ constant, i.e. that globular clusters were formed in these galaxies in direct proportion to the original gas mass of the protogalaxy. BCG galaxies (see text) are the *solid symbols* at high luminosity, while normal E's are plotted as *open symbols*. At the low-luminosity end, nucleated dE's are the *solid symbols* and non-nucleated dE's are the *starred symbols*. The *dashed line* assumes the Dekel/Silk model of mass loss, which strongly affects the smaller dwarfs. See text for discussion

high-temperature gas in the halo. As suggested above, in this first-order picture the gas is assumed to have belonged to the protogalaxy, but was heated at an early stage during star formation (by an energetic galactic wind, or tidal stripping of the SGMCs?), and left to occupy the dark-matter potential well in a shallower distribution than the halo stars.

An intriguing implication – and requirement – of this overall view would be that the main epoch of globular cluster formation must be *early* in the star-forming stage: that is, the clusters form in numbers that are in direct proportion to the total gas supply, and do so ahead of most of the stars. Then, in the most massive protogalaxies, the star formation is interrupted before it can run to completion, leaving behind lots of clusters as well as a considerable amount of hot, diffuse gas surrounding the visible galaxy.

It is probably not unreasonable to suppose that the star formation would proceed soonest and fastest within the densest clumps of gas (i.e., the protoclusters). For example, observational evidence from the color-magnitude diagram for R136 (in the LMC, and certainly the nearest example of a young globular cluster) indicates that it has taken $\lesssim 3$ Myr to form, from the lowest-mass stars to the highest-mass ones (Massey & Hunter 1998). It would be premature to claim that this result would be typical of all massive star clusters, but it favors the view that the densest clusters can form rapidly, over times far shorter than the $\sim 10^8$ y dynamical timescale of the protogalaxy.

8.5 Intergalactic Globular Clusters: Fact or Fancy?

White (1987) raised the intriguing idea that the BCGs in rich clusters of galaxies might be surrounded by populations of globular clusters that are not gravitationally bound to the BCG itself, but instead belong to the general potential well of the whole cluster. West et al. (1995) have pursued this concept in detail, suggesting that the huge globular cluster populations around most BCGs might be dominated by such "intergalactic" clusters.

At some level, free-floating intergalactic clusters must be present: strong evidence now exists for intergalactic Population II stars in the Fornax and Virgo clusters (e.g., Theuns & Warren 1997; Ferguson et al. 1998; Ciardullo et al. 1998), and some globular clusters should accompany these stars if they have been tidally stripped from the cluster galaxies. It is only a matter of time before individual cases are detected in nearby clusters of galaxies (accidentally or otherwise). Systematic searches are also underway for "orphan" collections of globular clusters near the centers of clusters of galaxies that do not have central BCGs.

It is, however, still totally unclear whether or not such objects would exist in sufficient numbers to affect the BCGs noticeably. Harris et al. (1998a) use the Virgo giant M87 as a detailed case study to reveal a number of critical problems with this scenario. To *create* the high S_N values seen in the BCGs, the putative intergalactic globulars would have to be present in large numbers *without adding contaminating field-star light of their own*; that is, the intergalactic material would itself need to possess a specific frequency in the range $S_N \sim 100$ or even higher. Furthermore, the "extra" clusters in M87 and the other BCGs are not just distributed in the outermost halo where the larger-scale intergalactic population might be expected to dominate; M87 has more clusters at *all* radii, even in the central few kpc. If all of these are intergalactic, then they would need to be concentrated spatially like the central galaxy, which is contradictory to the original hypothesis.

A promising new way to search for intergalactic material is through radial velocity measurement: such clusters (or planetary nebulae, among the halo stars) would show up as extreme outliers in the velocity histogram. Preliminary velocity surveys of the GCS around the Fornax cD NGC 1399 (Kissler-Patig 1998; Minniti et al. 1998; Kissler-Patig et al. 1999) are suggestive of either an intergalactic component with high velocity dispersion, or possibly contamination from neighboring galaxies. By comparison, few such objects

Table 17. Cooling-flow galaxies in Abell clusters

Abell Cluster	BCG	Young GCs ?	dM/dt
Virgo	M87	N	15
A426	N1275	Y	200
A496		N	38
A1060	N3311	Y?	10
A1795		Y	200
A2029	IC1011	N	260
A2052	U9799	(N)	55
A2107	U9958	(N)	8
A2199	N6166	(N)	70
A2597		N	135
MKW4	N4073	(N)	10

appear in the M87 cluster velocity data (see Cohen & Ryzhov 1997; Harris et al. 1998a). Larger statistical samples of velocities are needed.

8.6 The Relevance of Cooling Flows

An idea proposed some time ago by Fabian et al. (1984) was that the large numbers of "extra" clusters in M87 could have condensed out of the cooling flow from the X-ray gas. By inference, other high-S_N galaxies – mainly the central BCGs with large X-ray halos – should be ones with high cooling flows.

This hypothesis has become steadily less plausible. The biggest *a priori* difficulty is that to produce an increased S_N this way, we would have to invoke particularly efficient globular cluster formation (relative to field-star formation) out of the hot, dilute X-ray gas – exactly the type of situation that we would expect should be least likely to do this. However, probably the strongest argument against such a scheme is that there is not the slightest observational evidence that young, massive star clusters exist in any of the pure cooling-flow galaxies (for more extensive discussion, see Harris et al. 1995; Bridges et al. 1996; Holtzman et al. 1996). The true mass dropout rates from these cooling flows remain uncertain, and may in any case be considerably less than was thought in the early days of the subject.

A summary of the observed cases is given in Table 17: the Abell cluster designation and central galaxy NGC number (if any) are given in the first two columns (here "BCG" denotes brightest cluster galaxy), the presence or absence of young clusters in the third column, and the deduced cooling flow rate ($M_\odot$ per year) in the last column. GCS data are taken from the three papers cited above, and the cooling flow rates from Allen & Fabian

(1997), McNamara & O'Connell (1992), and Stewart et al. (1984). The only two cases which are seen to contain young globular clusters in their central regions (the BCGs in Abell 426 and 1795) are also the two which represent accreting or interacting systems, with large amounts of cooler gas present as well. It seems more probable that it is the cooler, infalling gas that has given rise to the recent starbursts in these giant galaxies, rather than the hot X-ray gas halo.

It is interesting to note that the cooling-flow scenario can, in some sense, be viewed historically as exactly the opposite of the view discussed above involving early mass loss from the BCGs. Rather than suggesting that large number of clusters formed at later times out of the hot X-ray gas, we now suggest that the clusters formed in their large numbers at an early stage, in direct proportion to the original reservoir of gas; but that considerable unused gas was ejected or left out in the halo to form the X-ray halo – which is now the source of the cooling flow.

8.7 Dwarf Ellipticals

The dwarf elliptical galaxies present a special puzzle. We see that extreme S_N values are also found among the lowest-luminosity dwarf ellipticals, at the opposite end of the galaxy size scale; but ones with low S_N are present too (Sect. 6). It seems likely that *early mass loss* is responsible for the high S_N values in these dwarfs. As discussed in Durrell et al. (1996a) and McLaughlin (1999), these tiny and isolated systems are the objects most likely to have suffered considerable mass loss from the first round of supernovae, leaving behind whatever stars and clusters had managed to form before then. Using this picture, Dekel & Silk (1986) argue that the expected gas vs. stellar mass scalings for dwarfs should go as $(M_{\rm gas}/M_\star) \sim L^{-0.4}$. But if the efficiency of cluster formation ε is constant (as was discussed above), then we should expect $S_N \sim L^{-0.4}$ for dwarf ellipticals. In deducing the role of the gas, note that there is one important difference between the dE's and the BCGs discussed above: in the dwarfs, the ejected fraction of the initial supply $M_{\rm gas}$ does not stay around in their halos. Only the stellar contribution $M_\star$ remains in their small dark-matter potential wells.

This scaling model ($S_N \sim L^{-0.4}$, starting at $M_T^V \simeq -18.4$ or about $2 \times 10^9 L_\odot$), is shown in Fig. 56. It does indeed come close to matching the observed trend for *nucleated* dE's but not the non-nucleated ones. At low luminosity ($M_V^T \sim -12$), we would expect from this scaling model that dE's should have S_N values approaching 20, much like what is seen in (e.g.) the Local Group dwarfs Fornax and Sagittarius, or some of the small nucleated dwarfs in Virgo. On the other hand, the non-nucleated dE's fall closer to the scaling model curve (60) which is simply the low-luminosity extension of the giant ellipticals. Intermediate cases are also present. Why the huge range between the two types?

The two brands of dwarf E's have other distinctive characteristics. The dE,N types have central nuclei which, in their spatial sizes and colors, resemble giant globular clusters (Durrell et al. 1996a; Miller et al. 1998) and indeed, some of the smaller nuclei may simply be single globular clusters drawn in to the center of the potential well of the dwarf by dynamical friction (the cluster NGC 6715 at the center of the Sagittarius dE may be the nearest such example). The most luminous nuclei, however, far exceed even the brightest known globular clusters; these may represent true nuclei formed by strongly dissipative gaseous infall at a moderately early stage (e.g., Caldwell & Bothun 1987; Durrell et al. 1996a). It is also well known that the dE,N and dE types exhibit different spatial distributions in the Virgo and Fornax clusters (Ferguson & Sandage 1989): the dE,N types follow a more centrally concentrated distribution resembling the giant E's, while the dE types occupy a more extended distribution resembling the spirals and irregulars. Differences in shape have also been noted; the dE (non-nucleated) types have more elongated isophotes on average (Ryden & Terndrup 1994; Binggeli & Popescu 1995).

A plausible synthesis of this evidence (see Durrell et al. 1996a; Miller et al. 1998) is that the dE,N types represent "genuine" small ellipticals in the sense that they formed in a single early burst. Many of them clearly formed near the giant BCGs, and thus may have been in denser, pressure-confined surroundings which allowed some to keep enough of their gas to build a nucleus later (e.g., Babul & Rees 1992). The dE types, by contrast, may represent a mixture of gas-stripped irregulars, some genuine ellipticals, or even quiescent irregulars that have simply age-faded. For many non-nucleated dE's, the scaling model $S_N \sim L^{-0.4}$ involving early mass loss may not apply, and if there was much less mass loss then something closer to $S_N \sim$ const should be more relevant.

If this interpretation of the dwarfs has merit, then once again we must assume that the main era of globular cluster formation went on at a very early stage of the overall starburst, before the supernova winds drove out the rest of the gas. More direct evidence in favor of this view, such as from contemporary starburst dwarfs (see below), would add an important consistency test to this argument.

For the complete range of elliptical galaxies, the pattern of specific frequency with luminosity is shown in Fig. 56, with the McLaughlin model interpretation. It is encouraging that a plausible basis for interpreting the high-S_N systems at both the top and bottom ends of the graph now exists, and that we have at least a partial answer to the classic "S_N problem". Nevertheless, individual anomalies remain at all levels, with cluster numbers that are too "high" or "low" for the mean line. Will we have to conclude from the high-S_N, non-BCG cases that genuinely high-efficiency cluster formation can indeed occur? Are all the low-S_N cases just instances of simple gas-poor mergers of spirals, which had few clusters to begin with? There is much still to be done to understand these cases, as well as to fill in the complete story of early star formation in the central giant ellipticals.

9 Formation: Mergers, Accretions, and Starbursts

The way to get good ideas is to get lots of ideas and throw the bad ones away.
Linus Pauling

When we begin reviewing the issues relating galaxy formation to cluster formation, we are plunging into much more uncharted territory than in the previous sections. The flavor of the discussion must now shift to material that is less quantitative, and less certain. It is fair to say that in the past decade especially, we have isolated the *processes* that need to be understood in more detail (gas dynamics within the clumpy structure of protogalaxies, and later processes such as galaxy mergers, satellite accretions, galactic winds, tidal stripping, and dynamical evolution). However, we are not always able to say with confidence *which* of these mechanisms should dominate the formation of the GCS in any one galaxy. In this exploratory spirit, let us move ahead to survey the landscape of ideas as they stand at present.

The formation scenarios discussed in Sect. 8 can be classified as *in situ* models: that is, the galaxy is assumed to be formed predominantly out of an initial gas supply that is "on site" from the beginning. In such models,later modifications to the population from outside influences are regarded as unimportant. From the limited evidence now available, the *in situ* approach may well be a plausible one for many ellipticals, but it cannot be the whole story. We see galaxies in today's universe undergoing major *mergers*; *starbursts* from large amounts of embedded gas; and *accretions* of smaller infalling or satellite galaxies. Can these processes have major effects on the globular cluster populations within large spirals and ellipticals?

9.1 Mergers and the Specific Frequency Problem

Considerable evidence now exists that merging of smaller already-formed galaxies occurs at all directly visible redshifts. In a high fraction of these cases, the outcome of repeated mergers is expected to be an elliptical galaxy, and a traditional question is to ask whether all ellipticals might have formed this way. Considerable enthusiasm for the idea can be found in the literature over the past two decades and more. However, a primary nagging problem has to do with the specific frequencies of the relevant galaxies. Disk galaxies are, in this view, postulated to be the progenitors from which larger E galaxies are built. But disks or spirals consistently have specific frequencies in a rather narrow range $S_N \lesssim 2$, while (as discussed in Sect. 6 above) specific frequencies for ellipticals occupy a much larger range up to several times higher. This "S_N problem" is not the same one discussed in the previous section (which applied to the BCGs vs. normal ellipticals); instead, it addresses the offset in S_N between two very different types of galaxies. Trying to circumvent this problem – that is, making a high-S_N galaxy by combining low-S_N ones – has generated an interesting and vivid literature.

A brief review of the key papers in historical sequence will give the flavor of the debate. The seminal paper of Toomre (1977) first showed convincingly from numerical simulations that direct mergers of disk galaxies could form large ellipticals. Toomre speculated that a large fraction of present-day gE's originated this way. Not long after that, the first surveys of GCSs in Virgo and in smaller galaxy groups came available (Hanes 1977; Harris & van den Bergh 1981). Using this material, Harris (1981) suggested that the *low-*S_N ellipticals, which are found characteristically in small groups and the field, might reasonably be argued to be the products of spiral mergers. Harris' paper concludes with the statement "The merger process cannot *increase* the specific frequency, unless vast numbers of extra clusters were somehow stimulated to form *during* a major collision early in its history, when substantial amounts of gas were still present". (It should be noted, however, that this last comment was a throwaway remark which the author did not really take as a serious possibility at that time!) Subsequently, van den Bergh (1982 and several later papers) repeatedly emphasized the difficulty of using disk-galaxy mergers to form "normal" (that is, Virgo-like) ellipticals in the range $S_N \sim 5$.

For descriptive purposes, let us call a *passive* merger one in which the stellar populations in the two galaxies are simply added together with no new star formation. In a passive merger, one would expect the specific frequency of the product to be the simple average of the two progenitors. Even after considerable age-fading of the Population I disk light, such a combination of low-S_N disk galaxies would never yield a sufficiently high S_N to match the Virgo-like ellipticals.

The debate gained momentum when Schweizer (1987) emphasized that mergers of spirals could be *active* rather than passive: that is, the progenitors could contain considerable gas, and globular clusters could form *during the merger*, thus changing the specific frequency of the merger product. In Schweizer's words, "What better environment is there to produce massive clusters than the highly crunched gas in [merging] systems? ... I would predict that remnants of merged spirals must have more globular clusters per unit luminosity than the spirals had originally". Many subsequent authors took this statement as a signal that the "specific frequency problem" had therefore been solved. But it had not.

In two influential papers, Ashman & Zepf (1992) and Zepf & Ashman (1993) published a more quantitative model for the merger of gas-rich galaxies and active cluster formation during the merger, and used it to predict other characteristics (metallicities, spatial distributions) for the resulting GCS. Their formalism emphasized single merger events of two roughly equal spirals, though it could be extended to multiple events. A considerable stimulus for this model was the growing awareness that the metallicity distributions of the globular clusters in many giant ellipticals had a bimodal form, suggesting (in the Ashman-Zepf view) that the metal-richer population formed during the merger. In their argument, during the collision the gas from both galaxies would dissipate, funnel in toward the center of the new proto-E galaxy, and form new stars and clusters there. Many Gyr after the merger had finished, we would then see an elliptical with a bimodal MDF and with the metal-richer component more centrally concentrated.[11]

The Ashman/Zepf model also predicts that the metallicity gradient should be steeper for the GCS than for the halo light (the "younger" merger-produced MRC clusters would have formed preferentially in the core regions and with greater efficiency). This overall picture became additionally attractive with the discovery of "young globular clusters" (compact, cluster-sized star forming regions with masses extending up past the $\gtrsim 10^5$ $M_\odot$ range) in gas-rich interacting galaxies such as NGC 1275 (Holtzman et al. 1992) and NGC 3597 (Lutz 1991). Many similar cases are now known (discussed below), in which considerable gas seems to have collected by merger or accretion events.

The community response to this merger scenario was initially enthusiastic and somewhat uncritical. However, counterarguments were also raised. Even if globular clusters form during mergers, a higher specific frequency is *not* necessarily the result: field-star formation goes on at the same time as cluster formation, and the final S_N could be either higher *or* lower depending on the efficiency of cluster formation (Harris 1995).

It may seem attractive to assume that the highly shocked and compressed gas generated during disk mergers would be a good place for high-efficiency cluster formation. But it is not clear that these shocks would be any more extreme than in the range of collisions that took place in the protogalactic era, when much more gas was present and the random motions were comparably high. There seems no need to automatically assume that the cluster formation efficiency in *present-day mergers* would be higher than in the protogalactic era. There is, in addition, a problem of sheer numbers: in a normal Virgo-like elliptical there are thousands of MRC clusters, and the quantitative demands

[11] Since Ashman & Zepf's original discussions, many authors have conventionally taken observations of bimodal MDFs as "supporting" the merger model. The correct statement is that bimodality is "consistent" with Ashman/Zepf. In general, observational evidence can be said to be *consistent* with a particular model if it falls within the expected results of that model. However, the same evidence might also be consistent with other models. To say that evidence *supports* a model is a much stronger statement: it requires the data to be consistent with that model *but inconsistent with other models*; that is, competing models are ruled out. We are fortunate indeed if our observations turn out to be strong enough to agree with only one model and to rule out competing ones! In this case, a bimodal MDF can equally well result from any *in situ* formation picture which involves at least two distinct epochs of star formation, with gaseous dissipation and infall occurring in between.

on the merger to create all of these are extreme (see Harris 1995 and the discussion below).

9.2 Observations of Merger Remnants

Clearly, what has been needed most of all to understand the characteristics of globular cluster formation during mergers is a series of new observations of merged galaxies. Whitmore, Schweizer, Zepf, and their colleagues have used the *HST* to carry out an important series of imaging studies of star clusters in galaxies that are clearly merger products, in an identifiable age sequence. Published cases include NGC 4038/39 where the young star formation is $\sim 10^8$ y old (Whitmore & Schweizer 1995); NGC 3256 at $\gtrsim 100$ Myr (Zepf et al. 1999); NGC 3921 (Schweizer 1996, Schweizer et al. 1996) and NGC 7252 (Miller et al. 1997), both at an age ~ 700 Myr; and NGC 1700 and 3610 (Whitmore et al. 1997), at ages of 3 to 4 Gyr. Other studies in this series are in progress. These galaxies are excellent testbeds for making quantitative measurements of young vs. old cluster subpopulations, asking where they are in the merging material, and deriving their specific frequencies and mass distributions.

An example of the data from the NGC 7252 study is shown in Fig. 57. Two groups of clusters are clearly visible, with the brighter, bluer and spatially more centrally concentrated population identified as the objects formed in the merger. In this case, the color difference between the bright, blue clusters and the fainter, redder ones is interpreted as primarily due to age (750 Myr vs. $\gtrsim 10$ Gyr) rather than metallicity. One can, however, make reasonable estimates of how the younger population would evolve in luminosity and color as the galaxy ages ("age fading"). In the Whitmore/Schweizer papers the stellar population models of Bruzual & Charlot (1993) are used, in which the clusters are assumed to form in a single burst and simply evolve passively by normal stellar evolution after that. Whitmore et al. (1997) present an interesting numerical simulation showing how a single-burst population of clusters would evolve progressively in a color-magnitude diagram such as Fig. 57. Intriguingly, for roughly solar abundance and ages near $\sim 1-3$ Gyr, the integrated colors of the clusters are near $(V-I) \simeq 1$, much like conventional old-halo globular clusters that are metal-poor ([Fe/H] $\simeq -1.7$). Thus for intermediate-age mergers, separating out the two types of cluster populations becomes extremely difficult.

For younger mergers, the mean age of the burst can be plausibly estimated by the color of the blue clusters and by the velocities and geometry of the progenitors, if they are still separate (Whitmore & Schweizer 1995; Fritze-von Alvensleben 1998). From the age-fading models, we can then estimate the LDF of the young clusters (number per unit luminosity) as it would look at a normal old-halo age. In Fig. 58, the age-faded LDFs for NGC 4038/39 and NGC 7252 are compared directly with that of M87. For the upper range $L \gtrsim 0.5 \times 10^5 L_\odot$, all three galaxies match extremely well, confirming our earlier

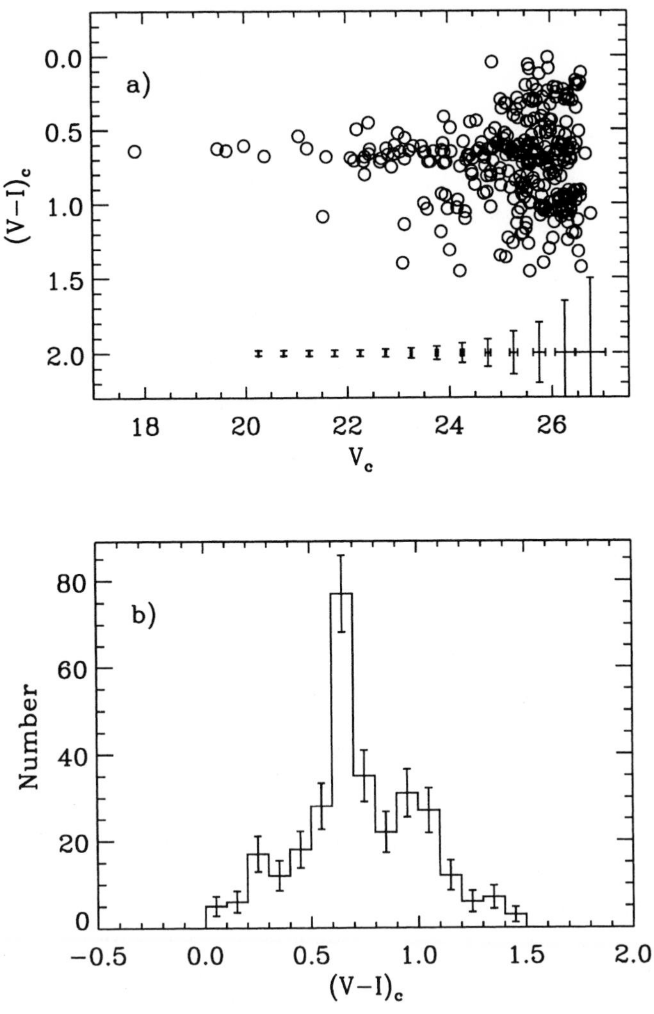

Fig. 57. (**a**) Integrated colors $(V - I)$ and apparent magnitudes V for the star clusters in the merger product NGC 7252 (Miller et al. 1997). The brighter, more numerous clusters centered at a mean color $(V - I) \simeq 0.65$ are presumed to be ones formed in the merger, while the redder, fainter population centered at $(V - I) \simeq 1.0$ is likely to be the old-halo cluster population from the two original galaxies. (**b**) Histogram of cluster colors. The two modes (young vs. old) are clearly visible. Figure courtesy Dr. B. Miller

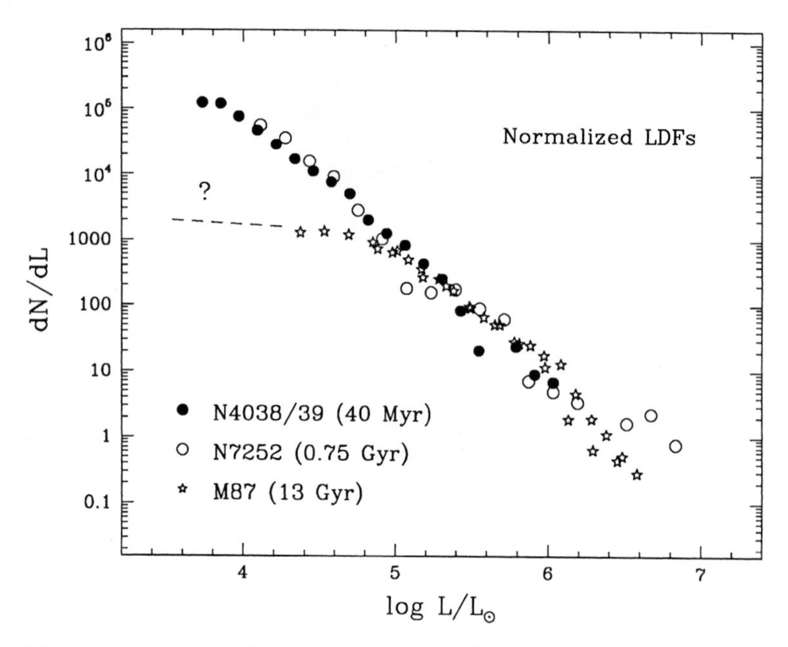

Fig. 58. Luminosity distribution functions (LDFs) for three galaxies: NGC 4038/39 (the Antennae merger system, at an age 40 Myr), NGC 7252 (a 750 Myr merger), and M87. The two merger remnants have been age-faded to an equivalent age of 13 Gyr with the Bruzual-Charlot models (see text) and then superposed on the M87 data for comparison

suggestions that the mass distribution function is not strongly affected by dynamical evolution for $M \gtrsim 10^5$ $M_\odot$. However, at lower masses, the M87 curve diverges strongly from the other two, falling well below them. Is this a signal that $\sim$ 10^{10} years of dynamical evolution have "carved away" this low-mass end of the LDF? Or have the numbers of low-mass clusters in the merger remnants been overestimated by observational selection effects? In Sect. 8 above, it was noted that a collisional growth model does predict an LDF shape continuing upward to low masses much like the observations, but this match is, perhaps, based on too simple a set of model assumptions.

In general, however, the LDF shapes in these obvious merger remnant galaxies closely match what is expected from the collisional-growth theory, $dn/dL \sim L^{-\alpha}$ with $\alpha = 1.8 \pm 0.2$ (assuming constant M/L). In NGC 4038/39 (the Antennae system), the slope is $\alpha = 1.78 \pm 0.05$; in NGC 3256, we have $\alpha = 1.8 \pm 0.1$; in NGC 3921, $\alpha = 2.12 \pm 0.22$; and in NGC 7252, $\alpha = 1.90 \pm 0.04$.

As for the sites of cluster formation, Whitmore & Schweizer (1995) note for the Antennae (the youngest merger) that "many of the clusters form tight groups, with a single giant HII region containing typically a dozen clusters". This clumpy distribution is exactly what we would expect from the SGMC formation picture, where large amounts of gas must be collected together to form local reservoirs from which cluster formation can proceed. Another intriguing feature of the Antennae system is that star formation is occurring at a high rate *well before the progenitor galaxies have completely merged.* The nuclei of the original disk galaxies are still clearly visible, and the disk gas has obviously not waited to "funnel" down in to the merged nucleus before starting star formation in earnest. Shocked, clumpy gas appears all over the merger region and even out along the tidal tails.

9.3 A Toy Model for Mergers and Specific Frequencies

Many of the issues surrounding the effect of mergers on specific frequencies can be clarified by building a simple quantitative model. Let us assume that two initial galaxies with luminosities (L_1, L_2) and specific frequencies (S_{N1}, S_{N2}) will merge to form an elliptical. (NB: these values are assumed to be "age-faded" ones, i.e. where all the light of the galaxy is reduced to the level it would have at an old-halo age $\sim 10-15$ Gyr.) Assume further that the galaxies bring in a total amount of gas M_g which is turned into new stars and clusters. There may be *additional* gas which is left unused; here we simply assume M_g is the amount actually turned into stars.

How many globular clusters do we get? During the merger we will form N_3 new clusters with a total mass $M_{\rm cl}$ at an efficiency (using our previous notation)

$$\varepsilon = \frac{M_{\rm cl}}{M_g} = 3 \times 10^5\,\mathrm{M}_\odot\,\frac{N_3}{M_g} \tag{61}$$

where the mean cluster mass is 3×10^5 $\mathrm{M}_\odot$, assumed the same as in the Milky Way. We will *also* form new stars, with a total (age-faded) luminosity

$$L_3 = \frac{M_g}{(M/L)_V} \tag{62}$$

where $(M/L)_V \simeq 8$ for old (Population II) stellar populations. Using $S_N = 8.55 \times 10^7 (N_{\rm cl}/L)$, we can quickly show that our "baseline" specific frequency $S_N^0 \simeq 3.5$ corresponds to an efficiency $\varepsilon_0 = 0.0015$. (This is about a factor of two less than for the Virgo and Fornax giant ellipticals analyzed by McLaughlin 1999, which have $S_N \sim 5$).

Now we add up the old and new clusters to get the final specific frequency in the product elliptical

$$S_N(\text{final}) = 8.55 \times 10^7 \left(\frac{N_1 + N_2 + N_3}{L_1 + L_2 + L_3}\right). \tag{63}$$

The L_3 term is forgotten by many writers.

Whether or not S_N ends up higher or lower than (S_{N1}, S_{N2}) clearly depends on both N_3 and L_3. Rewriting the result in terms of the formation efficiency ε and input gas mass M_g, we obtain

$$S_N = \left(\frac{S_{N1}L_1 + S_{N2}L_2 + 0.44 M_g(\varepsilon/\varepsilon_0)}{L_1 + L_2 + M_g/8}\right). \tag{64}$$

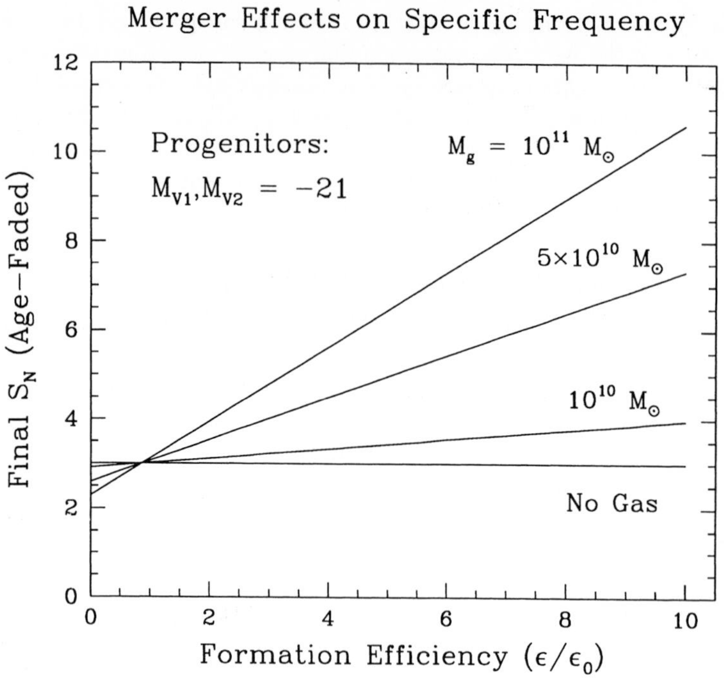

Fig. 59. Specific frequency S_N for an elliptical galaxy formed from the merger of two equal spirals. Here S_N is plotted versus the efficiency of cluster formation ε relative to the "normal" efficiency $\varepsilon_0 = 0.0015$ (see text). The four curves are labelled with the amount of gas M_g converted into stars during the merger

This is the general case for any two-galaxy merger. We can obtain a better idea of the effects if we look at specific cases. A particularly important one is for *equal progenitors*, that is, $L_1 = L_2$. If the merger is "passive" (gas-free; $M_g/M_{1,2} \ll 1$), then clearly

$$S_N = \frac{1}{2}(S_{N1} + S_{N2}) \ . \tag{65}$$

This latter situation is the limiting case discussed by Harris (1981). It was used to justify the suggestion that low-S_N ellipticals might be the end products of spiral mergers, especially in small groups where the galaxy-galaxy collision speeds are low, enhancing the probability of complete mergers.

The range of possibilities for "active" (gas-rich), equal-mass mergers is illustrated in Fig. 59. This graph shows an Antennae-like merger; both the incoming galaxies are adopted to have $M_1 \simeq M_2 = 1.7 \times 10^{11}$ $M_\odot$, equivalent to $M_V \simeq -21$. We assume them both to have age-faded specific frequencies $S_N \simeq 3$ (already an optimistically high value for spirals). For an input gas mass of 10^{10} $M_\odot$ (that is, about 3% of the total mass of the progenitors), the model shows only modest changes in the outcome S_N are possible even if the cluster formation during the merger is enormously efficient: there is simply not enough gas in this case to build a significant number of new clusters. Nevertheless, it is a plausible source for a low-S_N elliptical.

For input gas masses $M_g \gtrsim 5 \times 10^{10}$ M$_\odot$ (15% or more of the total galaxy mass, which corresponds to quite a gas-rich encounter), larger changes in S_N are possible *but only if the cluster formation efficiency is far above normal.* The essential prediction of this toy model is, therefore, that the expected result of a major merger will be an elliptical with $S_N \sim 3$, unless there is a *huge amount of input gas* and *a very high cluster formation efficiency*; both conditions must hold. Only then can we expect to build a Virgo-like (or, even more extreme, a BCG-like) elliptical this way.

For comparison, the Antennae merger has $\sim 2 \times 10^9$ M$_\odot$ of molecular gas (Stanford et al. 1990), while the extremely energetic starburst system Arp 220 has $\sim 10^{10}$ M$_\odot$ of H_2 (Scoville 1998). These are insufficient amounts of raw material to generate major changes in the specific frequencies. The most extreme case that may have been observed to date is in the giant starburst and cD galaxy NGC 1275. There, large amounts of gas have collected in its central regions, and many hundreds of young clusters or cluster-like objects have formed (Holtzman et al. 1992; Carlson et al. 1998). Carlson et al. estimate that, of the ~ 1180 young blue objects detected in their study (most of them of course at low luminosity), perhaps half would survive after 13 Gyr, assuming that they are all indeed globular clusters. Even this large new population, though, is not capable of changing the global specific frequency of this galaxy noticeably from its current level $S_N \sim 10$. Furthermore, spectra of five of the brighter young cluster candidates (Brodie et al. 1998) call into question their identification as globular clusters. Their integrated spectral properties are unlike those of young Magellanic or Galactic clusters, and can be interpreted as clusters with initial mass functions weighted strongly to the high-mass end. If so, a high fraction of them may self-disrupt or fade quickly away to extinction. This material emphasizes once again that we need to understand the first ~ 1 Gyr of evolution of a GCS before we can properly calculate the effects on the specific frequency and LDF.

The expected range of S_N can be compared with the actual merger products studied by Whitmore, Schweizer and their colleagues referred to above. In each case they have made empirical estimates of the global S_N in the end-product elliptical, and in each case it is at a level $S_N \lesssim 3$ consistent with the view that cluster formation efficiency during the merger is, in fact, not much different from the normal ε_0 level.

One of these calculations will illustrate the technique: for the Antennae, Whitmore & Schweizer (1995) find a total of ~ 700 young, blue objects (assumed to be mostly clusters); however, they identify only 22 to be brighter than the classic GCLF turnover point at $\sim 10^5$ M$_\odot$ and these are the important ones for calculating the specific frequency long after the merger is over, since the small or diffuse clusters will have largely been destroyed. Adding these to an estimated $\sim 50-70$ old-halo clusters already present in the progenitor galaxies then gives a total population of $\lesssim 100$ old clusters and a total (age-faded) E galaxy luminosity $M_V^T \simeq -20.7$ long after the merger is complete. The resulting specific frequency is $S_N \simeq 0.5$, which places it at the bottom end of the scale for observed E galaxies. The more recent analysis of the same material by Fritze-von Alvensleben (1998, 1999), who uses individual age estimates for each object to estimate their masses, suggests instead that the number of young clusters more massive than the mass distribution is already Gaussian in number per unit log mass, and that there may be as many as ~ 150 above the turnover. Adopting this higher total, however, only raises the final estimate to $S_N \simeq 1.2$. In short, the Antennae merger is producing a cluster-poor elliptical.

Similar calculations for the other galaxies in the series listed above (NGC 1700, 3610, 3921, 5018, 7252) are perhaps a bit more reliable since they are older remnants, and differences in internal reddening and age are less important. For these, the age-faded specific frequencies predicted empirically (see the references cited above) are all in the range $1.4 \lesssim S_N \lesssim 3.5$. The model displayed in Fig. 59 with normal production efficiencies appears to be an entirely tolerable match to these observations.

There is clear evidence from specific frequencies that E galaxies can be built, and are being built today, by mergers of pre-existing disk systems. But the type of elliptical being produced in such mergers resembles the low-S_N ones in small groups and in the field.

9.4 Other Aspects of the Merger Approach

The ellipticals in rich clusters provide a much stronger challenge to the simple merger scheme. The first and perhaps biggest barrier is connected with the sheer numbers of clusters in these gE's, and can be illustrated as follows (Harris 1995). Let us take NGC 4472 in Virgo as a testbed "normal" object ($S_N \simeq 5$). Its clusters display a bimodal MDF with about 3660 in the metal-poor population and 2400 in the metal-richer population. Now, *if* the entire galaxy formed by mergers, then by hypothesis all the MPC clusters must have come from the pre-existing disk systems. This would require amalgamating about 30 galaxies the size of the Milky Way, or an appropriately larger number of dwarfs (see below). Similarly, suppose we assume optimistically that all the MRC clusters formed during the mergers at rather high efficiency ($\varepsilon \simeq 0.003$); then the amount of input gas needed to do this would be at least 2.4×10^{11} M$_\odot$, or $\sim 10^{10}$ M$_\odot$ per merger. These are *very* gas-rich mergers. The amounts may be even larger if we account for wastage, i.e. gas lost to the system. In addition, over many different mergers it is not clear that a cleanly bimodal MDF could be preserved.

Alternately, one could assume that the galaxy was built by just one or two much larger mergers with almost all the gas ($\sim 3 \times 10^{11}$ M$_\odot$) coming in at once. The progenitor galaxies would, in fact, then have to be mostly gaseous, unlike any merger happening today. The only epoch at which such large supplies of gas were routinely available was the protogalactic one. It is then not clear how the merger scheme would differ in any essential way from the regular *in situ* (Searle/Zinn-like) model.

Specific frequencies are not the only outcome of a merger that is amenable to observational test. Detailed evaluations of other measurable features, particularly the metallicity distributions, have been given recently especially by Geisler et al. (1996), Forbes et al. (1997), and Kissler-Patig et al. (1998). Noteworthy problems that arise from these analyses are as follows:

- In the merger model, the highest-S_N ellipticals should have the great majority of their clusters in the metal-rich component, since by hypothesis these are created during the merger at high efficiency. A few of them do (NGC 3311 in Hydra I and IC 4051 in Coma; see Secker et al. 1995, Woodworth & Harris 1999), but many certainly do not (notably M87 and NGC 1399, where the metal-poor clusters are in a slight majority).
- If a sequence of mergers is required to build up a giant elliptical, then the gas – which is enriched further at each stage – should produce a multimodal or broad MDF, rather than the distinct bimodal (or even narrow, unimodal) MDFs that are observed.
- In most giant ellipticals, the metal-poor component is itself at *higher* mean metallicity (typically [Fe/H] $\simeq -1.2$; see Forbes et al. 1997) than the MPCs in spirals and dwarfs (at [Fe/H] $\simeq -1.6$), suggesting that they did not originate in these smaller systems after all.
- Radial metallicity gradients in the GCSs are usually *shallower* in ellipticals with *lower* S_N, the reverse of what is expected from the merger model.

A still newer line of attack, complementary to the specific frequencies and MDFs, is in the kinematics of the cluster systems. With the new 8- and 10-meter telescopes, it is now possible to accumulate large samples of accurate radial velocities for the globular clusters in the Virgo and Fornax ellipticals and thus to carry out kinematical analyses similar to the ones traditionally done for the Milky Way or M31. Data of this type for M87 (Kissler-Patig & Gebhardt 1998; Cohen & Ryzhov 1997) suggest, albeit from sketchy coverage of the halo at large radii, that the *outer* part of the halo ($R_{gc} \gtrsim 20$ kpc, mostly from the MPC clusters) shows a substantial net rotation of 200 $km\,s^{-1}$ or more, directed along the isophotal major axis. The MRC clusters by themselves show a more modest net rotation $V_{rot} \simeq 100$ km s^{-1} at all radii. Was this the result of a single merger between two galaxies that were already giants? This would be a simple way to leave large amounts of angular momentum in the outer halo. However, such a conclusion does not seem to be forced on us; even the merger product of several large galaxies can yield large outer-halo rotation (e.g., Weil & Hernquist 1996). In the other Virgo supergiant, NGC 4472, the cluster velocities (Sharples et al. 1998) confirm that the MRC has a distinctly lower velocity dispersion and is thus a dynamically cooler subsystem. In addition, there are hints from their still-limited dataset that the MPC has a distinctly higher net rotation speed, in analogy with M87. The interpretation of these systems is growing in complexity, and they present fascinating differences when put in contrast with the Milky Way.

In summary, mergers undoubtedly play a role in the formation of large ellipticals. They may be the dominant channel for forming ones in small groups, out of the S and Irr galaxies that are found in large numbers in such environments. However, the bigger and higher-S_N ellipticals in rich environments present several much more serious problems, which do not appear to be met by the Ashman/Zepf scenario in its initial form.

9.5 The Role of Accretions

The "merger" of a small galaxy with a much larger one is normally called an *accretion.* This is another type of event which must be fairly common for any large galaxy with significant numbers of satellites (again, the absorption of the Sagittarius dE by the Milky Way is the nearest and most well known example). The possibility of building up the observed GCS characteristics in giant ellipticals by accretions of many small satellites was pursued in some early numerical simulations by Muzzio (1986, 1988, and other papers; see Harris 1991 for an overview). It has been investigated anew by Côté et al. (1998) with the specific goal of reproducing the observed metallicity distribution functions. In brief, the assumption of their model is that an original central elliptical forms in a single major phase, giving rise to the MRC clusters. The correlation of the mean [Fe/H](MRC) with galaxy luminosity noted by Forbes et al. (1997) is laid down at this time (the bigger the galaxy, the more metal-rich the MRC component is).

Then, the "seed" galaxy – large, but not nearly at its present-day size – begins to accrete neighboring satellites. These are drawn randomly from a Schechter galaxy luminosity function, so the majority are dwarfs. Since all the accreted objects are smaller, their attendant globular clusters are more metal-poor, and over time, the entire MPC component builds up from the accreted material. In the sense used above, this process is assumed to be a *passive* one, with no new star formation. The Côté et al. model provides a valuable quantification of the results to be expected on the MDF from accretion. The attractive features are that it has the potential to explain the wide galaxy-to-galaxy differences in the MDF for the metal-poor component (Forbes et al.), while maintaining the similarity of the metal-rich clusters from one gE to another. In one sense, the accretion model is the reverse of the merger model: in the accretion scenario, the MRC clusters are the ones belonging to the "original" gE, while in the merger picture, the MRC clusters are formed actively during the buildup.

The Virgo giant NGC 4472 again is used as a template for specific comparisons. Some of these are shown in Fig. 60. Two obvious points to be drawn

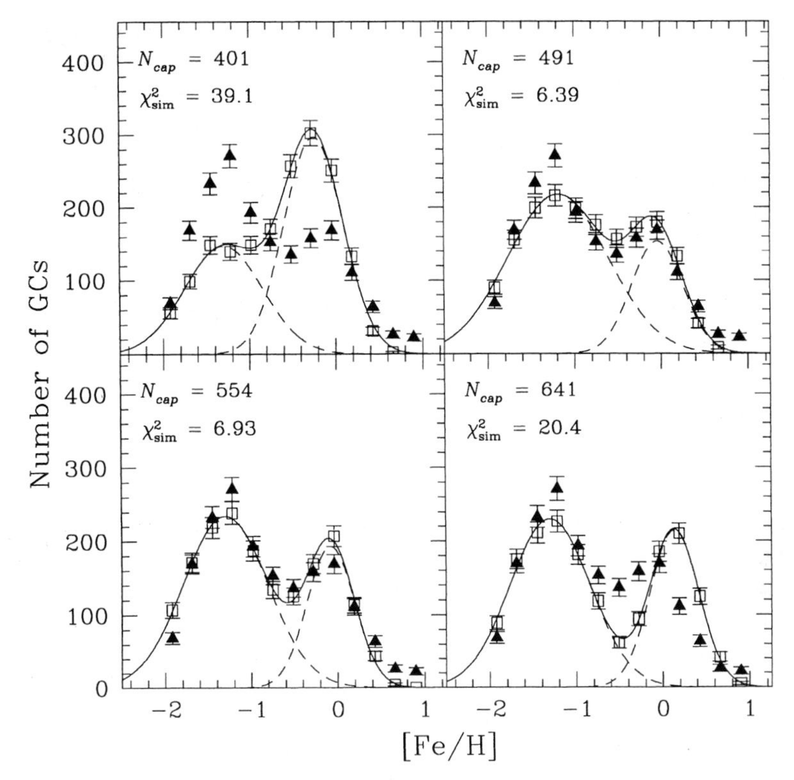

Fig. 60. Four synthetic metallicity distribution functions for globular clusters in a giant elliptical, taken from Côté et al. (1998). The metal-poor part of the bimodal MDF is produced from $N_{\rm cap}$ captured dwarf galaxies which were drawn randomly from a Schechter LF with slope −1.8. *Open squares* represent the final model galaxy; *closed triangles* show the observed MDF for NGC 4472 in Virgo; and the *lines* are the double-Gaussian combination best fitting the synthetic galaxy. Figure courtesy Dr. P. Côté

from the comparison (see also others shown in Côté et al.) are that the model has the flexibility to produce a wide range of bimodal-type MDFs; and that *large* numbers of dwarfs – many hundreds – need to be accreted to build up the metal-poor component sufficiently. The steeper the Schechter function slope, the larger the proportion of globular clusters accreted from small dwarfs and the more metal-poor the MPC appears.

The passive-accretion model has the distinct advantage of using a process that must surely happen at some level in an ongoing fashion. It does, however, have its share of characteristic problems:

- If $\sim$ 500 small galaxies were needed to build up (say) NGC 4472, then many thousands of them would have been absorbed over the whole Virgo cluster to generate the many giant ellipticals there today. This would require, in turn, that *most* (perhaps 90%) of the original dwarf population is now gone. Is this plausible? Perhaps. However, the collision cross sections for dwarf ellipticals are small, and full-scale dynamical simulations may be needed to test the expected accretion rates in detail (it is interesting to note that the earlier semi-analytical studies by Muzzio predicted rather small exchange effects from the dwarfs).

- The specific frequencies of spiral galaxies (suitably age-faded) and many dE's are in the range $S_N \sim 1-3$, whereas the present-day Virgo and Fornax gE's have $S_N \simeq 5$. Reconciling these figures would require us to assume that the initial metal-rich seed elliptical had $S_N \simeq 8$, which is in the BCG range – not a fatal objection, but one which does not necessarily favor the model. Côté et al. suggest that the gE may avoid diluting its cluster population this way if it accretes only the outer envelopes of the incoming galaxies by tidal stripping. Since the globular clusters in dE galaxies are found preferentially in the outskirts of their galaxies, the accreted material would be cluster-rich and S_N could stay nearly constant or even increase. But if this were the case, then the present-day gE's should be surrounded by hundreds (or thousands, in the case of M87) of stripped cores of former dE's. These remnants should be easily noticeable, but where are they?
- The most worrisome problem seems to me to be connected with the metallicity of the halo light. In this model, roughly *half* of the gE is accreted, low-[Fe/H] material. Thus, the mean color of the halo light should be roughly halfway between the red MRC and the bluer MPC. However, the measured color profiles of gE halos *match extremely well with the color of the MRC clusters* (see Geisler et al. 1996 and Fig. 47 for NGC 4472). How can such a galaxy accrete metal-poor clusters without accreting almost no field stars?

Another approach to the accretion scenario is taken by Kissler-Patig et al. (1999), who use the measured specific frequencies in the Fornax giant ellipticals to argue that the cD NGC 1399 might have accreted clusters and halo material predominantly from the neighboring *large* ellipticals in Fornax rather than dE's. This approach would, at least, circumvent the metallicity issue mentioned above. However, full-scale dynamical simulations will be needed to evaluate the plausibility of this level of stripping and halo redistribution among the large galaxies.

Passive accretion therefore leaves some fairly serious difficulties, just as did passive mergers. However, the conditions would change significantly if the accreted small galaxies were highly gaseous, so that (once again) new clusters could form in the process (see also Hilker 1998 for a similar view). This *active accretion* could then allow most of the field stars in the merged product to be at the necessary high metallicity, since most of them would have formed in the accretions. Clusters would also be added to both the MRC and the original MPCs that were present in the seed elliptical and the accreted

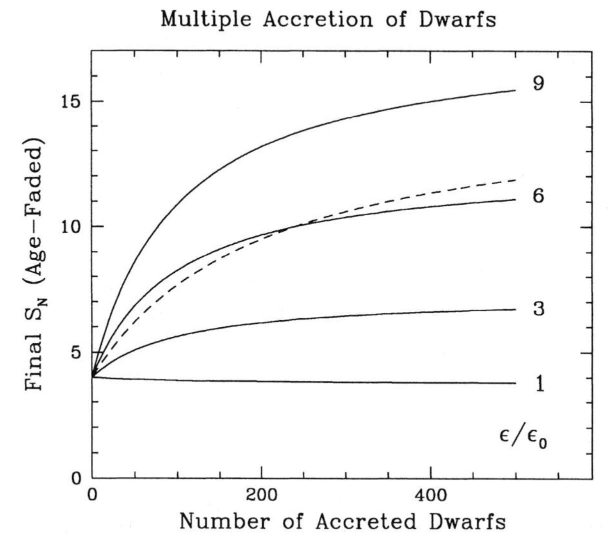

Fig. 61. Specific frequency S_N for an elliptical galaxy which grows by the accretion of many gas-rich dwarf galaxies. The initial E galaxy is assumed to have $M_V = -21.2$ ($M \sim 2 \times 10^{11}$ $M_\odot$) and has specific frequency $S_N(\text{init}) = 4$. The accreted dwarfs each have 10^9 $M_\odot$ of stellar mass and 10^9 $M_\odot$ of gas mass, all of which is converted to stars and globular clusters during the accretion. The relative efficiency of globular cluster formation $\varepsilon/\varepsilon_0$ is denoted at right. The *dashed line* indicates the result of "passive" accretion of gas-free dwarfs (no new star formation) which all have $S_N = 15$

dwarfs. However, just as for the merger picture, the entire formation model would once again begin to resemble a Searle/Zinn-like one whereby the galaxy builds up from many small gas clouds.

The "toy model" for mergers described in the previous section can be used to evaluate the effect of multiple accretions as well. In Fig. 61, the final S_N of the product gE is plotted as a function of the number of accreted dwarfs. The initial E galaxy is assumed to be a moderately luminous elliptical but not a giant ($M_V = -21.2$), while the accreted dwarfs have equal amounts of gas and stars (10^9 $M_\odot$ each; these are *very* gas-rich dwarfs). All the gas is assumed to be converted to stars during the accretions, with a cluster formation efficiency ε as defined previously. It is clear that, to attain final S_N values in the BCG range ($\gtrsim 10$), one needs to accrete many hundreds of dwarfs and convert their gas to clusters with abnormally high efficiency.

Alternately, one can assume that the accretions are "passive" (no new gas or star formation) and that the accreted dE's themselves have high $S_N \sim 15$. The result is shown as the dashed line in Fig. 61, and is roughly equivalent to the case for $\varepsilon/\varepsilon_0 = 6$.

9.6 Starburst Galaxies

To gain additional help in understanding what the early stage of galaxy formation from many small, dwarf-sized pieces may have looked like, we should take a closer look at the small galaxies in which large amounts of star formation are now taking place. Massive star clusters (young globular clusters) are seen to be forming in many individual galaxies which have large amounts of gas and are undergoing energetic star formation at the present time. Such galaxies are loosely called "starburst" systems, and their embedded young clusters have often been called "super star clusters" in the literature. Such clusters were known to exist more than 20 years ago (e.g., in NGC 1569 and 1705; see below), but were not connected until recently with globular clusters. This unfortunate communication gap persisted because the community of astronomers studying these young galaxies, and the community studying traditional globular clusters, had little contact with each other. The traditional and needlessly restrictive paradigm of globular clusters as exclusively old objects has taken a long time to fade away.

The small, dwarf-sized starburst galaxies are extremely interesting laboratories for our purposes. Many of these have high proportions of gas, and promise to give us our best direct view of what the protogalactic SGMCs (Searle/Zinn fragments) may have looked like. One of the nearest and best studied of these, NGC 5253 in the Centaurus group, is shown in Fig. 62 (Calzetti et al. 1997). If we could visualize many dozens of these dwarfs, sprinkled across a $\sim$ 50 kpc region of space and all undergoing their first starbursts, we might gain an image of what the early Milky Way galaxy looked like.

Many starburst dwarfs are found to have handfuls of young star clusters (or, at the very least, associations) that are massive enough to qualify as genuine young globular clusters. A summary of several of them, drawn from the recent literature, is shown in Table 18. Here, columns (1) and (2) give the galaxy name and the number of young, massive star clusters in it; and column (3) the age of the starburst in Myr, usually estimated from such factors as the integrated colors of the clusters, the luminosities and colors of the OB field stars present, and the presence or absence of HII regions and Wolf-Rayet stars. Columns (4-6) give the total mass $M_{\rm cl}$ in all the massive young clusters, the estimated mass $M(\rm H_2)$ in molecular hydrogen contained within the dwarf, and the mass $M(\rm HI)$ in neutral hydrogen. All masses are given in $\rm M_{\odot}$. (NB: The last entry, NGC 4449, is a fairly large system resembling the LMC in many respects and it is not easily possible to assign a single age to the star formation epoch.) Many others are known in addition to the ones listed, though with less complete material for the embedded star clusters (see, e.g., Meurer et al. 1995; Mayya & Prabhu 1996).

There are several common themes to be drawn out of this comparison:

- Massive cluster formation takes place in these little starburst systems preferentially in the *central, densest* regions of the collected gas.

Fig. 62. *HST* image of the starburst dwarf galaxy NGC 5253. The field of view shown is about 0.9 kpc across, with the assumption that the galaxy is 4 Mpc distant. Young OB stars are seen all across the face of the dwarf, along with half a dozen brighter clumps in the central regions, which appear to be massive young star clusters. Image courtesy of Dr. D. Calzetti

- Many of these dwarfs are rather isolated systems, and their recent starburst has not obviously been provoked by tidal encounters or other external stimuli. No outside "trigger" is apparently required to set off vigorous star formation.
- Many of the starbursts listed above are extremely young ($\lesssim 10$ Myr). Give or take a few Myr, the data indicate that the clusters form contemporary with the field stars (the "distributed" blue light), or at least during the leading edge of the burst.
- Large amounts of gas are present, and the burst does not appear to consume the entire supply. The residual amount of HI and H_2 gas not yet

Table 18. Starburst dwarf galaxies with young clusters

Galaxy	N_{cl}	Age (Myr)	M_{cl}(tot)	$M(H_2)$	M(HI)	Source
NGC 1140	12	10	8×10^5		$10^8 - 10^9$	1,2
NGC 1569	7	5	6.4×10^5	5×10^7	1.4×10^8	3,4,5,6
NGC 1705	2	13	4×10^5		1.2×10^8	5,7
NGC 4214	few	5	$\gtrsim 10^5$	1.0×10^8	1.1×10^9	8,9
NGC 5253	6	$\lesssim 10$	$\sim 10^6$	$\lesssim 2 \times 10^8$		10,11,12,13
He 2-10	19	$\simeq 5$	$10^6 - 10^7$	1.6×10^8	2×10^8	14,15,16
UGC 7636	18	10; 100	5×10^5		7×10^7	17
NGC 4449	dozens			8.7×10^8	5×10^9	18,19,20

Sources: (1) Hunter et al. 1994a (2) Hunter et al. 1994b (3) De Marchi et al. 1997 (4) Gonzalez Delgado et al. 1997 (5) O'Connell et al. 1994 (6) Waller 1991 (7) Meurer et al. 1992 (8) Leitherer et al. 1996 (9) Kobulnicky & Skillman 1996 (10) Calzetti et al. 1997 (11) Turner et al. 1997 (12) Beck et al. 1996 (13) Gorjian 1996 (14) Conti & Vacca 1994 (15) Kobulnicky et al. 1995 (16) Matthews et al. 1995 (17) Lee et al. 1997 (18) Bothun 1986 (19) Tacconi & Young 1985 (20) Israel 1997

converted into stars is always $\gtrsim 100$ times more than the total mass contained in the young clusters.

All of these factors are consistent with the characteristics of the formation model outlined in Sect. 8: the available reservoir of gas must be at least 10^8 $M_\odot$ to form globular-sized clusters: a handful of clusters form within one SGMC; their total masses use up only $\lesssim 1\%$ of the total gas supply; and they form at an early stage of the burst.

Massive star clusters are also seen forming in much larger galaxies – again, always as part of an energetic starburst event. In these cases, the source of the burst may be an accretion of a gas-rich satellite (thus not a "merger" in the restricted sense used in the previous section), a tidal shock from a close encounter, or the collection of gas into a central ring or bar, among other mechanisms. The most well known of these cases are probably NGC 1275 (Holtzman et al. 1992; Carlson et al. 1998) and NGC 3597 (Lutz 1991; Holtzman et al. 1996), thanks to the recent high resolution imaging of the *HST* cameras which has revealed many details of the nuclear star formation activity in these distant systems. A summary of parameters – numbers and masses of young clusters and total gas mass – for several large starburst galaxies is given in Table 19. As above, masses are given in $M_\odot$ units.

These larger galaxies present a much more heterogeneous collection than the simpler starburst dwarfs. Some (NGC 1097, 6951) show star formation along an inner ~ 1 kpc ring of gas; some (M82, NGC 253) may have been stimulated by tidal shocks; some (NGC 1275, 5128) are giant ellipticals which

Table 19. Large starburst galaxies with young clusters

Galaxy	N_{cl}	Age (Myr)	M_{cl}(tot)	$M(H_2)$	M(HI)	Source
NGC 1275	~ 1180	$\gtrsim 100$	$> 10^8$	1.6×10^{10}	$\gtrsim 5 \times 10^9$	1,2,3,4,5
M82	> 100	10:	$\gtrsim 10^6$	2×10^8	2×10^8	6,7,8
NGC 253	4	$10 - 50$	2×10^6	2×10^8	4×10^8	9,10,11
NGC 5128	dozens	50:	10^5:	3×10^8	3×10^8	12,13,14,15
NGC 1097	88	$\lesssim 10$	$\sim 10^6$			16
NGC 6951	24	$\lesssim 10$	$\sim 10^6$			16
Arp 220	> 8	$\lesssim 100$:		9×10^9		17,18
NGC 3597	$\simeq 70$	~ 200?	$10^7 - 10^8$:	3×10^9		19,20,21
NGC 7252	$\simeq 140$	700	4×10^7	3.5×10^9		22,23
NGC 3256	hundreds	$\gtrsim 100$	6×10^7	1.5×10^{10}		24

Sources: (1) Holtzman et al. 1992 (2) Carlson et al. 1998 (3) Lazareff et al. 1989 (4) Jaffe 1990 (5) Bridges & Irwin 1998 (6) O'Connell et al. 1995 (7) Lo et al. 1987 (8) Satypal et al. 1997 (9) Watson 1996 (10) Mauersberger et al. 1996 (11) Scoville et al. 1985 (12) Alonso & Minniti 1997 (13) Minniti et al. 1996 (14) Schreier et al. 1996 (15) Eckart et al. 1990 (16) Barth et al. 1995 (17) Scoville 1998 (18) Shaya et al. 1994 (19) Holtzman et al. 1996 (20) Wiklind et al. 1995 (21) Lutz 1991 (22) Miller et al. 1997 (23) Wang et al. 1992 (24) Zepf et al. 1999

appear to have undergone accretions of smaller gas-rich satellites; and some (NGC 3256, 3597, 7252) are suggested to be merger remnants in the sense used above, i.e. the collisions of two roughly equal disk galaxies. All of them have complex structures and morphologies. For example, the nearby and well studied elliptical NGC 5128 has star clusters in its inner few kpc which appear to be a broad mix of ages and metallicities (some from the original elliptical, some which may have been acquired from the disk-type galaxy it recently accreted, and some bluer objects recently formed out of the accreted gas; see, for example, Alonso & Minniti 1997; Schreier et al. 1996; Minniti et al. 1996). Its halo within $R_{gc} \lesssim 20$ kpc shows the characteristics of a fairly complex triaxial structure (Hui et al. 1995) revealed through the kinematics of both its planetary nebulae and globular clusters.

Nevertheless, these large galaxies display some important features in common with the starburst dwarfs: the massive young star clusters are forming preferentially in the densest, central regions of the collected gas; and their total masses are, once again, of order 1 percent of the residual gas (HI + H_2) present in the active regions. It does not appear to matter how the gas is collected; but there needs to be lots of gas collected into SGMC-sized regions before globular clusters can be built.

9.7 A Brief Synthesis

In this and Sect. 8, we have approached the discussion of galaxy formation from a number of different directions (*in situ*, mergers, accretions). How well do these different hypotheses score, as ways to build globular clusters?

By now, it should be apparent that each one of these generic pictures represents an extreme, or limiting, view of the way that galaxies must have assembled. Thus in some sense a "scorecard" is irrelevant: each approach in its extreme form would get a passing grade in some situations but an obvious failing grade in others. In other words, it is becoming increasingly clear that we need elements of *all* these approaches for the complete story. Mergers of disk systems are plainly happening in the present-day universe and should have happened at greater rates in the past. Accretions of small satellites are also happening in front of us, and will continue to take place in the ongoing story of galaxy construction. Yet there must also have been a major element of *in situ* formation, involving the amalgamation of many small gas clouds at early times while the first rounds of star formation were already going on within those pregalactic pieces.

The *in situ* approach imagines that a considerable amount of star formation took place at early times. For large elliptical galaxies, which have dominated much of the discussion in these chapters, there is now much evidence that their main epoch of formation was at redshifts $z \sim 3-5$ (see, e.g., Larson 1990b; Maoz 1990; Turner 1991; Whitmore et al. 1993; Loewenstein & Mushotzky 1996; Mushotzky & Loewenstein 1997; Steidel et al. 1996, 1998; Giavalisco et al. 1996; Bender et al. 1996; Ellis et al. 1997; Stanford et al. 1998; Baugh et al. 1998, for only a few examples of the extensive literature in this area). Evidence is also gathering that ellipticals in sparse groups may also have formed with little delay after the rich-cluster ellipticals (Bernardi et al. 1998; G. Harris et al. 1999), somewhat contrary to the expectations of hierarchical-merging simulations.

Our impression of what a large protogalaxy looked like at early times continues to be influenced strongly by the Searle-Zinn picture: the logical sites of globular cluster formation are $\sim 10^8 - 10^9\ M_\odot$ clouds, which are capable of building up the basic power-law mass spectrum of clusters that matches the observations (McLaughlin & Pudritz 1996). We can expect that vigorous star formation should be happening within these SGMCs at the same time as they are combining and spilling together to build up the larger galaxy. For dwarf E galaxies, perhaps a single SGMC or only a few of them combined, and one initial starburst truncated by early mass loss may spell out the main part of the formation history (Sect. 8 above). For giant E galaxies, dozens or hundreds of SGMCs would have combined (Harris & Pudritz 1994), and there could well have been at least two major epochs of star formation separated by a few Gyr, leaving their traces in the bimodal MDFs of the globular clusters (Forbes et al. 1997) and the halo field stars (G. Harris et al. 1999). The metal-poor part of the halo appears (from preliminary findings in NGC 4472 and NGC 5128, as discussed above) to be remarkably more "cluster-rich" than the metal-rich component, strongly suggesting that a great deal of protogalactic gas ended up being unconverted to stars until the second round or later (McLaughlin 1999); perhaps much of the gas in the original SGMCs was stripped away during infall (Blakeslee 1997) or driven out during the first starburst by galactic winds (Harris et al. 1998a). There are many potential influences to sort out here, and the sequence (and nature) of the events is still murky even without bringing in later influences from mergers and accretions.

The basic *accretion* model starts with a large initial galaxy which had already formed *in situ*, by hypothesis in a single major burst. Around this core, we then add a sequence of smaller galaxies and thus build up the metal-poor halo component. But there are two basic varieties of accretion: gas-free or gas-rich. If the satellites are gas-free dwarf ellipticals, then we should expect to build up a larger galaxy with a specific frequency in the normal range, and a halo MDF that is intermediate or moderately low metallicity. But if the accreted objects are gas-rich, then new clusters and halo stars can form in the process, drive the MDF increasingly toward the metal-rich end, and (if the total amounts of accreted gas are *very* large) possibly change the specific frequency. But in a fairly literal sense, this latter version of building a galaxy by adding together gas-rich dwarfs is quite close to the generic Searle-Zinn picture.

The *merger* model applies specifically for gE galaxy formation. If we amalgamate pre-existing galaxies of roughly equal size, the result should in most cases be an elliptical. But again, the amount of gas will play an important role in the outcome. If the mergers are taking place at high redshift (that is, at very early times), then we can expect the progenitors to be largely gaseous, in which case the majority of stars in the merged product would actually form during the merger of gas clouds. This, too, can be viewed as an extension of the basic Searle-Zinn formation. On the other hand, if the merging is happening at low redshift nearer to the present day, then the galaxies have smaller amounts of gas; much less star formation can happen during the merger, and the result will be a low-S_N elliptical such as we see in small groups.

The presence or absence of gas is therefore a critical factor in evaluating the success of any formation picture. It is only the presence – or the removal – of large amounts of gas during the most active cluster formation epochs which will permit significant changes in the total numbers of globular clusters (and thus the specific frequency), and the form of the metallicity distribution function. The challenge for any one galaxy is to identify the individual combination of events which led to its present-day structure.

Here – appropriately, in the confusion and uncertainty that characterizes the frontier of any active subject – we must end our overview of globular cluster systems. Many areas of this growing subject have been missed, or dealt with insufficiently, and I can only urge readers to explore the rich literature for themselves. The surest prediction is that new surprises await us.

To see the world for a moment as something rich and strange is the private reward of many a discovery.

Edward M. Purcell

Acknowledgements
It is a genuine pleasure to thank Lukas Labhardt, Bruno Binggeli, and their colleagues and staff for a superbly organized and enjoyable conference in a beautiful setting. Bruce Carney and Tad Pryor were the ideal fellow lecturers for this concentrated meeting on globular clusters. Finally, I am grateful to Dean McLaughlin and David Hanes, who gave me careful and constructive readings of the manuscript before publication.

10 Appendix: An Introduction to Photometric Measurement

Many of the issues in these lectures are strongly related to our ability to *measure* certain characteristics of globular clusters reliably and believably: luminosities, distances, chemical compositions, reddenings, ages ... the potential list is a long one. Many of these tasks boil down to the process of photometric data reduction from CCD images – simple in principle, but surprisingly intricate and challenging in practice.

10.1 An Overview of the Color-Magnitude Diagram

To start with, let us look at the essential features of a typical globular star cluster as they appear in the color-magnitude diagram (CMD), Fig. 63. Here, all the evolutionary stages of stars in this ancient object are laid out for us to see. First is the zero-age main sequence (ZAMS), where the low-mass stars are quietly undergoing core hydrogen burning (at the bottom end, the luminosity starts to decline steeply as the mass decreases toward the hydrogen-burning limit, and our best attempts at tracing them may be lost in the increasing scatter of photometric measurement uncertainty.) Core hydrogen exhaustion is marked by the turnoff point (MSTO), after which the stars move rapidly across the subgiant branch (SGB), then steadily up the red-giant branch (RGB) as the hydrogen-burning shell gradually moves outward in mass through the stellar interior and the inert helium core gradually increases in mass. Core helium ignition takes place at the RGB tip, and the star rapidly readjusts to a new equilibrium on the horizontal branch (HB). If the cluster has moderately high metallicity or low age, the HB stars will have large hydrogen envelopes and low surface temperatures and will thus all be on the red side of the horizontal branch (RHB). But if the cluster is very

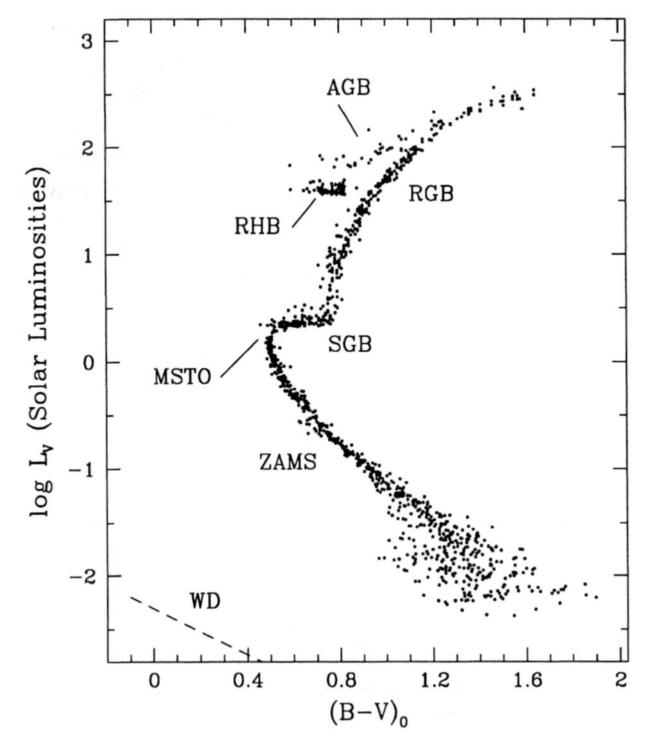

Fig. 63. Color-magnitude diagram for a typical globular cluster (data from Hesser et al. 1987, for 47 Tucanae). The axes are plotted as visual luminosity in Solar units, against the ratio of visual to blue luminosity (essentially, the color index $B - V$). The principal evolutionary stages in the stars' history are marked with the abbreviations defined in the text

old, or has low metallicity, or the stars have suffered high mass loss from their surfaces, then the residual hydrogen envelope will be small, the stars will have high surface temperatures, and the horizontal branch will extend far over to the blue (BHB) side of the CMD.

The last active stage of nuclear burning for globular cluster stars is the asymptotic giant branch (AGB) in which two fusion shell sources (hydrogen to helium, helium to carbon) sit on top of an inert core. When these shells approach too close to the stellar surface, the remaining envelope is ejected as a planetary nebula, the shell-burning sources are permanently extinguished, and the central dead core (a mixture of He/C/O in proportions determined by the original mass, with a tiny surface skin of hydrogen) settles into the white dwarf (WD) phase. The star is now supported mainly by the degeneracy pressure of the electron gas, and as it emits its residual heat, it gradually slides down the WD cooling line to the cold black-dwarf state.

The basic features of the various nuclear burning stages (ZAMS through AGB) have been recognized in the CMD for many years (for comprehensive reviews, see Renzini & Fusi Pecci 1988; Chiosi et al. 1992; and Carney's lectures in this volume). Current evolutionary stellar models and isochrones are now able to reproduce the observed CMDs in considerable detail (see, e.g., Sandquist et al. 1996; Harris et al. 1997b; Salaris et al. 1997; Stetson et al. 1999; Cassisi et al. 1999; Richer et al. 1997; Dorman et al. 1991; Lee et al. 1994, for just a few of the many examples to be found in the literature). However, the extremely faint WD sequence, and the equally faint bottom end of the ZAMS where the masses of the stars approach the hydrogen-burning limit, have come within reach of observation for even the nearest clusters only in very recent years (see Cool et al. 1996; Richer et al. 1997; King et al. 1998, for recent studies that delineate these limits). All in all, the CMDs for globular clusters provide one of the strongest and most comprehensive bodies of evidence that our basic understanding of stellar evolution is on the right track, even if many detailed steps still need work.

10.2 Principles of Photometry and the Fundamental Formula

High quality color-magnitude studies are obtained only after carefully designed observations and data reduction.

Figure 64 shows a pair of CCD images obtained with the *HST*/WFPC2 cameras. The first one is a single exposure with the PC1, after preprocessing.[12] It illustrates a number of problems that we need to attack, once we have our CCD image in hand:

- We need to eliminate *artifacts* from the image (bad pixels, cosmic rays, and so on).
- We need to *find* the real objects (stars, galaxies, asteroids, ...) in an objective and reproducible way.
- We need to *classify* the objects, i.e., divide our objects into separate lists of stars, galaxies, or other things.
- We need to *measure* the brightness and location of each object.
- We need to understand the *uncertainties* in the measurement, both random and systematic.

Eliminating unwanted artifacts from a single raw image such as that in Fig. 64 can be done, to some extent, with sophisticated rejection algorithms which look for sharp features (hot or very cold pixels, or cosmic-ray hits which affect only one or two pixels); or elongated or asymmetric features (bad columns, cosmic-ray streaks, bright stars which bleed along rows or columns). You can then attempt to smooth over these places with the hope

[12] The steps involved in preprocessing – bias subtraction and trim, dark current subtraction, flat fielding – are not discussed here. More detailed discussions of these operations can be found, e.g., in Walker (1987); Tyson (1989); Massey & Jacoby (1992); Howell (1992); Gullixson (1992); or Gilliland (1992).

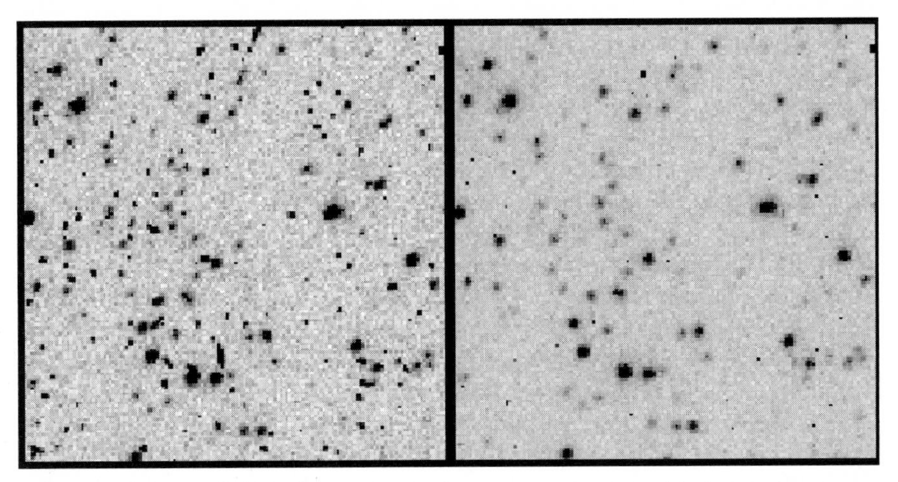

Fig. 64. *Left panel:* A segment of a single exposure in the field of globular cluster NGC 2419, taken with the *HST* (PC1 camera). Many star images are present, but the frame is heavily contaminated by numerous cosmic-ray hits and bad pixels. *Right panel:* The same field after re-registering and combining 8 individual exposures that were sub-pixel-shifted. Only the stars and a few easily distinguished bad pixels remain, on a much smoother background. Raw image data are taken from the study of Harris et al. (1997b)

that the adjacent pixels will allow you to reconstruct the information on these damaged pixels correctly. This approach will not work so well if the star images themselves are undersampled, such as in the WF frames of the *HST*, or if the artifact happens to be embedded somewhere in a real object (a star or galaxy). A much more effective process in any case is to take a series of images that are deliberately shifted in position between exposures (sub-pixel-shifted or "dithered") so that the pixel grid samples the field at several different positions.[13] When the frames are re-registered, and combined through a median or averaging algorithm which rejects extreme values, the artifacts will almost entirely drop out, leaving an enormously cleaner combined frame. Obviously, the more frames you can use to do this, the better. If several independent positions have been sampled differing by fractions of a pixel, a higher-resolution summed image with finer pixels can also be constructed (see, e.g., the 'drizzle' software of Hook & Fruchter 1997 and Fruchter & Hook 1999, or the ALLFRAME code of Stetson 1994), yielding gains in the ability to measure crowded objects or to determine image structure. Image reconstruction from images that have been carefully sub-pixel-shifted can yield powerful improvements especially in image structure or morphology studies (see Lauer 1999 for a review of reconstruction algorithms). However, recon-

[13] "Dither" literally means "to act nervously or indecisively; vacillate". This is an unfortunate choice of term to describe a strategy of sub-pixel-shifting which is quite deliberate!

structed images may not be suitable for photometry since some algorithms do not conserve flux; be cautious in making a choice of techniques.

Photometry of *stars* is a far easier job than photometry of nonstellar objects, for one predominant reason: on a given CCD image, all stars – which are point sources of light blurred by the atmosphere (for ground-based telescopes) and the telescope optics – have the same profile shape called the point spread function (PSF).[14] The PSF width can be characterized roughly by the full width at half-maximum height (FWHM) of the profile.

Finding stars in an objective manner is a straightforward job in principle: look for objects whose brightest pixels stand clearly above the pixel-to-pixel scatter of the sky background (see Fig. 65). If we define z_s as the mean sky brightness, and σ_s as the standard deviation of the sky pixels, then we can set a *detection threshold* for real objects by looking for any pixels with intensities $z > (z_s + k \cdot \sigma_s)$ for some threshold parameter k. Thresholds in the range $k \simeq 3.5 - 4.0$ are normal for faint stellar photometry; one can try to go lower, but choices $k \lesssim 3$ inevitably lead to lots of false detections and serious contamination problems (see below).

The sky characteristics (z_s, σ_s) are *local* quantities which may differ strongly across the image field. The all-important noise parameter σ_s, which directly fixes the faint limit of your photometry, is governed by (a) the raw sky brightness z_s through simple photon statistics (are you working in the wings of a large galaxy or nebula which covers much of the frame?); (b) the 'lumpiness' of the sky (are you working in a crowded field, or trying to find stars within a patchy nebula or a spiral arm of a distant galaxy?); (c) bad pixels and cosmic rays, if you were unable to remove those; (d) instrumental noise such as readout noise and dark current; (e) additional noise introduced in the preprocessing, such as flat-fielding (did your flat-field exposures have inadequate signal in them compared with the target exposures?).

The principles of simple aperture photometry will illustrate several of the basic limits inherent in photometric measurement. Suppose that we have a star located on a grid of pixels, as in Fig. 66, and that its intensity profile is given by the matrix $z(x, y)$. As before, denote the local sky and background noise as (z_s, σ_s). Suppose we now surround the star with a circular measuring aperture of radius r, which will then contain n_{px} pixels approximately given by $n_{px} \simeq \pi r^2$. (The finite pixellation of the image means that the ideal circular

[14] Strictly speaking, the assumption that all stars have the same PSF on a given image is only true to first order, since image scale or optical aberrations can differ subtly across the field of view in even the best situations. An important example is in the *HST*/WFPC2 cameras, but even here the variations can be fairly simply characterized; see the WFPC2 Instrument Handbook available through the StScI website. More generally, we can say that the PSF for star images should be a *known and slowly changing function of position on the frame*, and should be *independent of brightness* as long as the detector is linear. Neither of these statements will be true for nonstellar objects such as galaxies, whose profile shapes cover an enormously broader parameter space.

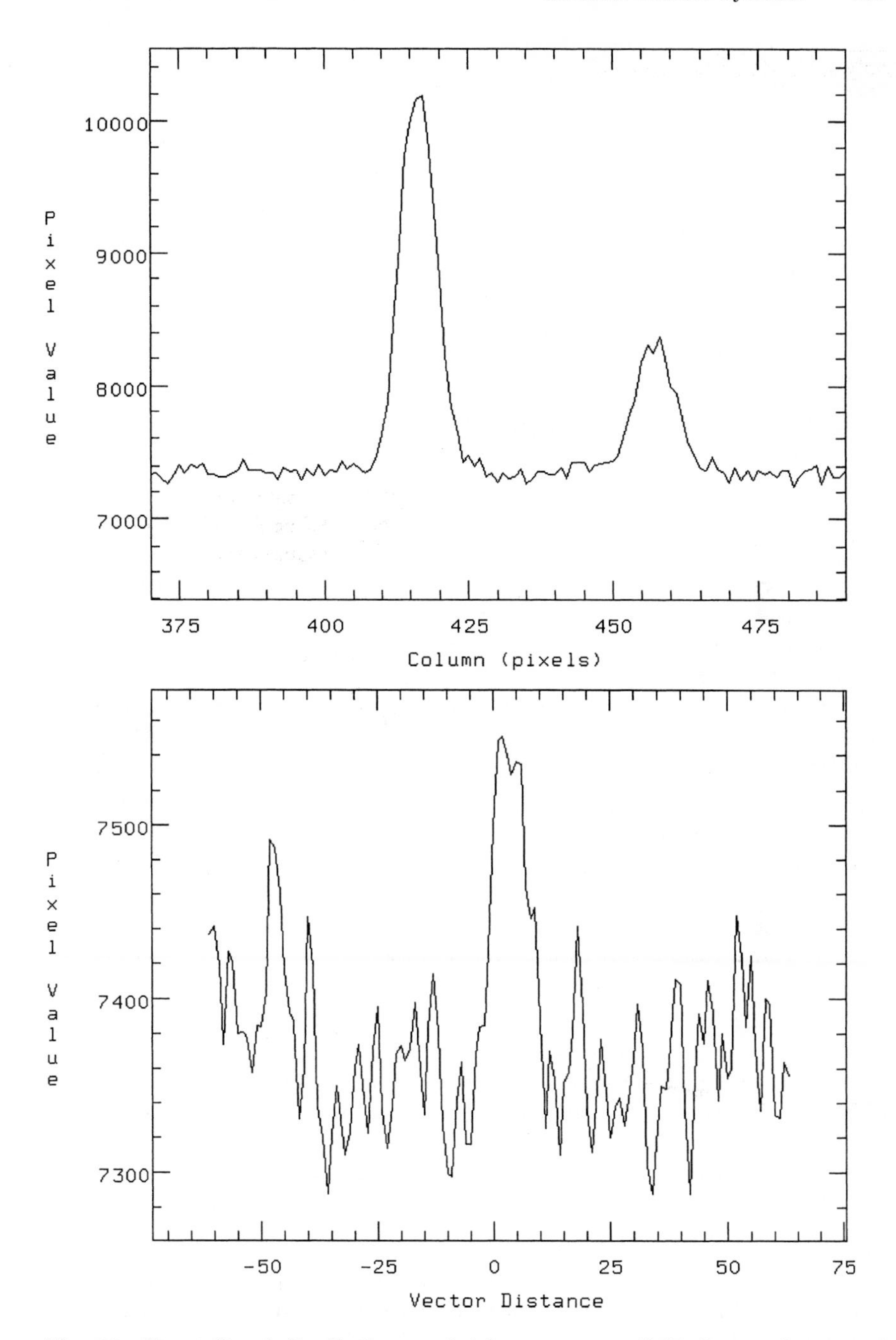

Fig. 65. *Upper Panel:* Profile for two bright stars on a CCD frame; these stand clearly above the sky background noise. *Lower Panel:* Profile for a much fainter star on the same frame. This star is near the limit of detectability relative to the standard deviation of the sky background noise

Fig. 66. CCD image of a bright star, surrounded by a digital 'aperture' (*white line*). The aperture boundary follows the quantized pixels and is only an approximation to a circle. Note three faint stars inside the aperture, which will contaminate the measurement of the central bright star

aperture boundary becomes a jagged line enclosing only whole pixels. In some photometry codes including DAOPHOT, second-order corrections are made to round off the boundary by adding fractional amounts of light from pixels along the rim of the circle.) Now let

$N_\star$ = number of collected e^- from the star image

N_s = number of collected e^- from the sky background light in the aperture

I = variance per pixel (in e^-) of the instrumental noise (including readout noise, dark current, quantization noise, and perhaps other factors).

These three quantities are normally uncorrelated, so we can add their variances to obtain the total variance of the random noise in the aperture. The *signal-to-noise ratio* of the measurement is then

$$\frac{S}{N} = \frac{N_\star}{\sqrt{N_\star + N_s + n_{px} I}} \; . \tag{66}$$

This ratio represents the internal uncertainty in the measured brightness of the star. The *instrumental magnitude* of the star (for example, in the V filter) is normally expressed as

$$v_{inst} \equiv -2.5 \log \left(N_\star / t \right) + \text{const}$$

where t is the exposure time and thus $N_\star / t \equiv r_\star$ is the "count rate" (e^- per second) from the star. The last term is an arbitrary constant. The instrumental magnitudes can be transformed to standard V magnitudes by measurements of photometric standard stars with known magnitudes and colors, taken during the same sequence of observations as the program exposures. For a thorough outline of precepts for standard-star transformations, see Harris et al. (1981).

What governs each of the factors in the uncertainty?

- $N_\star$ is proportional to the photon collection rate $r_\star$ from the star; the exposure time t; and the detection efficiency Q (conversion ratio of incident photons to stored electrons, characteristic of the detector). In turn, $r_\star$ is proportional to the brightness of the star $b_\star$ and the telescope aperture size D^2 (the collecting area).
- N_s is proportional to t, Q, D^2, n_px, and μ_s where μ_s is the sky brightness (number of photons per second per unit area). Also, we have $n_\mathrm{px} \sim r^2 \sim \alpha^2$ where α denotes the star image size (FWHM). Smaller seeing disks can be surrounded by smaller apertures, thus lowering the amount of contaminating skylight and improving the signal-to-noise.
- I (instrumental noise per pixel) is approximately constant if dark current is negligible; if dark current dominates, however, then $I \sim t$.

For most broad-band imaging applications and modern low-noise CCD detectors (which characteristically have $I \lesssim 5\ e^-/\mathrm{px}$), the instrumental noise is not important. (A notable exception is, again, *HST*/WFPC2 where the readnoise $n_\mathrm{px} I$ is roughly equal to the sky noise N_s even for full-orbit exposures and broad-band filters.) In addition, we are usually interested in how well we can do at the faint limit where $N_\star \ll N_\mathrm{s}$. So the most interesting limit of the S/N formula is the "sky-limited" case,

$$\frac{S}{N} \simeq \frac{N_\star}{\sqrt{N_\mathrm{s}}} \,. \tag{67}$$

If we put in our scaling laws $N_\star \sim b_\star\, t\, Q\, D^2$ and $N_\mathrm{s} \sim \mu_\mathrm{s}\, t\, Q\, D^2 \alpha^2$, we obtain

$$\left(\frac{S}{N}\right)^2 \sim \frac{b_\star^2\, t\, Q\, D^2}{\alpha^2 \mu_\mathrm{s}} \,. \tag{68}$$

Clearly there are several ways we can achieve deeper photometric limits. We can:

- Increase the exposure time.
- Use a bigger telescope.
- Find darker sky.
- Improve the seeing quality.
- Employ a better detector (lower noise or higher Q).

The circumstances will determine which routes are possible at any given time. Modern optical CCD detectors have detection efficiencies Q approaching 1, and readnoise levels I of only a few e^-, so they can be "improved" substantially only by making them in physically larger arrays or extending their wavelength coverage. Much effort has been directed toward techniques for improving image quality (seeing), either by space-based observations or by adaptive optics from the ground. It is important to note that better seeing

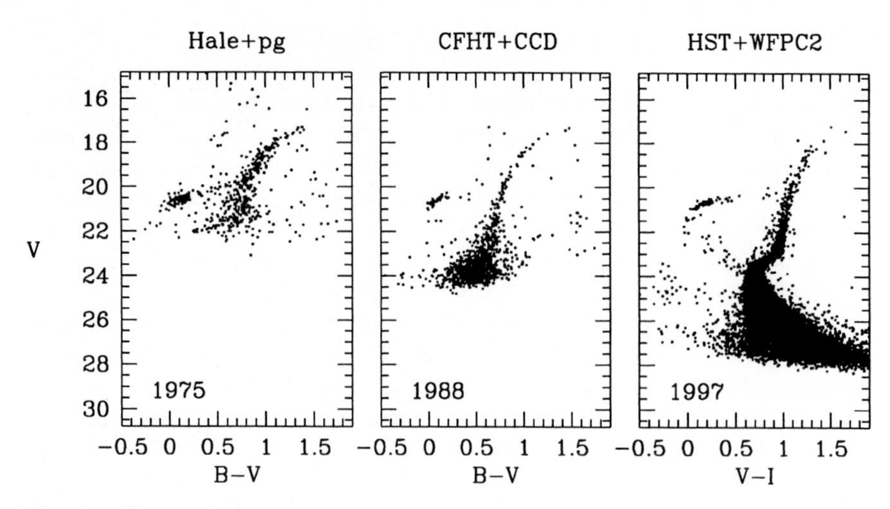

Fig. 67. Three color-magnitude diagrams for the remote halo cluster NGC 2419, showing the gain in depth from increased DQE and spatial resolution. *Left panel*: CMD containing 547 stars (Racine & Harris 1975), obtained via photographic plates with the Hale 5-m telescope. *Center panel*: CMD containing 1316 stars (Christian & Heasley 1988), obtained with CCD photometry from the CFHT. *Right panel*: CMD containing 17275 stars (Harris et al. 1997b), obtained with the WFPC2 CCD camera on the *HST*. The total exposure times in all three cases were similar

yields other gains beyond the formal improvement in S/N: a narrower stellar profile will also reduce contamination from image crowding, and allow us to see more detailed structure within nonstellar objects.

The gains in limiting magnitude in astronomical photometry over the past two decades have been spectacular (see Fig. 67): more than an order of magnitude of depth was gained in the early 1980's with the appearance of CCD detectors, which were ~ 100 times higher in quantum efficiency than photographic plates. Another decade later, *HST* was able to reach another order of magnitude deeper because of the jump in spatial resolution (FWHM $\simeq 0\farcs1$ as opposed to $\gtrsim 0\farcs5$ from the ground) and the considerably darker sky as observed from space. Still further gains in the post-*HST* era will require larger apertures and longer exposure times.

A useful illustration of the way the factors in (68) complement each other is to compare two similar photometric experiments carried out in completely different eras. Recently, Harris et al. (1998b) used the *HST* cameras to resolve the brightest old-halo RGB stars in a dwarf elliptical galaxy in the Virgo cluster, at a distance of $d \simeq 16$ Mpc (see Fig. 68). Half a century earlier, Walter Baade (1944) achieved exactly the same thing for the Local Group dwarf ellipticals NGC 185 and 205 ($d = 0.8$ Mpc), in a classic experiment which first showed that these galaxies were built of old-halo stars and were companions to M31. Baade used a total exposure time of 4 hours with the

Fig. 68. *HST* image in the I-band of the Virgo dwarf elliptical VCC 1104, from Harris et al. (1998b). Almost one magnitude of the galaxy's old red giant branch is clearly resolved

then-new red-sensitized photographic plates, on the Mount Wilson 2.5-meter telescope. The seeing was probably $\alpha \simeq 1''$; the efficiency Q of the emulsion is hard to guess but would certainly have been less than 1%. By comparison, the *HST* is a 2.4-meter telescope, the DQE of WFPC2 is $Q \simeq 30\%$, the "seeing" is $\alpha = 0\farcs1$, and the total exposure on the Virgo dwarf was 9 hours. Both experiments measured stars of the same absolute magnitude, so the relative photon collection rate scales just as the inverse square of the distance, $b_\star \sim d^{-2}$. Putting these factors into (68), we have as our comparison for the two experiments

$$\frac{\frac{S}{N}(\mathrm{Baade})}{\frac{S}{N}(HST)} \simeq \frac{\left(\frac{4}{9} \cdot \frac{1}{50}\right)^{1/2} \cdot 1}{\left(\frac{1}{20}\right)^2 \cdot 10 \cdot 10^{1/2}} \simeq 1.2\,.$$

The photometric limits of both experiments turn out to be similar – both of them reach just about one magnitude below the RGB tip of their respective target galaxies, just as the quantitative comparison would predict. The distance ratio is obviously the biggest factor in the equation; it makes the same types of stars in the Virgo dwarf 400 times fainter in apparent brightness than those in the Local Group ellipticals.

10.3 Aperture and PSF Measurement

We have already discussed the initial process of determining the sky background noise σ_s and finding stars across the field. Now we need to decide how to measure them. Most of all, this decision depends on whether the frame is *crowded* or *uncrowded.* Suppose the picture has a total area of A pixels and there are n detected objects, so that the average area per object is $\simeq A/n$ px. A useful guideline is that if $(A/n) \gtrsim \pi(4\,FWHM)^2 \simeq 50\,(FWHM)^2$, then the image is not very crowded in absolute terms.

For **uncrowded images**, straightforward aperture photometry (Fig. 66) will work well, and is conceptually the simplest measurement technique. To understand the empirical image profile, it will help to take several bright stars and plot up the *curve of growth* – i.e., the apparent magnitude of the star as a function of aperture radius – to find out how large an aperture one needs to enclose virtually all the light of the star. Choose a radius $r_{\rm max}$ which contains, e.g., $\gtrsim 97\%$ of the asymptotic total, but which is not so large that sky noise starts to affect the internal uncertainty. This is the best aperture size to use for calibration purposes, since it will be unaffected by minor differences in seeing (which affects the core image structure) from place to place on the frame, or between frames taken at different times during the same observing session. We measure known standard stars on other frames through the same aperture and thus define the transformation between the "instrumental" magnitude scale v into true magnitude V.

However, to compare the relative magnitudes of all the stars on one frame, both bright and faint, we should instead choose an aperture radius which maximizes S/N. For the bright stars where $N_\star \gg N_s$, we have $(S/N) \simeq \sqrt{N_\star}$ and a large aperture is best. But the vast majority of the stars on the frame will be faint ($N_\star \ll N_s$), and for them we have $(S/N) \simeq N_\star/\sqrt{N_s}$. Suppose the star intensity profile is given by $I(r)$. Then

$$N_\star = \int_0^r I(\varrho) \cdot 2\pi\,\varrho\,d\varrho \tag{69}$$

and thus the signal-to-noise scales as

$$\frac{S}{N} \sim \frac{1}{r} \int \varrho\, I\, d\varrho\ . \tag{70}$$

Once we know the profile shape $I(r)$, we can solve (70) numerically to find the optimum radius r_0. For a Gaussian profile, we have $I(r) = I_0\ \exp(-r^2/2\sigma^2)$

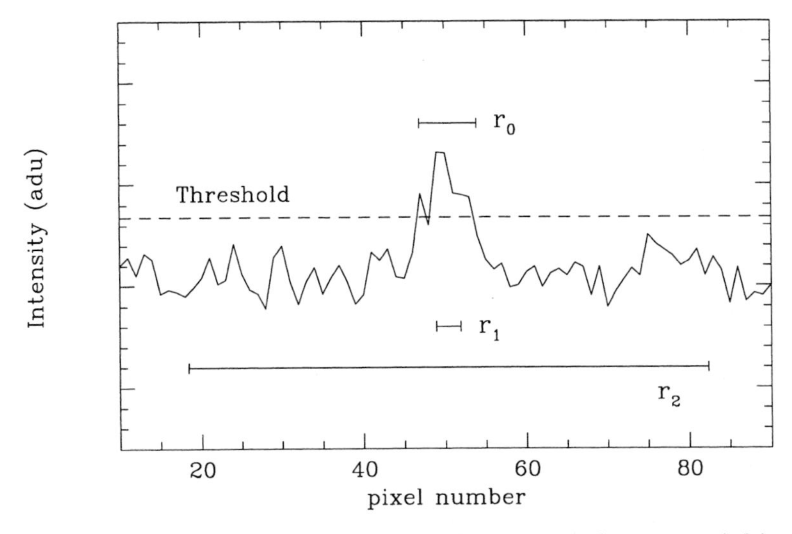

Fig. 69. Definition of the optimum radius for aperture photometry. A faint-star image profile superimposed on sky noise is shown, with its central few pixels extending up past the detection threshold (*dashed line*). A small aperture (r_1) contains too small a fraction of the star light to produce adequate S/N, while a large one (r_2) is dominated by sky noise. The intermediate aperture r_0 maximizes the signal-to-noise ratio

and the maximum S/N occurs for an intermediate radius $r_0 \simeq 1.6\,\sigma \simeq 0.67$ FWHM (see Pritchet & Kline 1981). Essentially, this optimum radius is big enough to include most of the starlight, but not so big that a large amount of sky background intrudes to dominate the scatter. The point is illustrated in Fig. 69.

Once you have all the magnitudes measured through the optimum radius r_0, the brighter stars can be used to find the mean correction from r_0 to r_{max}, so that the whole list is then calibrated.

For **crowded images**, we have to plunge all the way into the more formidable job of profile fitting. One good approach is to use the bright stars with high S/N to define the PSF shape empirically. (Pure analytic approximations to the PSF, such as Gaussians or Moffat profiles, can also be effective and may work in situations where empirical PSFs are difficult to derive.) If the shape parameters (FWHM, noncircularity, orientation) depend on location (x, y) in the frame, then typically a few dozen stars spread evenly over the frame will be needed to map it out; the more the better. Once you have defined the PSF, then try to fit it to each object in the detection list. For *each* star there are at least four adjustable parameters: the object center (x_c, y_c), the brightness scale factor A, and the predicted local sky level $z_s(x_c, y_c)$ at the object center. To test the quality of the result, subtract off the fitted PSF at each star and look for anomalous or distorted residuals. In practice, any one star image will overlap to varying degrees with the wings of all its neighbors, so that the full-blown solution actually requires a simultaneous and highly nonlinear fit of the PSF model to all the stars at once.

The fitted quantity we are most interested in is the scale factor A (unless we are doing astrometry!). Fortunately, the exact form of the model PSF does not have a major effect on A, since both faint and bright stars have the same profile shape and we only need to know their relative brightness scale factors. A more accurate PSF will, however, allow you to do a better job of subtracting out neighboring stars that are crowding your target star, so it is worth spending time to get the best possible PSF. For very crowded fields, defining the PSF itself is an iterative and sometimes painful business: one must make a first rough PSF, use that to subtract out the neighboring objects around the stars that defined the PSF, then get a cleaner PSF from them and repeat the steps. Finding stars, too, is an iterative process; after the first pass of PSF fitting and subtraction, additional faint stars are often found that were hiding in the wings of the brighter ones or somehow lost in the first pass. These should be added to the starlist and the solution repeated.

The ability to fit *and subtract* stars from the frame is one of the most powerful and helpful features of digital photometry. For example, it can be used even to improve simple aperture photometry, by "cleaning" the area around each measured star even if it is not severely crowded in absolute terms (see, e.g., Stetson 1990; Cool et al. 1996). Nevertheless, in extreme situations the practice of PSF-fitting photometry may take a great deal of the photometrist's time and thought. The best work is still something of an art.

All the basic steps discussed above can be turned into automated algorithms, and several flexible and powerful codes of this type are available in the literature (DAOPHOT, DoPHOT, Romafot, and others; see Stetson 1987, 1994; Stetson et al. 1990; Schechter et al. 1993; Buonanno & Iannicola 1989; Mighell & Rich 1995). These papers, as well as other sources such as Stetson's DAOPHOT manual, supply more advanced discussions of the process of CCD stellar photometry.

10.4 Testing the Data

If stars can be subtracted from the frame, they can also be added. That is, we can put artificial stars (scaled PSFs) into the image at arbitrarily chosen brightnesses and locations, and then detect and measure these simulated objects in the same way that we did the real stars. Since these added stars are built from the actual PSF, and are put onto the real sky background, they resemble the real stars quite closely. The huge advantage is that we know beforehand just how bright they are and exactly where we put them. The ability to create simulated images that resemble the real ones in almost every respect is a powerful way to test the data and understand quantitatively our measurement uncertainties and systematic errors.

Three extremely important results emerge from the analysis of simulated images:

- We can determine the *completeness of detection* $f(m)$: that is, we can find out what fraction of the artificially added stars were picked up in the normal object-finding process, as a function of magnitude m.
- We can determine any *systematic bias* $\Delta(m)$ in the measured magnitudes, again as a function of magnitude.
- We can determine the *random uncertainty* $\sigma(m)$ in the measured magnitudes. Here $\sigma(m)$, in magnitude units, is related to our earlier S/N ratio approximately by

$$\sigma \simeq 2.5 \log \left(1 + \frac{1}{(S/N)} \right) .$$

The bright stars are easy to deal with: they will almost all be found ($f \simeq 1$), and measured without bias ($\Delta \simeq 0$) and with low random uncertainty ($\sigma \rightarrow 0$). The real problems show up at the opposite end of the scale:

First, at progressively fainter levels, more and more stars fall below the threshold of detection and the completeness fraction f becomes small. At the 50% completeness level, the brightest pixel in the object is nominally *just* at the detection threshold, so it has an equal chance of falling above or below it depending on photon statistics from both the star and the sky it is sitting on. Fainter stars would nominally never have a bright enough pixel to sit above threshold, but they will be found (though with lower probability) if they happen to fall on a brighter than average patch of sky pixels. Similarly, a star just a little brighter than the nominal threshold can be missed if its local sky level is lower than average. Thus in practice, the transition from $f \simeq 1$ to $f \simeq 0$ is a smooth declining curve (see Fig. 70). A simple analytic function due to C. Pritchet which accurately matches most real $f(m)$ curves is

$$f = \frac{1}{2} \left(1 - \frac{a(m - m_0)}{\sqrt{1 + a^2(m - m_0)^2}} \right) \tag{71}$$

where the two free parameters are m_0 (the magnitude at which f is exactly 1/2) and a (which governs the slope; higher values of a correspond to steeper downturns). Completeness also depends significantly on the degree of crowding, and the intensity level of the sky; that is, it is a lot harder to find objects in extreme crowding conditions, or on a noisier background.

Second, the systematic bias $\Delta(m)$ starts to grow dramatically at fainter magnitudes, primarily because the fainter stars will be found more easily if they are sitting on brighter patches of background, which produces a measured magnitude brighter than it should be. See Fig. 71, and notice that $\Delta(m)$ rises *exponentially* as we go fainter than the 50% completeness level.

Third, the random uncertainty in the measured magnitude increases at fainter levels as the signal declines toward zero and the surrounding noise

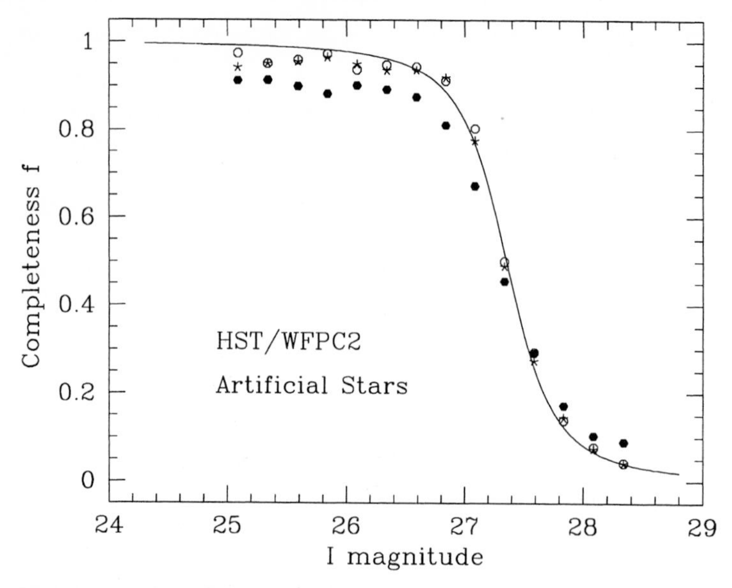

Fig. 70. Completeness of detection f, plotted as a function of magnitude, for images taken with the *HST*/WFPC2 camera. Data are from the I-band photometry of the Virgo dwarf elliptical shown in Fig. 68. The different symbols represent three different regions of the WF2 frame: *open* and *starred symbols* are from relatively uncrowded areas, whereas the *solid dots* are from areas closest to the galaxy center and thus most affected by crowding. The plotted points are binned means of several thousand artificial stars spread over all magnitudes. The model *line* is the Pritchet interpolation function defined in the text, with parameters $m_0 = 27.36$ and $a = 2.37$. Typically, f declines smoothly from nearly unity to nearly zero over roughly a one-magnitude run of image brightness

dominates more and more. Under a wide range of practical conditions, it can be shown that the 50% completeness level corresponds to a typical uncertainty $\sigma \sim 0.2$ mag or $S/N \sim 5$ (see Harris 1990a).

Examples of all three of these effects are illustrated in Figs. 70, 71, and 72. The messages from these simulation studies are clear. All aspects of your data will become seriously unreliable below the crossing point of $\sim 50\%$ completeness, and you need to know where that point is. Do not be tempted to believe any interesting features of your data that you think you see at still fainter levels; and above all, do not publish them!

10.5 Dealing with Nonstellar Objects

In most photometry projects, it is extremely helpful to be able to separate out starlike objects from nonstellar things. The former will include true stars, and also other objects that you may be trying to find in distant galaxies such as faint globular clusters, HII regions, galactic nuclei, and so on. The latter

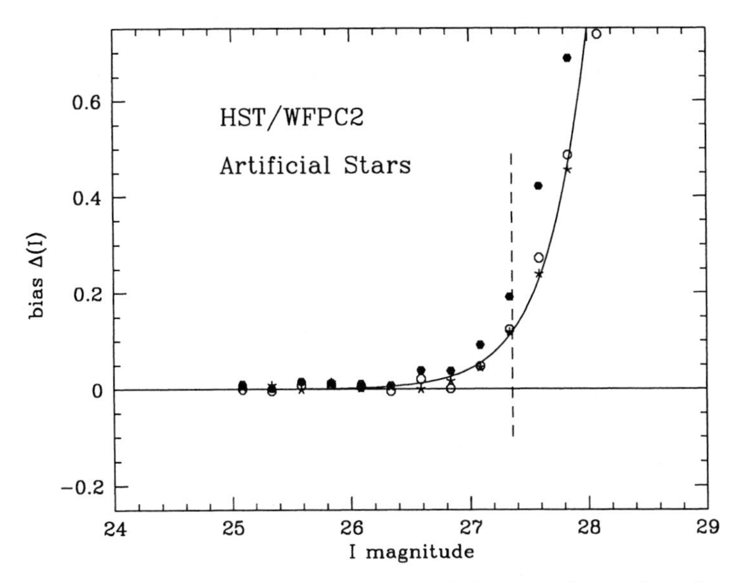

Fig. 71. Systematic measurement bias, defined as $\Delta = m(\text{input}) - m(\text{measured})$, plotted against magnitude. Data are from the same sample as in the previous figure. The 50% detection completeness level is marked with the *dashed line*, while the *solid curve* shows an exponential function in magnitude fitted to the data points

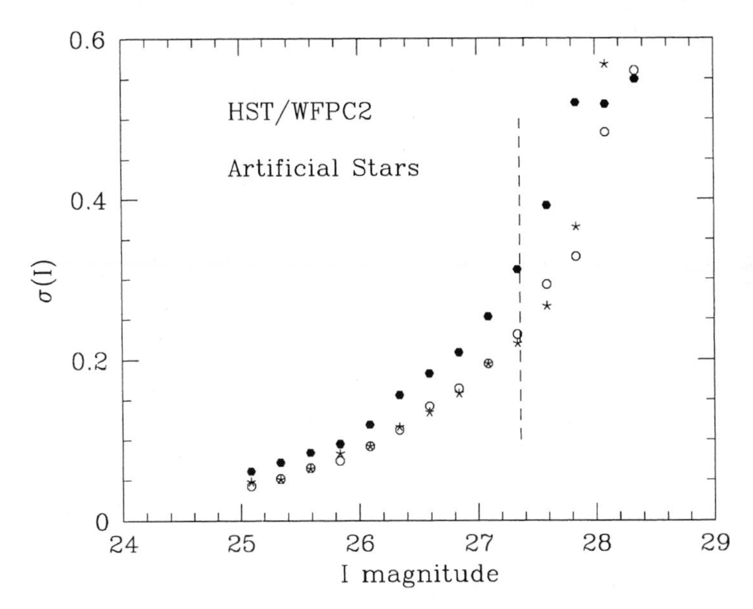

Fig. 72. Random measurement error $\sigma(m)$, defined as the root mean square scatter of the magnitude differences (input − measured). Data are from the same sample as in the previous figures. The 50% detection completeness level is marked with the *dashed line*

category will include anything that does not match the stellar PSF: small, faint background galaxies, resolved nebulae and clusters, unresolved clumps of stars, or even artifacts on the image.

A variety of algorithms can be constructed to *classify* the objects that you find on the image. The rather restricted question we ask during the process of stellar photometry is: how well does the PSF fit a given object? We now replace it by a subtly different and more general one: what parameters can we construct to *maximize the difference between starlike and nonstellar objects?*. And since the range of parameter space occupied by nonstellar objects is much larger than for starlike ones, the "best" answer may depend on the situation.

The answer to this question generally depends on using the fact that nonstellar objects have more extended contours and radial shapes than the starlike objects on the image. To quantify the characteristic "size" or extent of the object, let us define a general *radial image moment* which is constructed from the pixels within the object (e.g., Tyson & Jarvis 1979; Kron 1980; Harris et al. 1991):

$$r_n = \left(\frac{\sum r_i^n \cdot z_i}{\sum z_i} \right)^{1/n} . \tag{72}$$

Here z_i is the intensity of the ith pixel above sky background; r_i is the radial distance of each pixel from the center of the object; and there are $i = (1, \ldots, N)$ pixels in the image brighter than some chosen threshold. The object can have any arbitrary shape, so that the sum is simply taken over all the connected pixels making up the object rather than within any fixed aperture. Clearly, r_n represents a *characteristic radius* for the object in pixel units, calculated from the nth radial moment of the intensity distribution. Nonstellar objects (as well as some random clumps of noise) will have larger wings and asymmetric shapes, and thus have larger r_n values than stars do at the same total brightness. A simple plot of r_n against magnitude then effectively separates out the stars from other types of objects.

The choice of weighting exponent n is usually not crucial: positive n-values will weight the more extended wings of nonstellar objects more highly, while negative n gives more weight to the sharper cores of starlike objects. The simplest nontrivial moment, r_1, turns out in most situations to work well at separating out a high fraction of the distinguishable nonstellar images (cf. Harris et al. 1991). Other types of moments can be constructed that are sensitive to (for example) image asymmetry, or are linear combinations of various radial and nonradial moments (e.g., Jarvis & Tyson 1981; Valdes et al. 1983).

Though these quantities can be effective at picking out nonstellar objects, they are also good at finding crowded stars which one may not want to eliminate. Choose what works best for your situation. In some cases, it is important to distinguish between *crowded* pairs of objects (star/star, star/galaxy, galaxy/galaxy) and *single* galaxies with complex, lumpy structures; develop-

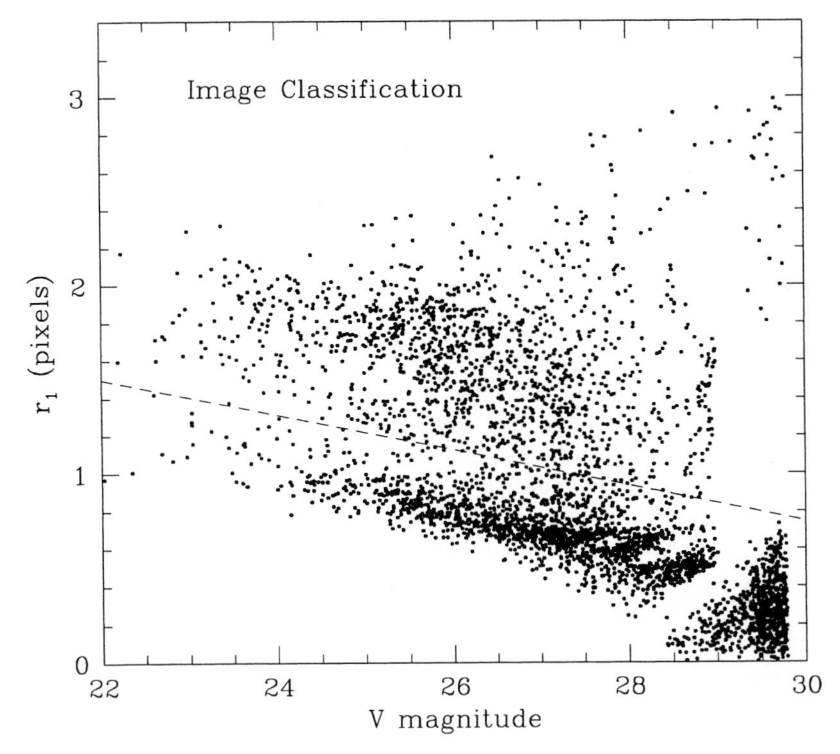

Fig. 73. Image structure diagnostic graph for a sample of measurements taken from deep *HST*/WFPC2 exposures of the remote elliptical galaxy IC 4051 (adapted from Woodworth & Harris 2000). The radial image moment r_1 defined in (72) is plotted against V magnitude for $\simeq$ 4500 objects measured on the WF2,3,4 CCD fields

ing an unbiased algorithm to separate out these cases is a nontrivial exercise (e.g., Jarvis & Tyson 1981; Bertin & Arnouts 1996).

A simple example of one of these diagnostic graphs is shown in Fig. 73. The datapoints in the figure represent a mixture of small nonstellar objects (the points falling above the dashed line, which are mostly faint background galaxies) and starlike objects (the points below the dashed line, which are mostly foreground stars and globular clusters belonging to the target galaxy IC 4051). Clearly, the great majority of the nonstellar objects can be separated out cleanly. (Note in this example that all the starlike objects do not have the same characteristic "size" r_1; instead, r_1 decreases for fainter objects. This is because the moment sum defining r_n is taken only over all pixels above a certain threshold, so that the fainter objects have fewer included pixels and smaller characteristic moments. For this reason, the points at the faint end (lower right corner of the figure) fall rather noticeably into quantized groups which represent the small numbers of pixels defining the intensity sum. Also, the V magnitude against which the radial moment is plotted is actually not a true fixed-aperture magnitude, but is instead something resembling an "isophotal" magnitude defined from the intensity sum of all pixels brighter than the threshold used in the calculation of r_1. That is, we have $V = \text{const} -2.5 \log(\sum z_i)$. See Harris et al. (1991) for additional discussion.)

It should be stressed that the image moment quantities described here are used *only* for classification purposes and not for actual measurement of the total magnitudes and colors (see below). Thus, they can be defined in whatever way will maximize the difference between stellar and nonstellar objects. Once they have served their purpose of separating out the two kinds of objects, they can be put aside.

The last stage of the classification process is to draw appropriate boundary lines between stellar and nonstellar regions of your chosen diagnostic diagram, and extract the unwanted ones from your object lists. In the simple case defined above, we would use the single diagram of r_1 versus total magnitude and define one empirical boundary line (shown in Fig. 73). But in principle, we could simultaneously use many more parameters, such as the aperture growth curve, peak intensity, and nonradial moments. Bertin & Arnouts (1996) nicely describe this step as mapping out the *frontier* between stellar and nonstellar objects in the multi-dimensional parameter space. Where to define the boundary of the frontier is always a matter of judgement, and artificial-star tests can be extremely helpful here for deciding where to place it. The codes of Valdes et al. (1983) and Bertin & Arnouts (1996) combine several image parameters in a manner equivalent to a neural network, and employ simulated images as training sets for the neural net.

At the faint end of the photometry, random noise eventually overwhelms the ability of even the most advanced decision-making algorithms to discriminate between stellar and nonstellar objects. Nevertheless, image classification is worth doing. The eventual payoff is that the contamination "noise" in your selected sample of objects can be tremendously reduced, and in some cases it is critical to the ability to define the sample at all.

In addition, the noise for one experiment can literally be the signal for another: for example, in a deep high-latitude field we might want to study the population of faint galaxies, in which case the foreground stars would be the contaminants. But if we want to study the Milky Way halo stars, exactly the opposite is true.

Measuring the *total magnitudes* of nonstellar objects correctly and consistently is a nontrivial job with a whole new set of special problems: unlike stellar images, there is no "PSF" to refer to, and the parameter space of image properties is vastly larger. One defensible and widely applicable approach is to measure the total magnitude within an aperture which encloses some large fraction (say 90%) of the object's asymptotic total flux, while not becoming so large that the enclosed light is too sensitive to sky noise. Since no two nonstellar objects have the same shape, this optimum aperture will have a different numerical value for each object and must be determined for each one. Under fairly general conditions, the first-order radial moment r_1 defined above is nearly equal to the half-light radius of a faint galaxy; a radius $2r_1$ turns out to enclose $\simeq 90\%$ of the total light and gives a reasonable choice for the optimum aperture magnitude. For detailed discussions of this method, see, for example, Kron (1980); Infante (1987); Infante & Pritchet (1992); Bershady et al. (1994); Bertin & Arnouts (1996); or Secker & Harris (1997), among others. Readers can refer to these same papers for an entry to the extensive literature on this subject.

References

1. Abraham, R. G., & van den Bergh, S. 1995, ApJ, 438, 218
2. Ajhar, E. A., Blakeslee, J. P., & Tonry, J. L. 1994, AJ, 108, 2087
3. Ajhar, E. A. et al. 1996, AJ, 111, 1110
4. Alcock, C. et al. 1997, ApJ, 482, 89
5. Allen, S. W., & Fabian, A. C. 1997, MNRAS, 286, 583
6. Alonso, M. V., & Minniti, D. 1997, ApJS, 109, 397
7. Armandroff, T. E. 1989, AJ, 97, 375
8. Armandroff, T. E. 1993, in The Globular Cluster–Galaxy Connection, ASP Conf. Series 48, ed. G. Smith & J. P. Brodie (San Francisco: ASP), 48
9. Armandroff, T. E., Olszewski, E. W., & Pryor, C. 1995, AJ, 110, 2131
10. Arp, H. C. 1965, in Galactic Structure, ed. A. Blaauw & M. Schmidt (Chicago: Univ. Chicago Press), 401
11. Ashman, K. M., Bird, C. M., & Zepf, S. E. 1994, AJ, 108, 2348
12. Ashman, K. M., & Zepf, S. E. 1992, ApJ, 384, 50
13. Ashman, K. M., & Zepf, S. E. 1998, Globular Cluster Systems (New York: Cambridge Univ. Press)
14. Baade, W. 1944, ApJ, 100, 147
15. Baade, W. 1958, in Stellar Populations, Proceedings of a Conference at Vatican Observatory, ed. D. J. K. O'Connell (New York: Interscience)
16. Babul, A., & Rees, M. J. 1992, MNRAS, 255, 346
17. Bahcall. N. A., Fan, X., & Cen, R. 1997, ApJ, 485, L53
18. Barbuy, B., Bica, E., & Ortolani, S. 1998, A&A, 333, 117
19. Barnes, J. E. 1988, ApJ, 331, 699
20. Barth, A. J., Ho, L. C., Filippenko, A. V, & Sargent, W. L. W. 1995, AJ, 110, 1009
21. Battistini, P. et al. 1987, A&AS, 67, 447
22. Baugh, C. M., Cole, S., Frenk, C. S., & Lacey, C. G. 1998, ApJ, 498, 504
23. Baum, W. A. 1952, AJ, 57, 222
24. Baum, W. A. 1955, PASP, 67, 328
25. Baum, W. A. et al. 1995, AJ, 110, 2537
26. Baum, W. A. et al. 1997, AJ, 113, 1483
27. Baumgardt, H. 1998, A&A, 330, 480
28. Beck, S. C., Turner, J. L., Ho, P. T. P., Lacy, J. H., & Kelly, D. M. 1996, ApJ, 457, 610

29. Bender, R., Ziegler, B., & Bruzual, G. 1996, ApJ, 463, L51
30. Bernardi, M. et al. 1998, ApJ, 508, L143
31. Bershady, M. A., Hereld, M., Kron, R. G., Koo, D. C., Munn, J. A., & Majewski, S. R. 1994, AJ, 108, 870
32. Bertin, E., & Arnouts, S. 1996, A&AS, 117, 393
33. Binggeli, B., & Popescu, C. C. 1995, A&A, 298, 63
34. Binggeli, B., Popescu, C. C., & Tammann, G. A. 1993, A&AS, 98, 275
35. Binney, J., & Tremaine, S. 1987, Galactic Dynamics (Princeton: Princeton Univ. Press)
36. Blakeslee, J. P. 1997, ApJ, 481, L59
37. Blakeslee, J. P., & Tonry, J. L. 1996, ApJ, 465, L19
38. Blakeslee, J. P., Tonry, J. L., & Metzger, M. R. 1997, AJ, 114, 482
39. Bothun G. D. 1986, AJ, 91, 507
40. Brewer, J. P., Richer, H. B., & Crabtree, D. R. 1995, AJ, 109, 2480
41. Bridges, T. J., Carter, D., Harris, W. E., & Pritchet, C. J. 1996, MNRAS, 281, 1290
42. Bridges, T. J., & Hanes, D. A. 1992, AJ, 103, 800
43. Bridges, T. J., Hanes, D. A., & Harris, W. E. 1991, AJ, 101, 469
44. Bridges, T. J. et al. 1997, MNRAS, 284, 376
45. Bridges, T. J., & Irwin, J. A. 1998, MNRAS, 300, 967
46. Brodie, J. P., & Huchra, J. P. 1991, ApJ, 379, 157
47. Brodie, J. P., Schroder, L. L., Huchra, J. P., Phillips, A. C., Kissler-Patig, M., & Forbes, D. A. 1998, AJ, 116, 691
48. Bruzual, G., & Charlot, S. 1993, ApJ, 405, 538
49. Buonanno, R., Corsi, C. E., Zinn, R., Fusi Pecci, F., Hardy, E., & Suntzeff, N. B. 1998, ApJ, 501, L33
50. Buonanno, R., & Iannicola, G. 1989, PASP, 101, 294
51. Burkert, A., & Smith, G. H. 1997, ApJ, 474, L15
52. Caldwell, N., & Bothun, G. D. 1987, AJ, 94, 1126
53. Calzetti, D. et al. 1997, AJ, 114, 1834
54. Capriotti, E. R., & Hawley, S. L. 1996, ApJ, 464, 765
55. Carlberg, R. G., & Pudritz, R. E. 1990, MNRAS, 247, 353
56. Carlberg, R. G. et al. 1996, ApJ, 462, 32
57. Carlson, M. N. et al. 1998, AJ, 115, 1778
58. Carney, B. W., & Seitzer, P. 1986, AJ, 92, 23
59. Carney, B. W., Storm, J., & Jones, R. V. 1992, ApJ, 386, 663
60. Carretta, E., & Gratton, R. G. 1997, A&AS, 121, 95
61. Carretta, E., Gratton, R. G., Clementini, G., & Fusi Pecci, F. 1999, ApJ, 533, 215
62. Carroll, S. M., & Press, W. H. 1992, ARAA, 30, 499
63. Cassisi, S., Castellani, V., Degl'Innocenti, S., Salaris, M., & Weiss, A. 1999, A&AS, 134, 103
64. Catelan, M., & de Freitas Pacheco, J. A. 1995, A&A, 297, 345
65. Cesarsky, D. A., Lequeux, J., Laustsen, S., Schuster, H.-E., & West, R. M. 1977, A&A, 61, L31
66. Chaboyer, B., Demarque, P., Kernan, P. J., & Krauss, L. M. 1998, ApJ, 494, 96
67. Chaboyer, B., Demarque, P., & Sarajedini, A. 1996, ApJ, 459, 558
68. Chiosi, C., Bertelli, G., & Bressan, A. 1992, ARAA, 30, 235

69. Christian, C. A., & Heasley, J. N. 1988, AJ, 95, 1422
70. Christian, C. A., & Heasley, J. N. 1991, AJ, 101, 848
71. Ciardullo, R., Jacoby, G. H., Feldmeier, J. J., & Bartlett, R. E. 1998, ApJ, 492, 62
72. Ciardullo, R., Jacoby, G. H., & Ford, H. C. 1989, ApJ, 344, 715
73. Ciardullo, R., Jacoby, G. H., & Tonry, J. L. 1993, ApJ, 419, 479
74. Cohen, J. G. 1992, ApJ, 400, 528
75. Cohen, J. G., Blakeslee, J. P., & Ryzhov, A. 1998, ApJ, 496, 808
76. Cohen, J. G., & Ryzhov, A. 1997, ApJ, 486, 230
77. Colless, M., & Dunn, A. M. 1996, ApJ, 458, 435
78. Conti, P. S., & Vacca, W. D. 1994, ApJ, 423, L97
79. Cool, A. M., Piotto, G., & King, I. R. 1996, ApJ, 468, 655
80. Côté, P. 1999, AJ, 118, 406
81. Côté, P., Marzke, R. O., & West, M. J. 1998, ApJ, 501, 554
82. Couture, J., Harris, W. E., & Allwright, J. W. B. 1990, ApJS, 73, 671
83. Couture, J., Racine, R., Harris, W. E., & Holland, S. 1995, AJ, 109, 2050
84. Cowan, J. J., McWilliam, A., Sneden, C., & Burris, D. L. 1997, ApJ, 480, 246
85. Crampton, D., Cowley, A. P., Schade, D., & Chayer, P. 1985, ApJ, 288, 494
86. Cudworth, K. M. 1985, AJ, 90, 65
87. Cudworth, K. M., & Hanson, R. B. 1993, AJ, 105, 168
88. Da Costa, G. S. 1984, ApJ, 285, 483
89. Da Costa, G. S., & Armandroff, T. E. 1995, AJ, 1090, 2533
90. Dekel, A., & Silk, J. 1986, ApJ, 303, 39
91. De Marchi, G., Clampin, M., Greggio, L., Leitherer, C., Nota, A., & Tosi, M. 1997, ApJ, 479, L27
92. de Vaucouleurs, G. 1970, ApJ, 159, 435
93. de Vaucouleurs, G. 1977, Nature, 266, 126
94. Dinescu, D. I., Girard, T. M., & van Altena, W. F. 1999, AJ, 117, 1792
95. Dorman, B., Lee, Y.-W., & VandenBerg, D. A. 1991, ApJ, 366, 115
96. Durrell, P. R., Harris, W. E., Geisler, D., & Pudritz, R. E. 1996a, AJ, 112, 972
97. Durrell, P. R., McLaughlin, D. E., Harris, W. E., & Hanes, D. A. 1996b, ApJ, 463, 543
98. Eckart, A. et al. 1990, ApJ, 363, 451
99. Eggen, O. J., Lynden-Bell, D., & Sandage, A. 1962, ApJ, 136, 748 (ELS)
100. Ellis, R. S. et al. 1997, ApJ, 483, 502
101. Elmegreen, B. G., & Efremov, Y. N. 1997, ApJ, 480, 235
102. Elmegreen, B. G., Efremov, Y. N., Pudritz, R. E., & Zinnecker, H. 1999, in Protostars and Planets IV, ed. A. P. Boss, S. S. Russell, & V. Mannings (Tucson: Univ. Arizona Press), in press
103. Elmegreen, B. G., & Falgarone, E. 1996, ApJ, 471, 816
104. Elson, R. A. W., Grillmair, C. J., Forbes, D. A., Rabban, M., Williger, G. M., & Brodie, J. P. 1998, MNRAS, 295, 240
105. Fabian, A. C., Nulsen, P. E. J., & Canizares, C. R. 1984, MNRAS, 201, 933
106. Fall, S. M., & Rees, M. J. 1977, MNRAS, 181, 37P
107. Fall, S. M., & Rees, M. J. 1985, ApJ, 298, 18
108. Feast, M. 1998, MNRAS, 293, L27
109. Feldmeier, J. J., Ciardullo, R., & Jacoby, G. H. 1997, ApJ, 479, 231
110. Ferguson, H. C., & Sandage, A. 1989, ApJ, 346, L53

111. Ferguson, H. C., Tanvir, N. R., & von Hippel, T. 1998, Nature, 391, 461
112. Fernley, J. et al. 1998a, A&A, 330, 515
113. Fernley, J. et al. 1998b, MNRAS, 293, L61
114. Ferrarese, L. et al. 1996, ApJ, 464, 568
115. Ferraro, F. R. et al. 1997, A&A, 320, 757
116. Field, G. B., & Saslaw, W. C. 1965, ApJ, 142, 568
117. Fleming, D. E. B., Harris, W. E., Pritchet, C. J., & Hanes, D. A. 1995, AJ, 109, 1044
118. Forbes, D. A. 1996a, AJ, 112, 954
119. Forbes, D. A. 1996b, AJ, 112, 1409
120. Forbes, D. A., Brodie, J. P., & Grillmair, C. J. 1997, AJ, 113, 1652
121. Forbes, D. A., Brodie, J. P., & Huchra, J. 1996a, AJ, 112, 2448
122. Forbes, D. A., Franx, M., Illingworth, G. D., & Carollo, C. M. 1996b, ApJ, 467, 126
123. Forbes, D. A., Grillmair, C. J., Williger, G. M., Elson, R. A. W., & Brodie, J. P. 1998, MNRAS, 293, 325
124. Ford, H. C., Hui, X., Ciardullo, R., Jacoby, G. H., & Freeman, K. C. 1996, ApJ, 458, 455
125. Forte, J. C., Strom, S. E., & Strom, K. M. 1981, ApJ, 245, L9
126. Freedman, W. L., & Madore, B. F. 1990, ApJ, 365, 186
127. Frenk, C. S., & White, S. D. M. 1980, MNRAS, 193, 295
128. Fritze-von Alvensleben, U. 1998, A&A, 336, 83
129. Fritze-von Alvensleben, U. 1999, A&A, 342, L25
130. Frogel, J. A., Cohen, J. G., & Persson, S. E. 1983, ApJ, 275, 773
131. Fruchter, A. S., & Hook, R. N. 1999, preprint (astro-ph/9808087)
132. Fusi Pecci, F., Bellazzini, M., Cacciari, C., & Ferraro, F. R. 1995, AJ, 110, 1664
133. Fusi Pecci, F. et al. 1996, AJ, 112, 1461
134. Geisler, D., & Forte, J. C. 1990, ApJ, 350, L5
135. Geisler, D., Lee, M. G., & Kim, E. 1996, AJ, 111, 1529
136. Giavalisco, M., Steidel, C. C., & Macchetto, F. D. 1996, ApJ, 474, 223
137. Gieren, W. P., Fouqué, P., & Gómez, M. 1998, ApJ, 496, 17
138. Gilliland, R. L. 1992, in Astronomical CCD Observing and Reduction Techniques, ASP Conf. Series 23, ed. S. B. Howell (San Francisco: ASP), 68
139. Girardi, M., Biviano, A., Giuricin, G., Mardirossian, F., & Mezzetti, M. 1993, ApJ, 404, 38
140. Glazebrook, K., Abraham, R., Santiago, B., Ellis, R., & Griffiths, R. 1998, MNRAS, 297, 885
141. Gnedin, O. Y. 1997, ApJ, 487, 663
142. Gnedin, O. Y., & Ostriker, J. P. 1997, ApJ, 474, 223
143. Gonzalez Delgado, R. M., Leitherer, C., Heckman, T., & Cervino, M. 1997, ApJ, 483, 705
144. Gorjian, V. 1996, AJ, 112, 1886
145. Gould, A. 1994, ApJ, 426, 542
146. Gould, A., & Popowski, P. 1998, ApJ, 508, 844
147. Graham, J. A. 1977, PASP, 89, 425
148. Graham, J. A. et al. 1997, ApJ, 477, 535
149. Gratton, R. G. 1998, MNRAS, 296, 739
150. Gratton, R. G., Fusi Pecci, F., Carretta, E., Clementini, G., Corsi, C. E., & Lattanzi, M. 1997, ApJ, 491, 749

151. Grillmair, C. J. 1998, in Galactic Halos: A UC Santa Cruz Workshop, ASP Conf. Series 136, ed. D. Zaritsky (San Francisco: ASP), 45
152. Grillmair, C. J. et al. 1994, ApJ, 422, L9
153. Grillmair, C. J., Forbes, D. A., Brodie, J. P., & Elson, R. A. W. 1999, AJ, 117, 167
154. Grundahl, F., VandenBerg, D. A., & Andersen, M. I. 1998, ApJ, 500, L179
155. Gullixson, C. A. 1992, in Astronomical CCD Observing and Reduction Techniques, ASP Conf. Series 23, ed. S. B. Howell (San Francisco: ASP), 130
156. Hamuy, M., Phillips, M. M., Suntzeff, N. B., Schommer, R. A., Maza, J., & Avilés, R. 1996, AJ, 112, 2398
157. Hanes, D. A. 1977, MNRAS, 180, 309
158. Hanes, D. A., & Whittaker, D. G. 1987, AJ, 94, 906
159. Hanson, R. B. 1979, MNRAS, 186, 875
160. Harris, G. L. H., Geisler, D., Harris, H. C., & Hesser, J. E. 1992, AJ, 104, 613
161. Harris, G. L. H., Harris, W. E., & Poole, G. B. 1999, AJ, 117, 855
162. Harris, G. L. H., Poole, G. B., & Harris, W. E. 1998, AJ, 116, 2866
163. Harris, H. C. 1988, in Globular Cluster Systems in Galaxies, IAU Symposium 126, ed. J. E. Grindlay & A. G. D. Philip (Dordrecht: Kluwer), 205
164. Harris, W. E. 1976, AJ, 81, 1095
165. Harris, W. E. 1981, ApJ, 251, 497
166. Harris, W. E. 1988a, in Globular Cluster Systems in Galaxies, IAU Symposium 126, ed. J. E. Grindlay & A. G. D. Philip (Dordrecht: Kluwer), 237
167. Harris, W. E. 1988b, in The Extragalactic Distance Scale, ASP Conf. Series 4, ed. S. van den Bergh & C. J. Pritchet (San Francisco: ASP), 231
168. Harris, W. E. 1990a, PASP, 102, 949
169. Harris, W. E. 1990b, PASP, 102, 966
170. Harris, W. E. 1991, ARAA, 29, 543
171. Harris, W. E. 1993, in The Globular Cluster–Galaxy Connection, ASP Conf. Series 48, ed. G. H. Smith & J. P. Brodie (San Francisco: ASP), 472
172. Harris, W. E. 1995, in Stellar Populations, IAU Symposium 164, ed. P. C. van der Kruit & G. Gilmore (Dordrecht: Kluwer), 85
173. Harris, W. E. 1996a, AJ, 112, 1487
174. Harris, W. E. 1996b, in Formation of the Galactic Halo...Inside and Out, ASP Conf. Series 92, ed. H. Morrison & A. Sarajedini (San Francisco: ASP), 231
175. Harris, W. E. 1998, in Galactic Halos: A UC Santa Cruz Workshop, ASP Conf. Series 136, ed. D. Zaritsky (San Francisco: ASP), 33
176. Harris, W. E. 1999, in Globular Clusters, Tenth Canary Islands Winter School of Astrophysics, ed. C. Martinez Roger, I. Perez Fournon & F. Sanchez (Cambridge: Cambridge Univ. Press), 325
177. Harris, W. E., Allwright, J. W. B., Pritchet, C. J., & van den Bergh, S. 1991, ApJS, 76, 115
178. Harris, W. E., & Canterna, R. 1979, ApJ, 231, L19
179. Harris, W. E. et al. 1997a, AJ, 113, 688
180. Harris, W. E. et al. 1997b, AJ, 114, 1030
181. Harris, W. E., Durrell, P. R., Petitpas, G. R., Webb, T. M., & Woodworth, S. C. 1997c, AJ, 114, 1043
182. Harris, W. E., Durrell, P. R., Pierce, M. J., & Secker, J. 1998b, Nature, 395, 45
183. Harris, W.E., FitzGerald, M. P., & Reed, B. C. 1981, PASP, 93, 507

184. Harris, W. E., Harris, G. L. H., & McLaughlin, D. E. 1998a, AJ, 115, 1801
185. Harris, W. E., Pritchet, C. J., & McClure, R. D. 1995, ApJ, 441, 120
186. Harris, W. E., & Pudritz, R. E. 1994, ApJ, 429, 177
187. Harris, W. E., & Racine, R. 1979, ARAA, 17, 241
188. Harris, W. E., & Smith, M. G. 1976, ApJ, 207, 1036
189. Harris, W. E., & van den Bergh, S. 1981, AJ, 86, 1627
190. Hartwick, F. D. A. 1976, ApJ, 209, 418
191. Hartwick, F. D. A. 1996, in Formation of the Galactic Halo...Inside and Out, ASP Conf. Series 92, ed. H. Morrison & A. Sarajedini (San Francisco: ASP), 444
192. Hartwick, F. D. A., & Sargent, W. L. W. 1978, ApJ, 221, 512
193. Heasley, J. N., Christian, C. A., Friel, E. D., & Janes, K. A. 1988, AJ, 96, 1312
194. Hesser, J. E., Harris, W. E., VandenBerg, D. A., Allwright, J. W. B., Shott, P., & Stetson, P. B. 1987, PASP, 99, 739
195. Hilker, M. 1998, PhD thesis, University of Bonn
196. Hodge, P. W. 1974, PASP, 86, 289
197. Holland, S. 1998, AJ, 115, 1916
198. Holland, S., Fahlman, G. G., & Richer, H. B. 1997, AJ, 114, 1488
199. Holtzman, J. A. et al. 1992, AJ, 103, 691
200. Holtzman, J. A. et al. 1996, AJ, 112, 416
201. Hook, R. N., & Fruchter, A. S. 1997, in Astronomical Data Analysis Software and Systems VI, ASP Conf. Series 125, ed. G. Hunt & H. E. Payne (San Francisco: ASP), 147
202. Howell, S. B. 1992, in Astronomical CCD Observing and Reduction Techniques, ASP Conf. Series 23, ed. S. B. Howell (San Francisco: ASP), 105
203. Hubble, E. 1932, ApJ, 76, 44
204. Huchra, J. P. 1988, in The Extragalactic Distance Scale, ASP Conf. Series 4, ed. S. van den Bergh & C. J. Pritchet (San Francisco: ASP), 257
205. Huchra, J. P., & Brodie, J. P. 1987, AJ, 93, 779
206. Huchra, J. P., Brodie, J. P., & Kent, S. M. 1991, ApJ, 370, 495
207. Hui, X., Ford, H. C., Freeman, K. C., & Dopita, M. A. 1995, ApJ, 449, 592
208. Hunter, D. A., O'Connell, R. W., & Gallagher, J. S. 1994a, AJ, 108, 84
209. Hunter, D. A., van Woerden, H., & Gallagher, J. S. 1994b, ApJS, 91, 79
210. Infante, L. 1987, A&A, 183, 177
211. Infante, L., & Pritchet, C. J. 1992, ApJS, 82, 237
212. Irwin, M. J., Demers, S., & Kunkel, W. E. 1995, ApJ, 453, L21
213. Israel, F. P. 1997, A&A, 328, 471
214. Jablonka, P., Alloin, D., & Bica, E. 1992, A&A, 260, 97
215. Jablonka, P., Bica, E., Pelat, D., & Alloin, D. 1996, A&A, 307, 385
216. Jacoby, G. H., Ciardullo, R., & Ford, H. C. 1990, ApJ, 356, 332
217. Jacoby, G. H., Ciardullo, R., & Harris, W. E. 1996, ApJ, 462, 1
218. Jacoby, G. H. et al. 1992, PASP, 104, 599
219. Jaffe, W. 1990, A&A, 240, 254
220. Jarvis, J. F., & Tyson, J. A. 1981, AJ, 86, 476
221. Jerjen, H., & Tammann, G. A. 1993, A&A, 276, 1
222. Johnson, J. A., & Bolte, M. 1998, AJ, 115, 693
223. Johnson, J. A., Bolte, M., Stetson, P. B., Hesser, J. E., & Somerville, R. S. 1999, ApJ, 527, 199

224. Johnston, K. V. 1998, in Galactic Halos: A UC Santa Cruz Workshop, ASP Conf. Series 136, ed. D. Zaritsky (San Francisco: ASP), 365
225. Jones, B. F., Klemola, A. R., & Lin, D. N. C. 1994, AJ, 107, 1333
226. Kavelaars, J. J. 1998, PASP, 110, 758
227. Kavelaars, J. J. 1999, in Galaxy Dynamics, ASP Conf. Series 182, ed. D. R. Merritt, M. Valluri, & J. A. Sellwood (San Francisco: ASP), 437
228. Kavelaars, J. J., & Gladman, B. 1998, in Proc. 5th CFHT Users Meeting, ed. P. Martin & S. Rucinski (Kamuela: CFHT), 155
229. Kavelaars, J. J., & Hanes, D. A. 1997, MNRAS, 285, L31
230. Kavelaars, J. J., Harris, W. E., Hanes, D. A., Hesser, J. E., & Pritchet, C. J. 2000, ApJ, 533, 125
231. King, I. R., Anderson, J., Cool, A., & Piotto, G. 1998, ApJ, 492, L37
232. Kinman, T. D. 1959a, MNRAS, 119, 538
233. Kinman, T. D. 1959b, MNRAS, 119, 559
234. Kinman, T. D. et al. 1991, PASP, 103, 1279
235. Kissler, M., Richtler, T., Held, E. V., Grebel, E. K., Wagner, S. J., & Capaccioli, M. 1994, A&A, 287, 463
236. Kissler-Patig, M. 1998, in Galactic Halos: A UC Santa Cruz Workshop, ASP Conf. Series 136, ed. D. Zaritsky (San Francisco: ASP), 63
237. Kissler-Patig, M., Ashman, K. M., Zepf, S. E., & Freeman, K. C. 1999, AJ, 118, 197
238. Kissler-Patig, M., Forbes, D. A., & Minniti, D. 1998, MNRAS, 298, 1123
239. Kissler-Patig, M., & Gebhardt, K. 1998, AJ, 116 2237
240. Kissler-Patig, M., Grillmair, C. J., Meylan, G., Brodie, J. P., Minniti, D., & Goudfrooij, P. 1999, AJ, 117, 1206
241. Kissler-Patig, M., Kohle, S., Hilker, M., Richtler, T., Infante, L., & Quintana, H. 1997a, A&A, 319, 470
242. Kissler-Patig, M., Richtler, T., Storm, J., & Della Valle, M. 1997b, A&A, 327, 503
243. Kobulnicky, H. A., Dickey, J. M., Sargent, A. I., Hogg, D. E., & Conti, P. S. 1995, AJ, 110, 117
244. Kobulnicky, H. A., & Skillman, E. D. 1996, ApJ, 471, 211
245. Kohle, S., Kissler-Patig, M., Hilker, M., Richtler, T., Infante, L., & Quintana, H. 1996, A&A, 309, L39
246. Kron, G. E., & Mayall, N. U. 1960, AJ, 65, 581
247. Kron, R. 1980, ApJS, 43, 305
248. Kroupa, P., & Bastian, U. 1997a, New Astron., 2, 77
249. Kroupa, P., & Bastian, U. 1997b, in Proc. ESA Symposium "Hipparcos – Venice '97", ESA SP-402, 615
250. Kundu, A., & Whitmore, B. C. 1998, AJ, 116, 2841
251. Kundu, A., Whitmore, B. C., Sparks, W. B., Macchetto, F. D., Zepf, S. E., & Ashman, K. M. 1999, ApJ, 513, 733
252. Kunkel, W. E., & Demers, S. 1975, RGO Bulletin No. 182
253. Kunkel, W. E., & Demers, S. 1977, ApJ, 214, 21
254. Kwan, J. 1979, ApJ, 229, 567
255. Larson, R. B. 1988, in Globular Cluster Systems in Galaxies, IAU Symposium No. 126, ed. J. E. Grindlay & A. G. D. Philip (Dordrecht: Kluwer), 311
256. Larson, R. B. 1990a, in Physical Processes in Fragmentation and Star Formation, ed. R. Capuzzo-Dolcetta, C. Chiosi, & A. DiFazio (Dordrecht: Kluwer), 389

257. Larson, R. B. 1990b, PASP, 653, 704
258. Larson, R. B. 1993, in The Globular Cluster–Galaxy Connection, ASP Conf. Series 48, ed. G. H. Smith & J. P. Brodie (San Francisco: ASP), 675
259. Larson, R. B. 1996, in Formation of the Galactic Halo...Inside and Out, ASP Conf. Series 92, ed. H. Morrison & A. Sarajedini (San Francisco: ASP), 241
260. Lauberts, A. 1976, A&A, 52, 309
261. Lauer, T. R. 1999, PASP, 111, 227
262. Lauer, T. R., Tonry, J. R., Postman, M., Ajhar, E. A., & Holtzman, J. A. 1998, ApJ, 499, 577
263. Layden, A. C., Hanson, R. B., Hawley, S. L., Klemola, A. R., & Hanley, C. J. 1996, AJ, 112, 2110
264. Lazareff, B., Castets, A., Kim, D. W., & Jura, M. 1989, ApJ, 336, L13
265. Lee, M. G., Freedman, W. L., & Madore, B. F. 1993a, ApJ, 417, 553
266. Lee, M. G., Freedman, W. L., Mateo, M., Thompson, I., Roth, M., & Ruiz, M. T. 1993b, AJ, 106, 1420
267. Lee, M. G., Kim, E., & Geisler, D. 1997, AJ, 114, 1824
268. Lee, M. G., Kim, E., & Geisler, D. 1998, AJ, 115, 947
269. Lee, Y.-W., Demarque, P., & Zinn, R. 1994, ApJ, 423, 248
270. Leitherer, C., Vacca, W. D., Conti, P. S., Filippenko, A. V., Robert, C., & Sargent, W. L. W. 1996, ApJ, 465, 717
271. Lin, D. N. C., Jones, B. F., & Klemola, A. R. 1995, ApJ, 439, 652
272. Lineweaver, C. H., Barbosa, D., Blanchard, A., & Bartlett, J. G. 1998, A&A, 322, 365
273. Little, B., & Tremaine, S. 1987, ApJ, 320, 493
274. Liu, T., & Janes, K. A. 1990, ApJ, 360, 561
275. Lo, K. Y., Cheung, K. W., Masson, C. R., Phillips, T. G., Scott, S. L., & Woody, D. P. 1987, ApJ, 312, 574
276. Loewenstein, M. & Mushotzky, R. F. 1996, ApJ, 466, 695
277. Long, K., Ostriker, J. P., & Aguilar, L. 1992, ApJ, 388, 362
278. Lutz, D. 1991, A&A, 245, 31
279. Lutz, T. E., & Kelker, D. H. 1973, PASP, 85, 573
280. Lynden-Bell, D. 1976, MNRAS, 174, 695
281. Lynden-Bell, D., Cannon, R. D., & Godwin, P. J. 1983, MNRAS, 204, 87P
282. Lynden-Bell, D., & Lynden-Bell, R. M. 1995, MNRAS, 275, 429
283. Madau, P., Ferguson, H. C., Dickinson, M.E., Giavalisco, M., Steidel, C. C., & Fruchter, A. 1996, MNRAS, 283, 1388
284. Madore, B. F., & Arp, H. C. 1979, ApJ, 227, L103
285. Madore, B. F. et al. 1998, Nature, 395, 47
286. Majewski, S. R. 1993, ARAA, 31, 575
287. Majewski, S. R. 1994, ApJ, 431, L17
288. Majewski, S. R., Hawley, S. L., & Munn, J. A. 1996, in Formation of the Galactic Halo...Inside and Out, ASP Conf. Series 92, ed. H. Morrison & A. Sarajedini (San Francisco: ASP), 119
289. Maoz, E. 1990, ApJ, 359, 257
290. Marsakov, V. A., & Suchkov, A. A. 1976, Sov. Astron. Letters, 2, 148
291. Massey, P., & Hunter, D. A. 1998, ApJ, 493, 180
292. Massey, P., & Jacoby, G. H. 1992, in Astronomical CCD Observing and Reduction Techniques, ASP Conf. Series 23, ed. S. B. Howell (San Francisco: ASP), 240

293. Mateo, M. 1996, in Formation of the Galactic Halo...Inside and Out, ASP Conf. Series 92, ed. H. Morrison & A. Sarajedini (San Francisco: ASP), 434
294. Mateo, M., Fischer, P, & Krzeminski, W. 1995, AJ, 110, 2166
295. Mateo, M., Olszewski, E. W., Pryor, C., Welch, D. L., & Fischer, P. 1993, AJ, 105, 510
296. Mateo, M., Olszewski, E. W., Welch, D. L., Fischer, P., & Kunkel, W. 1991, AJ, 102, 914
297. Mathews, G. J., Kajino, T., & Orito, M. 1996, ApJ, 456, 98
298. Matthews, L. D., Gallagher, J. S., & Littleton, J. E. 1995, AJ, 110, 581
299. Mauersberger, R., Henkel, C., Wielebinski, R., Wiklind, T., & Reuter, H.-P. 1996, A&A, 305, 421
300. Mayall, N. U. 1946, ApJ, 104, 290
301. Mayall, N. U., & Eggen, O. J. 1953, PASP, 65, 24
302. Mayya, Y. D., & Prabhu, T. P. 1996, AJ, 111, 1252
303. McKee, C.F., Zweibel, E. G., Goodman, A. A., and Heiles, C. 1993, in Protostars and Planet III, ed. M. Matthews & E. Levy (Tucson: Univ. Arizona Press), 327.
304. McLaughlin, D. E. 1994, PASP, 106, 47
305. McLaughlin, D. E. 1999, AJ, 117, 2398
306. McLaughlin, D. E., Harris, W. E., & Hanes, D. A. 1994, ApJ, 422, 486
307. McLaughlin, D. E., & Pudritz, R. E. 1996, ApJ, 457, 578
308. McMillan, R., Ciardullo, R., & Jacoby, G. H. 1993, ApJ, 416, 62
309. McNamara, B. R., & O'Connell, R. W. 1992, ApJ, 393, 579
310. Meurer, G. R. 1995, Nature, 375, 742
311. Meurer, G. R., Freeman, K. C., Dopita, M. A., & Cacciari, C. 1992, AJ, 103, 60
312. Meurer, G. R., Heckman, T. M., Leitherer, C., Kinney, A., Robert, C., & Garnett, D. R. 1995, AJ, 110, 2665
313. Mighell, K. J., & Rich, R. M. 1995, AJ, 110, 1649
314. Mighell, K. J., & Rich, R. M. 1996, AJ, 111, 777
315. Miller, B. W., Lotz, J. M., Ferguson, H. C., Stiavelli, M., & Whitmore, B. C. 1998, ApJ, 508, L133
316. Miller, B. W., Whitmore, B. C., Schweizer, F., & Fall, S. M. 1997, AJ, 114, 2381
317. Minniti, D. 1995, AJ, 109, 1663
318. Minniti, D., Alonso, M. V., Goudfrooij, P., Jablonka, P., & Meylan, G. 1996, ApJ, 467, 221
319. Minniti, D., Kissler-Patig, M, Goudfrooj, P., & Meylan, G. 1998, AJ, 115, 121
320. Morgan, W. W. 1956, PASP, 68, 509
321. Morris, P. W., & Shanks, T. 1998, MNRAS, 298, 451
322. Mould, J. R., Oke, J. B., De Zeeuw, P. T., & Nemec, J. M. 1990, AJ, 99, 1823
323. Mould, J. R., Oke, J. B., & Nemec, J. M. 1987, AJ, 93, 53
324. Mould, J. et al. 1995, ApJ, 449, 413
325. Murali, C., & Weinberg, M. D. 1997a, MNRAS, 288, 749
326. Murali, C., & Weinberg, M. D. 1997b, MNRAS, 288, 767
327. Murali, C., & Weinberg, M. D. 1997c, MNRAS, 291, 717
328. Mushotzky, R. F., & Loewenstein, M. 1997, ApJ, 481, L63
329. Muzzio, J. C. 1986, ApJ, 306, 44
330. Muzzio, J. C. 1988, in Globular Cluster Systems in Galaxies, IAU Symposium 126, ed. J. Grindlay & A. G. D. Philip (Dordrecht: Reidel), 543

331. Navarro, J. F., Frenk, C., & White, S. D. M. 1996, ApJ, 462, 563 (NFW)
332. Neilsen, E. H. Jr, Tsvetanov, Z. I., & Ford, H. C. 1997 ApJ, 483, 745
333. Nemec, J. M., Wehlau, A., & de Oliviera, C. M. 1988, AJ, 96, 528
334. Norris, J. E. 1993, in The Globular Cluster–Galaxy Connection, ASP Conf. Series 48, ed. G. Smith & J. P. Brodie (San Francisco: ASP), 259
335. Norris, J. E., & Hawkins, M. R. S. 1991, ApJ, 380, 104
336. O'Connell, R. W., Gallagher, J. S., & Hunter, D. A. 1994, ApJ, 433, 65
337. O'Connell, R. W., Gallagher, J. S., Hunter, D. A., & Colley, W. N. 1995, ApJ, 446, L1
338. Odenkirchen, M., Brosche, P., Geffert, M., & Tucholke, H.-J. 1997, New Astron., 2, 477
339. Okazaki, T., & Tosa, M. 1995, MNRAS, 274, 48
340. Olsen, K. A. G., et al. 1998, MNRAS, 300, 665
341. Oort, J. H. 1977, ApJ, 218, L97
342. Ostriker, J. P., & Gnedin, O. Y. 1997, ApJ, 487, 667
343. Ostrov, P. G., Forte, J. C., & Geisler, D. 1998, AJ, 116, 2854
344. Pahre, M. A., & Mould, J. R. 1994, ApJ, 433, 567
345. Pascarelle, S. M., Windhorst, R. A., Keel, W. C., & Odewahn, S. C. 1996, Nature, 383, 45
346. Peebles, P. J. E. 1993, Principles of Physical Cosmology (Princeton: Princeton Univ. Press)
347. Peebles, P. J. E., & Dicke, R. H. 1968, ApJ, 154, 891
348. Perelmuter, J.-M., & Racine, R. 1995, AJ, 109, 1055
349. Perlmutter, S. et al. 1997, ApJ, 483, 565
350. Pierce, M. J., Welch, D. L., McClure, R. D., van den Bergh, S., Racine, R., & Stetson, P. B. 1994, Nature, 371, 385
351. Pont, F., Mayor, M., Turon, C., & VandenBerg, D. A. 1998, A&A, 329, 87
352. Pritchet, C. J., & Kline, M. I. 1981, AJ, 86, 1859
353. Pritchet, C. J., & van den Bergh, S. 1987, ApJ, 316, 517
354. Pritchet, C. J., & van den Bergh, S. 1994, AJ, 107, 1730
355. Puzia, T. H., Kissler-Patig, M., Brodie, J. P., & Huchra, J.P. 1999, AJ, 118, 2734
356. Queloz, D., Dubath, P, & Pasquini, L. 1995, A&A, 300, 31
357. Racine, R. 1968, JRASC, 62, 367
358. Racine, R. 1980, in Star Clusters, IAU Symposium 85, ed. J. E. Hesser (Dordrecht: Reidel), 369
359. Racine, R. 1991, AJ, 101, 865
360. Racine, R., & Harris, W. E. 1975, ApJ, 196, 413
361. Racine, R., & Harris, W. E. 1989, AJ, 98, 1609
362. Racine, R., & Harris, W. E. 1992, AJ, 104, 1068
363. Racine, R., Oke, J. B., & Searle, L. 1978, ApJ, 223, 82
364. Reed, L., Harris, G. L. H., & Harris, W. E. 1992, AJ, 103, 824
365. Reed, L., Harris, G. L. H., & Harris, W. E. 1994, AJ, 107, 555
366. Rees, R. F. 1996, in Formation of the Galactic Halo...Inside and Out, ASP Conf. Series 92, ed. H. Morrison & A. Sarajedini (San Francisco: ASP), 289
367. Rees, R. F., & Cudworth, K. M. 1991, AJ, 102, 152
368. Reid, I. N. 1997, AJ, 114, 161
369. Reid, I. N., & Freedman, W. 1994, MNRAS, 267, 821
370. Renzini, A., & Fusi Pecci, F. 1988, ARAA, 26, 199

371. Renzini, A. et al. 1996, ApJ, 465, L23
372. Rich, R. M., Mighell, K. J., Freedman, W. L., & Neill, J. D. 1996, AJ, 111, 768
373. Richer, H. B. et al. 1995, ApJ, 451, L17
374. Richer, H. B. et al. 1997, ApJ, 484, 741
375. Richtler, T. 1992, Habilitationsschrift, Sternwarte der Universität Bonn
376. Richtler, T., Grebel, E. K., Domgorgen, H., Hilker, M., & Kissler, M. 1992, A&A, 264, 25
377. Riess, A. G. et al. 1998, AJ, 116, 1009
378. Rodgers, A. W., & Paltoglou, G. 1984, ApJ, 283, L5
379. Roman, N. G. 1952, ApJ, 116, 122
380. Rutledge, G. A., Hesser, J. E., & Stetson, P. B. 1997, PASP, 109, 907
381. Ryden, B. S., & Terndrup, D. M. 1994, ApJ, 425, 43
382. Saha, A., Sandage, A., Labhardt, L., Tammann, G. A., Macchetto, F. D., & Panagia, N. 1996a, ApJ, 466, 55
383. Saha, A., Sandage, A., Labhardt, L., Tammann, G. A., Macchetto, F. D., & Panagia, N. 1996b, ApJS, 107, 693
384. Sakai, S., Madore, B. F., Freedman, W. L., Lauer, T. R., Ajhar, E. A., & Baum, W. A. 1997, ApJ, 478, 49
385. Salaris, M., Degl'Innocenti, S., & Weiss, A. 1997, ApJ, 479, 665
386. Sandage, A. 1968, ApJ, 152, L149
387. Sandage, A. 1970, ApJ, 162, 841
388. Sandquist, E. L., Bolte, M., Stetson, P. B., & Hesser, J. E. 1996, ApJ, 470, 910
389. Sarajedini, A., Chaboyer, B., & Demarque, P. 1997, PASP, 109, 1321
390. Sargent, W. L. W., Kowal, S. T., Hartwick, F. D. A., & van den Bergh, S. 1977, AJ, 82, 947
391. Satypal, S. et al. 1997, ApJ, 483, 148
392. Schechter, P. L., Mateo, M., & Saha, A. 1993, PASP, 105, 1342
393. Schreier, E. J., Capetti, A., Macchetto, F. D., Sparks, W. B., & Ford, H. J. 1996, ApJ, 459, 535
394. Schweizer, F. 1987, in Nearly Normal Galaxies, Eighth Santa Cruz Summer Workshop in Astronomy and Astrophysics, ed. S. M. Faber (New York: Springer), 18
395. Schweizer, F. 1996, AJ, 111, 109
396. Schweizer, F., Miller, B. W., Whitmore, B. C., & Fall, S. M. 1996, AJ, 112, 1839
397. Scoville, N. Z. 1998, ApJ, 492, L107
398. Scoville, N. Z., Soifer, B. T., Neugebauer, G., Young, J. S., Matthews, K., & Yerka, J. 1985, ApJ, 289, 129
399. Searle, L. 1977, in The Evolution of Galaxies and Stellar Populations, ed. B. M. Tinsley & R. B. Larson (New Haven: Yale Univ. Observatory), 219
400. Searle, L., & Zinn, R. 1978, ApJ, 225, 357 (SZ)
401. Secker, J. 1992, AJ, 104, 1472
402. Secker, J., Geisler, D., McLaughlin, D. E., & Harris, W. E. 1995, AJ, 109, 1019
403. Secker, J., & Harris, W. E. 1993, AJ, 105, 1358
404. Secker, J., & Harris, W. E. 1997, PASP, 109, 1364
405. Seyfert, C. K., & Nassau, J. J. 1945, ApJ, 102, 377

406. Shapley, H. 1918, ApJ, 48, 154
407. Sharov, A. S. 1976, Sov. Astron., 20, 397
408. Sharples, R. M. et al. 1998, AJ, 115, 2337
409. Shaya, E. J., Dowling, D. M., Currie, D. G., Faber, S. M., & Groth, E. J. 1994, AJ, 107, 1675
410. Simard, L., & Pritchet, C. J. 1994, AJ, 107, 503
411. Smecker-Hane, T. A., Stetson, P. B., Hesser, J. E., & Lehnert, M. D. 1994, AJ, 108, 507
412. Smith, E. O., Neill, J. D., Mighell, K. J., & Rich, R. M. 1996, AJ, 111, 1596
413. Sodemann, M., & Thomsen, B. 1996, AJ, 111, 208
414. Stanek, K. Z., & Garnavich, P. M. 1998, ApJ, 503, L131
415. Stanford, S. A., Eisenhardt, P. R., & Dickinson, M. 1998, ApJ, 492, 461
416. Stanford, S. A., Sargent, A. I., Sanders, D. B., & Scoville, N. Z. 1990, ApJ, 349, 492
417. Steidel, C. C., Adelberger, K. L., Dickinson, M., Giavalisco, M., Pettini, M., & Kellogg, M. 1998, ApJ, 492, 428
418. Steidel, C. C., Giavalisco, M., Pettini, M., Dickinson, M., & Adelberger, K. L. 1996, ApJ, 462, L17
419. Stetson, P. B. 1987, PASP, 99, 191
420. Stetson, P. B. 1990, PASP, 102, 932
421. Stetson, P. B. 1994, PASP, 106, 250
422. Stetson, P. B. 1998, private communication
423. Stetson, P. B. et al. 1999, AJ, 117, 247
424. Stetson, P. B., Davis, L. E., & Crabtree, D. R. 1990, in CCDs in Astronomy, ASP Conf. Series 8, ed. G. H. Jacoby (San Francisco: ASP), 289
425. Stetson, P. B., & Harris, W. E. 1988, AJ, 96, 909
426. Stetson, P. B., Vandenberg, D. A., & Bolte, M. 1996, PASP, 108, 560
427. Stewart, G. C., Fabian, A. C., Jones, C., & Forman, W. 1984, ApJ, 285, 1
428. Storm, J., Carney, B. W., & Latham, D. W. 1994b, A&A, 291, 121
429. Storm, J., Nordstrom, B., Carney, B. W., & Andersen, J. 1994a, A&A, 290, 443
430. Strom, S. E., Forte, J. C., Harris, W. E., Strom, K. M., Wells, D. C., & Smith, M. G. 1981, ApJ, 245, 416
431. Suntzeff, N. B., Mateo, M., Terndrup, D. M., Olszewski, E. W., Geisler, D., & Weller, W. 1993, ApJ, 418, 208
432. Surdin, V. G. 1979, Sov. Astron., 23, 648
433. Tacconi, L. J., & Young, J. S. 1985, ApJ, 290, 602
434. Theuns, T., & Warren, S. J. 1997, MNRAS, 284, L11
435. Tonry, J. L., Blakeslee, J. P., Ajhar, E. A., & Dressler, A. 1997, ApJ, 475, 399
436. Toomre, A. 1977, in The Evolution of Galaxies and Stellar Populations, ed. B. M. Tinsley & R. B. Larson (New Haven: Yale Univ. Observatory), 401
437. Tremaine, S. D., Ostriker, J. P., & Spitzer, L. Jr 1975, ApJ, 196, 407
438. Turner, E. L. 1991, AJ, 101, 5
439. Turner, J. L., Beck, S. C., & Hunt, R. L. 1997, ApJ, 474, L11
440. Tyson, A. 1989, in CCDs in Astronomy, ASP Conf. Series 8, ed. G. H. Jacoby (San Francisco: ASP), 1
441. Tyson, J. A., & Jarvis, J. F. 1979, ApJ, 230, L153
442. Valdes, F., Tyson, J. A., & Jarvis, J. F. 1983, ApJ, 271, 431

443. van Altena, W. F., Lee, J. T., & Hoffleit, E. D. 1995, General Catalogue of Trigonometric Stellar Parallaxes, Fourth Edition (New Haven: Yale Univ. Observatory)
444. VandenBerg, D. A., Stetson, P. B., & Bolte, M. 1996, ARAA, 34, 461
445. van den Bergh, S. 1969, ApJS, 19, 145
446. van den Bergh, S. 1977, Vistas Astron., 21, 71
447. van den Bergh, S. 1982, PASP, 94, 459
448. van den Bergh, S. 1993a, AJ, 105, 971
449. van den Bergh, S. 1993b, ApJ, 411, 178
450. van den Bergh, S. 1995, ApJ, 446, 39
451. van den Bergh, S., Abraham, G. G., Ellis, R. S., Tanvir, N. R.,, Santiago, B. X., & Glazebrook, K. G. 1996, AJ, 112, 359
452. van den Bergh, S., Pritchet, C. J., & Grillmair, C. 1985, AJ, 90, 595
453. Vesperini, E. 1997, MNRAS, 287, 915
454. Vesperini, E. 1998, MNRAS, 299, 1019
455. Vesperini, E., & Heggie, D. C. 1997, MNRAS, 289, 898
456. Veteŝnik, M. 1962, Bull. Astron. Inst. Czech., 13, 180 and 218
457. Vogt, S. S., Mateo, M., Olszewski, E. W., & Keane, M. J. 1995, AJ, 109, 151
458. von Hoerner, S. 1955, Zs. f. Astrophys., 35, 255
459. Walker, A. R. 1989, AJ, 98, 2086
460. Walker, G. 1987, Astronomical Observations (Cambridge: Cambridge Univ. Press)
461. Waller, W. H. 1991, ApJ, 370, 144
462. Wang, Z., Schweizer, F., & Scoville, N. Z. 1992, ApJ, 396, 510
463. Watson, A. M. 1996, AJ, 112, 534
464. Webb, T. 1998, MSc Thesis, McMaster University
465. Weil, M. L., & Hernquist, L. 1996, ApJ, 460, 101
466. West, M. J., Côté, P., Jones, C., Forman, W., & Marzke, R. O. 1995, ApJ, 453, L77
467. White, R. E. 1987, MNRAS, 227, 185
468. Whitmore, B. C. 1997, in The Extragalactic Distance Scale, ed. M. Livio, M. Donahue, & N. Panagia (Baltimore: StScI), 254
469. Whitmore, B. C., Gilmore, D. M., & Jones, C. 1993, ApJ, 407, 489
470. Whitmore, B. C., Miller, B. W., Schweizer, F., & Fall, M. 1997 AJ, 114, 797
471. Whitmore, B. C., & Schweizer, F. 1995, AJ, 109, 960
472. Whitmore, B. C., Sparks, W. B., Lucas, R. A., Macchetto, F. D., & Biretta, J. A. 1995, ApJ, 454, L73
473. Wiklind, T., Combes, F., & Henkel, C. 1995, A&A, 297, 643
474. Woodworth, S., & Harris, W. E. 2000, AJ, 119, 2699
475. Zaritsky, D., Olszewski, E. W., Schommer, R. A., Peterson, R. C., & Aaronson, M. 1989, ApJ, 345, 759
476. Zepf, S. E., & Ashman, K. M. 1993, MNRAS, 264, 611
477. Zepf, S. E., Ashman, K. M., English, J., Freeman, K. C., & Sharples, R. M. 1999, AJ, 118, 752
478. Zepf, S. E., Ashman, K. M., & Geisler, D. 1995, ApJ, 443, 570
479. Zinn, R. 1980, ApJ, 241, 602
480. Zinn, R. 1985, ApJ, 293, 424
481. Zinn, R. 1993a, in The Globular Cluster–Galaxy Connection, ASP Conf. Series 48, ed. G. Smith & J. P. Brodie (San Francisco: ASP), 38

482. Zinn, R. 1993b, in The Globular Cluster–Galaxy Connection, ASP Conf. Series 48, ed. G. Smith & J. P. Brodie (San Francisco: ASP), 302
483. Zinn, R. 1996, in Formation of the Galactic Halo...Inside and Out, ASP Conf. Series 92, ed. H. Morrison & A. Sarajedini (San Francisco: ASP), 211
484. Zinn, R., & West, M. J. 1984, ApJS, 55, 45
485. Zinnecker, H., McCaughrean, M. J., & Wilking, B. A. 1993, in Protostars and Planets III, ed. E. H. Levy & J. I. Lunine (Tucson: Univ. Arizona Press), 429

Index

Druck: Strauss Offsetdruck, Mörlenbach
Verarbeitung: Schäffer, Grünstadt

Math